ADVANCES IN Chromatography

VOLUME 34

ADVANCES IN
Chromatography

VOLUME 34

Edited by

Phyllis R. Brown
UNIVERSITY OF RHODE ISLAND
KINGSTON, RHODE ISLAND

Eli Grushka
THE HEBREW UNIVERSITY OF JERUSALEM
JERUSALEM, ISRAEL

Marcel Dekker, Inc.   New York•Basel•Hong Kong

ISBN: 0-8247-9087-1

The publisher offers discounts on this book when ordered in bulk quantities. For more information, write to Special Sales/Professional Marketing at the address below.

This book is printed on acid-free paper.

Copyright © 1994 by Marcel Dekker, Inc. All Rights Reserved.

Neither this book nor any part may be reproduced or transmitted in any form or by any means, electronic or mechanical, including photocopying, microfilming, and recording, or by any information storage and retrieval system, without permission in writing from the publisher.

Marcel Dekker, Inc.
270 Madison Avenue, New York, New York 10016

Current printing (last digit):
10 9 8 7 6 5 4 3 2 1

PRINTED IN THE UNITED STATES OF AMERICA

Contributors to Volume 34

W. M. Coleman III, Ph.D. Master Chemist, Department of Analytical Chemistry, R.J. Reynolds Tobacco Company, Winston-Salem, North Carolina

Joe M. Davis, Ph.D. Associate Professor, Department of Chemistry and Biochemistry, Southern Illinois University at Carbondale, Carbondale, Illinois

Ruth Freitag, Ph.D. Institut für Technische Chemie, University of Hannover, Hannover, Germany

Bert M. Gordon, Ph.D. Master Chemist, Department of Analytical Chemistry, R.J. Reynolds Tobacco Company, Winston-Salem, North Carolina

Yuzuru Hayashi, Ph.D. Senior Researcher, Department of Medical Devices, National Institute of Hygienic Sciences, Setagaya, Tokyo, Japan

Norman E. Hoffman, Ph.D. Professor, Department of Chemistry, Marquette University, Milwaukee, Wisconsin

Ralf Lausch Institut für Technische Chemie, University of Hannover, Hannover, Germany

Rieko Matsuda, Ph.D. Senior Researcher, Department of Food, National Institute of Hygienic Sciences, Setagaya, Tokyo, Japan

Mary Ellen P. McNally, Ph.D. Senior Research Chemist, Agricultural Products Department, E. I. DuPont de Nemours & Co., Inc., Wilmington, Delaware

Leah J. Mulcahey, Ph.D. Section Research Chemist, Agricultural Products Department, E. I. DuPont de Nemours & Co., Inc., Wilmington, Delaware

Christine L. Rankin, Ph.D. Section Research Chemist, Agricultural Products Department, E. I. DuPont de Nemours & Co., Inc., Wilmington, Delaware

Ziad El Rassi, Ph.D. Associate Professor, Department of Chemistry, Oklahoma State University, Stillwater, Oklahoma

Oscar W. Reif Institut für Technische Chemie, University of Hannover, Hannover, Germany

Contents of Volume 34

Contributors to Volume 34   iii
Contents of Other Volumes   ix

1. **High-Performance Capillary Electrophoresis of Human Serum and Plasma Proteins**   1

   Oscar W. Reif, Ralf Lausch, and Ruth Freitag

   I. Introduction
   II. Basic Considerations in Protein Separation
   III. Use of CE in Serum and Plasma Protein Separation
   IV. Summary
          Appendix A. Register of Human Proteins Separated by CE
          Appendix B. Physical and Chemical Properties of Selected Human Serum and Plasma Proteins
          References

2. **Analysis of Natural Products by Gas Chromatography/Matrix Isolation/Infrared Spectrometry**   57

   W. M. Coleman III and Bert M. Gordon

   I. Introduction
   II. Hardware
   III. Spectral Interpretation
   IV. Applications
   V. Summary
   References

3. **Statistical Theories of Peak Overlap in Chromatography**   109

   Joe M. Davis

   I. Introduction
   II. Statistical Modeling of Overlap in One-Dimensional Separations
   III. Statistical Modeling of Overlap in Multidimensional Separations
   IV. Conclusions
   Glossary of Symbols
   References

4. **Capillary Electrophoresis of Carbohydrates**   177

   Ziad El Rassi

   I. Introduction
   II. Electrophoretic System
   III. Separation Methodologies and Applications
   References

5. **Environmental Applications of Supercritical Fluid Chromatography**   251

   Leah J. Mulcahey, Christine L. Rankin, and Mary Ellen P. McNally

   I. Introduction
   II. Polychlorinated Biphenyls

- III. Pesticides and Herbicides
- IV. Phenols
- V. Polynuclear Aromatic Hydrocarbons
- VI. Conclusions
  References

6. HPLC of Homologous Series of Simple Organic Anions and Cations   309

Norman E. Hoffman

- I. Introduction
- II. Reversed-Phase Chromatography of Organic Ions
- III. Ion-Exchange Chromatography of Organic Ions
- IV. Ion-Exclusion Chromatography of Organic Ions
  References

7. Uncertainty Structure, Information Theory, and Optimization of Quantitative Analysis in Separation Science   347

Yuzuru Hayashi and Rieko Matsuda

- I. Introduction
- II. Stochastic Properties of Signals
- III. Uncertainty Structure of Quantitative Analysis
- IV. Information Theory and Quantitative Analysis
- V. Fundamentals of Optimization
- VI. Applications of FUMI and MEI
- VII. Total Chromatographic Optimization
- VIII. Factors Affecting Precision and Throughput
- IX. Outlook
  References

Index   425

Contents of Other Volumes

Volumes 1-10 out of print

Volume 11

Quantitative Analysis by Gas Chromatography *Josef Novák*
Polyamide Layer Chromatography *Kung-Tsung Wang, Yau-Tang Lin, and Iris S. Y. Wang*
Specifically Adsorbing Silica Gels *H. Bartels and P. Prijs*
Nondestructive Detection Methods in Paper and Thin-Layer Chromatography *G. C. Barrett*

Volume 12

The Use of High-Pressure Liquid Chromatography in Pharmacology and Toxicology *Phyllis R. Brown*
Chromatographic Separation and Molecular-Weight Distributions of Cellulose and Its Derivatives *Leon Segal*
Practical Methods of High-Speed Liquid Chromatography *Gary J. Fallick*
Measurement of Diffusion Coefficients by Gas-Chromatography Broadening Techniques: A Review *Virgil R. Maynard and Eli Grushka*
Gas-Chromatography Analysis of Polychlorinated Diphenyls and Other Nonpesticide Organic Pollutants *Joseph Sherma*
High-Performance Electrometer Systems for Gas Chromatography *Douglas H. Smith*

Steam Carrier Gas-Solid Chromatography   *Akira Nonaka*

Volume 13

Practical Aspects in Supercritical Fluid Chromatography   *T. H. Gouw and Ralph E. Jentoft*
Gel Permeation Chromatography: A Review of Axial Dispersion Phenomena, Their Detection, and Correction   *Nils Friis and Archie Hamielec*
Chromatography of Heavy Petroleum Fractions   *Klaus H. Altegelt and T. H. Gouw*
Determination of the Adsorption Energy, Entropy, and Free Energy of Vapors on Homogeneous Surfaces by Statistical Thermodynamics   *Claire Vidal-Madjar, Marie-France Gonnord, and Georges Guiochon*
Transport and Kinetic Parameters by Gas Chromatographic Techniques   *Motoyuki Suzuki and J. M. Smith*
Qualitative Analysis by Gas Chromatography   *David A. Leathard*

Volume 14

Nutrition: An Inviting Field to High-Pressure Liquid Chromatography   *Andrew J. Clifford*
Polyelectrolyte Effects in Gel Chromatography   *Bengt Stenlund*
Chemically Bonded Phases in Chromatography   *Imrich Sebestian and István Halász*
Physicochemical Measurements Using Chromatography   *David C. Locke*
Gas-Liquid Chromatography in Drug Analysis   *W. J. A. VandenHeuvel and A. G. Zacchei*
The Investigation of Complex Association by Gas Chromatography and Related Chromatographic and Electrophoretic Methods   *C. L. de Ligny*
Gas-Liquid-Solid Chromatography   *Antonio De Corcia and Arnaldo Liberti*
Retention Indices in Gas Chromatography   *J. K. Haken*

Volume 15

Detection of Bacterial Metabolites in Spent Culture Media and Body Fluids by Electron Capture Gas-Liquid Chromatography   *John B. Brooks*
Signal and Resolution Enhancement Techniques in Chromatography   *Raymond Annino*
The Analysis of Organic Water Pollutants by Gas Chromatography and Gas Chromatography-Mass Spectrometry   *Ronald A. Hites*
Hydrodynamic Chromatography and Flow-Induced Separations   *Hamish Small*
The Determination of Anticonvulsants in Biological Samples by Use of High-Pressure Liquid Chromatography   *Reginald F. Adams*
The Use of Microparticulate Reversed-Phase Packing in High-Pressure Liquid Chromatography of Compounds of Biological Interest   *John A. Montgomery, Thomas P. Johnston, H. Jeanette Thomas, James R. Piper, and Carroll Temple Jr.*
Gas-Chromatographic Analysis of the Soil Atmosphere   *K. A. Smith*
Kinematics of Gel Permeation Chromatography   *A. C. Ouano*
Some Clinical and Pharmacological Applications of High-Speed Liquid Chromatography   *J. Arly Nelson*

Volume 16

Analysis of Benzo(a)pyrene Metabolism by High-Pressure Liquid Chromatography *James K. Selkirk*
High-Performance Liquid Chromatography of the Steroid Hormones *F. A. Fitzpatrick*
Numerical Taxonomy in Chromatography *Desire L. Massart and Henri L. O. De Clercq*
Chromatography of Oligosaccharides and Related Compounds on Ion-Exchange Resins *Olof Samuelson*
Applications and Theory of Finite Concentrations Frontal Chromatography *Jon F. Parcher*
The Liquid-Chromatography Resolution of Enantiomers *Ira S. Krull*
The Use of High-Pressure Liquid Chromatography in Research on Purine Nucleoside Analog *William Plunkett*
The Determination of Di- and Polyamines by High-Pressure Liquid and Gas-Chromatography *Mahmoud M. Abdel-Monem*

Volume 17

Progress in Photometric Methods of Quantitative Evaluation in TLO *V. Pollak*
Ion-Exchange Packings for HPLC Separations: Care and Use *Fredric M. Rabel*
Micropacked Columns in Gas Chromatography: An Evaluation *C. A. Cramers and J. A. Rijks*
Reversed-Phase Gas Chromatography and Emulsifier Characterization *J. K. Haken*
Template Chromatography *Herbert Schott and Ernst Bayer*
Recent Usage of Liquid Crystal Stationary Phases in Gas Chromatography *George M. Janini*
Current State of the Art in the Analysis of Catecholamines *Anté M. Krstulovic*

Volume 18

The Characterization of Long-Chain Fatty Acids and Their Derivatives by Chromatography *Marcel S. F. Lie Ken Jie*
Ion-Pair Chromatography on Normal- and Reversed-Phase Systems *Milton T. W. Hearn*
Current State of the Art in HPLC Analyses of Free Nucleotides, Nucleosides, and Bases in Biological Fluids *Phyllis R. Brown, Anté M. Krstulovic, and Richard A. Hartwick*
Resolution of Racemates by Ligand-Exchange Chromatography *Vadim A. Danankov*
The Analysis of Marijuana Cannabinoids and Their Metabolites in Biological Media by GC and/or GC-MS Techniques *Benjamin J. Gudzinowicz, Michael J. Gudzinowicz, Joanne Hologgitas, and James L. Driscoll*

Volume 19

Roles of High-Performance Liquid Chromatography in Nuclear Medicine *Steven How-Yan Wong*
Calibration of Separation Systems in Gel Permeation Chromatography for Polymer Characterization *Josef Janča*

Isomer-Specific Assay of 2,4-D Herbicide Products by HPLC: Regulaboratory Methodology   *Timothy S. Stevens*
Hydrophobic Interaction Chromatography   *Stellan Hjertén*
Liquid Chromatography with Programmed Composition of the Mobile Phase   *Pavel Jandera and Jaroslav Churáček*
Chromatographic Separation of Aldosterone and Its Metabolites   *David J. Morris and Ritsuko Tsai*

Volume 20

High-Performance Liquid Chromatography and Its Application to Protein Chemistry   *Milton T. W. Hearn*
Chromatography of Vitamin D_3 and Metabolites   *K. Thomas Koshy*
High-Performance Liquid Chromatography: Applications in a Children's Hospital   *Steven J. Soldin*
The Silica Gel Surface and Its Interactions with Solvent and Solute in Liquid Chromatography   *R. P. W. Scott*
New Developments in Capillary Columns for Gas Chromatography   *Walter Jennings*
Analysis of Fundamental Obstacles to the Size Exclusion Chromatography of Polymers of Ultrahigh Molecular Weight   *J. Calvin Giddings*

Volume 21

High-Performance Liquid Chromatography/Mass Spectrometry (HPLC/MS)   *David E. Grimes*
High-Performance Liquid Affinity Chromatography   *Per-Olof Larsson, Magnus Glad, Lennart Hansson, Mats-Olle Månsson, Sten Ohlson, and Klaus Mosbach*
Dynamic Anion-Exchange Chromatography   *Roger H. A. Sorel and Abram Hulshoff*
Capillary Columns in Liquid Chromatography   *Daido Ishii and Toyohide Takeuchi*
Droplet Counter-Current Chromatography   *Kurt Hostettmann*
Chromatographic Determination of Copolymer Composition   *Sadao Mori*
High-Performance Liquid Chromatography of K Vitamins and Their Antagonists   *Martin J. Shearer*
Problems of Quantitation in Trace Analysis by Gas Chromatography   *Josef Novák*

Volume 22

High-Performance Liquid Chromatography and Mass Spectrometry of Neuropeptides in Biologic Tissue   *Dominic M. Desiderio*
High-Performance Liquid Chromatography of Amino Acids: Ion-Exchange and Reversed-Phase Strategies   *Robert F. Pfeifer and Dennis W. Hill*
Resolution of Racemates by High-Performance Liquid Chromatography   *Vadium A. Davankov, Alexander A. Kurganov, and Alexander S. Bochkov*
High-Performance Liquid Chromatography of Metal Complexes   *Hans Veening and Bennett R. Willeford*

Chromatography of Carotenoids and Retinoids *Richard F. Taylor*
High Performance Liquid Chromatography *Zbyslaw J. Petryka*
Small-Bore Columns in Liquid Chromatography *Raymond P. W. Scott*

Volume 23

Laser Spectroscopic Methods for Detection in Liquid Chromatography
  Edwards S. Yeung
Low-Temperature High-Performance Liquid Chromatography for Separation of Thermally Labile Species *David E. Henderson and Daniel J. O'Connor*
Kinetic Analysis of Enzymatic Reactions Using High-Performance Liquid Chromatography *Donald L. Sloan*
Heparin-Sepharose Affinity Chromatography *Akhlaq A. Farooqui and Lloyd A. Horrocks*
Chromatopyrography *John Chih-An Hu*
Inverse Gas Chromatography *Seymour G. Gilbert*

Volume 24

Some Basic Statistical Methods for Chromatographic Data *Karen Kafadar and Keith R. Eberhardt*
Multifactor Optimization of HPLC Conditions *Stanley N. Deming, Julie G. Bower, and Keith D. Bower*
Statistical and Graphical Methods of Isocratic Solvent Selection for Optimal Separation in Liquid Chromatography *Haleem J. Issaq*
Electrochemical Detectors for Liquid Chromatography *Ante M. Krstulović, Henri Colin, and Georges A. Guiochon*
Reversed-Flow Gas Chromatography Applied to Physicochemical Measurements *Nicholas A. Katsanos and George Karaiskakis*
Development of High-Speed Countercurrent Chromatography *Yoichiro Ito*
Determination of the Solubility of Gases in Liquids by Gas-Liquid Chromatography *Jon F. Parcher, Monica L. Bell, and Ping J. Lin*
Multiple Detection in Gas Chromatography *Ira S. Krull, Michael E. Swartz, and John N. Driscoll*

Volume 25

Estimation of Physicochemical Properties of Organic Solutes Using HPLC Retention Parameters *Theo L. Hafkenscheid and Eric Tomlinson*
Mobile Phase Optimization in RPLC by an Iterative Regression Design *Leo de Galan and Hugo A. H. Billiet*
Solvent Elimination Techniques for HPLC/FT-IR *Peter R. Griffiths and Christine M. Conroy*
Investigations of Selectivity in RPLC of Polycyclic Aromatic Hydrocarbons *Lane C. Sander and Stephen A. Wise*
Liquid Chromatographic Analysis of the Oxo Acids of Phosphorus *Roswitha S. Ramsey*

HPLC Analysis of Oxypurines and Related Compounds *Katsuyuki Nakano*
HPLC of Glycosphingolipids and Phospholipids *Robert H. McCluer, M. David Ullman, and Firoze B. Jungalwala*

Volume 26

RPLC Retention of Sulfur and Compounds Containing Divalent Sulfur
  Hermann J. Möckel
The Application of Fleuric Devices to Gas Chromatographic Instrumentation
  Raymond Annino
High Performance Hydrophobic Interaction Chromatography *Yoshio Kato*
HPLC for Therapeutic Drug Monitoring and Determination of Toxicity *Ian D. Watson*
Element Selective Plasma Emission Detectors for Gas Chromatography *A. H. Mohamad and J. A. Caruso*
The Use of Retention Data from Capillary GC for Qualitative Analysis: Current Aspects
  Lars G. Blomberg
Retention Indices in Reversed-Phase HPLC *Roger M. Smith*
HPLC of Neurotransmitters and Their Metabolites *Emilio Gelpi*

Volume 27

Physicochemical and Analytical Aspects of the Adsorption Phenomena Involved in GLC
  Victor G. Berezkin
HPLC in Endocrinology *Richard L. Patience and Elizabeth S. Penny*
Chiral Stationary Phases for the Direct LC Separation of Enantiomers *William H. Pirkle and Thomas C. Pochapsky*
The Use of Modified Silica Gels in TLC and HPTLC *Willi Jost and Heinz E. Hauck*
Micellar Liquid Chromatography *John G. Dorsey*
Derivatization in Liquid Chromatography *Kazuhiro Imai*
Analytical High-Performance Affinity Chromatography *Georgio Fassina and Irwin M. Chaiken*
Characterization of Unsaturated Aliphatic Compounds by GC/Mass Spectrometry
  Lawrence R. Hogge and Jocelyn G. Millar

Volume 28

Theoretical Aspects of Quantitative Affinity Chromatography: An Overview
  Alain Jaulmes and Claire Vidal-Madjar
Column Switching in Gas Chromatography *Donald E. Willis*
The Use and Properties of Mixed Stationary Phases in Gas Chromatography
  Gareth J. Price
On-Line Small-Bore Chromatography for Neurochemical Analysis in the Brain
  William H. Church and Joseph B. Justice, Jr.
The Use of Dynamically Modified Silica in HPLC as an Alternative to Chemically
  Bonded Materials *Per Helboe, Steen Honoré Hansen, and Mogens Thomsen*
Gas Chromatographic Analysis of Plasma Lipids *Arnis Kuksis and John J. Myher*
HPLC of Penicillin Antibiotics *Michel Margosis*

Volume 29

Capillary Electrophoresis *Ross A. Wallingford and Andrew G. Ewing*
Multidimensional Chromatography in Biotechnology *Daniel F. Samain*
High-Performance Immunoaffinity Chromatography *Terry M. Phillips*
Protein Purification by Multidimensional Chromatography *Stephen A. Berkowitz*
Fluorescence Derivitization in High-Performance Liquid Chromatography
    Yosuke Ohkura and Hitoshi Nohta

Volume 30

Mobile and Stationary Phases for Supercritical Fluid Chromatography
    Peter J. Schoenmakers and Louis G. M. Uunk
Polymer-Based Packing Materials for Reversed-Phase Liquid Chromatography
    Nobuo Tanaka and Mikio Araki
Retention Behavior of Large Polycyclic Aromatic Hydrocarbons in Reversed-Phase
    Liquid Chromatography *Kiyokatsu Jinno*
Miniaturization in High-Performance Liquid Chromatography *Masashi Goto,*
    Toyohide Takeuchi, and Daido Ishii
Sources of Errors in the Densitometric Evaluation of Thin-Layer Separations with Special
    Regard to Nonlinear Problems *Viktor A. Pollak*
Electronic Scanning for the Densitometric Analysis of Flat-Bed Separations
    Viktor A. Pollak

Volume 31

Fundamentals of Nonlinear Chromatography: Prediction of Experimental Profiles
    and Band Separation *Anita M. Katti and Georges A. Guiochon*
Problems in Aqueous Size Exclusion Chromatography *Paul L. Dubin*
Chromatography on Thin Layers Impregnated with Organic Stationary Phases
    Jiri Gasparic
Countercurrent Chromatography for the Purification of Peptides *Martha Knight*
Boronate Affinity Chromatography *Ram P. Singhal and S. Shyamali M. DeSilva*
Chromatographic Methods for Determining Carcinogenic Benz(c)-acridine
    Noboru Motohashi, Kunihiro Kamata, and Roger Meyer

Volume 32

Porous Graphitic Carbon in Biomedical Applications *Chang-Kee Lim*
Tryptic Mapping by Reversed Phase Liquid Chromatography *Michael W. Dong*
Determination of Dissolved Gases in Water by Gas Chromatography *Kevin Robards,*
    Vincent R. Kelly, and Emilios Patsalides
Separation of Polar Lipid Classes into Their Molecular Species Components by Planar
    and Column Liquid Chromatography *V. P. Pchelkin and A. G. Vereshchagin*
The Use of Chromatography in Forensic Science *Jack Hubball*
HPLC of Explosives Materials *John B. F. Lloyd*

Volume 33

Planar Chips Technology of Separation Systems: A Developing Perspective in Chemical Monitoring   *Andreas Manz, D. Jed Harrison, Elizabeth Verpoorte, and H. Michael Widmer*

Molecular Biochromatography: An Approach to the Liquid Chromatographic Determination of Ligand–Biopolymer Interactions   *Irving W. Wainer and Terence A. G. Noctor*

Expert Systems in Chromatography   *Thierry Hamoir and D. Luc Massart*

Information Potential of Chromatographic Data for Pharmacological Classification and Drug Design   *Roman Kaliszan*

Fusion Reaction Chromatography: A Powerful Analytical Technique for Condensation Polymers   *John K. Haken*

The Role of Enantioselective Liquid Chromatographic Separations Using Chiral Stationary Phases in Pharmaceutical Analysis   *Shulamit Levin and Saleh Abu-Lafi*

ADVANCES IN Chromatography

VOLUME 34

1
High-Performance Capillary Electrophoresis of Human Serum and Plasma Proteins

Oscar W. Reif, Ralf Lausch, and Ruth Freitag *Institut für Technische Chemie, University of Hannover, Hannover, Germany*

I.	INTRODUCTION	2
	A. Capillary Electrophoresis	2
	B. Serum and Plasma Proteins	7
II.	BASIC CONSIDERATIONS IN PROTEIN SEPARATION	10
	A. Sample Preparation	10
	B. Sample Injection	11
	C. Protein Detection	15
	D. Capillaries and Coatings	22
	E. Separation of Serum and Plasma Proteins	25
III.	USE OF CE IN SERUM AND PLASMA PROTEIN SEPARATION	35
	A. Capillary Zone Electrophoresis	35
	B. Capillary Isotachophoresis	39
	C. Capillary Isoelectric Focusing	40
	D. Capillary Gel Electrophoresis	42
	E. Combined Separation Techniques	44
IV.	SUMMARY	44
	Appendix A. Register of Human Proteins Separated by CE	46
	Appendix B. Physical and Chemical Properties of Selected Human Serum and Plasma Proteins	47

1

I. INTRODUCTION

The analysis of serum and plasma proteins is one of the most important tasks in clinical analysis, for a wide range of human diseases can be related to changes in the composition of the blood proteins. Furthermore, the increasing therapeutic application of single-serum proteins requires close process and quality control of the process of human serum fractionation and the biotechnological production with recombinant culture organisms, respectively.

Capillary electrophoresis (CE) is a comparatively new technique for the analysis of biological samples such as blood and serum proteins. It combines the separation principles of conventional electrophoresis with the advanced instrumental design of high-performance liquid and gas chromatography. In fact, CE has often proven to be superior to chromatographic techniques in terms of resolution, analysis time, cost, and ease of handling, thus even extending the range of separations that can be performed on a day-to-day laboratory routine basis. Combined with an analysis of the analytes by high-performance liquid chromatography (HPLC) and gas chromatography (GC), CE provides additional information as the separation is based on a different mechanism.

The topic of this chapter is the potential use of CE in the analysis of human serum and plasma proteins, with a strong focus on applications. Due to the complex composition of serum and plasma, however, experience gained in the separation of blood proteins will certainly be useful for any kind of protein separations. In this section the various modes of CE are discussed; in Section II, basic considerations for the separation of proteins; and in Section III, applications of CE for serum and plasma protein separations.

A. Capillary Electrophoresis

Electrophoretic separations are based on differences in the velocities of the analytes in an electric field. In CE the electric field is generated over a long, narrow tubing filled with the electrophoretic medium. The wide variety of possible media is one of the reasons for the versatility of CE. The components of a sample mixture, introduced as a narrow zone into the electrophoresis system, are forced to move under the influence of an electric field. Based on the pioneering work of Hjertén [1–3], Jorgenson and Lukacs [4–7], Mikkers [8], Terabe [9], and Cohen and Karger [10,11], different modes of capillary electrophoretic separations were developed, often based on the experience gained in conventional slab gel electrophoresis and chromatography. These modes are:

Capillary zone electrophoresis (CZE)
Capillary isotachophoresis (CITP)

Capillary isoelectric focusing (CIEF)
Capillary gel electrophoresis (CGE)
Micellar electrokinetic capillary chromatography (MECC)

In Section II, these techniques and their mechanisms are discussed briefly with regard to protein separations. A more detailed discussion of the various separation methods can be found in a number of recently published review articles [2,12–18] and books [19,20].

The fundamental feature for each electrophoretic separation is the nature of the electrophoresis buffer. Both continuous or discontinuous buffer systems can be used in CE. In the continuous buffer the electrolytes form a continuum along the migration path, resulting in the generation of an electric conducting medium. Since the composition of the electrolyte is constant along the migration path, the electric potential and the electrophoretic mobilities of the analytes remain constant. Separation is based solely on differences in velocities. CZE, CGE, and MECC are examples of this type of electrophoretic method. This kinetic separation process can be changed to a steady-state process by modifying the properties of the electrolyte system. In discontinuous systems the composition of the electrolyte changes along the migration path. One or more of the buffer's properties known to influence mobility (e.g., the pH value) are varied along the capillary. In CIEF, analyte molecules move to the spot where the local pH values correspond to their isoelectric points (IEPs), where no further movement is likely. Discontinuous buffer systems are also used in capillary isotachophoresis, where the analytes migrate as discrete zones between a leading electrolyte and a terminating electrolyte. These electrolytes are different in their electrophoretic mobilities.

Capillary Zone Electrophoresis

Capillary zone electrophoresis (CZE), also termed free zone capillary electrophoresis (FZCE), is the most commonly used technique in the CE of proteins. The method is inexpensive, fast, and easy to use. Fused silica and Teflon constitute the most common material for capillaries in CE. The silanol groups on the inner surface of the fused silica capillaries are ionized in liquid media above a pH value of 2. Thus a negatively charged silica surface is formed to which an immobile layer of electrolyte cations is afixed, named the Stern layer. Adjacent to the Stern layer is a diffuse layer of cations. In an electrical field these cations migrate to the cathode, causing a concomitant migration of the liquid through the capillary. This flow is called electroendosmotic flow (EOF). In contrast to the forced convective flow of, for example, the mobile phase in liquid chromatography, the EOF runs with a plug flow rather than a parabolic flow profile. This phenomenon is of great consequence for the width of the sample zones and it is the main reason for the extraordinarily high plate numbers and separation efficiencies found in CE. However, as Grushka [21] pointed out, in real separations the migration velocity may differ across the capillary, for the flat migration velocity profile may be somewhat distorted, owing to the Joule heat generated within the capillary.

Direction and speed of movement of analytes in CZE will depend largely on their mass/charge ratio and the resulting electrophoretic mobility. In fused silica capillaries, the EOF will overlay the electrophoretic movement. Due to the fact that the velocity of the EOF will always be higher than the electrophoretic velocity of any protein, all proteins regardless of presence or sign of a net charge will be detected in CZE. Cationic analytes with a low mass/charge ratio will be detected first, followed by the cations with a higher ratio, neutral analytes, anions with a high mass/charge ratio, and finally, the anionic molecules with the lowest mass/charge ratios. The electrophoretic mobility can be influenced by varying the pH of the electrophoresis buffer, which will modify the mass/charge ratio. Furthermore, the separation of the sample molecules can be influenced by controlling the EOF, which depends on the pH, the viscosity, the ionic strength, and the dielectric constant of the electrophoresis buffer, as well as on the applied voltage. A wide range of additives and coating of the inner surface can be applied to control or to suppress the EOF.

Capillary Isotachophoresis

Another promising method for protein separation in capillary electrophoresis is capillary isotachophoresis (CITP), called high-performance displacement electrophoresis by Hjertén [30]. The basic theory of this technique has been taken from slab gel electrophoresis. Isotachophoresis is performed in a discontinuous buffer system. The sample molecules are injected into the capillary between a leading and a terminating electrolyte solution. When the electrical field is applied, the ions in the leading buffer, chosen for their high mobility, move rapidly toward the electrode, thus reducing the ion concentration at the interface of the sample zone. As a result, the strength of the electric field increases in this area, and the sample molecules are sped up and condense in narrow zones of the specific analytes. In the steady state the sample molecules are arranged according to their electrophoretic mobilities, which often correspond to their isoelectric points. The consecutive sample zones move with identical speed between the leading and terminating electrolytes. To improve the detection of the single sample zones, spacers such as ampholytes or amino acids are added to the sample. These spacers move between the sample zones according to their isoelectric points and electrophoretic velocities, increasing the distance between the sample zones. By choosing an appropriate detection wavelength (e.g., 254 or 280 nm), sample components such as proteins or peptides can be detected, whereas the spacers are not detected. Modern CITP, which used capillaries with internal diameters of less than 100 μm, has a precurser in a number of isotachophoresis systems, in which capillaries with an inner diameter of up to 500 μm were used [31–33]. This type of isotachophoresis is used as a standard method for analysis of serum, plasma, urine, and cerebrospinal fluid, thereby inspiring putative applications in high-performance capillary electrophoresis [34–37].

Capillary Isoelectric Focusing
Isoelectric focusing is a powerful technique in CE. This method had long been used in agarose or polyacrylamide slab gel electrophoresis, when Hjertén et al. [25–27] succeeded in transferring the principle to CE. In capillary isoelectric focusing (CIEF) the capillary ends are placed in two different electrolytes, specified as anolyte and catholyte buffer. The anolyte contains an acidic, the catholyte a basic solution. The capillary is filled with sample and a carefully designed mixture of ampholytes with varied IEPs, which in addition show a high buffer capacity at their IEP. As the electric field is applied, the ampholytes align themselves along the capillary according to their IEP, thus forming a steady pH gradient. Coincidentally, the sample molecules migrate toward the attracting electrode until they reach a pH region corresponding to their isoelectric points, where they become electrically neutral and stop migrating. The analytes are thus focused into narrow zones until a steady state is reached between zone broadening by molecular diffusion and zone narrowing by isoelectric focusing. Afterward the sample zones have to be mobilized toward the detector. One option for mobilization is the addition of salt (e.g., NaCl) to the anolyte or catholyte buffer. The sodium and chloride ions substitute for the protons or hydroxyl ions [26]. Thus the pH gradient is forced into an imbalance, which causes a migration of the zones toward the capillary end, where the ion exchange took place. The same effect is observed by replacing the acidic anolyte by a basic solution or the basic catholyte by an acidic solution after focusing. A further option is the pressure-driven mobilization of the focused zones by the application of a vacuum or pressure to one end of the capillary or simply by generating a height difference between the endings of the capillary tube. The reproducibility of the entire separation depends mainly on the reproducibility of the mobilization conditions. Therefore, mobilizing by salt addition or solute exchange is often superior to pressure-driven options which result in forced convection.

Although CIEF was demonstrated in uncoated capillaries [28,29], the high resolving power of this method is accomplished primarily in coated capillaries in which the EOF is minimized. The resolution can be optimized by changing the range of the pH gradient or the length of the capillary. Diffusion can be minimized by working at low temperatures. The separation time can be reduced by applying higher electric field strengths, by working at low temperatures, or by reducing the capillary length. The technique is one of the most important in the analysis of human serum and plasma proteins.

Capillary Gel Electrophoresis

The separation of sample molecules in capillary gel electrophoresis (CGE) is based on a molecular sieving effect. In CGE the analyte molecules migrate through a porous gel or a solution of an entangled polymer as a sieving matrix and separate according to size. Various gels have been used, although agarose, cross-linked and non-cross-linked polyacrylamide, dextran, or poly(ethylene glycol) are by far the most common. Suitable gels have to be temperature and pH stable. The sieving effect must be well defined and controllable, the polymers should not interact with analytes, and the viscosity should be high enough to counteract sample zone broadening by minimizing convection and diffusion. Charged analytes can be separated in a gel in their native state, although the use of a detergent such as sodium dodecyl sulfate (SDS) to solubilize and concomitantly denature all proteins is more common. The SDS interacts with the peptide bonds of the sample proteins. The amount of SDS needed to solubilize 1 g of protein is often given as 1.4 to 2.2 g. The proteins thus treated appear negatively charged regardless of their charge before solubilization. Since the SDS–protein complexes have nearly similar electrophoretic mobilities, all proteins are separated solely by size. Since the complexes are negatively charged, the injection is performed at the cathode and the detector installed on the anodic side of the capillary.

Agarose and polyacrylamide are used most often in the analysis of nucleotides, peptides, and proteins. Acrylamide is polymerized in the capillary in the presence (cross-linked gel) or absence of a cross-linking agent such as bisacrylamide. The resulting gel has a coiled gel structure with a definite pore size distribution [38,39]. The distribution of the pore size can be controlled by the ratio of the concentration of the gel and the concentration of the cross-linker [3,25]. Still, the application of gels for the separation by size of proteins faces some serious difficulties. The average pore sizes achievable in cross-linked acrylamide gels were found to be too small for large proteins, resulting in inefficient and prolonged separation times. However, non-cross-linked gels also show a sieving effect, presumably due to the physical entanglement of the long polymer chains. Because of their comparatively low viscosity such gels can be injected into the capillary and removed from it by pressure (i.e., the gel can be replaced after each run) [40]. Therefore, irreproducible results caused by contamination of the gel by the sample are omitted. Other promising non-cross-linked gels are based on dextran, poly(ethylene oxide), poly(ethylene glycol), or a mixture of polyacrylamide and poly(ethylene glycol) [42], but separations of native proteins with a molecular weight of more than 800,000 (such as IgM or lipoproteins) have not yet been demonstrated, even with these sieving gels.

In CGE the EOF is suppressed primarily by coating of the inner capillary surface or chemical attachment of the sieving matrix to the column to prevent the gel from being carried out of the capillary by EOF. However, at high viscosities of the sieving gel, a coating proved not to be necessary [159]. Various coatings and their influence on the separation in CGE have been discussed in the literature [38,41]. This is discussed in detail in Section II.

Micellar Electrokinetic Capillary Chromatography

The separation of molecules by micellar electrokinetic capillary chromatography (MECC) is based on the partitioning of the analytes between a micellar phase and the electrophoresis buffer. If a critical micellar concentration is exceeded, micelles are formed by aggregation of charged surfactants in a liquid solution. In case of an anionic surfactant, these micelles are small circular drops with a negatively charged surface and a hydrophobic inner volume. MECC is usually performed in uncoated fused silica capillaries. Because of the EOF, even negatively charged micelles move toward the anode, albeit at low flow rates. The sample molecules partition in and out of the micelles and the average dwelling time of each analyte species within the micelles will be related to their hydrophobicity. Therefore, the more hydrophilic molecules are eluted first, while the hydrophobic molecules are eluted later. The most commonly applied surfactant is sodiumdodecyl sulfate (SDS), which is an anionic surfactant [22]. A large number of other micelle-forming surfactants, such as cetyltrimethylammonium bromide (CTAB) [23] or dodecyltrimethylammonium chloride (DTAC) [24], are known. Furthermore, nonmicellar systems such as cyclodextrins were found suitable for chiral separations (CD–MECC). The resolution of separation can be enhanced by the choice of buffer additive. The MECC is commonly used in the separation of small molecules even in a complex mixture. A common clinical application of MECC is the analysis of metal ions or pharmaceutically important drugs such as barbiturates, thiols, or apoxicillins in human serum or plasma. Since most proteins are hydrophilic, MECC is not suitable for protein separations. So far there have been only a few applications of MECC to protein separations, and none for serum or plasma proteins. Therefore, this method will not be discussed further here.

B. Serum and Plasma Proteins

Circulating human blood consists of a fluid portion (plasma) and the formed elements. Plasma constitutes 55 to 60% of the volume of whole blood. The formed elements are the erythrocytes, leucocytes, and blood platelets. If blood is removed from the body, it clots (coagulates) in a few minutes and becomes gelatinous. If this process is left undisturbed, a clear straw-colored fluid, the serum, is gradually squeezed out. The serum can be separated by centriguation or filtration. Blood plasma, on the other hand, is obtained by centrifugation of blood to which an anticoagulant was added immediately after donation. Serum and plasma are similar in appearance and composition [43,44]: Plasma contains all blood substances besides the formed elements; serum contains all blood substances besides the formed elements and the clotting factors. Furthermore, serum and plasma contain large amounts of various salts (e.g., 0.1 M sodium chloride). Total plasma or serum contains approximately 90 to 92% water and 6 to 8% proteins [45]. Even in healthy donors the concentration of the various serum proteins may vary from person to person as well as from day to day by as much as 10% of the average value. The variation caused by certain diseases may be considerably higher. This must be taken into account in choosing the standard control sample, which should reflect concentrations normally present in human serum. Figure 1 compares the CZE separation of a healthy patient, a NIST standard serum, and bovine serum, showing the differences in areas and retention times [46]. Special care must be taken when choosing and standardizing the sample preparation procedure to ensure the reliability of the analytical result obtained. The major protein types include albumins and globulins. The approximate concentrations of these in human plasma or serum, as determined by electrophoresis, are shown in Table 1. Albumin presents up to 60% of the whole plasma proteins and is therefore the main contaminant of other protein fractions in electrophoresis or liquid chromatography [45]. Albumin consists of three different albumin proteins, varying in their molecular weights and the amount of carbohydrates: albumin and pre- and postalbumin. At present, over 100 different plasma proteins have been described [47]. They have been isolated mainly by electrophoretic fractionation and identified or quantified by immunoelectrophoresis. The majority of the proteins in serum and plasma seem to be conjugated, especially the glycoproteins and lipoproteins.

The γ-globulin fraction of the plasma proteins contains a number of components that are of great medical and pharmaceutical importance. Some of the best investigated serum proteins are the immunoglobulins, also referred to as antibodies, which are found in the electrophoretic γ-globulin zone. The γ-globulins represent 9 to 12% of the total plasma proteins. The immunoglobulins play a key role in the immune defense of the body, and quantitative changes in the amounts of immunoglobulins in the plasma or serum are an indicator of various human diseases. A more detailed analysis of the subfractions of the immunoglobulins offers even more information about certain physical disorders.

Table 1 Concentrations of Albumins and Globulins in Human Plasma or Serum

Component	Total protein (%)	Concentration (mg/mL)
Albumins	50–60	40–50
Globulins	40–50	20–30
α_1	5–13	3–6
α_2	7–12	4–9
β	10–14	6–11
γ	12–19	7–15
Total protein		60–80

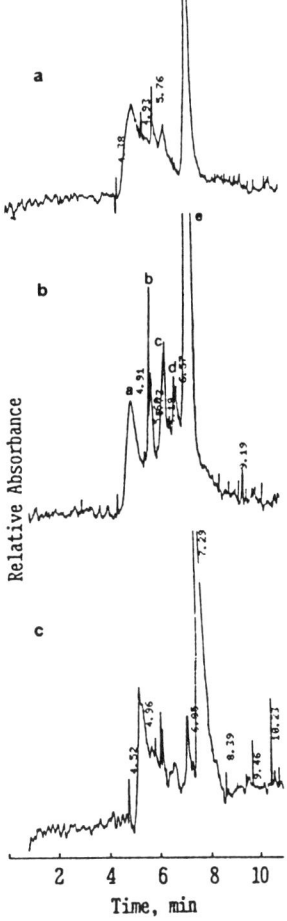

Fig. 1 Separation of proteins with CZE from bovine serum (a), from NIST standard reference material (b), and human serum from a patient (c). Conditions: capillary, 100 cm length, 50 μm ID; applied voltage, 30 kV; buffer, 50 mM NaOH pH 9.5; detection, 200 nm. (From Ref. 46.)

In addition to the traditional clinical applications of albumin and whole plasma, in recent years an increasing number of plasma and serum proteins have been used for therapeutic purposes. Monoclonal and polyclonal antibodies especially are gaining great importance as a new generation of drugs. As the majority of these antibodies are produced by recombinant organisms in cell cultures, sophisticated quality and purity control becomes essential. At the same time, the large-scale fractionation of human serum was optimized in recent years to meet the growing demand of specific proteins, such as immunoglobulins, clotting factors, and so on. The extremely low concentration of some of these blood proteins gave rise to the need for improved fractionation and process control techniques. Furthermore, verification of the product's identity and composition, especially in the case of therapeutically important substances (e.g., factor VIII, interleukins) is essential, for even small impurities can lead to a total loss of therapeutic quality of the substance or cause an immune reaction.

Analysis of blood proteins is performed by electrophoresis. The applied techniques range from simple zone electrophoresis to crossed immunoelectrophoresis. Agarose gel electrophoresis is commonly used in clinical laboratories as a screening technique for the qualitative determination of abnormalities of major human plasma and serum proteins. The results offer a fast indication for the pathologist to seek additional conformation, thus simplifying the final diagnosis. Often, immunoelectrophoresis or isoelectric focusing are performed in addition. Still, the quantitation of protein bands in a gel is difficult and time consuming. CE overcomes these disadvantages of conventional electrophoresis with respect to use in clinical and industrial laboratories. Most electrophoretic separation modes can be transformed directly to CE. With CE the handling is easier if no gel preparation is necessary. The labor-intensive staining procedures are replaced by on-line detection and quantitation of the sample zones. With the exception of some tasks in sample preparation, the entire analytical procedure, including aquisition and interpretation of the data, can be automated. Capillary electrophoresis can replace slab gels for many procedures in the clinical laboratory; however, for some analytical problems, immunoelectrophoresis carried out in slab gels is still required since immunoelectrophoresis has extremely high resolution power.

II. BASIC CONSIDERATIONS IN PROTEIN SEPARATION

The separation of serum and plasma proteins presents several problems, related primarily to the fact that the total protein concentration is about 60 to 80 mg/mL, while the concentration of single proteins in the plasma may range from 0.1 μg/mL to 50 mg/mL. Therefore, the electrophoretic separation method, as well as the mode of the injection and detection, have to be suitable for trace analysis on the one hand, and for the screening of the entire protein spectrum present in plasma samples, on the other. Coating of the capillary or the addition of surfactants has to be performed with respect to the hydrophilic and hydrophobic character of the blood proteins; otherwise, protein adsorption on the inner surface of the capillary might disturb analysis significantly. Furthermore, the large amounts of salts present in serum and plasma may cause a decline in resolution or performance.

A. Sample Preparation

For all applications in capillary electrophoresis the composition of the protein sample has to be considered. If commercial human standard proteins are used, the procedure through which the protein was isolated must be taken into account. Most of these standard proteins are isolated from donor blood by ethanol and salt precipitation [49,50] or by chromatographic procedures such as affinity or ion-exchange chromatography [48]. The salt content of the protein fractions and consequently, the conductivity of a sample of a given protein concentration may differ considerably, depending on which method was used. If possible, standards and samples should be dialyzed against the electrophoresis buffer. Especially in CIEF, a desalting of a biological sample to approximately 10 mM salt concentration prior to analysis is recommended [171]. Interfering substances such as detergents should always be removed. Most protein fractions isolated from blood are contaminated by albumin and other proteins with related physical and chemical properties. Furthermore, the protein content is given solely in activity per unit (e.g., in the case of antibodies). Depending on the CE technique used, serum and plasma are generally diluted from 1:1 [25] up to 1:40 [51] or 1:100 [52] with an appropriate buffer. A dilution ratio of 1:50 was found to be preferable with whole serum or plasma [52]. To avoid degradation of the serum and plasma proteins, samples should be stored at 4°C and analyzed within 24 h. Furthermore, samples should never be left at room temperature for a prolonged time, as this will lead to a degradation of the sample and evaporation of the solvent, which in turn will influence the apparent sample concentration [53]. If possible, the autosampler should be cooled. If samples have to be stored over a longer period, these samples should be frozen and stored at $-4°C$ and thawed only once before use. Figure 2 shows the reproducibility from two CZE runs made 2 months apart on two different samples from the same person. No sample degradation can be observed and the retention times and even the peak areas are reproducible [51]. All buffers and samples should be filtered through a 0.45-μm filter before use. Each technique in capillary electrophoresis requires special pretreatment of the biological sample. Details are given in the following sections.

B. Sample Injection

The introduction of the sample into the capillary is one of the crucial points in CE. For example, the resolution can be reduced substantially by simple overloading. Additionally, the reliability of the quantitation of samples depends on exact and reproducible injection. The majority of commercially available CE systems use one of two injection modes: electrokinetic and hydrostatic. In both cases the amount of sample applied can be regulated by changing the injection time. In the case of electrokinetic injection the applied voltage, and in case of the hydrostatic method the applied pressure, is of great importance [57]. Generally, the composition of the sample diluent should be carefully adjusted with regard to the conductivity of the final sample. Conductivities higher or lower than that of the electrophoresis buffer cause sample-zone broadening or focusing. Often, distilled water is the most suitable diluent.

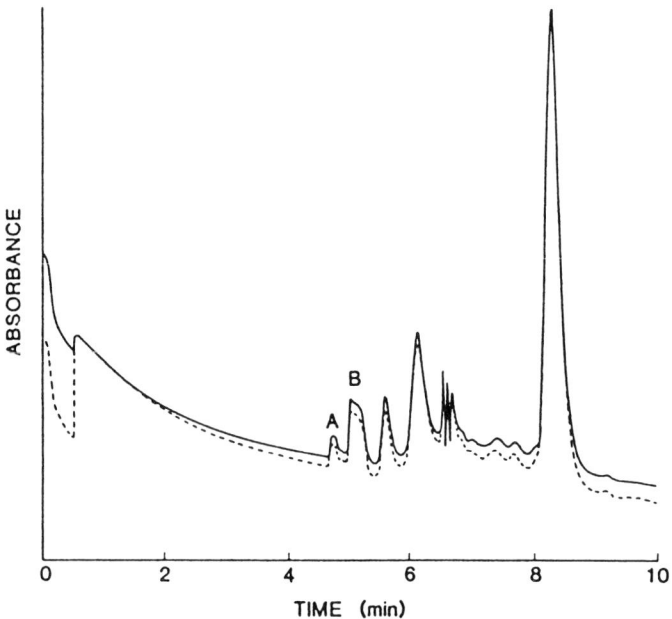

Fig. 2 Overlay of two CZE runs of the same human serum protein sample. Peaks A and B are ethylene glycol. Capillary, 37.5 cm length, 75 μm ID; sample, 40:1 with 1 mM boric acid pH 4.5, 20% ethylene glycol; separation, 50 mM sodium borate pH 10; applied voltage, 10 kV; detection, 200 nm. (From Ref. 51.)

Electrokinetic injection or electromigration is based on the migration of charged molecules into the capillary in an electric field [4,54]. Assuming an EOF is directed toward the cathode [5,7], the anodic end of the capillary is placed in the sample while the cathode is placed in the electrophoresis buffer. By applying an electrical field, the sample is forced into the capillary [55]. After the time span chosen for injection, the anodic end of the capillary is replaced in the electrophoresis buffer and electrophoretic separation of the sample proceeds. In coated capillaries, where the EOF is minimized or suppressed, only charged analytes are introduced by electromigration. In both cases electrokinetic injection is biased towards molecules of low mass/charge ratio. Since the migration velocity of such molecules is higher than that of molecules with lower electrophoretic mobilities, they are injected in larger relative quantities [56]. Enrichment of the electrophoretically slower-moving components in the sample vial may present a problem when repeated injections of the same sample are required.

Hydrodynamic injection, also called hydrostatic injection, is the second common injection mode in CE. In this case the sample is introduced into the capillary by the application of pressure to the sample vial or vacuum to the opposite vial, thereby pushing or sucking a defined sample volume into the capillary. Siphoning, another hydrodynamic variant, is performed by raising the sample vial with the capillary end to a certain height level above the grounded capillary end (detector end) [58].

As opposed to hydrodynamic flow injection, electrokinetic injection has no effect on the separation efficiency, even though a plug flow profile [54] can be assumed in the former. Although the hydrodynamic injection has the advantage of not affecting the sample composition, it is not suitable in all CE methods; for example, in CGE only electrokinetic injection can be used. A comparison of the two injection modes [54] found a relative standard deviation (RSD) of 4.1% for electrokinetic injection and of 2.9% for the hydrodynamic mode.

The detection limit is another aspect to be considered in sample injection. It was shown that the detection limit of human growth hormone (hGH) is a function of the sample concentration [60]. Based on the analysis of a 0.02-mg/mL hGH sample solution and a detector rise time of 1.0 s, the detection limit was found to be 19 pg of hGH (ca. 850 attomol). While the mass sensivity of ultraviolet (UV) detection is generally higher in HPCE than in HPLC, the concentration sensitivity is poorer, as the sample in CE is detected in small detection cells. To overcome this problem, sample concentration or injection time can be increased. Figure 3 shows the relationship between sample concentration and peak height or peak area divided by retention time. As peak area is inversely proportional to the velocity of the analyte zone when it passes the detector, division of peak area by retention time (R_t) is necessary to account for the difference. Linearity between the sample concentration and the normalized peak area was observed in the range

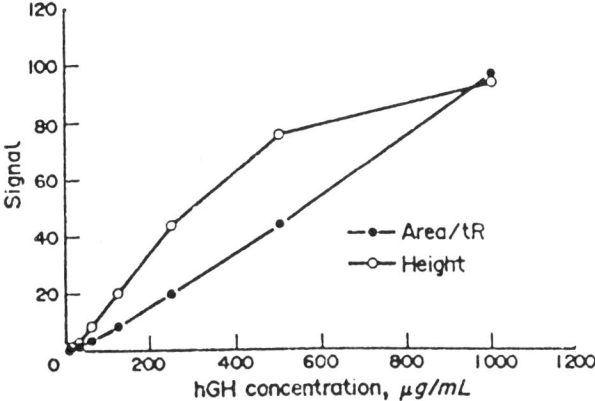

Fig. 3 Peak area/t_R and peak height versus hGH concentration. Capillary, 109 cm length, 50 μm ID; sample, 7.8 to 1000 μg/mL hGH, injected for 5 s by 16.8 kPa vacuum; separation, 10 mM tricine pH 8.0, 20 mM NaCl, 30 kV; detection, 200 nm. (From Ref. 60.)

16 to 1000 μg/mL, while linearity between the peak height and the protein concentration was observed only in the range 16 to 250 μg/mL [60]. This difference is caused by overloading effects, which have shown to increase the dispersion of the sample zone, thus influencing the maximum height recorded but not the overall peak area [31,62]. Concomitantly, separation performance, especially the theoretical plate number, decreases exponentially with increasing injection time, for the contribution of the sample plug to the total variance of the signal can be neglected only at small sample plug lengths [59,61]. If the injection time is varied for a given sample (Fig. 4), linearity between injection time and peak height was observed only in the range 0.2 to 1 s, while linearity between normalized peak area and injection time was observed over the much wider range 0.1 to 20 s [60]. This has been attributed to the fact that the sample plug constitutes the major contributor to the total peak width [63–65].

In Fig. 5, electrokinetic and hydrodynamic injection are compared in regard to the relationship between the peak parameters and the injection time. Excellent linearity is achieved in the case of hydrodynamic injection, whereas the correlation is nonlinear in electrokinetic injection. This is due to the continuous change of concentration and ionic strength in the sample zone during electrokinetic injection, which results in changes in the conductivity and electrical field over this zone [8,66].

Another way to improve the concentration sensivity of CE separations is the stacking of the injected sample zone [60,67,68]. This effect occurs whenever a sample zone is surrounded by a electrophoresis buffer with a higher conductivity. The electrophoretic velocity is proportional to the field strength, which is inversely proportional to the conductivity of the zone. When the velocity increases,

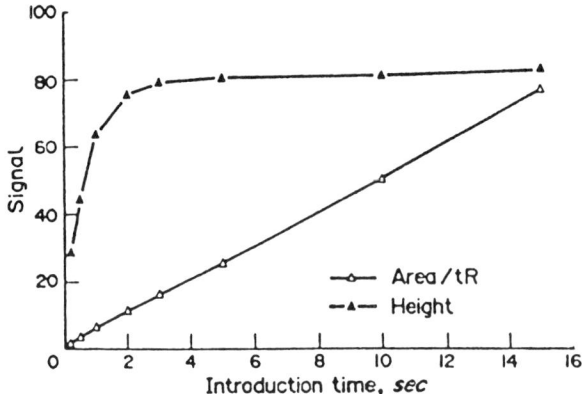

Fig. 4 Peak area/t_R and peak height versus sample injection time in seconds. Capillary, 100 cm length, 50 μm ID; sample, 0.1 mg of hGH injected by 16.8 kPa vacuum; separation, 10 mM tricine pH 8.0, 15 kV; detection, 200 nm. (From Ref. 60.)

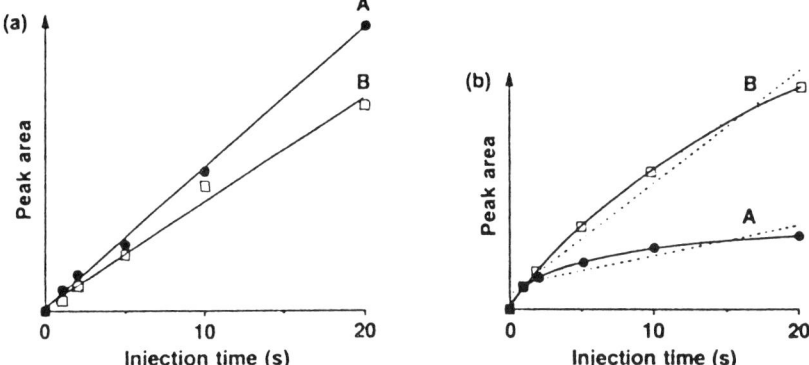

Fig. 5 Linearity of (a) vacuum and (b) electrokinetic injection as a function of injection time. (a) Capillary, 122 cm length, 50 μm ID; injection vacuum 16.9 kPa; buffer, 20 mM bicine–tetraethylammonium pH 8.5; applied voltage, 25 kV; detection, 200 nm. (b) Capillary, 72 cm length, 50 μm ID; injection voltage, 5 kV. Curves A, β-lactoglobulin B; curve B, moyglobin (horse heart). (From Ref. 66.)

the sample zone is "stacked" (i.e., concentrated). In this way the sensitivity and theoretical plate numbers are increased [60] (Fig. 6). A similar effect was exploited systematically in systems using an on-line coupled column [69,70] and on-column isotachophoretic preconcentration of the protein samples [71]. Despite the excellent results, no commercially available CE system features these methods.

Other sample injection modes include the electric sample splitter, where the sample migrates in two electrical circuits in which the ratio of the corresponding electric currents gives the splitting ratio [72], and rotary-type injectors, comparable to standard HPLC injectors [73,74]. Scale-down of the latter type of injector to the dimensions of a CE system still poses difficulties. More promising are microinjectors, able to handle ultralow sample volumes such as single cells or discrete tissue regions. Several microinjector systems have been developed for CE. They all employ heated glass capillaries that are pulled to form an extremely small tip area, which is connected to the sample end of the capillary [75,76]. However, microinjectors are still under development and commercial equipment is not yet available.

C. Protein Detection

To our knowledge, UV and to a lesser extent fluorescent detection can be regarded as the main detection modes in CE. However, other techniques, such as Raman spectroscopy [78], electrochemical detection [79], and amperometric detection [80,81], have been employed successfully. Recently, protein detection has even

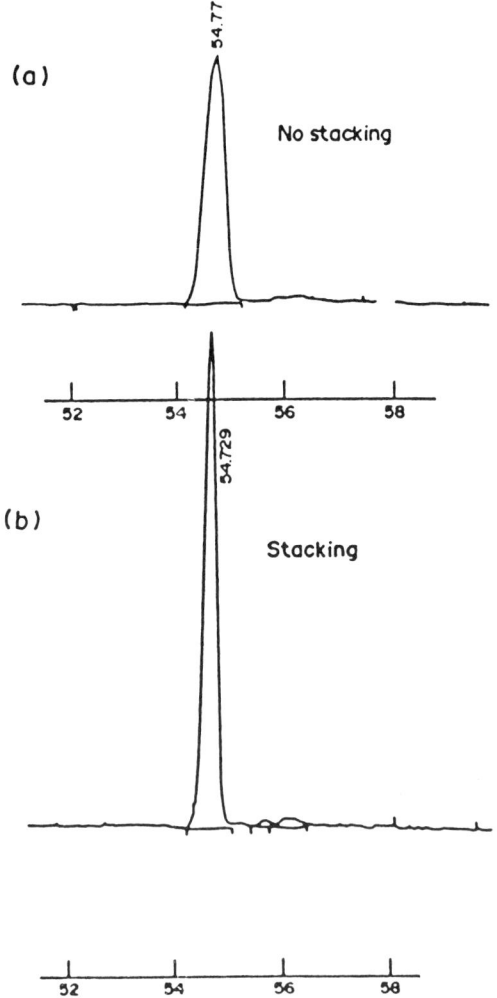

Fig. 6 Influence of sample stacking on the number of theoretical plates (x-axis: minimum; y-axis: arbitrary AU). Capillary, 100 cm length, 50 μm ID; sample, 0.1 mg of hGH in (a) running buffer (no stacking) or (b) distilled water (stacking), injected for 3 s by 16.8 kPa vacuum; separation, 10 mM tricine pH 8.0, 25 mM NaCl, 5 kV; detection, 200 nm. (From Ref. 60.)

been performed by mass spectrometry [82–84]. For these detection modes, the dimensions of the capillary constitute a major drawback, as the path length available for UV detection may be as short as 25 μm. Detection limits of 10^{-5} to 10^{-6} mol/L are found in the case of proteins as a result of the shortness of the light path [91,92]. Despite this disadvantage, UV detection is clearly the most popular detection mode for protein samples, partly because of its universal nature and applicability, partly because of the extensive data basis available for UV detection of various substances in HPLC. Consequently, all commercial CE systems are equipped with UV detectors. A large number of publications concerning the theory and practical application of UV detection are currently available [2,12–20,85–89]. Usually, a small detection window is created near one end of the capillary, but whole-column UV detection has also been reported [90], where the entire capillary was moved past the detector window after CIEF was performed. Various modifications of the detectors and the flow cell have been attempted to improve the detection sensitivity for proteins. In studies of light refraction and reflection in cylindrical cells it was found that the reflection losses are very small [2]. Cells with an adjustable aperture, a focusing lens, or a U-shaped cell were used [93]. Z-shaped and bubble cells have also been tested. The appropriate wavelength for the UV detection of proteins ranges from 190 to 280 nm. If possible, protein detection should be performed at low wavelengths, where the sensitivity is much higher [94]. The CE method chosen, however, may place some restrictions on the detection wavelength. Figure 7 shows the adsorbance of the spacer molecules used in a CIEF at 200 nm. Obviously, these spacers would be detected at this wavelength, while no such interference would occur at a detection wavelength of 280 nm. However, the protein signal of human hemoglobin A is decreased significantly at 280 nm compared to 200 nm [95]. Although most UV detectors employ only a single wavelength at a time, detectors such as a photodiode array and a multiwavelength UV detector have also been used to advantage in CE (Fig. 7). The spectral information can be used to identify unknown compounds, especially of contaminants of the protein fractions, or to confirm whether there are any overlapping peaks in a single electropherogram [96,97].

Fluorescence detectors used for on-column fluorescence detection in CE are usually adapted from similar HPLC detectors. As only a few of the fluorescence detectors are suitable for modification to capillary dimensions, custom-built detectors are the rule [4,5,98–103]. Both fluorescence and UV detection are handicapped by small capillary dimensions, but promising results are obtained with laser-induced fluorescence (LIF) detectors. Still, only a few publications deal with protein samples and even fewer with serum proteins; most publications concentrate on the analysis of amino acids [105], peptides [106], vitamins [107], thiols [108], and antibiotics [109]. However, under optimal conditions, as little as 4.2×10^{-8} M of myoglobin has been detected in CE [104].

Another drawback of fluorescence detection is the fact that most analytes show no natural activity. Therefore, pre- or postcolumn derivatization of the sample with some type of fluorophore must be performed. The fluorophores for amino acid and peptide labeling common in HPLC have been used in CE as well, including 3-(4-carboxybenzoyl)-2-quinoline-carboxaldehyde (CBQCA), dansyl

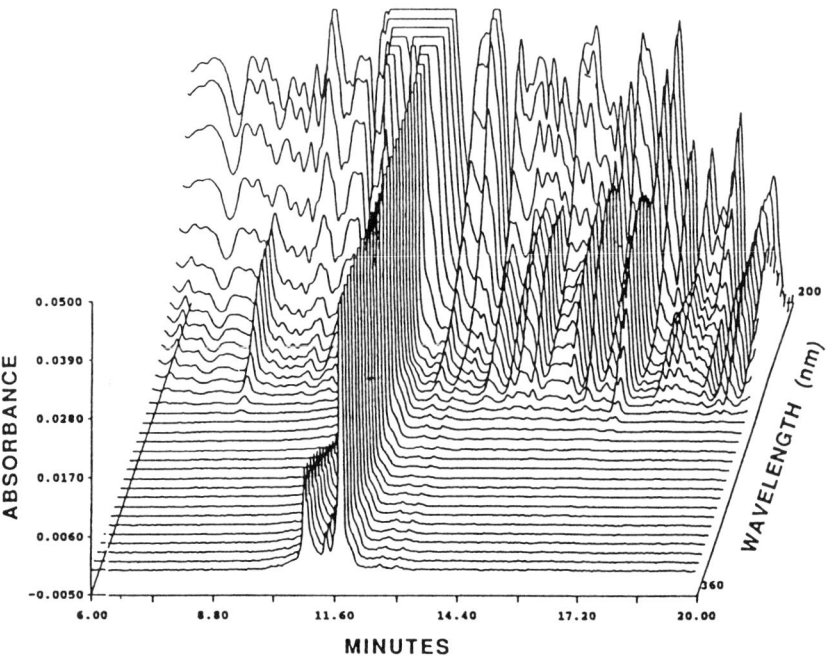

Fig. 7 Separation of hemoglobin A using scanning detection in the 200 to 360-nm UV region. Sample, 2% ampholytes pH 3 to 10, 0.1% Triton X-100; capillary, 17 cm length, 25 μm ID; focusing conditions, 20 mM phosphoric acid (anolyte), 40 mM NaOH (catholyte), 7 kV; mobilization conditions, zwitterionic solution, 8 kV; detection, 280 nm. (From Ref. 95.)

(DNS) [110], o-phthaldialdehyde (OPA) [111], naphthalenedialdehyde (NDA) [112], fluorescein isothiocyanate (FITC) [113], fluorescamine [114], and 4-chlor-7-nitrobenzofuran (NBD) [115]. Precolumn labeling of the compounds is simple to perform but causes problems with the detection of large peptides and proteins in CZE, as a single protein can give rise to a number of overlapping peaks [7]. The labeling of each amine group changes the charge of the protein and thereby the electrophoretic mobility. The sensitivity of CZE is so high that two protein molecules differing only in the number of fluorecence labels attached will be separated [104]. An attempt to overcome this problem was the formation of "supramolecular" complexes. the fluorescent dye eosine was found to migrate together with the sample proteins in a capillary tube as a supramolecular complex, overcoming the problems and increasing the sensitivity [119]. Human serum albumin, human γ-globulins, human hemoglobin, myoglobin, and other proteins were analyzed (Fig. 8).

Postcolumn derivatization of the sample proteins is generally performed in a T reactor connected to two coaxial fused silica capillaries where labeling agent is introduced through the third capillary [54,116]. However, the coaxial capillaries have to be matched closely in respect to their inner and outer diameters to avoid excessive band broadening [104]. Furthermore, the labeling agent had to be forced through the T reactor by gravity or low pressure, complicating flow rate control. Modifications of the cell by Jorgenson [104] overcame some of these problems by using pressure to drive the labeling solution and by placing one coaxial capillary into the other, thereby using the second capillary volume as a reactor cell. The fluorescent derivative was detected further down the capillary. Using an integrated microscope detection system [117] and a postcolumn reactor, FITC-labeled monoclonal antibodies could be analyzed by capillary electrophoresis at the pg/mL level (Fig. 9) [118]. This tagging of proteins to chromophores and the separation of these complexes by capillary electrophoresis might provide an excellent protein purity test.

Different from direct fluorescence detection, indirect fluorescence detection constitutes a widely applicable, sensitive detection mode. Here a fluorophore (e.g., 1.0 mM salicylate) [120] is added to the electrolysis buffer to obtain a

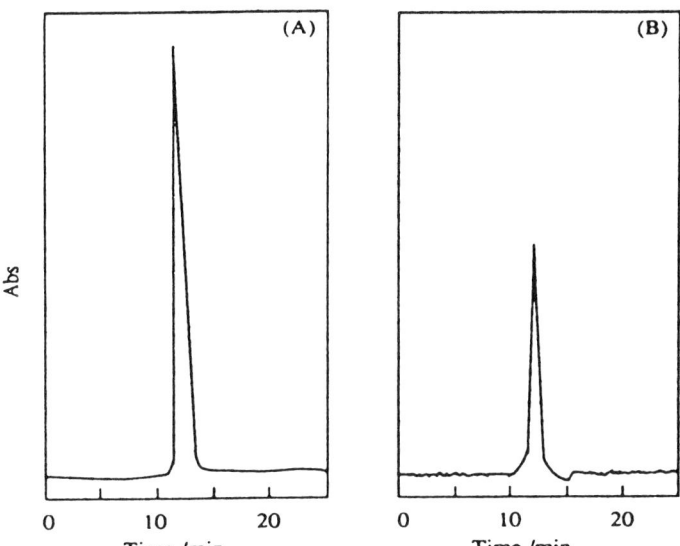

Fig. 8 Detection of the EY–HSA system. Capillary, 60 cm length, 50 μm ID; sample, EY 3 × 10^{-4} M, HSA 6.6 g/L; applied voltage, 20 kV; buffer, 50 mM phosphate pH 2.5; detection: (A) 210 nm (the peak corresponds to HSA), (B) 532 nm (the peak corresponds to EY–HSA), no peak was observed for EY alone up to 60 min. (From Ref. 119.)

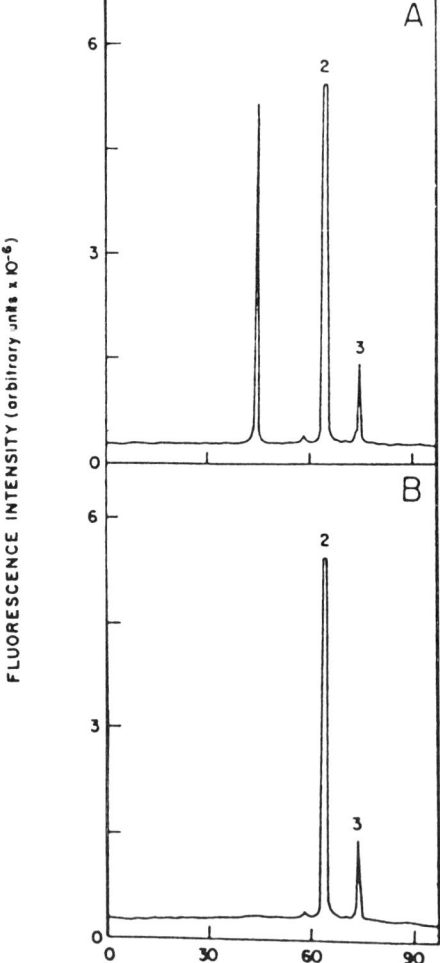

Fig. 9 Electropherogram of (A) monoclonal antibody conjugated to FITC and (B) a FITC solution control. Capillary, 55 cm length, 75 μm ID; separation, 50 mM sodium borate pH 8.3, 20 kV; detection, mercury lamp, bandpass filter 450 to 490 nm, emission filter 515 nm; peaks: 1, monoclonal antibody; 2, FITC isomer; 3, either a second isomer of FITC or a contaminatant. (From Ref. 118.)

steady background signal. Analytes are detected by the fluorescence detector as a negative signal [120]. No applications for serum proteins have yet been reported.

Other detection techniques, such as Raman spectroscopy, have been demonstrated as a potentially useful detection system. The detection is based on monitoring of the intensity and frequency of the scattered light, stemming from a monochromatic light source that irradiates the analyte [121]. However, at present protein analysis has not been performed using this detection mode. Potentiometric detection involves measurement of the potential between two indicator electrodes when a small constant current is applied [122]. Applications of capillary electrophoresis with potentiometric as well as with amperometric detectors [80] (using microelectrodes) have been limited to small ions such as sodium or potassium ions or hydroquinone, and few applications for proteins have been reported. Other detection techniques, such as radioisotope detection [133], refractive index detection [134], and capillary vibration detection [135], are still in the early stages of development. Again, no protein applications have been reported.

A great deal of interest has been paid to the development of capillary electrophoresis coupled to mass spectrometry (MS). Taking advantage of the best features of capillary electrophoresis and mass spectrometry, fast, efficient separation is combined with a selective and reliable detection method capable of analyzing all analytes, ranging from small ions to proteins [123–126]. Furthermore, the flow rates of CE are suited to the on-line application of the mass sensitive detector [84]. The first coupling attempts, which demonstrated the potential of MS with CE, used a valve to connect the detector. Today, various interfaces for CE–MS connection have been developed. The ionization and transfer of the analyte ions into the MS is performed by electrospray ionization (ESI) and fast-atom bombardment (FAB) [128]. Commonly employed mass spectrometers coupled to CE are quadrupole [123,127], time of flight (TOF) [129], and ion trap [130]. For the analysis of analytes with higher molecular masses, such as proteins, ESI has the advantage of forming multiple-charged ions. Exact molar masses could be calculated from the observed distribution of charge states of the molecule [131]. Detection limits in CZE–MS were found to be in the high-femtomole range, but considering that the concentration of the original protein sample was 10^{-5} M, the detection must be regarded as rather insensitive [125]. Karger et al. [84] used capillary isotachophoresis (CITP) or a combination of CITP and CZE for sample preconcentration, thereby increasing the detection limits for proteins to a concentration of 10^{-7} M of protein. In this case the identification of standard proteins could be accurately performed even in complex mixtures. Since the application of CE–MS to protein analyses is relatively new, few reports on the analysis of human serum or plasma proteins can be found in the literature [132]. The molecular heterogeneity of the glycoprotein bovine serum apotransferrin, a plasma protein involved in iron transport, was analyzed by CZE–MS (Fig. 10) [125]. In the future, CE–MS could very likely become the method of choice for purity and product heterogenity control of drugs, based on human proteins. At the moment, however, such a system is too expensive to be used as a standard method, and application of CE–MS is limited to research laboratories.

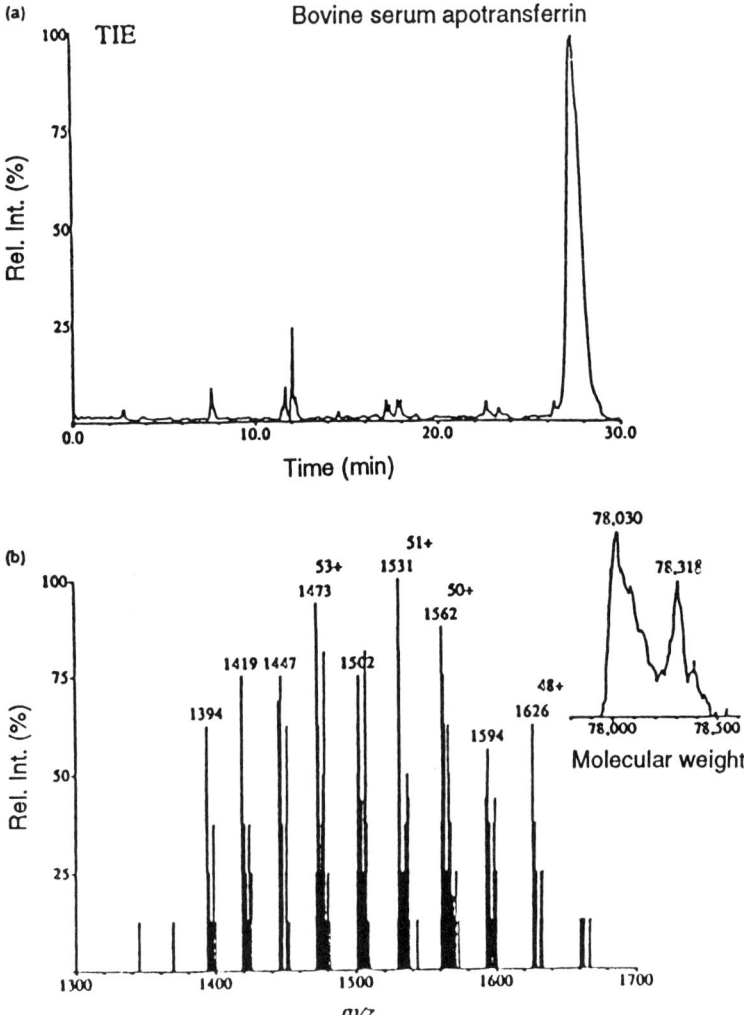

Fig. 10 CE–MS analysis of serum apotransferrin (bovine): (a) capillary zone electrophoresis; (b) mass spectrum of peak migrating at 27.3 min together with its reconstructed molecular weight profile. (From Ref. 125.)

D. Capillaries and Coatings

Capillaries in CE are generally made of fused silica with an polyimide coating on the outside to improve the mechanical stability of the material. Other materials, such as Pyrex, polytetrafluoroethylene (PTFE) [55], and poly(ethylene-propylene) [138], were investigated but were found unsuitable for the production of capillaries of inner diameters as small as 10 μm [136]. So far, circular capillaries have been the standard in CE. In the future, however, noncircular cross sections may have to be considered, for any increase in the path length of the UV light beam through the detection chamber will increase the sensitivity of an optical detector. Furthermore, circular capillaries are a disadvantage in heat dissipation, due to their low surface area/volume ratio. The performance of rectangular capillaries was investigated for borosilicate glass capillaries [137] and flattened poly(ethylene-propylene) tubing [138]. Analyses of immunglobulin G by isotachophoresis using rectangular and cylindrical capillaries were compared. A poly(ethylene-propylene) capillary with an inner-diameter of 500 μm was used, as smaller inner diameters were not available. Similar peak patterns were achieved; however, since there was superior heat dissipation of the rectangular tubing, higher voltages could be applied, resulting in shorter analysis times. Even better results may be achieved using this kind of capillary with smaller inner diameters.

Fused silica capillaries are popular in CE, due to their optical properties in the UV range and thermal characteristics. In addition, the silanol groups in the internal surface are essential with regard to the EOF in case of CZE. For other separation methods in CE, the EOF would interfere with the separation and has to be supressed (e.g., by coating of the capillary walls).

Protein adsorption on the capillary inner surface constitutes another serious problem in CE. In uncoated capillaries, positively charged proteins tend to stick to the deprotonated silanol groups. This effect is well known in the case of the separation of serum and plasma proteins. This complex mixture, containing hydrophylic and hydrophobic, charged and uncharged proteins, has a great variety of proteins capable of adsorption. The separation efficiency and the reproducibility are thus reduced from run to run, which has an adverse effect on the reliability of the analysis. Buffer additives to reduce protein adsorption are discussed in Chapter 2. For all CE methods except CZE, where the EOF is necessary, a reproducible and stable coating of the silanol groups on the inner surface of the capillary constitutes an efficient means of minimizing both the EOF and protein adsorption.

Various coatings have been developed and shown to suppress the EOF efficiently. In addition, most succeed in reducing protein adsorption. Still, not all coatings are suitable for serum and plasma protein separations, for either the pH stability was insufficient or some proteins were found to interact with the coating itself. Most separations are based on the polyacrylamide (PAA) coating with siloxane bonding introduced by Hjertén [30,139]. This coating can be considered as a standard coating, and the procedure is easy to perform. The coating is stable in the pH range 2 to 10.5 and no affinity to proteins could be observed. The coating of capillaries with small inner diameters has some problems. As the viscosity of the acrylamide used for coating increases with reaction time, due to the formation of more and larger polymers, the solution becomes difficult to withdraw from the narrow capillaries afterward. If the time allowed for the reaction is reduced, the problem was alleviated, but the lifetime of the capillary decreased as well [136,140]. Another common coating used in protein separation is based on poly(ethylene glycol) (PEG). It was found that PEG coating greatly reduces the EOF and is stable for long periods of time. PEG is bound to the capillary wall either by an aminopropyl sublayer and 3-aminopropyltriethoxysilane [141] or by γ-methylacryloxypropyltrimethoxysilane [142]. PEG-coated capillaries performed adequately in the separation of serum and plasma proteins, but just as with PAA, PEG was found to lack hydrolytic stability at extreme pH values. The pH of the buffer used for the separation and the separation itself are thereby limited, a drawback especially in CIEF, where focusing over the entire pH range is desired. A coating of improved hydrolytic stability was achieved by binding PAA to the fused silica via Si—C bonds [143] or by using a polyethyleneimine coating [144], which inverses the electroendosmotic flow. Although the performance of these coatings was excellent, they are not commonly used because the coating procedure is complicated and time consuming.

If protein adsorption has to be prevented without a concomitant elimination of the EOF (e.g., in CZE), coulombic repulsion is exploited by using a buffer with a pH value above the isoelectric points of the proteins in question [89,145]. Several human serum, plasma, and blood proteins have been investigated by that method. For example α-, β-, and γ-chains of human fetal hemoglobin were analyzed with a variation coefficient of between-day runs of 5.7% (Fig. 11) [146]. However, extreme pH values may have an inverse and irreversible effect on protein stability. If protein adsorption to the capillary walls is regarded simply as an ion-exchange mechanism, the addition of salts to the buffer should reduce the effect [145]. This kind of high salt buffer was used successfully for rapid CZE of whole serum or plasma proteins [147]. Yet the addition of salt to the electrophoresis buffer also increases the conductivity, and thus more Joule heat is produced when voltage is applied. As discussed above, dissipation of the heat produced often presents a problem in CE that should not be aggravated unnecessarily. Protein adsorption can also be prevented by using nonionic surfactants such as Tween or BRIJ. Such coatings showed high protein recovery rates and a pH stability over the range 4 to 10.5. Other coatings, such as dextrane, methylcellulose [141], arylpentafluoro [148], LC and GC types [149,150], or "dynamic coatings" [151,152] have been developed, but their physical and chemical stabilities were found to be inferior to those of PAA or PEG coatings. Nevertheless, depending on the application, for some protein samples these coatings may be the method of choice.

At this time, coating has to be done by the individual user, as commercially available precoated capillaries are expensive and tend to be of low capillary-to-capillary reproducibility. This is the main reason why, at present, application of

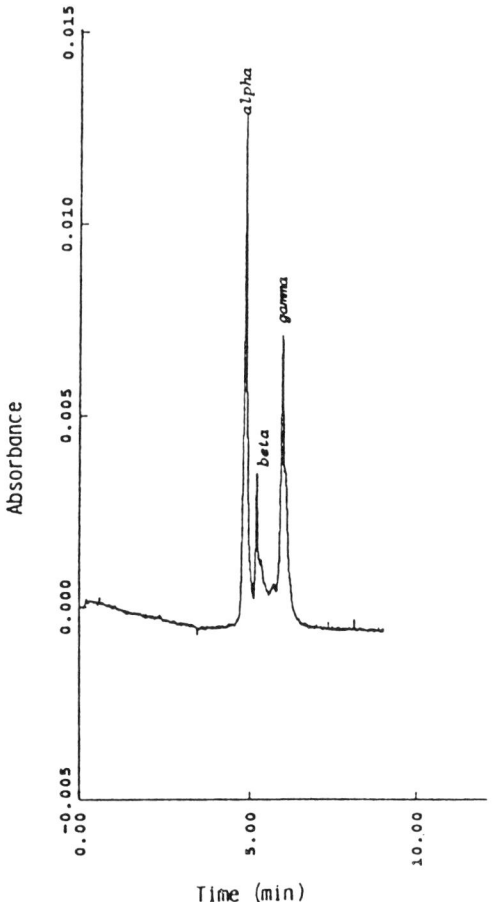

Fig. 11 Analysis of human fetal hemoglobin. Capillary, 42 cm length, 75 μm ID; buffer, 25 mM sodium phosphate pH 11.8; applied voltage, 20 kV, detection, 214 nm. (From Ref. 146.)

CE as a routine method is mainly reduced to CZE in uncoated capillaries, while promising high-resolution techniques such as CITP, CIEF, and CGE are not used to their full potential.

E. Separation of Serum and Plasma Proteins

This section deals with the separation of serum and plasma proteins by CE. No systematic proposition for optimizing the separation conditions is available, for the development of CE methodology is still done empirically.

Capillary Zone Electrophoresis

Although electrophoresis buffers strongly influence the separation of proteins, only a few guidelines for choosing an appropriate buffer are available, as most separation conditions were developed by empirical approaches. The electrophoretic buffer system should be chosen with regard to the stability, solubility, and degree of ionization of the analytes. The influence of the buffer ion and its pH on the electromigration of the proteins and the generation of heat should also be considered. The most commonly used buffers for capillary zone electrophoresis (CZE) are phosphate (pH 1.0 to 3.0, 6.0 to 8.5), acetate (pH 3.6 to 6.0), borate (pH 8.0 to 10.5), and zwitterionic buffers [169], such as Tris (pH 7.2 to 9.5), 2-[N-cyclohexylamino] ethanesulphonic acid (CHES) (7.0 to 8.5), and 2-[N-Morpholino] ethanesulphonic acid MES (pH 5.5 to 7.0) using salt concentrations ranging from 20 to 500 mM [10,56,165,166]. If CZE is performed in uncoated capillaries, the ionic strength and the pH of the buffer affects primarily the EOF and thereby the separation efficiency, for high pH and low ionic strength result in high velocities of the EOF [55,167]. On the other hand, high salt concentrations reduce the EOF and thus increase the retention times. The addition of zwitterionic additives may prove beneficial in this context, since these ions do not influence the conductivity of the buffer [168,169].

As discussed before, the buffer plays an important role in suppressing protein adsorption to the capillary walls. Optimal results were achieved by choosing an electrolyte system of high pH values and salt concentrations (Fig. 12), although this precludes analysis in the physiological range. The tendency of silica to dissolve at high pH values also limits this approach. For regeneration of the capillary and repeated etching of the capillary surface, 1 N NaOH and distilled water is rinsed through the capillary at least for 30 s. In the analysis of some proteins by CZE, such as purification of human IgM, other solvents (methanol, acids, etc.) have to be used. The capillary length and inner diameter restrict the applicable voltage in CE. By using longer or narrower capillaries, a higher potential can be used. Yet, as mentioned before, smaller diameters reduce the detection sensitivity, while longer capillaries will obviously increase the analysis time.

Through a variation of the parameters mentioned above, serum and plasma protein separation can be optimized. Most recent serum and plasma protein separations in CZE utilized borate and phosphate buffers in concentrations of 20 to 100 mM and a pH value of 6.5 to 7.8 or 9.0 to 10.5. The applied voltage is generally kept constant at 10 to 20 kV, but recently a voltage ramping for enhanced separation of several serum proteins was introduced [166]. The problem of small electrophoretic mobility differences masked by the EOF caused by high salt buffers and the resulting poor resolution of the peaks can be prevented by voltage ramping, as shown in Fig. 13.

Capillary Isotachophoresis

The separation of serum and plasma proteins with capillary isotachophoresis (CITP) is related to CIEF, with regard to the spacers, detection, coatings, and

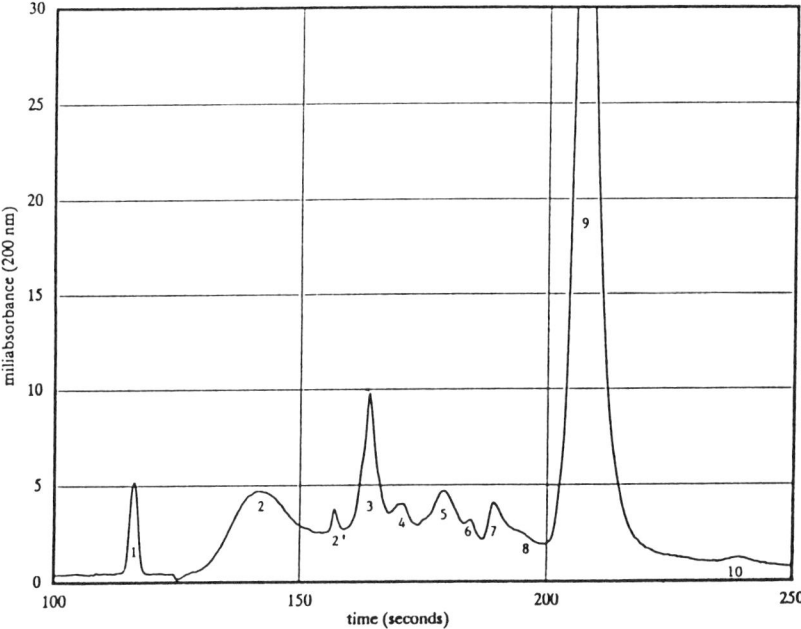

Fig. 12 Rapid separation of human serum proteins. Capillary, 25 cm length, 25 μm ID; applied voltage, 10 kV; high-ionic-strength borate buffer pH 10; detection, 200 nm; peaks: 1, dimethylformamide; 2, γ-globulin; 2', complements; 3, transferrin; 4, β-lipoproteins; 5, haptoglobin; 6, α_2-macroglobulin; 7, α_1-antitrypsin; 8, α_1-lipoproteins; 9, albumin; 10, prealbumin. (From Ref. 147.)

buffer additives used. More detailed discussion is given in the following section on CIEF. When the steady state in CITP is reached, a train of sample zones stack between the leading and terminating electrolytes, all migrating with the same velocity [178]. For the last 20 years, isotachophoretic analysis was performed in plastic tubes with an inner diameter of 200 to 500 μm, which inherently minimizes the EOF. The experience was transferred to modern CE using coated and uncoated fused silica capillaries [30], thus resolving human serum proteins into about 30 components (Fig. 14).

The common leading electrolyte is an acid such as phosphoric acid, HCl, or formic acid, titrated with ammediol or Tris to a pH value between 9.1 and 9.8. Most terminating electrolytes consists of a zwitterionic base such as valine or β-alanine, titrated to pH values between 9.4 and 10.4 with ammediol [179,180]. The spacer mixtures and buffer additives are mostly identical to the ampholytes used in CIEF and are discussed below. In several laboratories the application of discrete spacers such as amino acids is being investigated [181]. CITP has also been performed in uncoated capillaries using the same conditions as in coated

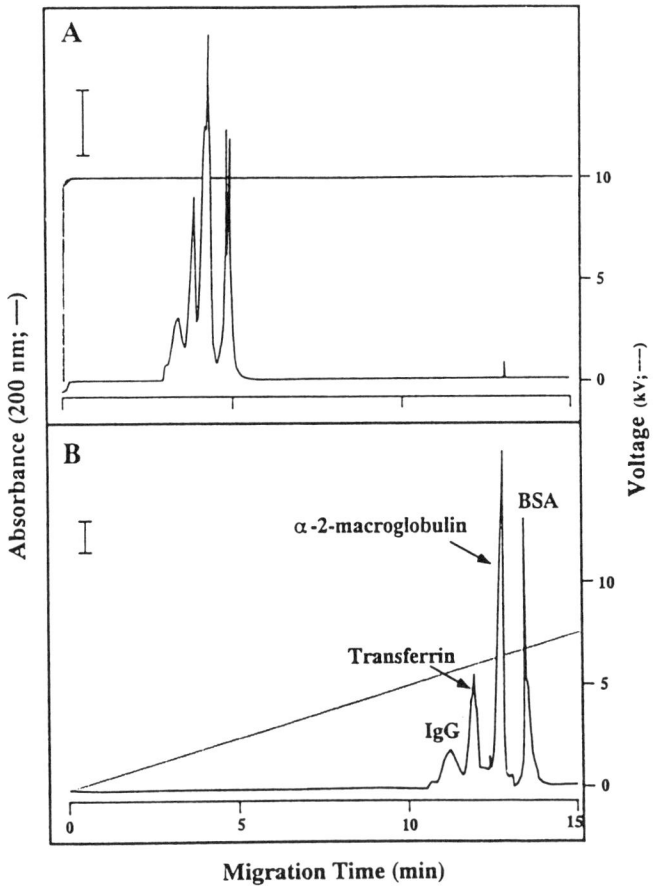

Fig. 13 Voltage ramping for enhanced separation of proteins. Capillary, 27 cm length, 50 μm ID; buffer, 500 mM sodium borate; detection, 200 nm; applied voltage: (A) 10 kV constant, (B) voltage ramp of 0 to 10 kV over 20 min. (From Ref. 166.)

columns [178,181], but no applications for the analysis of serum or plasma proteins have been reported. The resolving power of CITP for human serum and plasma samples is demonstrated in a number of publications, and further improvement of the spacers will probably enhance the selectivity of the separations [182,183].

Capillary Isoelectric Focusing

Capillary isoelectric focusing (CIEF) is a CE technique used frequently for the separation of serum and plasma proteins. A number of applications are presented in this section, proving CIEF equal to the related slab gel technique with respect to

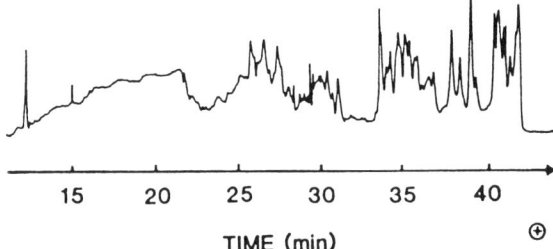

Fig. 14 High-performance displacement electrophoresis of human serum. Spacer, ampholytes pH 3 to 10; capillary, 35 cm length, 50 μm ID; leading solution (anolyte), 10 mM HCl–Tris pH 8.3; terminating solution (catholyte), 100 mM β-alanine/barium hydroxide pH 9.2; applied voltage, 10 kV; detection, 280 nm. (From Ref. 30.)

the resolution of complex mixtures in serum samples, and even superior regarding separation time and handling. Compared to isoelectric focusing in slab gels, CIEF is fast. Analysis is normally finished after 15 min [171]. The reproducibility is high, even with complex serum samples such as γ-globulins (Fig. 15). Most applications use 10 or 20 mM phosphoric acid as the anolyte and 20 or 40 mM sodium hydroxide as the catholyte [172]. The most common mobilization technique is performed by addition of 80 mM sodium chloride to either the anolyte or the catholyte. Other mobilization techniques, such as the application of zwitterionic salts, are still under development [171]. The stability of the pH gradient is increased by the low electrolyte concentration, for the buffer capacity of the ampholytes is not exceeded even at longer focusing times. The ampholytes, responsible for the pH gradient, are identical to the ampholytes used in slab gel isoelectric focusing. The available mixtures cover pH ranges from 3 to 10, or more discrete ranges such as 4 to 6, 5 to 7, 6 to 8, or 7 to 9. Usually, 1 to 5% v/v of ampholyte is added to the samples [173]. Since excess salt in the sample changes the pH gradient during the focusing step, the focusing time is prolonged. Optimal performance is achieved in narrow capillaries, despite the decreasing sensitivity if optical detection is used. A further loss in sensitivity is caused by the fact that detection is limited to wavelengths of more than 240 nm, as UV absorption of the ampholytes biases against detection below this wavelength [136]. For this purpose a gradient detector was invented, showing high sensitivity and, in combination with extremely short capillaries, excellent separation efficiencies [176]. Another problem is based on the fact that most commercial CE instruments are not designed to hold short capillaries, whereas the best separations in CIEF are achieved in these capillaries [171,175].

Since the EOF of uncoated capillaries would prevent the formation of a stable pH gradient, CIEF requires a coating of the inner surface of the capillary. The coating has to be stable at the extreme pH values present at the ends of the

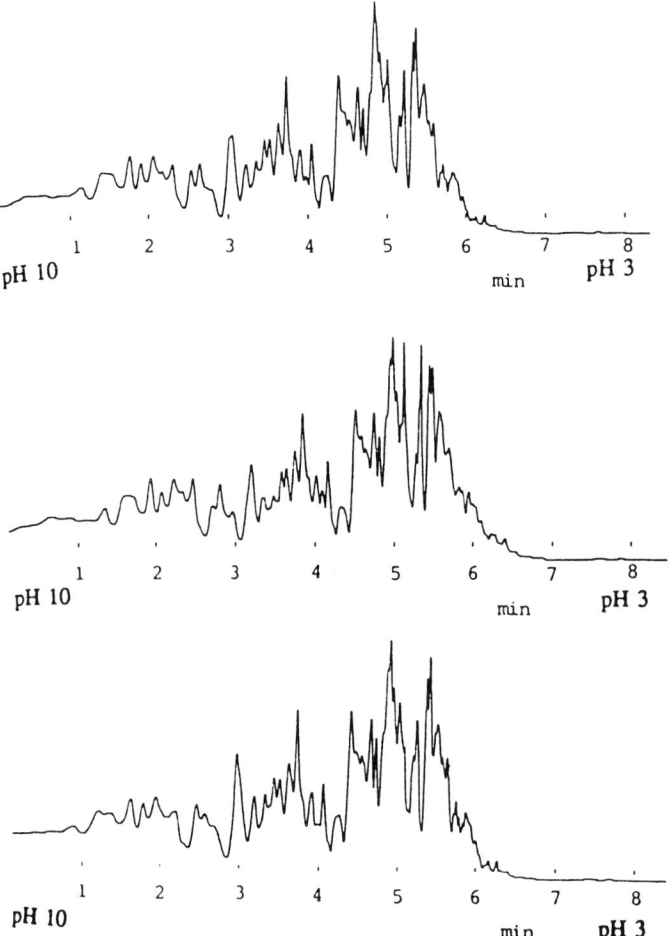

Fig. 15 Replicate separations of γ-globulins. Sample, 2% ampholytes pH 3 to 10, 1% Triton X-100; capillary, 20 cm length, 25 μm ID; focusing conditions, 10 mM phosphoric acid (anolyte), 20 mM NaOH (catholyte), 6 kV; mobilization conditions, zwitterionic solution pI 3.22, 8 kV. (From Ref. 171.)

capillaries. PAA is not truly suited since it hydrolyzes under these conditions. Yet it is commonly used in CIEF and the short lifetime of the capillary is accepted. Other materials, such as methylcellulose or dimethylpolysiloxane, have been suggested [172]. The addition of hydroxymethycellulose to the sample buffer was found to suppress any remaining EOF, resulting from a putatively incomplete coating under these conditions [163]. Generally speaking, the lack of truly reproducible coatings as well as the time required for coating constitute a drawback in CIEF.

The fact that in commercially available CE systems the detector is placed some distance from the capillary end poses another difficulty in CIEF, since this prevents the detection of proteins located between the monitor and the capillary end during mobilization. This problem is specific only for capillary isoelectric focusing. Through the addition of TEMED to the sample, this blind segment can be blocked by shifting the pH gradient to the point of detection [171].

Protein precipitation is a problem generic to isoelectric focusing. The proteins tend to aggregate and precipitate at their isoelectric points, when their net charge is zero. Precipitation can cause poor peak shapes, for in this case the counterions are stripped from the protein, resulting in zones of higher salt concentration and flow-disturbing heat generation (Fig. 16). Protein precipitation can be suppressed by the addition of nonionic detergents [27], ethylene glycol [177], and Triton X-100 [170,171].

Capillary Gel Electrophoresis

Capillary gel electrophoresis (CGE) is widely used as a protein-separating method in day-to-day routine analysis. Capillary electrophoresis overcomes the difficulties of the commonly used slab gels, for automation and easy quantification are introduced in the instrumental approach. Nevertheless, the experience gained in protein separation in slab gels can normally not be simply transferred to capillary electrophoresis.

As mentioned before, prerequisite to separation performance and gel stability, total elimination of the electroendosmotic flow by an appropriate coating should be considered. In case the gel or the entangled polymer itself interacts with the capillary wall and thereby suppresses the EOF, the coating can be neglected. Otherwise, inactivation of the inner capillary surface is recommended. Furthermore, the detergent sodium dodecyl sulfate (SDS), combined with mercaptoethanol as a reducing agent, dissociates the proteins into their subunits and creates protein–SDS complexes of constant mass/charge ratio. The electrophoresis buffers used in CGE are usually phosphate, borate, tricine, or glycine buffers of moderate pH (6.7 to 8.7) and salt concentrations in the range of 0.02 to 0.25 M.

In SDS gel electrophoresis, proteins are separated according to their size, in a polymer matrix that represents a sieve. The selectivity of the gel is controlled by the ratio of the concentration of the monomer to the concentration of the crosslinker. Pore sizes considered suitable in slab gel electrophoresis proved to be too narrow to allow large proteins to migrate along a capillary. Standard proteins with a molecular weight of 20,000 to 100,000 Da were separated by a cross-linked gel [11,153], but a complete separation of serum proteins lasted longer than 5 h for a single run [154]. These problems were solved by applying viscous solutions of an entangled polymer network functioning as a liquid sieving matrix [155–157], also

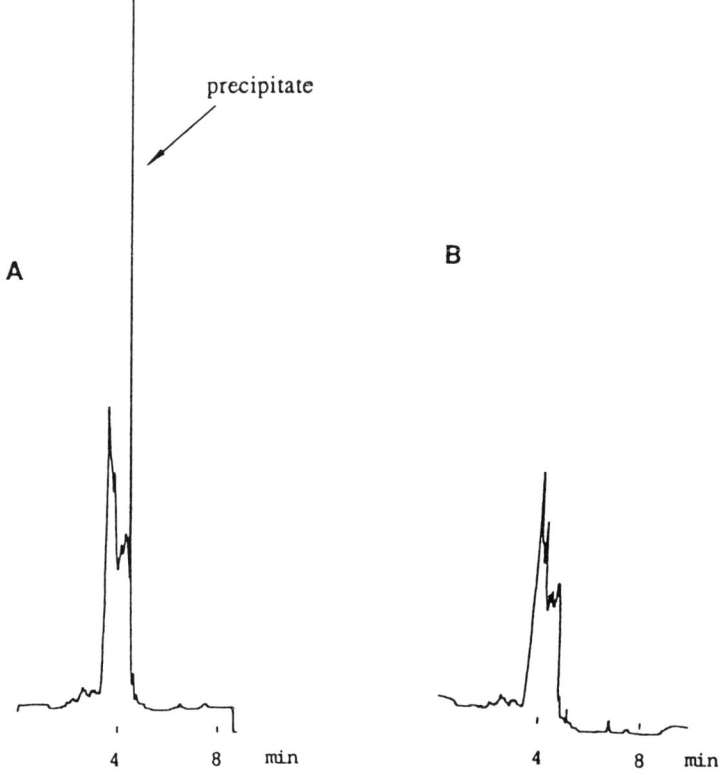

Fig. 16 Separation of monoclonal antibodies (A) with and (B) without precipitation. Sample, 2% ampholytes pH 3 to 10; capillary, 20 cm length, 25 μm ID; focusing conditions, 10 mM phosphoric acid (anolyte), 20 mM sodium NaOH (catholyte), 6 kV; mobilization conditions, zwitterionic solution pI 3.22, 8 kV; detection, 280 nm. (From Ref. 171.)

termed "linear gel." In contrast to the cross-linked polymer gels, where higher temperatures lead to deterioration of the gel, linear polymers are more stable with regard to temperature changes. However, both the migration time and the peak shape of the SDS–protein complex were found to decrease with increasing temperature [158]. Non-cross-linked PAA showed an excellent performance by separating proteins with relative molecular masses of 14,000 to 205,000 when the concentration of linear PAA was higher than 4%. In addition, large molecules such as IgG were already separated in gels containing 2% linear polyacrylamide (Fig. 17). At lower gel concentrations, non-cross-linked PAA is easily displaced from the capillary [159]. Nevertheless, even with linear gels, the lifetime of the column is limited to 40 to 60 separations per gel, especially if high voltages are employed. Additionally, the absorbance of polyacrylamide below 230 nm restricts detection to higher wavelengths, thereby causing a decrease in sensitivity. Karger et al. overcame these limitations by applying a transparent polymer matrix of dextran or PEG of low-to-moderate viscosity [160]. Even complex biological samples such as serum could be analyzed with a migration time reproducibility of 0.5% RSD or less when these matrices were used (Fig. 18). Using short separation distances, rapid migration of IgG in less than 2 min was achieved [161]. Other sieving matrices for protein separations are poly(ethylene oxide) [162], methylcellulose [163], or hydroxyethyl cellulose [156,164]. Further development in this area can be expected in the near future.

Entangled polymers as a sieving matrix overcome most problems of cross-linked gels, for sieving gels can easily be replaced from run to run. Problems of the cross-linked gels in CGE include the limited lifetime of the cross-linked gels due to air bubbles that occur at the end of the capillary as a result of drying, as well

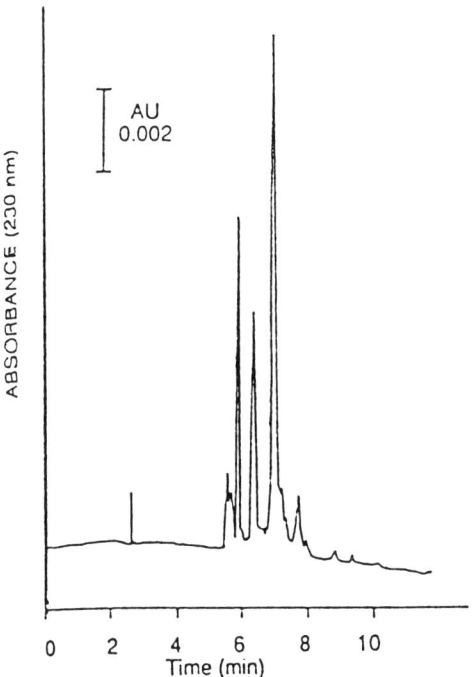

Fig. 17 Analysis of IgG (bovine) by CGE. Sample, SDS/protein mass ratio 2.5:1; capillary, 45 cm length, 75 μm ID; separation, 2% linear polyacrylamide, 0.1% SDS, Tris–borate buffer pH 8.1; applied voltage, 12 kV; detection, 280 nm. (From Ref. 159.)

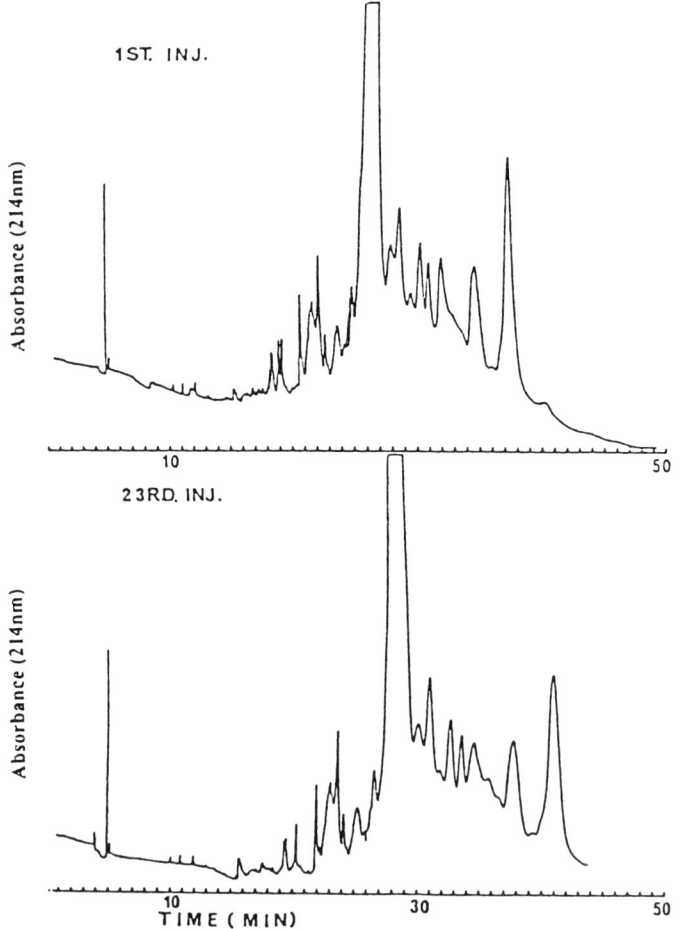

Fig. 18 Multiple direct injection of plasma (rat) using UV-transparent polymer networks (SDS–CGE). Capillary, 30 cm length, 100 μm ID; separation, 10% dextran (MW 2,000,000) in 50 mM AMPD–CACO pH 8.8, 0.1% SDS; sample, 60 mM Tris–HCl pH 6.6, 1% SDS, 5% 2-mercaptoethanol; applied electric field, 300 V/cm; detection, 214 nm. (From Ref. 160.)

as their low thermal stability, low flexibility in regard to buffer changes, and with regard to protein separations, the inappropriate small pore sizes. Since the systematic development of linear sieving gels started only 2 years ago, there are only a few applications to serum and plasma proteins. However, the development of these gels offered the first real size-dependent and reproducible protein separations in CE.

III. USE OF CE IN SERUM AND PLASMA PROTEIN SEPARATION

In this section a detailed overview of the published separations of serum and plasma proteins is given. These separations are grouped into CE method categories. In Appendix B, the grouping into serum and plasma proteins is omitted. The primary aim is to furnish information on the applicability of CE in clinical or pharmaceutical application. In addition, all information gained from such a complex mixture of proteins as serum can be the basis for developing analytical procedures for protein samples.

A. Capillary Zone Electrophoresis

Using 50 mM sodium borate at a pH value of 9.6, 10.0, 10.2, and 11.0, whole serum, diluted 40:1 in 1 mM boric acid containing 20% ethylene glycol, was separated into albumin and α_1-, α_2-, β-, and γ-globulins within 10 min. The applied voltage was 10 kV at 20°C. The detection was performed by UV absorption at 200 nm. Serum of a multiple-myeloma patient was analyzed and compared with the results obtained with agarose gels [51]. Similar results were achieved within 8 min using 50 mM borate buffer at pH 9.5, an applied voltage of 30 kV, and UV absorption at 200 nm. NIST serum, albumin, and IgG-depleted serum (Fig. 19) were analyzed [46].

Comparable results were achieved by Chen et al. [184,205] analyzing IgG kappa myeloma serum and hemoglobin (A, S, and C) in normal serum (diluted 1:20) with a borate buffer of 80 mM at pH 10.0 and a column voltage gradient of 200 V/cm. Detection was at 214 nm. The separation time could be reduced to 4 min using 150 mM borate buffer pH 10.5 and 10 kV at 22°C. The proteins, γ-globulin, transferrin, β-lipoproteins, haptoglobin, α_2-macroglobulin, α_1-antitrypsin, α_1-lipoproteins, albumin, and prealbumin were analyzed [147,189]. With these electrophoretic conditions, abnormal serums with elevated IgG, IgM, and IgA levels were investigated [52]. The proteins were monitored at 200 or 206 nm. The separation could be shortened to 90 s by using lower buffer strength and a voltage of 10 kV [147].

Using a coated capillary, a 100 mM Tris–HCl electrophoresis buffer (pH 8.5), and a potential of 10 kV, human serum proteins could be separated and monitored at 230 mn. The peak patterns were similar to those achieved with uncoated capillaries [7]. Rabbit IgG, bovine serum albumin (BSA), transferrin, and α_2-macroglobulin in 25 mM PBS were separated in 500 mM sodium borate buffer run at 10 kV constant voltage or a voltage ramp of 0 to 10 kV over 20 min at 25°C [166].

The optimization and validation of analytical CE conditions for BSA was investigated by varying the borate buffer from 50 to 300 mM and the pH value between 8.0 and 9.0 at a constant voltage of 12 kV and a temperature of 20°C.

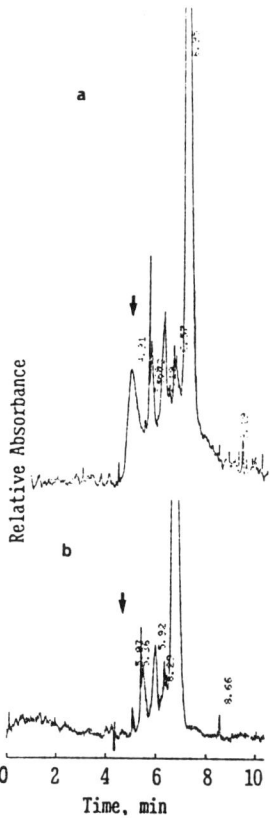

Fig. 19 Comparison between normal human serum (a) and IgG-depleted human serum (b). Conditions: capillary, 100 cm length, 50 μm ID; applied voltage, 30 kV; buffer, 50 mM sodium borate, pH 9.5; detection, 200 nm. (From Ref. 46.)

The detection was carried out at 214 nm. Additives such as Triton X-100, triethylamine, urea, and ethylene glycol were applied with coated and uncoated capillaries [185]. Additionally, BSA and BSA dimer were determined by CZE–MS [132].

The separation of immune complexes from free unreacted antibody and antigen was performed using 100 mM tricine electrophoresis buffer at pH 8.0 and a voltage of 30 kV at a temperature of 25°C. Human insulin and human growth hormone (hGH) together with their antibodies (IgG class) were separated and characterized. The UV absorbance wavelength was 200 nm [186]. The antibody–antigen complexes of hGH and the antibody (monoclonal immunglobulin G) were investigated further [187]. At a temperature of 30°C and a UV detection at

200 nm, a 100 mM tricine buffer pH 8.0 and a voltage of 30 kV were applied. The complexes and the unreacted proteins could be separated in various mixtures within 12 min (Fig. 20).

CZE was employed for the separation of unpurified enzyme-labeled monoclonal antibody conjugates from the conjugate, unreacted alkaline phosphatase and IgG as a fast in-process control [188]. Methylcellulose (0.5%) and 0.5 mM SDS in a running buffer of 100 mM borax at pH 10.0 was used. The separation was performed at a temperature of 15°C and a voltage of 5 kV, with the detection at 280 nm. Monoclonal antibody microheterogeneity analysis was performed at 200 nm in phosphate buffer and an applied voltage of 12 kV, and the immunoglobulin G isoforms could be resolved [190].

A monoclonal immunoglobulin G directed as an antibody against human placental prolyl 4-hydrolase was derivatized with fluorescein isothiocyanate

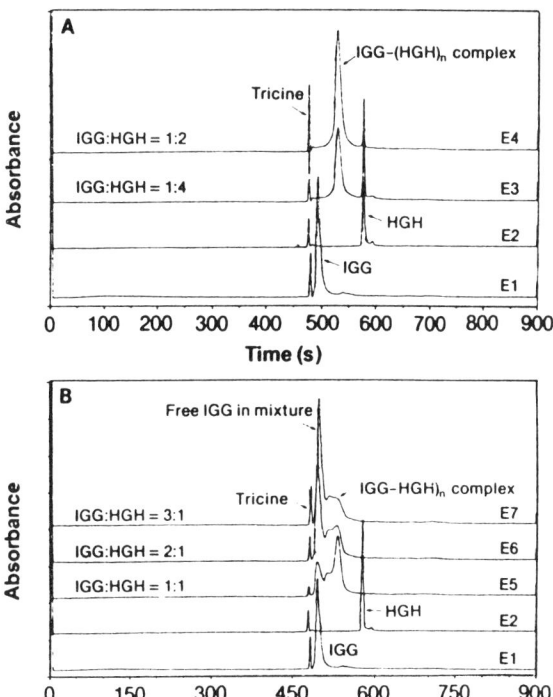

Fig. 20 Separation of IgG, hGH, and IgG–(hGH)$_n$ complexes by CZE: (A) electropherograms of IgG, hGH, and mixtures containing an excess of hGH; (B) electropherograms of IgG, hGH, and mixtures containing an excess of IgG. Capillary, 100 cm length, 50 μm ID; separation, 100 mM tricine buffer pH 8.0; applied voltage, 30 kV; detection, 200 nm; temperature, 30°C. (From Ref. 187.)

(FITC). The sample was separated into labeled antibody and unreacted FITC at 25°C and 20 kV using 50 mM borate buffer (pH 8.3). Fluorescence detection was performed using a mercury lamp. The bandpass filter was a 450- to 490-nm filter and the emission filter was a 515-nm interference filter [118]. A monoclonal antibody directed against the human receptor for interleukin-2 was derivatized with fluorescamine, and the effects of buffer constituents were studied. The applied voltage was 29 kV. The capillary temperature was maintained at 25°C and the labeled and unlabeled analytes were monitored at 214 nm. Various pH values, constituents, concentrations, and salt additives were studied [204].

The purification of the isoforms of human monoclonal antibodies (IgG) against the gp-41 of AIDS virus and of human recombinant superoxide dismutase (r-SOD) has been monitored by free zone capillary electrophoresis (FZCE). The electrophoretic conditions were 10 kV with 100 mM phosphate buffer (pH 5.8) for the monoclonal antibody or 21 kV with 20 mM citrate buffer (pH 4.0) for r-SOD. Quantification was satisfactory with detection at 200 nm [202].

Biosynthetic human insulin and biosynthetic human growth hormone were examined and compared to results achieved by reversed-phase high-performance liquid chromatography (RP-HPLC). Excellent correlation was observed between these techniques, applying approximately 30 kV with FZCE. The buffer consisted of 20 mM tricine, 5.8 mM morpholine, and 20 mM NaCl. Detection was at 200 nm [191]. Human growth hormone and its digests [192] were investigated further at selected pH values of 4.0, 6.0, and 9.0 by FZCE for quality control [193] or as a protein standard in CZE–MS [201]. Human growth hormone with a molecular weight of 22,125 and an isoelectric point of 5 was used as a model protein for a practical approach to CZE to examine various injection conditions, buffers, and voltages. Reproducibility and relation between peak area and sample concentration were studied [60]. High-density (HDL) and low-density (LDL) plasma apolipoproteins were analyzed within 12 min using 30 mM borax buffer containing 0.1% SDS (pH 9.0) in coated and uncoated capillaries. The applied voltage was 333 V/cm for all experiments, using a wavelength of 220 nm for detection [194].

Analysis of carbohydrate-mediated heterogeneity and characterization of N-linked oligosaccharides of recombinant tissue plasminogen activator (rt-PA) and α_1-acid glycoprotein was performed with 20 min, using 100 mM phosphate (pH 6.6) or 100 mM tricine buffer (pH 8.2) containing 1.25 mM putrescine [195]. The analyses of microheterogenity of glycoproteins were investigated further with several modifiers of the EOF [196]. CZE was used to fractionate the rt-PA glycoforms. The sialic acids were removed from the carbohydrate chains by treating the sample with neuraminidase; desialyated and untreated rt-PA were compared. Electrophoresis was performed with 100 mM ammonium phosphate buffer at pH 4.6, containing 0.01% Triton X-100 and 200 mM ϵ-aminocaproic acid [170]. Furthermore, these conditions were employed for monitoring the charge heterogenity of human rt-PA and r-hGH [203]. Recombinant interleukin-3 [197] and related proteins [198] such as human serum albumin were separated and collected using 10 mM bicarbonate (pH 7.2) or 20 mM CHES containing 10 mM KCl (pH 9.0). The detection was at 220 nm and a voltage of 15 kV was used.

Globin chains of human hemoglobin were analyzed by FZCE under optimized conditions and the reproducibility was investigated; α-, β-, and γ-globin chains were separated using 25 mM phosphate buffer at pH 11.8, 22°C, and 20 kV with excellent reproducibility in between-day runs [146]. Abnormal hemoglobins were also examined at a lower pH of 9.98 (100 mM borate buffer, 30 kV) and an elevated temperature of 30°C [199]. Globin chain analysis of adult human hemoglobin A, fetal hemoglobin F, and hemoglobin variants S and C was performed at a low pH value of 3.2 (100 mM phosphate) in the presence of 7 M urea and 1% Triton X-100 [93]. The globin chains were separated at 8 kV constant voltage within 20 min. The detection wavelength was 210 nm. Several recombinant human interleukins, interferons, and the tumor necrosis factor were analyzed using a polyimide-coated and an uncoated capillary. The temperature of the separation was elevated from 15°C to 30°C. Phosphate buffers with pH values of 6.0, 7.0, and 8.0 were employed with a voltage of 17 kV and a detection wavelength of 200 nm [209].

Human transferrin isoforms were also investigated by CZE. The di-, tri-, tetra-, penta-, and hexasialo transferrins were separated either with or without a neuraminidase treatment in a running buffer of 18 mM boric acid–0.3 mM ethylenediaminetetraacetic acid (EDTA) (pH 8.4) using a voltage of 8 kV and a detection wavelength of 280 nm [172].

B. Capillary Isotachophoresis

Hjertén et al. used a leading buffer of 10 mM Tris–HCl (pH 8.3) and 100 mM β-alanine/barium hydroxide (pH 9.2) as a terminating buffer. The separation of human plasma was performed in coated (polyacrylamide) and uncoated capillaries, and the proteins were monitored at 280 nm. Discrete and continuous spacers were used and the influence of the voltage was investigated [30]. A large number of discrete spacers for serum protein separation were investigated and are discussed [181,207]. Human plasma and the human growth hormone was separated with 50 mM phosphoric acid–Tris (pH 6.6) as leading buffer and 23 mM glycine–Tris (pH 9.3) as terminating buffer. The sample contained pharmalytes (pH 4.5 to 6 and 3 to 10). A voltage of 3 kV was used with detection at 280 nm [39].

Various IgG preparations were investigated using a hydroxypropyl methylcellulose (HPMC)-coated flattened poly(ethylene-propylene) capillary. The leading electrolyte was 5 mM HCl–9.3 mM 2-amino-2-methyl-1-propanol (pH 9.9), and the terminating solution was 50 mM transexamic acid–15 mM potassium hydroxide (pH 10.8). The sample solution contained 0.3% ampholytes (pH 7 to 9 and 9 to 11) [138].

Serum lipoproteins were investigated using hydroxypropyl methylcellulose as an additive in Teflon capillaries. Various amino acids were applied as spacers, optimizing the separation of the hydrophobic proteins (HDL, VDL, IDL, and LDL). The leading electrolyte consisted of 5 mM phosphoric acid, 0.25% HPMC, and 20 mM ammediol (pH 9.2). The terminating electrolyte was 100 mM valine, adjusted to pH 9.4 with ammediol. Detection was performed at 280 nm or, after prestaining with Sudan Black B, at 570 nm [180].

Capillary isotachophoresis of human serum proteins was investigated thoroughly with respect to specially designed isotachophoretic equipment. This isotchophoretic method used capillaries with an inner diameter of 200 to 600 μm. Despite the fact that this technique is somewhat different from high-performance CE, the experimental conditions can generally be transferred. For detailed information the publications of Everaerts et al. [31] and Delmotte [32] are recommended.

C. Capillary Isoelectric Focusing

Isoelectric focusing of γ-globulins was performed by Hjertén using 20 mM NaOH as catholyte and 10 mM phosphoric acid as anolyte [25]. The sample was dissolved in a 1% solution of polybuffer (pH 4 to 9) containing 1% v/v Triton X-100 R as a sample additive. For mobilization the anolyte was replaced with 10 mM phosphoric acid, containing 80 mM NaCl. Both focusing and mobilization were conducted at 6 kV. The capillary was coated with methylcellulose or polyacrylamide. Hemoglobin A$_{1c}$ was dissolved in 1% biolyte (pH 6 to 8) and separated applying the experimental conditions as described for γ-globulins above [25]. Using this method, transferrin was analyzed [26]. A mixture of hemoglobin and transferrin containing a 2% v/v solution of pharmalyte (pH 3 to 10) was focused at 6 kV with 20 mM phosphoric acid as anolyte and 20 mM sodium hydroxide as catholyte. Mobilization was performed by replacing the catholyte with 10 mM glycine (pH $\approx$ 6) and with the application of a voltage of 6 kV, the capillary was emptied by pumping anolyte from the cathodic side into the tube. On-line detection was performed at 280 nm [25].

Separation of transferrin and hemoglobin was performed in coated (non-cross-linked polyacrylamide and methylcellulose) and uncoated capillaries. The sample was solved in 1% pharmalyte (pH 3 to 10). Focusing was carried out with on-line detection at 280 nm at 2 kV with 20 mM phosphoric acid as anolyte and 20 mM sodium hydroxide as catholyte. Mobilization was achieved by replacing the anolyte with 20 mM sodium hydroxide [139]. Isoelectric focusing of recombinant human interleukins, tumor necrosis factor, and interferon was performed with ampholytes (1% pH 3 to 10) and a focusing and mobilization voltage of 8 kV in polyimide-coated capillaries [200]. Under these experimental conditions, human growth hormone (hGH) was separated in a sample mixture with 2% ampholyte (pH 3 to 10 or 4 to 6) in a polyacrylamide-coated capillary [192].

Recombinant tissue plasminogen activator (rt-PA) glycoforms were analyzed in a sample mixture of 0.4% TEMED and 1% ampholyte solution (pH 3 to 10). Focusing conditions were 10 mM phosphoric acid as anolyte, 20 mM NaOH as catholyte, and a voltage of 12 kV. Cathodic mobilization was achieved by adding 80 mM NaCl to the catholyte. The mobilizing voltage was set at 8 kV. Detection was at 280 nm [170]. γ-Globulins were separated in various sample mixtures. The influence of Triton X-100, TEMED, and 100 mM NaCl in 2% ampholyte solution on the separation efficiency was investigated. The mobilization through salt addition or application of zwitterions in basic solution (pI 3.22) replacing the catholyte was examined. The mobilization voltage was 8 kV and the proteins were monitored at 280 nm. Focusing was achieved using 10 mM phosphoric acid as anolyte and 20 mM NaOH as catholyte at a voltage of 6 kV [171]. Under identical experimental conditions, human monoclonal antibodies were investigated [171]. Based on this antibody analysis, anti-TAC (IgG1 class humanized hybrid antibody against human interleukin-2 receptor) was investigated and the results were compared to slab gel isoelectric focusing methods [206].

The application of CIEF with a concentration gradient detector to the analysis of human blood serum, serum albumin, hemoglobin, and monoclonal antibodies was investigated [136]. The anolyte and catholyte were 10 mM phosphoric acid and 20 mM NaOH. The applied focusing voltage was 8 or 6 kV. Cathodic mobilization was performed replacing the catholyte with a solution containing 20 mM NaOH and 80 mM NaCl (10 or 12 kV). Samples were mixed with pharmalytes (pH 3 to 10) to a final concentration of 2% ampholytes. Using these conditions, hemoglobin chains were separated within 9 min, including the focusing time [163].

CIEF was used to separate iron-free and iron-containing transferrin using 20 mM phosphoric acid (focusing) or 20 mM NaOH (mobilization) as anolyte and 20 mM NaOH as catholyte (Fig. 21). The voltage was 5 kV and the proteins were monitored at 280 nm. The sample mixture contained 2% ampholytes (pH 5 to 7) [173]. Using identical experimental conditions, asialo, mono-, di-, tri-, tetra-, penta-, and hexasialo transferrin components could be separated [172].

Isoelectric focusing in capillaries was evaluated for hemoglobin A, F, S, and C and globin chains. The samples, containing 2% ampholytes (pH 3 to 10), were focused at 7 kV, using 40 mM NaOH as catholyte and 20 mM phosphoric acid as anolyte. Cathodic mobilization was performed by replacing the catholyte with a zwitterionic solution and application of 8 kV mobilization voltage. Single-wavelength mode detection was at 280 nm; in the scanning mode spectra were acquired at 5-nm intervals [95].

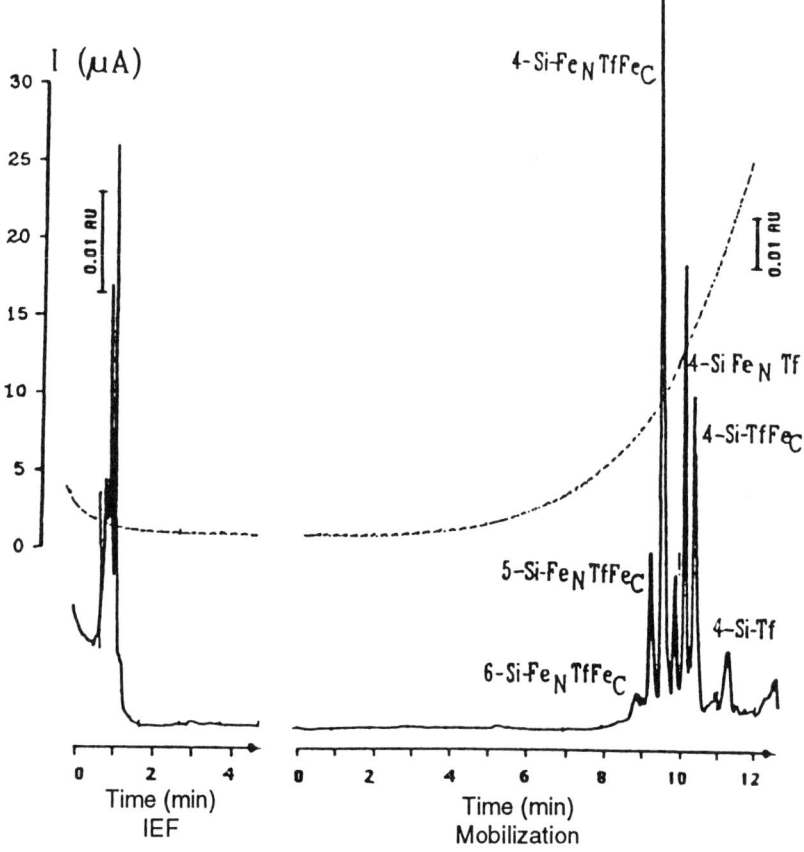

Fig. 21 CIEF of transferrin sample. Sample, 2% ampholytes pH 5 to 7; capillary, 18.5 cm length, 25 μm ID; focusing conditions, 20 mM phosphoric acid (anolyte), 20 mM NaOH (catholyte), 5 kV; mobilization conditions, 20 mM NaOH (anolyte), 5 kV; detection, 280 nm. (From Ref. 173.)

D. Capillary Gel Electrophoresis

Using 10% liquid linear polymer in a 50 mM phosphate buffer at pH 5.5 containing 0.5% SDS, conalbumin, ovalbumin, bovine albumin, and myoglobin (treated with 0.5% SDS at 100°C for 5 min) were separated within 60 min. A constant voltage of 20 kV, resulting in a current of 120 to 160 μA, was applied. UV absorption was measured at 254 nm [157]. The purity of IgG (bovine) was examined by SDS–CGE. The IgG sample was treated with SDS in a mass ratio of SDS to protein of 2.5:1. UV detection was performed at 280 nm. Using a 2%

polyacrylamide linear gel containing 0.1% SDS in a Tris–borate buffer (8.1), the heavy chain, light chain, and a complex of both could be separated from impurities within 10 min [149].

Karger et al. used dextran as a UV-transparent polymer network. Detection was performed at 214 nm. Dextran (10%; MW 2,000,000) was dissolved in 50 mM AMPD–CACO (AMPD: 2-amino-2-methyl-1,3-propanediol; CACO: cacodylic acid) at pH 8.8, containing 0.1% SDS. Protein samples were treated with a solution of 60 mM Tris–HCl (pH 6.6), 1% SDS, and 5% 2-mercaptoethanol for 10 min at 100°C. The heavy and light chains of human IgG were separated [160]. Whole plasma (rat) could be analyzed within 50 min. The human growth hormone was separated from its dimer using a 8% linear polyacrylamide gel in 120 mM Tris–120 mM histidine (pH 8.8) buffer, containing 0.1% SDS. Detection was at 280 nm [160].

Using a capillary with an effective path length of 7 cm to the detector, the light and heavy chains of human IgG were separated within 2 min (Fig. 22) [161].

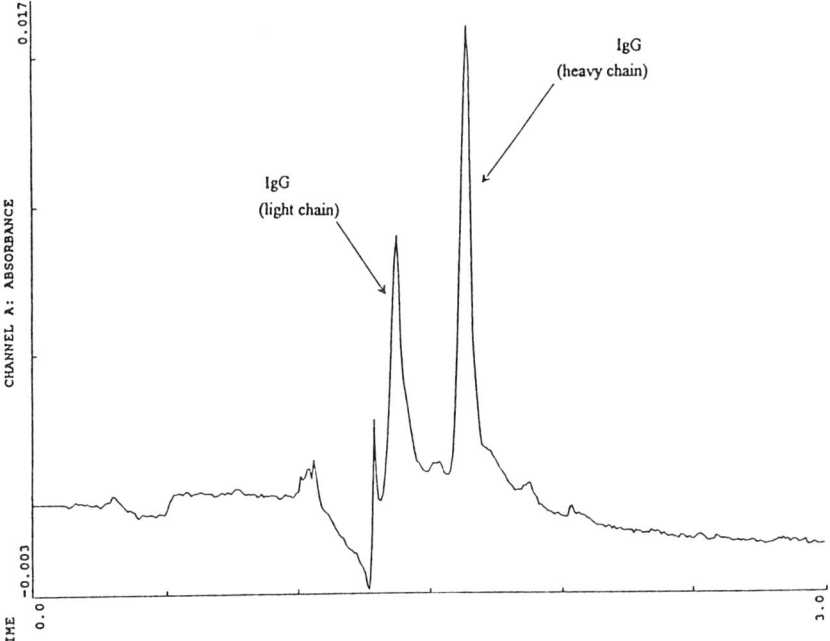

Fig. 22 Rapid SDS–CGE separation of human IgG. Capillary, 50 μm ID, 27 cm length—effective length 7 cm; separation, 10% dextran (MW 2,000,000), 50 mM Tris–CHES pH 8.6, 0.1% SDS; applied field strength, 740 V/cm; sample, 100 mM Tris–CHES pH 8.6, 1% SDS, 5% 2-mercaptoethanol; detection, 200 nm. (From Ref. 161.)

The proteins were monitored at 200 nm. The gel consisted of 10% dextran (MW 2,000,000) containing 100 mM Tris–CHES (pH 8.6) and 0.1% SDS. The sample was dissolved in 50 mM Tris–CHES (pH 8.6), 1% SDS, and 5% 2-mercaptoethanol. The sample mixture was heated to 95°C for 15 min.

E. Combined Separation Techniques

An apparatus for coupled high-performance liquid chromatography and capillary electrophoresis was used for analysis of serum proteins. First, the serum was separated by a gel permeation chromatography (GPC). The fractions were injected into the capillary via an injection port connected to the HPLC and CE system. The GPC fractions containing the serum proteins were further separated by capillary isotachophoresis [208]. This method is also termed two-dimensional capillary electrophoresis. In the first dimension, the serum proteins are separated by size, and in the second, separation is based on the electrophoretic mobility. Based on this high-resolution method, serum proteins such as α_2-macroglobulin, haptoglobulin, transferrin, albumin, and IgG could be analyzed.

Another analytical and preparative technique is based on the adsorption of the proteins onto a moving blotting membrane as they migrate out of the capillary. The proteins adsorbed are visualized by conventional staining techniques or by specific fluorescent labeling. Human serum proteins were separated by capillary isotachophoresis and immobilized on a moving PCDF [poly(vinylidene difluoride)] membrane. The adsorbed serum proteins were first labeled with FITC and visualized under UV light. For specific identification of transferrin, the membrane was treated with alkaline phosphatase antitransferrin [209]. Furthermore, human growth hormone labeled with radioactive iodine was separated with this method and the radioactivity was measured on the membrane.

IV. SUMMARY

In recent years, capillary electrophoresis has proved extremely useful for separation of biomolecules such as nucleotides, amino acids, peptides, and proteins. The advanced instrumental design of commercial CE equipment has offered several advantages to conventional electrophoresis. Precise sample application, temperature regulation, and stable, reproducible instrument control with respect to voltage, current, pressure, and run-to-run reproducibility is routinely achieved in contemporary commercially available equipment. Based on these instrumental features, CE is already used as a routine method for the analysis of small molecules.

In contrast, serum and plasma proteins represent a difficult challenge for capillary electrophoresis. Routine analysis of serum and plasma proteins is reduced to the application of capillary zone electrophoresis, which proved to be a reliable method for rapid screening of whole serum, purity control of serum proteins and fractions, as well as on-line quantification of the components. With regard to the simplicity of the method and the low cost of operation, CZE proved to be an attractive alternative to comparable conventional methods.

Other separation techniques, such as CITP, CIEF, and CGE, provide higher resolution of the sample proteins than that provided by CZE. Furthermore, information about physicochemical properties of the analyte, such as isoelectric point or molecular weight, can be gained by CIEF or CGE. Even though this information is essential for protein analysis and characterization, these methods are not commonly used at present because reliable methods for controlling the EOF are not commercially available. Coated capillaries are expensive, and it is still difficult to purchase a set of coated capillaries with identical properties. As long as the reproducibility from capillary to capillary cannot be guaranteed, the high resolving methods will not be applied in routine analysis. The development of reliable regulation of the EOF, either by an appropriate column material or coating or an external field control, will certainly lead to a breakthrough for the capillary electrophoresis of human serum and plasma proteins.

APPENDIX A: REGISTER OF HUMAN PROTEINS SEPARATED BY CE

Protein	Method	Refs.
Whole serum, serum standard	CZE	7, 46, 51, 52, 147, 184, 205
	CITP	30, 39, 181, (31, 32)
	CIEF	136
	CGE	160
Abnormal serum	CZE	46, 51, 52, 184, 205
Albumin, prealbumin	CZE	46, 132, 166, 185, 198
	CIEF	136
	CGE	157
Antitrypsin	CZE	147, 189
γ-Globulins	CZE	147
	CIEF	25, 171
Glycoproteins	CZE	195, 196
Haptoglobulins	CZE	147, 189, 208
Hemoglobins, globin chains (A, S, C)	CZE	184, 193, 199, 205
	CIEF	25, 26, 95, 136, 139, 163
Human growth hormone	CZE	60, 186, 187, 192, 193, 201, 203
	CIEF	192
	CITP	39
Immunglobuin G, antibodies	CZE	52, 118, 147, 186–188, 190, 202, 204, 208
	CITP	138
	CIEF	136, 171
	CGE	149, 160, 161
Insuline	CZE	191
Interleukin-2, -3, interferon	CZE	197, 198, 204, 209
	CIEF	200
Lipoproteins	CZE	147, 194
	CITP	147, 180
Macroglobulins	CZE	147, 166, 208
Tissue plasminogen activator	CZE	170, 195, 203
	CIEF	170
Transferrin	CZE	166, 172, 208, 209
	CIEF	25, 26, 139, 172, 173

APPENDIX B: PHYSICAL AND CHEMICAL PROPERTIES OF SELECTED HUMAN SERUM AND PLASMA PROTEINS

Fraction	Compound	Molecular weight	Isoelectric point	Amount in normal plasma (mg/mL)	Electrophoretic mobility, pH 8.6, Barbital $I = 0.1$
	Prealbumin	61,000	4.7	0.28–0.35	7.6
	Albumin	69,000	4.9	35–45	5.92
α_1	Antitrypsin	45,000	4.0	2.1–4.0	5.42
	Lipoprotein	50,000			
	High-density lipoproteins				
	HDL_2	435,000	—	0.37–1.17	—
	HDL_3	195,000	—	2.17–2.7	—
	$HDL_{1,185}$	166,000–175,000	—	—	—
	$VHDL_1$	148,000–153,800	—	—	—
	Orosomucoid	44,100	2.7	0.75–1.0	5.2
	α_1-Acid glycoprotein				
	Transcortin	51,700	—	0.041 ± 0.004	4.9
	Corticosteroid binding globulin (CBG)				
	α_1X-Glycoprotein	68,000	—	0.4–0.6	—
	α_1-Antichymotrypsin				
	α_1B-Glycoprotein	50,000	—	0.19–0.25	—
	α_1-Easily precipitable glycoprotein				
α_2	Zn-α_2-Glycoprotein	41,100	3.8	0.042–0.054	4.2
	α_2HS-Glycoprotein	49,000	4.1–4.3	0.30–0.90	4.2
	Mucoproteins	—	4.9	5	—
	Haptoglobin	100,000			
	Type 1-1		4.1	1.0–2.2	—
	2-1, 2-2		4.5	1.2–2.6	
	α_2-Lipoproteins	$5 \times 10^6 – 20 \times 10^6$	—	1.5–2.3	—

APPENDIX B Continued

Fraction	Compound	Molecular weight	Isoelectric point	Amount in normal plasma (mg/mL)	Electrophoretic mobility, pH 8.6, Barbital $I = 0.1$
	Low-density lipoproteins (LDLs)				
	Ceruloplasmin	134,000	4.4	0.27–0.39	4.6
	Prothrombin	68,700	4.2	0.1	—
β_1	Lipoproteins (LDLs)	3.2×10^6	—	2.8–4.4	3.1
	Hemopoxin	57,000	—	0.5–1.0	3.1
	Heme-binding β-globulin				
	Plasminogen	81,000	5.6	0.48 ± 0.09	3.7
	Profibrinolysin		6.3–8.6		
	Fibrinogen	341,000	5.8	2–6	2.1
$\beta_1\beta_2$	Complement	80,000–400,000	—	0.01–1.7	—
	Transferrin	76,000	5.2	2–4	3.63
β_2	Glycoprotein III	35,000	—	0.05–0.15	—
	Glycoprotein I	40,000	6.2	0.15–0.30	1.6
γ	Blood group globulins and immunoglobulins	150,000–1,000,000	6.3–7.3	7–15	—
	Immunoglobulin E IgE, γE	190,000	—	6×10^{-5}–10^{-3}	—
	Immunoglobulin D IgD, γD	172,000–148,000	—	0.003–0.4	—
	Immunoglobulin A IgA, γA	162,000	—	3.28	2.1
	Immunoglobulin M IgM, γM	1,000,000	5.1–7.8	0.8–0.9	2.1
	Immunoglobulin G IgG, γG 7 S-globulin	153,000	5.8–7.3	12–18	1.2

Source: Refs. 210 to 212.

REFERENCES

1. S. Hjertén, S. Jerstedt, and A. Tiselius, *Anal. Chem., 11:* 211 (1965).
2. S. Hjertén, *Chromatogr. Rev., 9:* 122 (1967).
3. S. Hjertén, *J. Chromatogr., 270:* 7 (1983).
4. J. W. Jorgenson and K. D. Lukacs, *Anal. Chem., 53:* 1298 (1981).
5. J. W. Jorgenson and K. D. Lukacs, *J. Chromatogr., 218:* 209 (1981).
6. J. W. Jorgenson and K. D. Lukacs, *J. High Resolut. Chromatogr. Chromatogr. Commun., 4:* 230 (2981).
7. J. W. Jorgenson and K. D. Lukacs, *Science, 222:* 266 (1983).
8. F. E. P. Mikkers, F. M. Everaerts, and T. P. E. M. Verheggen, *J. Chromatogr., 169:* 11 (1979).
9. S. Terabe, K. Otsuka, K. Ichikawa, A. Tsuchiya, and T. Ando, *Anal. Chem., 56:* 111 (1984).
10. B. L. Karger, A. S. Cohen, and A. Guttman, *J. Chromatogr., 492:* 585 (1989).
11. A. Cohen and B. L. Karger, *J. Chromatogr., 397:* 409 (1987).
12. T. Dülffer, R. Herb, H. Herrmann, and U. Kobold, *Chromatographia,* 30(11/12): 675 (1990).
13. B. L. Karger, *Curr. Opin. Biotechnol., 3:* 59 (1992).
14. B. L. Karger, A. S. Cohen, and A. Guttman, *J. Chromatogr., 492:* 585 (1989).
15. S. W. Compton and R. G. Brownlee, BioTechniques, 6(5): 432 (1988).
16. M. V. Novotny, K. A. Cobb, and J. Liu, *Electrophoresis, 11:* 735 (1990).
17. M. Lederer, *J. Chromatogr., 488:* 5 (1989).
18. B. L. Karger, *Nature, 339:* 641 (1989).
19. R. A. Wallingford and A. G. Ewing, in *Advances of Chromatography,* Vol. 29, (J. C. Giddings, E. Grushka, and P. Brown, Eds.), Marcel Dekker, New York, 1989.
20. S. F. Y. Li, *Capillary Electrophoresis,* Elsevier, Amsterdam, 1992.
21. E. Grushka, R. M. McGormick, and J. J. Kirkland, *Anal. Chem., 61:* 241 (1989).
22. M. Martin, G. Guiochon, Y. Warbroehl, and J. Jorgenson, *Anal. Chem., 57:* 559 (1985).
23. T. Tsuda, *J. High Resolut. Chromatogr. Chromatogr. Commun., 10:* 622 (1987).
24. D. E. Burton, M. J. Sepaniak, and M. P. Maskarinec, *J. Chromatogr. Sci., 25:* 514 (1987).
25. S. Hjertén, K. Elenbring, F. Kilar, J. L. Liao, A. J. L. Chen, C. J. Siebert, and M. D. Zhu, *J. Chromatogr., 403:* 47 (1987).
26. S. Hjertén and M. D. Zhu, *J. Chromatogr., 346:* 265 (1985).
27. S. Hjertén, J. L. Liao, and K. Yao, *J. Chromatogr., 387:* 127 (1987).

28. J. R. Mazzeo and I. S. Krull, *Anal. Chem., 63:* 2852 (1991).
29. J. R. Mazzeo and I. S. Krull, *J. Chromatogr., 606:* 291 (1992).
30. S. Hjertén and M. Kiessling Johansson, *J. Chromatogr., 550:* 811 (1991).
31. F. M. Everaerts, J. L. Beckers, and T. P. Verheggen, *Isotachophoresis: Theory, Instrumentation and Application,* Elsevier, Amsterdam, 1976.
32. P. Delmotte, *Electrofocusing and Isotachophoresis,* Walter de Gruyter, New York, 1977.
33. K. G. Kjellin, U. Moberg, and L. Hallander, *Sci. Tools, 22:* 3 (1975).
34. L. Arlinger and R. J. Routs, *Sci. Tools, 17:* 21 (1970).
35. P. Delmotte, *Z. Klin. Chem. Klin. Biochem., 9:* 334 (1971).
36. E. Borris and S. Husmann-Holloway, *Z. Anal. Chem., 311:* 467 (1982).
37. P. Delmotte, *Sci. Tools, 24:* 33 (1977).
38. S. Hjertén and M. D. Zhu, *J. Chromatogr., 327:* 157 (1987).
39. S. Hjertén and M. D. Zhu, *Protides Biol. Fluids, 33:* 537 (1985).
40. A. Widhalm, C. Schwer, D. Blaas, and E. Kenndler, *J. Chromatogr., 549:* 446 (1991).
41. H. F. Yin, J. A. Lux, and G. Schomburg, *J. High Resolut. Chromatogr., 13:* 624 (1990).
42. P. G. Righetti, S. Caglio, M. Saracchi, and S. Quaroni, *Electrophoresis, 13:* 587 (1992).
43. L. B. Dotti and J. M. Orten, *Laboratory Instructions in Biochemistry,* C.V. Mosby, St. Louis, Mo., 1971.
44. W. S. Beck, *N. Engl. J. Med., 290:* 695 (1974).
45. J. L. Gamble, *Chemical Anatomy, Physiology and Pathology of the Extracellular Fluid,* Harvard University Press, Cambridge, Mass., 1954.
46. K. J. Lee and G. S. Heo, *J. Chromatogr., 559:* 317 (1991).
47. A. Schulze and C. Heremans, *Molecular Biology of Human Proteins with Special Reference to Plasma Proteins,* Vol. 1, Elsevier, Amsterdam, 1966.
48. J. M. Curling, Ed., *Methods of Plasma Protein Fractionation,* Academic Press, London, 1980.
49. E. J. Cohn, in *Blood Cells and Plasma Proteins,* J. L. Tullis, Ed., Academic Press, London, 1953.
50. P. Kistler and H. Nitschmann, *Vox Sang., 7:* 414 (1962).
51. M. J. Gordon, K. J. Lee, A. A. Arias, and R. N. Zare, *Anal. Chem., 63:* 69 (1991).
52. F. T. A. Chen, U.S. patent 5,139,630A, Aug. 18, 1992.
53. M. A. Strege and A. L. Lagu, *J. Liq. Chromatogr., 16:* 51 (1993).
54. D. J. Rose and J. W. Jorgenson, *Anal. Chem., 60:* 642 (1988).
55. K. D. Lukacs and J. W. Jorgenson, *J. High Resolut. Chromatogr. Chromatogr. Commun., 8:* 407 (1985).
56. X. Huang, M. J. Gordon, and R. N. Zare, *Anal. Chem., 60:* 375 (1988).
57. Z. Deyl and R. Struzinsky, *J. Chromatogr., 569:* 63 (1991).

58. S. Fjuiwara and S. Honda, *Anal. Chem., 58:* 1811 (1986).
59. X. Huang, W. F. Coleman, and R. N. Zare, *J. Chromatogr., 480:* 95 (1989).
60. A. Vinther, H. Søeberg, H. H. Sørensen, and A. M. Jespersen, *Talanta, 38:* 1369 (1991).
61. T. Tsuda, K. Nomura, and G. Nagawa, *J. Chromatogr., 264:* 385 (1983).
62. F. Kohlrausch, *Ann. Phys. Chem., 62:* 209 (1897).
63. A. Vinther and H. Søberg, *J. Chromatogr., 559:* 3 (1991).
64. S. Hjertén, *Electrophoresis, 11:* 665 (1990).
65. K. Otsuka and S. Terabe, *J. Chromatogr., 480:* 91 (1989).
66. S. E. Moring, J. C. Colburn, P. D. Grossman, and H. H. Lauer, *Liq. Chromatogr. Gas Chromatogr., 8:* 34 (1991).
67. R. L. Chien and D. S. Burgi, *J. Chromatogr., 559:* 141 (1991).
68. P. Jandik and W. R. Jones, *J. Chromatogr., 546:* 431 (1991).
69. D. Kaniansky and J. Marák, *J. Chromatogr., 498:* 191 (1990).
70. V. Dolnik, K. A. Cobb, and M. V. Novotny, *J. Microcolumn,* Sept. 2, p. 127 (1990).
71. F. Foret, E. Szoko, and B. L. Karger, *J. Chromatogr., 608:* 3 (1992).
72. M. Deml, F. Foret, and P. Boček, *J. Chromatogr., 320:* 159 (1985).
73. T. Tsuda, T. Mizuno, and J. Akiyama, *Anal. Chem., 59:* 678 (1987).
74. T. Tsuda and R. N. Zare, *J. Chromatogr., 559:* 103 (1991).
75. R. A. Wallingford and A. G. Ewing, *Anal. Chem., 59:* 678 (1987).
76. R. A. Wallingford and A. G. Ewing, *Anal. Chem., 60:* 1972 (1988).
77. W. G. Kuhr, *Anal. Chem., 62:* 403 (1990).
78. C. Y. Chen and M. D. Morris, *Appl. Spectrosc., 42:* 515 (1988).
79. A. J. Bard and L. R. Faulkner, *Electrochemical Methods,* Wiley, New York, 1980.
80. C. M. St. Claire III and J. W. Jorgenson, *J. Chromatogr. Sci., 23:* 186 (1985).
81. R. A. Wallingford and A. G. Ewing, *J. Chromatogr., 441:* 299 (1988).
82. R. D. Smith and H. R. Udseth, *Nature, 331:* 638 (1988).
83. C. M. Whitehouse, R. N. Dreyer, M. Yamashita, and M. Fenn, *Anal. Chem., 57:* 675 (1985).
84. T. J. Thomson, F. Foret, P. Vouros, and B. Karger, *Anal. Chem., 65:* 900 (1993).
85. T. Tsuda, G. Nakgawa, M. Sato, and K. Yagi, *J. Appl. Biochem., 5:* 330 (1983).
86. K. H. Row, W. H. Griest, and M. P. Maskarinec, *J. Chromatogr., 409:* 193 (1987).
87. M. A. Firestone, J. P. Michaud, R. N. Carter, and W. Thorman, *J. Chromatogr., 407:* 363 (1987).

88. A. S. Cohen, S. Terabe, J. A. Smith, and B. L. Karger, *Anal. Chem., 59:* 1021 (1987).
89. H. H. Lauer and D. McManigill, *Anal. Chem., 58:* 165 (1986).
90. T. Wang and R. A. Hartwick, *Anal. Chem., 64:* 1745 (1992).
91. T. Wang, R. H. Hartwick, and P. B. Champlin, *J. Chromatogr., 462:* 147 (1989).
92. Y. Wahlbroel and J. Jorgenson, *J. Chromatogr., 315:* 135 (1984).
93. G. J. M. Bruin, G. Stegeman, A. C. van Asten, X. Xu, J. C. Kraak, and H. Poppe, *J. Chromatogr., 559:* 163 (1991).
94. O. W. Reif, R. Lausch, and R. Freitag, submitted.
95. M. Zhu, R. Rodriguez, T. Wehr, and C. Siebert, *J. Chromatogr., 608:* 225 (1992).
96. S. Kobayashi, T. Ueda, and M. Kikomoto, *J. Chromatogr., 480:* 179 (1989).
97. G. A. Ross and D. Perrett, *Biochem. Soc. Trans., 21:* 19s (1992).
98. J. Jorgenson and K. D. Lukacs, *Clin. Chem., 27:* 1551 (1981).
99. E. J. Guthrie and J. W. Jorgenson, *Anal. Chem., 56:* 483 (1984).
100. J. S. Green and J. W. Jorgenson, *J. Chromatogr., 352:* 337 (1986).
101. E. Gassmann, J. E. Kuo, and R. N. Zare, *Science, 230:* 813 (1985).
102. A. T. Balchunas and M. J. Sepaniak, *Anal. Chem., 60:* 617 (1988).
103. M. C. Roach, P. Gozel, and R. N. Zare, *J. Chromatogr. Biomed. Appl., 426:* 129 (1988).
104. B. Nickerson and J. W. Jorgenson, *J. Chromatogr., 480:* 157 (1989).
105. W. Kuhr and E. Yeung, *Anal. Chem., 60:* 1832 (1988).
106. J. Y. Zhao, K. C. Waldron, J. Miller, J. Z. Zhang, H. Harke, and N. J. Dovichi, *J. Chromatogr., 608:* 239 (1992).
107. D. E. Burton, M. J. Sepaniak, and M. P. Mascarinec, *J. Chromatogr. Sci., 24:* 347 (1986).
108. B. L. Ling and W. R. G. Baeyens, *Anal. Chim. Acta, 255:* 283 (1991).
109. S. K. Yeo, H. K. Lee, and S. F. Y. Li, *J. Chromatogr., 585:* 133 (1991).
110. A. Guttman, A. Paulus, A. Cohen, N. Grinberg, and B. Karger, *J. Chromatogr., 448:* 41 (1988).
111. J. Liu, K. A. Cobb, and M. Novotny, *J. Chromatogr., 468:* 55 (1989).
112. E. Gurthrie, J. Jorgenson, and P. Dluzneski, *J. Chromatogr., 22:* 171 (1984).
113. Y. F. Cheng and N. J. Dovichi, *Science, 242:* 562 (1989).
114. B. Wright, G. A. Ross, and R. D. Smith, *J. Microcolumn,* Sept. 1, p. 85 (1989).
115. A. T. Balchunas and M. J. Sepaniak, *Anal. Chem., 59:* 1466 (1987).
116. T. Tsuda, Y. Kobayashi, A. Hori, T. Matsumoto, and O. Suzuki, *J. Chromatogr., 456:* 375 (1988).
117. N. A. Guzman, L. Hernandez, and S. Terabe, in *Analytical Biotechnology:*

Capillary Electrophoresis and Chromatography, C. Horváth and J. G. Nikelly, Eds., American Chemical Society, Washington, D.C., 1990.
118. N. A. Guzman, M. A. Trebilock, and J. P. Advis, *Anal. Chim. Acta, 249:* 247 (1991).
119. T. Hara, S. Okumura, S. Kato, J. Yokogi, and R. Nakajima, *Anal. Sci., 7:* 261 (1991).
120. W. Kuhr and E. Yeung, *Anal. Chem., 60:* 1832 (1988).
121. C. Y. Chen and M. D. Morris, *Appl. Spectrosc., 42:* 515 (1988).
122. A. J. Bard and L. R. Faulkner, *Electrochemical Methods,* Wiley, New York, 1980.
123. E. D. Lee, W. Muck, J. D. Henion, and T. R. Covey, *J. Chromatogr., 458:* 313 (1988).
124. F. Garcia and J. D. Henion, *Anal. Chem., 64:* 985 (1992).
125. P. Thibault, C. Paris, and S. Pleasance, *Rapid Commun. Mass Spectrom., 5:* 484 (1991).
126. M. A. Moseley, L. J. Deterding, K. B. Tomer, and J. W. Jorgenson, *Anal. Chem., 63:* 109 (1991).
127. J. A. Olivares, N. T. Nguyen, C. R. Yonker, and R. D. Smith, *Anal. Chem., 59:* 1232 (1987).
128. M. A. Moseley, L. J. Detering, K. B. Tomer, and J. W. Jorgenson, *J. Chromatogr., 3:* 87 (1989).
129. R. W. Hallen, C. B. Shumante, W. F. Siems, T. Tsuda, and H. H. Hill, *J. Chromatogr., 480:* 233 (1989).
130. J. C. Schwartz and I. Jardine, poster presented at the *40th ASMS Conference on MS and Allied Topics,* Washington, D.C., 1992.
131. J. B. Fenn, M. Mann, C. K. Meng, S. F. Wong, and C. M. Whitehouse, *Science, 246:* 64 (1989).
132. R. D. Smith, J. A. Loo, C. J. Baringa, C. G. Edmonds, and H. R. Udseth, *J. Chromatogr., 480:* 211 (1989).
133. S. Pentoney, Jr. and R. Zare, *Anal. Chem., 61:* 1643 (1989).
134. C. Chen, T. Demana, S. Huang, and M. Morris, *Anal. Chem., 61:* 1593 (1986).
135. J. Wu, T. Odake, T. Katimori, and T. Sawada, *Anal. Chem., 63:* 2216 (1991).
136. J. Wu and J. Pawliszyn, *J. Chromatogr., 608:* 121 (1992).
137. T. Tsuda, J. V. Sweedler, and R. N. Zare, *Anal. Chem., 62:* 2149 (1990).
138. T. Istumi, T. Nagahori, and T. Okugama, *J. High Resolut. Chromatogr. Chromatogr. Commun., 14:* 352 (1991).
139. S. Hjertén, *J. Chromatogr., 347:* 191 (1985).
140. F. Kilár and S. Hjertén, *Electrophoresis, 10:* 23 (1989).
141. B. J. Herren, S. G. Shafer, S. V. Alstine, J. M. Harris, and R. S. Snyder, *J. Colloid Interface Sci., 115:* 46 (1987).

142. G. J. M. Bruin, J. Chang, R. Kuhlman, K. Zegers, J. Kraak, and H. Poppe, *J. Chromatogr., 471:* 429 (1987).
143. K. A. Cobb, V. Dolnik, and M. Novotny, *Anal. Chem., 62:* 2478 (1990).
144. J. K. Towns and F. E. Regnier, *J. Chromatogr., 516:* 69 (1990).
145. H. H. Lauer and D. McManigill, *Trends Anal. Chem., 5:* 11 (1986).
146. C. N. Ong, L. S. Liau, and H. Y. Ong, *J. Chromatogr., 576:* 346 (1992).
147. F. T. A. Chen, *J. Chromatogr., 559:* 445 (1991).
148. S. A. Swedberg, *Anal. Biochem., 185:* 51 (1990).
149. G. J. M. Bruin, P. Tock, J. Kraak, and H. Poppe, *J. Chromatogr., 517:* 557 (1990).
150. A. M. Dougherty, C. L. Wolley, D. L. Williams, D. F. Swaile, R. D. Cole, and M. J. Sepaniak, *J. Liq. Chromatogr., 14:* 907 (1991).
151. J. E. Wiktorowicz and J. C. Colburn, *Electrophoresis, 11:* 769 (1990).
152. K. Tsuji and R. J. Little, *J. Chromatogr., 594:* 317 (1992).
153. K. Tsuji, *J. Chromatogr., 550:* 823 (1991).
154. O. W. Reif, K. Hebenbrock, and K. Schügerl, *HPCE'92*, Amsterdam, 1992.
155. J. Sudor, F. Foret, and P. Boček, *Electrophoresis, 12:* 1056 (1991).
156. S. Nathakarnkitkoll, P. J. Öffner, G. Bartsch, M. A. Chin, and G. K. Bonn, *Electrophoresis, 13:* 18 (1992).
157. A. Widhalm, C. Schwer, D. Blaas, and E. Kenndler, *J. Chromatogr., 549:* 446 (1991).
158. A. Guttman, J. Horváth, and N. Cooke, *Anal. Chem., 65:* 199 (1993).
159. D. Wu and F. E. Regnier, *J. Chromatogr., 608:* 349 (1992).
160. K. Ganzler, K. S. Greve, A. S. Cohen, A. Guttman, N. C. Cooke, and B. L. Karger, *Anal. Chem., 64:* 2665 (1992).
161. R. Lausch, O. W. Reif, J. Schlösser, J. Fleischer, R. Freitag, and T. Scheper, in press.
162. H. J. Bode, *FEBS Lett., 65:* 56 (1976).
163. M. Zhu, D. L. Hansen, S. Burd, and F. Gannon, *J. Chromatogr., 480:* 311 (1989).
164. M. Strege and A. Lagu, *Anal. Chem., 63:* 1233 (1991).
165. H. Lauer and D. McManigill, *Trends Anal. Chem., 5:* 11 (1986).
166. J. P. Landers, R. P. Oda, T. C. Spelsberg, J. A. Nolan, and K. J. Ulfelder, *BioTechniques, 14:* 98 (1993).
167. K. D. Altria and C. F. Simpson, *Chromatographia, 24:* 527 (1987).
168. M. Bushey and J. Jorgenson, *J. Chromatogr., 480:* 301 (1989).
169. W. J. Ferguson and N. E. Good, *Anal. Biochem., 104:* 300 (1980).
170. K. W. Yim, *J. Chromatogr., 559:* 401 (1991).
171. M. Zhu, R. Rodriguez, and T. Wehr, *J. Chromatogr., 559:* 479 (1991).
172. S. M. Chen and J. E. Wiktorowicz, *Anal. Biochem., 206:* 84 (1992).
173. F. Kilár and S. Hjertén, *J. Chromatogr., 480:* 351 (1989).

174. F. Kilar, *J. Chromatogr., 545:* 403 (1991).
175. T. Wehr, M. Zhu, R. Rodriguez, D. Burke, and K. Duncan, *Am. Biotechnol. Lab., 8:* 22 (1990).
176. J. Wu and J. Pawliszyn, *J. Chromatogr., 608:* 121 (1992).
177. R. E. Jones, W. A. Hemmings, and W. Page Faulk, *Immunochemistry, 8:* 299 (1971).
178. W. Thormann, *J. Chromatogr., 516:* 211 (1990).
179. T. Izumi, T. Nagahori, and T. Okuyama, *J. High Resolut. Chromatogr., 14:* 351 (1991).
180. Dj. Josic, A. Bottcher, and G. Schmitz, *Chromatographia, 30:* 703 (1990).
181. F. S. Stover, *J. Chromatogr., 470:* 201 (1989).
182. J. Beckers and F. Everaerts, *J. Chromatogr., 508:* 3, 19 (1990).
183. S. Hjalmarsson and A. Baldsten, *Crit. Rev. Anal. Chem., 11:* 264 (1981).
184. F. T. A. Chen and J. C. Sternberg, U.S. patent 5,120,413, June 9, 1992.
185. D. G. Pande, R. V. Nellore, and H. R. Bhagat, *Anal. Chem., 204:* 103 (1992).
186. P. D. Grossman, J. C. Colburn, H. H. Lauer, R. G. Nielsen, R. M. Riggin, G. S. Sittampalam, and E. C. Rickard, *Anal. Chem., 61:* 1186 (1989).
187. R. G. Nielsen, E. C. Rickard, P. F. Santa, D. A. Sharknas, and G. Sittampalam, *J. Chromatogr., 539:* 177 (1991).
188. S. J. Harrington, R. Varro, and T. M. Li, *J. Chromatogr., 559:* 385 (1991).
189. Z. K. Shihabi, *Ann. Clin. Lab. Sci., 22:* 399 (1992).
190. B. J. Compton, *J. Chromatogr., 559:* 357 (1991).
191. R. G. Nielsen, G. S. Sittampalam, and E. C. Rickard, *Anal. Biochem., 177:* 20 (1989).
192. J. Frenz, S. L. Wu, and W. S. Hancock, *J. Chromatogr., 480:* 379 (1989).
193. R. G. Nielsen and E. C. Rickard, *ACS Symp. Ser., 434:* 36 (1990).
194. T. Tadey and W. C. Purdy, *J. Chromatogr., 583:* 111 (1992).
195. M. Taverna, A. Baillet, D. Biou, M. Schlüter, R. Werner, and D. Ferrier, *Electrophoresis, 13:* 359 (1992).
196. J. P. Landers, R. P. Oda, B. J. Madden, and T. C. Spelsberg, *Anal. Biochem., 205:* 115 (1992).
197. R. I. Hecht, J. F. Coleman, J. C. Morris, F. S. Stover, and C. Demarest, *Prep. Biochem., 19:* 363 (1989).
198. R. I. Hecht, J. C. Morris, F. S. Stover, L. Fossey, and C. Demarest, *Prep. Biochem., 19:* 201 (1989).
199. N. Ishioka, N. Iyori, J. Noli, and S. Kurioka, *Biomed. Chromatogr., 6:* 224 (1992).
200. T. M. Phillips, *Liq. Chromatogr. Gas Chromatogr. Int., 6:* 290 (1993).
201. R. M. Caprioli, W. T. Moore, M. Martin, B. N. DaGue, K. Wilson, and S. Moring, *J. Chromatogr., 480:* 247 (1989).

202. E. Wenisch, C. Tauer, A. Jungbauer, H. Katinger, M. Faupel, and P. G. Righetti, *J. Chromatogr., 516:* 133 (1990).
203. S. L. Wu, G. Teshima, J. Cacia, and W. S. Hancock, *J. Chromatogr., 516:* 115 (1990).
204. N. A. Guzman, J. Moschera, K. Iqbal, and A. W. Malick, *J. Chromatogr., 608:* 197 (1992).
205. F. T. Chen, C. M. Liu, Y. Z. Hsieh, and J. C. Sternberg, *Clin. Chem., 37:* 14 (1991).
206. M. A. Costello, C. Woititz, J. De Feo, D. Stremlo, L. F. L. Wen, D. J. Palling, K. Iqbal, and N. A. Guzman, *J. Liq. Chromatogr., 15:* 1081 (1992).
207. T. Yagi, K. Kojima, M. Yagi, and Y. Kajita, in *Electrophoresis '93,* H. Hirai, Ed., Walter DeGruyter, Berlin, 1984.
208. H. Yamamoto, T. Manabe, and T. Okuyama, *J. Chromatogr., 515:* 659 (1990).
209. K. O. Eriksson, A. Palm, and S. Hjertén, *Anal. Biochem., 201:* 211 (1992).
210. G. D. Fasman, Ed., *Handbook of Biochemistry and Molecular Biology,* Vol. II, CRC Press, Cleveland, Ohio, 1976.
211. E. Seligson, Ed., *Clinical Laboratory Science,* Vol. I, CRC Press, Cleveland, Ohio, 1977.
212. R. C. Weast, Ed., *Handbook of Clinical Laboratory Data,* CRC Press, Cleveland, Ohio, 1968.

2
Analysis of Natural Products by Gas Chromatography/Matrix Isolation/Infrared Spectrometry

W. M. Coleman III and Bert M. Gordon *R.J. Reynolds Tobacco Company, Winston-Salem, North Carolina*

I.	INTRODUCTION	58
	A. Chromatography/Detector Interfaces	59
	B. Chromatography/IR Interfaces	59
II.	HARDWARE	60
	A. Gas Chromatography/Matrix Isolation/Fourier Transform Infrared Spectrometry	60
	B. Gas Chromatography/Light Pipe/Fourier Transform Infrared Spectrometry	64
	C. Multidimensional Gas Chromatography	66
	D. GC/MI/FT-IR Methodologies	68
III.	SPECTRAL INTERPRETATION	71
	A. Matrix Isolation Spectrometry	71
	B. Characteristics of Matrix Isolation	72
	C. IR Spectra of Matrix-Isolated Organic Compounds and Applications to Natural Products	73
	D. Matrix Shifts	75
	E. MI/FT-IR Spectra of Selected Compound Types	77
	F. IR Spectral Libraries	90
	G. Matrix Versus Light Pipe	91
IV.	APPLICATIONS	94
V.	SUMMARY	103

I. INTRODUCTION

Complex mixtures such as natural product essential oils have been separated successfully and analyzed both qualitatively and quantitatively based on advances in several types of high-resolution chromatography. Among the more recent chromatographic advances that have been realized are fields such as supercritical fluid chromatography [1], microbore liquid chromatography [2], capillary electrophoresis [3], and gas chromatography [4]. Collectively, all of these separation techniques have addressed selected segments of sample types and have employed arrays of methods to detect the presence of separated components. These arrays of methods have been developed as a means to differentiate between the components of the samples from those of the separation process; for example, the components eluting from a gas chromatograph (GC) can be detected in the presence of the carrier gas He. In a number of cases, information-rich detectors based on Fourier transform infrared spectrometry (FT-IR) and mass spectrometry (MS) were employed. These types of detectors have advantages over mass flux (flame ionization) and concentration-dependent (thermal conductivity) detectors in that structural information can be obtained on the components in unknown mixtures as well as quantitative information. These two approaches complement each other in that FT-IR can establish functional group identities and MS can provide molecular weight and structural information.

The powerful structural elucidation capability realized by wedding these two spectroscopic techniques with chromatographic separations has stimulated advancements in the technology. Since the successful combination of gas chromatography and mass spectrometry (GC/MS) and gas chromatography and infrared spectrometry (GC/IR), approximately 30 years has passed with astonishing improvements. These advances have culminated in part in the commercial availability of instruments such as GC/IR/MS. Parallel advances in both chromatography and spectrometry lay at the foundation of the breakthroughs realized during the evolution of these hyphenated techniques [5]. With the advent of high-resolution fused silica capillary column technologies and the miniaturization of mass spectrometers, resolution improved dramatically and detection limits decreased to the nanogram level with accompanying structural information. Similar advances in light pipe design for the infrared spectrometer, Fourier transform (FT) spectral acquisition capabilities, and computer acquisition capabilities aided in its use as an information-rich detector [6]. The historical development of GC/IR has been recorded by timely reviews of the subject [7,8] and the reader is referred to these reviews for detailed developments.

Insofar as natural product essential oil GC/FT-IR analyses are concerned, tremendous benefits in minor component identification have been realized due to the increased sensitivity of the newer systems. Since the first GC/FT-IR system in 1967 [9], the sensitivity has dropped from the initial nanomole levels to pico- and femtomole levels. GC/matrix isolation (MI)/FT-IR has been responsible primarily for these lower detection limits.

A. Chromatography/Detector Interfaces

When discussing the technology involved with hyphenated [5] techniques, it is necessary to weigh the merits of each technique separately as opposed to them joined together. In most cases some sacrifice is required on both parts to produce the hyphenated technique. Thus the hyphen in hyphenated techniques entails an essential and real device that links the two technologies: that is, the interface.

The decision must be made as to who adapts to whom in order to design the appropriate interface. For instance, should a lower-cost instrument such as a GC be adapted to the requirements of a more expensive "detector" such as an IR or MS, or vice versa? Once this decision is made, instrument integration by means of computer control must be established. The computer would then control not only the chromatography but also the detector. In the case of GC/matrix isolation/Fourier transform infrared (Section II.A), the computer controls the GC, the IR, the rotation and position of the cryodisk, collection of the IR spectra, and when requested, provides searches of spectral databases.

B. Chromatography/IR Interfaces

At the heart of the innovation in chromatography/IR has been the development of the interface between the chromatograph and the IR spectrometer. This is particularly significant for the MI case (Section III.A). Specific interfaces depend on many factors, among them the physical state of matter being analyzed. For infrared spectrometry, several types of phases, liquids, vapors, supercritical fluids, and solids have been employed, each with its own unique set of demands for an interface when coupled with a separation process. Three types of interfaces have dominated applications for IR spectrometry in conjunction with sample separation by gas chromatography. The first successful interfaces for GC/FT-IR were comprised of a heated flow-through cell, termed a light pipe. With this interface the effluent (in the vapor phase) from the GC column passes through the pipe and the presence of separated components is detected and their IR spectra gathered by absorption of infrared radiation. With this light pipe interface, the critical parameters shown to be important are (1) geometry, length/diameter relationships, (2) wall coating, (3) window materials, and (4) temperature.

Second, interfaces for subambient trapping of GC effluents have been demonstrated in conjunction with FT-IR detection. It has been disclosed [10] that eluants from a GC could be trapped at low temperatures followed by detection with infrared microscopy. This particular interface required that the trapping medium be cooled and transparent to IR irradiation. Among the critical parameters of this interface are (1) positioning the GC column restrictor, (2) temperature of the medium, (3) volatility of effluent, (4) distance between trapping medium and restrictor, and (5) anhydrous chamber atmosphere. Commercial instruments based on this low-temperature interface design and light pipe technologies have been introduced.

Third, interfaces for gas chromatographs based on detection of the effluents trapped in a matrix have been designed [11–14]. This interface was constructed such that the gas chromatographic effluent and argon were codeposited at 10 K on the surface of a gold-plated disk. Among the critical parameters of this interface are (1) deposition rate, (2) matrix gas/eluate ratios, (3) temperature, (4) distance between gold disk and column end, and (5) vacuum. The most recent version of this instrument incorporates a FT-IR spectrometer encased in a vacuum as well as a mass spectrometer.

II. HARDWARE

A. Gas Chromatography/Matrix Isolation/Fourier Transform Infrared Spectrometry

Matrix isolation/Fourier transform infrared (MI/FT-IR) spectra were obtained with a Mattson Cryolect/Sirius 100 GC/FT-IR system, and the GC/MI/FT-IR system is shown schematically in Fig. 1. The GC/MI/FT-IR consists of four individual modules appropriately interfaced: (1) gas chromatograph, (2) cryogenic matrix isolation apparatus, (3) FT-IR spectrometer, and (4) computer for instrument control and data collection and reduction. Each module is discussed below with specific attention to the interfacing.

The carrier gas for the GC was helium premixed with a small (ca. 1%) amount of argon. The outlet of the chromatographic column is attached to an outlet splitter (FSOS, SGE) adjusted with appropriate deactivated fused silica tubings to provide a split ratio of 10 to 20% (FID flow) and 80 to 90% of the flow to the open split interface. The zero dead volume cross (SGE) of the open split interface (Fig. 2) and the outlet splitter were housed inside the chromatographic oven. The flow controllers, pressure regulator, pressure gauge, and toggle on/off valves were housed inside a separate cabinet and used the same helium/argon gas mixture as that used by the chromatographic column. The ca. 1 m of 180 μm ID deactivated fused silica transfer line was inserted directly inside the high-vacuum area of the cryodisk and provided a constant flow of about 1 to 2 mL/min. Since the flow rate at the column outlet varied depending on many chromatographic properties, including column head pressure, column length, column diameter, and oven temperature, the makeup gas flow rate was adjusted so that the column flow plus makeup flow was slightly greater than the transfer line flow. The purge line flow was incorporated into the design to allow additional flow if required by the vacuum system. The chromatographic conditions and the column varied depend-

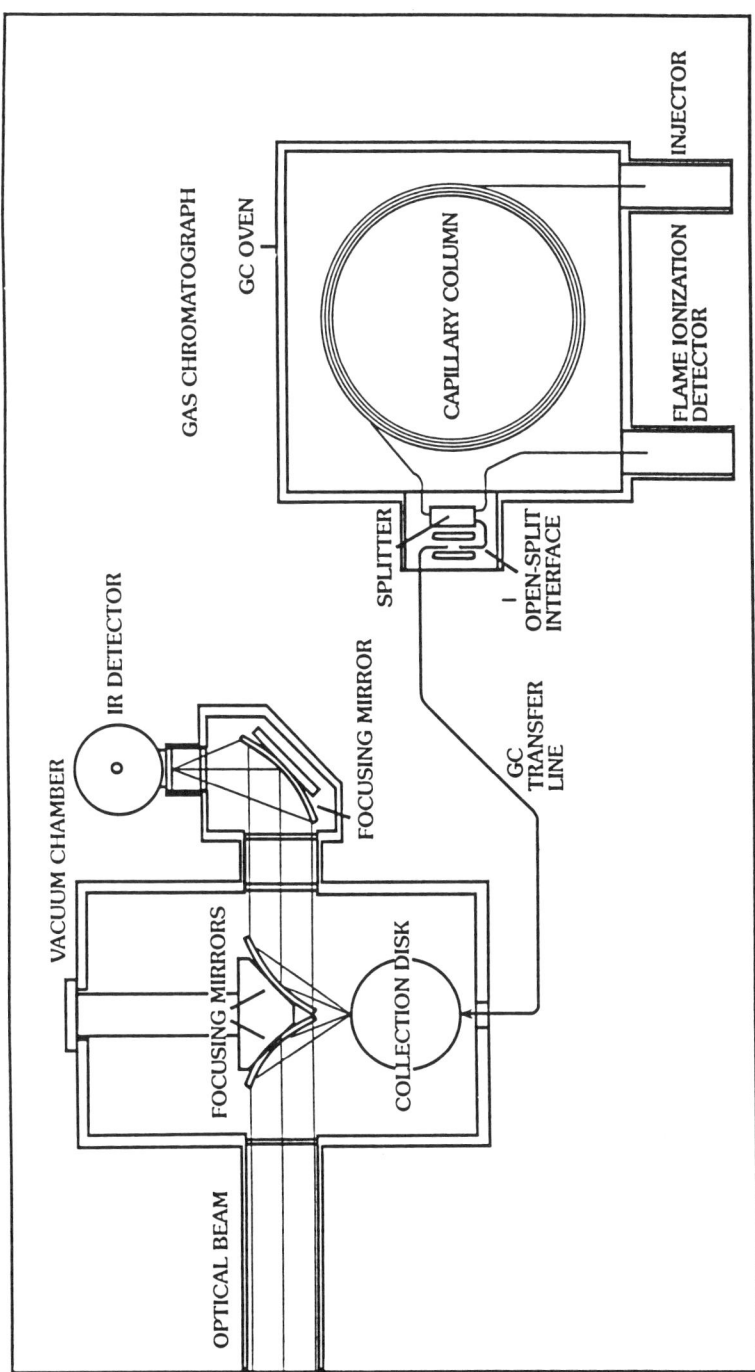

Fig. 1 Schematic of the gas chromatograph/matrix isolation/Fourier transform–infrared spectrometer instrumentation. (Courtesy of Mattson Instrument Corporation.)

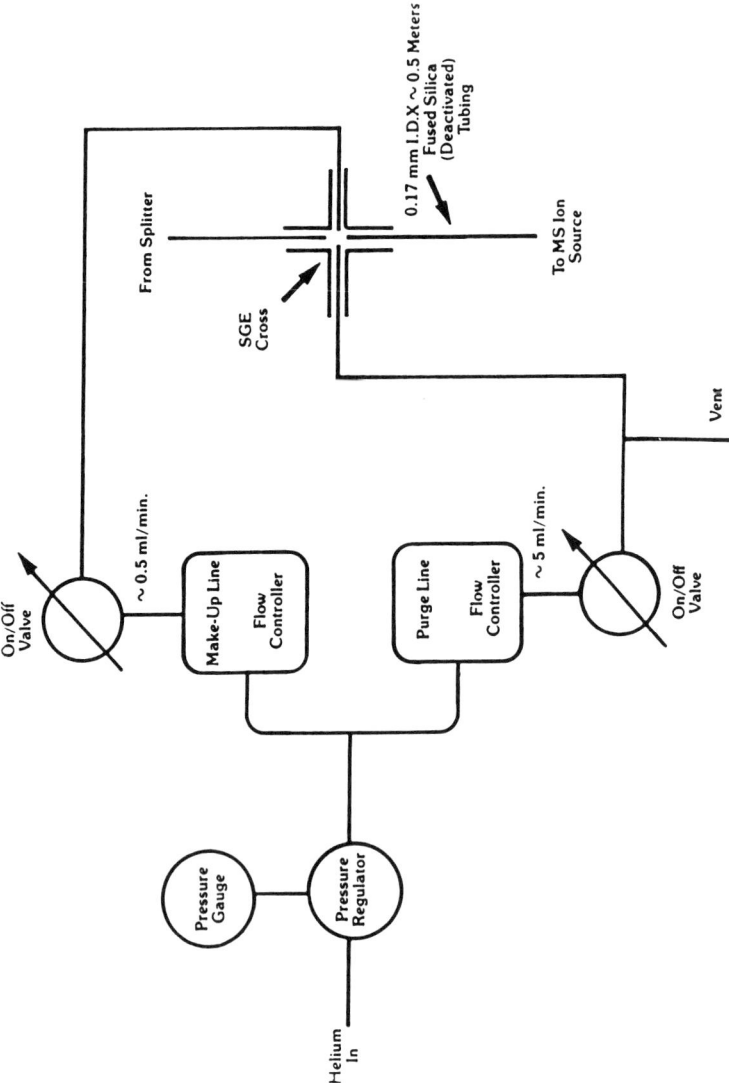

Fig. 2 Flow diagram of the open split interface. (From Ref. 16.)

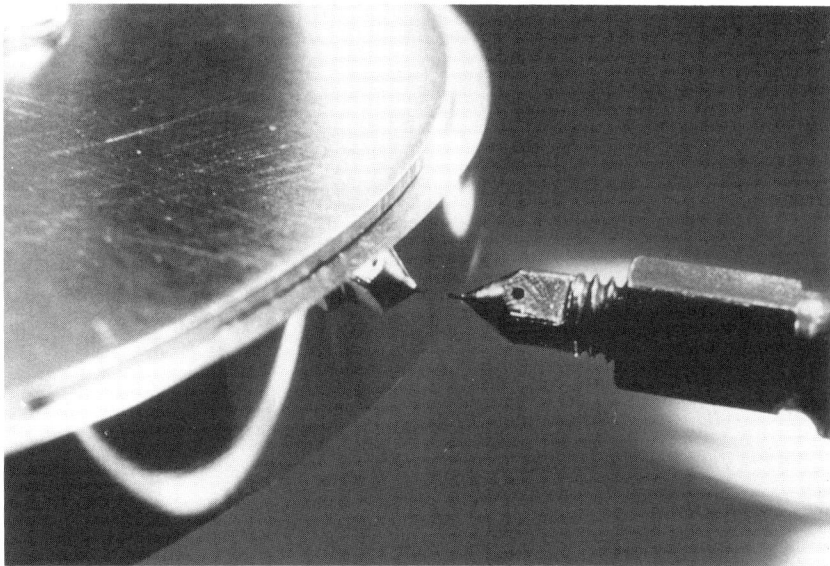

Fig. 3 Cryogenic collection disk, transfer line tip, and metal nozzle. (Courtesy of Mattson Instrument Corporation.)

ing on the requirements of the separation. In most cases the column was either a DB-Wax 60 m × 0.32 mm ID, 0.5-μm film or a DB-5 60 m × 0.32 mm ID, 0.25-μm film (J&W Scientific). The cryogenic matrix isolation apparatus consists of a vacuum chamber, a gold-plated collection disk approximately 10 cm in diameter, and a helium refrigeration unit. The vacuum chamber is maintained at pressures around 10 μm (Hg) via a combination of roughing and diffusion pumps. The GC transfer line is maintained at a temperature approximately 20°C higher than the maximum column temperature for that particular separation. The transfer line enters the vacuum chamber and protrudes 2 to 4 cm through a nozzle that is maintained at a temperature similar to that of the transfer line, to avoid condensation of analyte in the transfer line. The nozzle, exit end of the transfer line, and collection disk are shown in Fig. 3. The collection disk is maintained at a constant temperature of 10 K via the helium refrigeration unit (Airco). This very low temperature is required so that when the column effluent exits the hot (250°C) transfer line and strikes the collection disk:

1. The helium (boiling point = 4 K) remains in the vapor state and is swept away by the vacuum system.

2. The argon and the analyte are codeposited on the collection disk (producing a cryogram).

Experience has shown that this level of vacuum is necessary to avoid "clouding" of the disk with molecules present in the vacuum chamber. The collection disk rotates at a known rate controlled by the computer system and correlated to the GC/FID elution times via this system.

The FT-IR was a Mattson Sirius 100 operated at 1-cm^{-1} resolution coadding 128 scans per spectrum unless otherwise stated. When attached to the MI module, the collimated optical beam is directed external to the standard optical bench via a plain mirror and impinges on the first of a pair of off-axis parabolic focusing mirrors mounted inside the vacuum chamber. The beam is focused on the mirrored gold disk. As such, the infrared beam passes through the cryogram twice before being refocused via two parabolic mirrors onto the mercury cadmium telluride (MCT) detector. Two KCl windows are used to seal and allow the infrared beam to enter and exit the vacuum chamber.

The computer system that controlled the FT-IR spectrometer and the collection disk provided for data collection and reduction (FFT) were standard equipment provided by Mattson Instruments Inc. The start run on the GC system initiated the computer, which in turn controlled the turning of the cryodisk. After the GC run is complete, the computer positions the collection disk with the cryogram deposited in line with the infrared beam. The GC retention times obtained from the FID signal are entered into the computer manually for correlation with cryogram position times.

B. Gas Chromatography/Light Pipe/Fourier Transform Infrared Spectrometry

The gas chromatography/light pipe/Fourier transform infrared (GC/LP/FT-IR) system was a HP GC/IRD Model 5965A controlled by a Pascal ChemStation (Hewlett-Packard). This GC/FT-IR instrument is similar in design to other GC/LP/FT-IR instruments where the column is connected to a gold-plated small-volume flow cell termed a light pipe. Since the effluent from the GC column is in the vapor phase and hot, the light pipe must be heated to temperatures equal to or greater than the maximum temperature that the column reaches during the separation process. If any area of the interface is below column temperature, there is a significant probability that components in the eluant will condense inside the light pipe.

The IR radiation from a robust interferometer designed specifically for GC applications passes through KBr windows at the ends of the light pipe (Fig. 4). FT-IR spectra are acquired "on the fly" or as the sample passes through the light pipe. The impact of the on-the-fly data acquisition is very significant. The residence time of individual components inside the light pipe is essentially equal

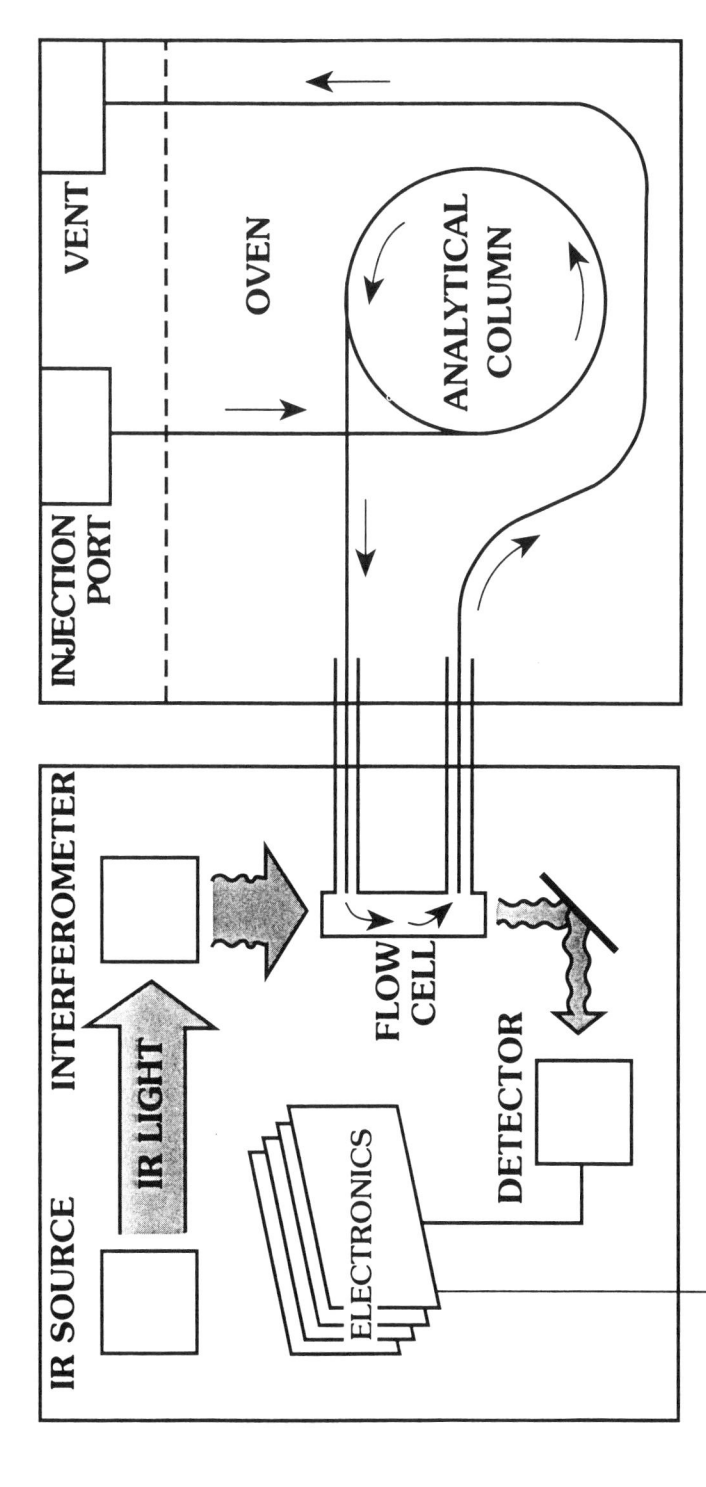

Fig. 4 Diagram of GC/LP/FT-IR system. (Courtesy of Hewlett-Packard Instrument Co.)

to (or slightly greater than) the peak width at the base of the peak. In high-resolution capillary gas chromatography, this is roughly 6 s. The best infrared spectra are usually acquired at the peak apex (i.e., the middle 3 s). Signal averaging over a 3-s period does not significantly increase the signal/noise ratio. If the chromatographic resolution is increased via a decrease in column internal diameter, the peak width becomes even shorter, allowing even less effect on the signal/noise ratio.

As stated above, the light pipe is gold plated. It is well documented in the gas chromatographic literature that hot metal is a very reactive environment and causes artifacts in the chromatographic analysis. The separation conditions used were:

Initial temperature, 40°C
Initial time, 1 min
Program rate, 5°C min^{-1}
Final temperature, 240°C
Final time, 20 min
Column, cross-linked OV-1701 (DB-1701 30 m × 0.32 mm ID, 1.0-μm film)
Injection, 1 μL

C. Multidimensional Gas Chromatography

Multidimensional gas chromatography (MDGC) is one of the most powerful techniques for separating complex mixtures. Previous work in our laboratories [15–17] compares various implementations of the technique and describes an experimental design for the analysis of natural products. The system designed, constructed, and used for the separation of natural products in this study is shown in Fig. 5. This is the same design previously described elsewhere and is reviewed below. The hardware is based on a multidimensional capillary switching device (Scientific Glass Engineering, Ringwood, Australia) modified for use with two separate gas chromatographs. All operating and column installation parameters were as described in the manual for the SGE capillary switching system except as stated below. The precolumn was located in the HP 5880 GC and the analytical column was located in an HP 5890 GC. The electronics side of the 5890 GC was physically removed from the right side of the chromatographic oven and placed behind the chromatograph. Numerous electrical connections were lengthened to accomplish this, thereby allowing the 5880 GC and 5890 GC to be placed closely together and connected by a heated interface. The SGE switching device was divided into two parts: The midpoint restrictor was mounted in the precolumn oven (5880 GC) and the liquid CO_2 cold trap/midpoint splitter assembly was mounted in the analytical column oven (5890 GC). A short length of uncoated

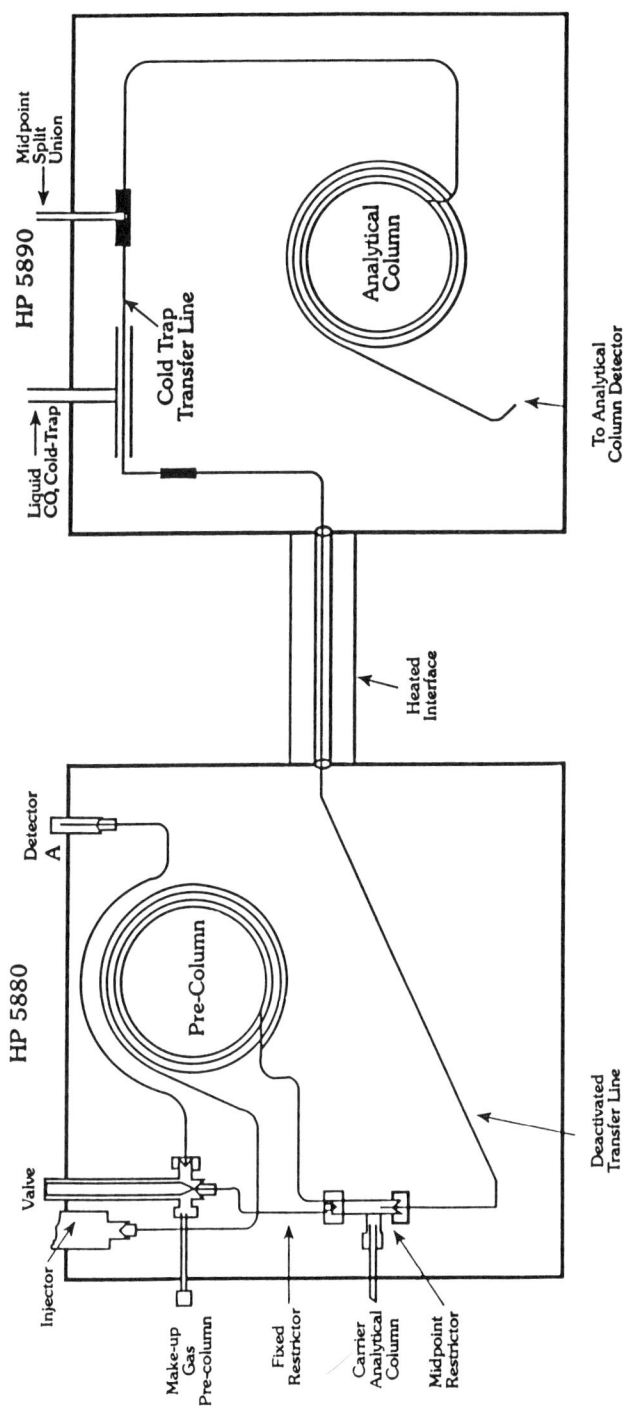

Fig. 5 Flow diagram of MDGC system. (From Ref. 16.)

deactivated fused silica (0.53 mm ID × ca. 0.5 m SGE) was used to connect the flows from the midpoint restrictor in the 5880 GC oven to the cold trap located in the 5890 GC oven (ca. 0.1 m × 0.53 mm ID, 0.25-μm film DB-1701, J&W Scientific). The precolumn separation conditions were:

Initial temperature, 45°C
Initial time, 1.0 min
Program rate, 5.0°C min^{-1}
Final temperature, 220°C
Final time, 60 min
Column, cross-linked Carbowax (DB-WAX 15 m × 0.53 mm ID, 1.0-μm film)
Injection, 5.0 μL, cold on-column

The analytical column separation conditions were:

Initial temperature, 45°C (until end of heartcut)
Initial time, 0.2 min
Program rate, 3.0°C min^{-1}
Final temperature, 240°C
Final time, 15 to 100 min
Column, cross-linked OV-1701 (DB-1701, 30 m × 0.25 mm ID, 0.25-μm film)

The end of the analytical column is then connected to a detector via the open split interface as described above.

D. GC/MI/FT-IR Methodologies

Since the operation of the cryolect is computer controlled, the deposition of the matrix onto the rotating gold disk and the collection of IR spectra post run are under software control. The user has the capability to manipulate the resolution of the scan, the number of scans taken, and the position of the IR beam relative to the run time of the chromatogram. However, to collect useful MI/FT-IR spectra, careful attention must be given to some very important points.

    1. The composition of the carrier gas must be held constant over time so that reproducible spectra can be obtained. It has been shown that variations in matrix gas/analyte ratios have significant impact on the spectra. For example, by altering the ratio of matrix gas to analyte by scanning a very broad peak of methylacetate, meaningful changes occurred not only in the number of absorptions found but in their ratio as well (Figs. 6 to 8). These results dictate that care must be taken in assessing the concentration of the analytes in a mixture so as to maintain acceptable matrix gas/analyte ratios.

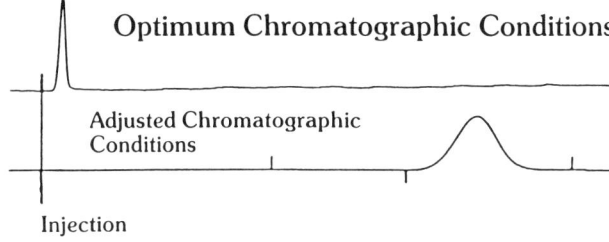

Fig. 6 Schematic drawing of optimum and adjusted chromatographic conditions for methylacetate. (From Ref. 30.)

2. The temperature of the deposited matrix must be held constant. It has been shown that the annealing of matrices results in a redistribution of the deposition sites, thereby altering the characteristics of the MI/FT-IR spectrum (Fig. 9). To maintain the rigidity of the matrix, the temperature must be held constant at less than one-half of the melting point of the matrix host, 10 K, in the

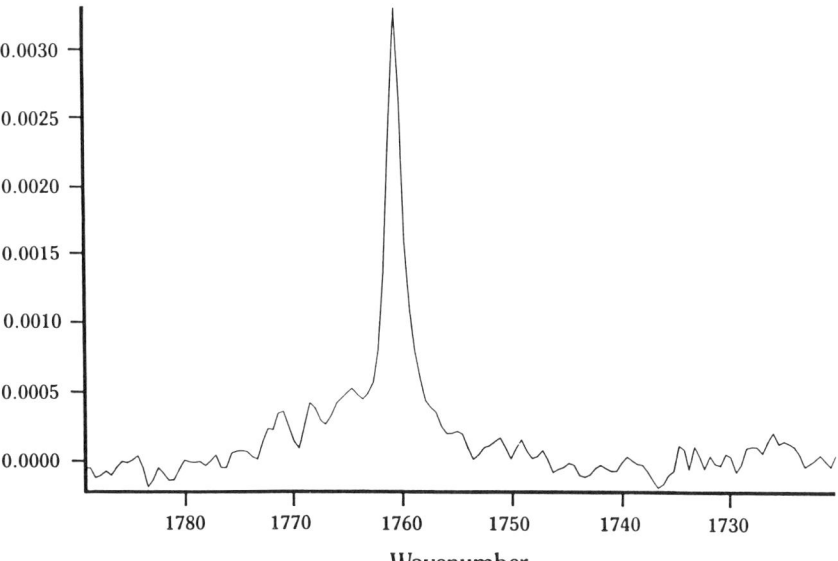

Fig. 7 MI/FT-IR spectrum, carbonyl region, gathered along the ascending slope of the methylacetate chromatographic peak, 0.6% Ar in He, in Fig. 6. (From Ref. 30.)

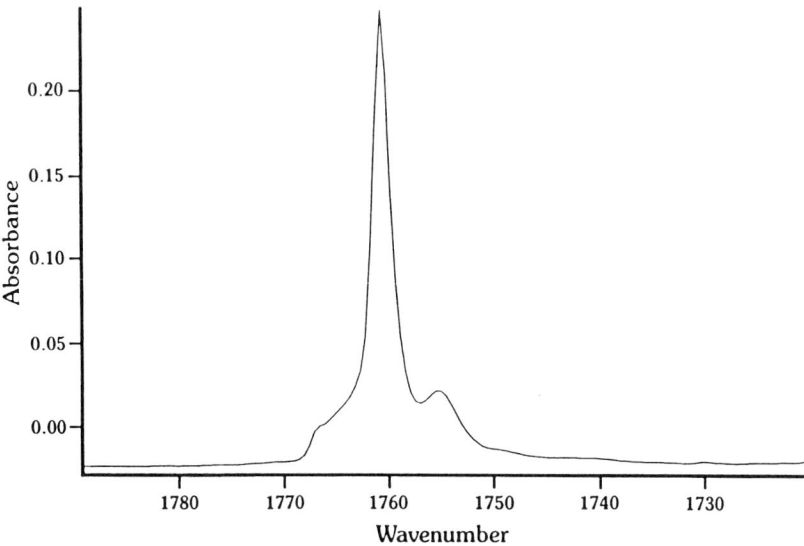

Fig. 8 MI/FT-IR spectrum, carbonyl region, gathered along the leading edge of the methylacetate chromatographic peak, 0.6% Ar in He, in Fig. 6. (From Ref. 30.)

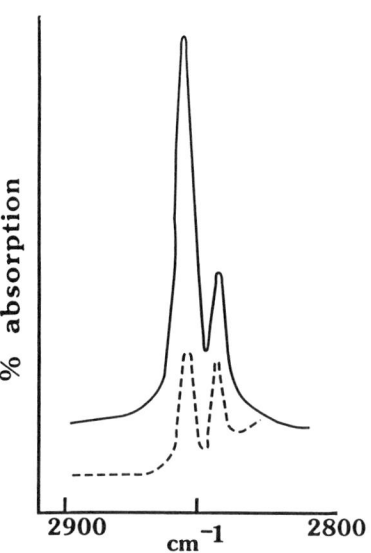

Fig. 9 Monomer IR absorption region of HCl in a N_2 matrix at 20 K; dashed spectrum is after annealing at 35 K for a few minutes. (From A. J. Barnes and W. J. Orville-Thomas, *Vibrational Spectroscopy: Modern Trends,* Elsevier, Amsterdam, 1977.)

case of Ar. Our experience has shown no measurable effects at 1 cm^{-1} resolution if the temperature is held to 11 ± 1 K.

3. The width of the collimated IR beam must be optimized to the width of the matrix line to obtain maximum sensitivity. This is accomplished by focusing the mirrors within the spectrometer, the dual parabolics in the vacuum chamber, and the position of the detector.

4. Since spectra are collected after the chromatographic run by rotating the disk, it is very important that the exact position of the analyte on the disk be established. When positioned directly underneath the IR beam, maximum sensitivity is achieved. This exact positioning is made possible by correlating the retention time of the FID detector with the position on the disk. All of the necessary information required to achieve this is captured by the computer for the specific run in question. Often, in practice, groups of spectra are collected on and around the selected retention time, so that maximum sensitivity is realized. This approach will allow for any small differences that may exist between the retention time measured at the FID detector and the retention time on the cryodisk. Experience has shown that with closely balanced flows, differences in FID retention time and deposition positions are less than 0.1 min.

5. The position of the transfer line relative to the collection disk should be within one column diameter of the gold disk surface. If the distance is too large, ineffective deposition occurs, resulting in a diffuse matrix and loss of sensitivity. On the other hand, if the distance is too close, the disk may become scratched by the end of the transfer line.

6. One of the subtle parameters under computer control is the speed of the rotating disk. The disk speed selected must be optimized so as to not destroy resolution of the components gained by the GC separation, yet maintain optimum IR sensitivity.

III. SPECIAL INTERPRETATION

A. Matrix Isolation Spectrometry

The wedding of infrared spectrometry and matrix isolation was first described by Whittle et al. in 1954 [18]. Since that time, many types of spectrometry have been allied with matrix isolation in the study of many different areas of chemistry [19]. Among the types of spectrometry employed were atomic absorption, electron spin resonance, infrared, microwave, Mössbauer, nuclear magnetic resonance, photoacoustic, Raman, luminescence, and ultraviolet-visible. As testament to the use of matrix isolation in combination with IR spectrometry, over 1650 articles appeared on the subject between the years 1954 and 1985 [19].

It was not until 1979 that matrix isolation/infrared spectrometry (MI/IR) was coupled successfully with gas chromatography, yielding gas chromatography/

matrix isolation/Fourier transform–infrared spectrometry (GC/MI/FT-IR) [11]. Nevertheless, the characteristics of the spectra obtained from organic and inorganic compounds trapped in matrix gases, prior to wedding the technique as a detector to a GC, proved to be a very important building block in the subsequent use of such data in structural determinations [20–22]. Thus accounts of the characteristics of the matrix itself as well as matrix-isolated species are necessary to reap the full benefits of the GC/MI/IR technique.

B. Characteristics of Matrix Isolation

In MI sample preparation, the sample of interest is mixed thoroughly with a diluent or matrix gas and the mixture is deposited on a cryogenic surface (Fig. 10). The large spheres represent the matrix gas, while the smaller spheres represent the isolated species. In most cases the diluent or matrix gases have been either nitrogen or argon. Among some of the characteristics of the matrix isolation experiment are (1) the random distribution of analytes through the host matrix such that they are unable to interact with each other; (2) constant temperature, usually 10 ± 1 K; (3) high-purity matrix host; and (4) constant high vacuum. Additional characteristics apply to unstable analytes and liquid host matrices. In essence, then, the method involves the trapping of the analyte of interest in a rigid cage of a chemically inert substance (matrix host) at a low temperature in a high

Fig. 10 Pictorial representation of matrix-isolated species.

vacuum [23]. Since the cage is rigid, diffusion of molecules is very limited. Due to the inertness of the cage atoms, the possibility of chemical reactions with the host is reduced significantly.

The formation of rigid matrices requires that the temperature of the matrix be held well below the melting point of the matrix material. For example, with a melting point of 29 K for argon, the matrix must be held around 10 K to maintain acceptable rigidity. Additionally, the formation and maintenance of the matrix cages require a high vacuum, so pressures in the range of 10 μm must be maintained consistently. Consequently, the matrix isolation experiment encompasses several distinct technologies, low temperature and high vacuum, each demanding and each interactive with the other. This observation does not address implications of the addition of a flow of hot gas from a gas chromatograph. All of these technologies are required to mesh before acceptable spectroscopic information can be obtained.

Conceptually, the matrix-isolated analyte is free of intermolecular interactions. Although this is not completely true, the severity of intermolecular interactions is reduced significantly compared with those of traditional liquid, vapor, or solid phases. A good indication of the intermolecular interactions is provided from the bandwidths found in IR spectra of compounds obtained in the solid state or in solution, since bandwidth differences often depend on the choice of solvent. To a first approximation, the spectroscopic properties of the analyte that are observed can be assumed to arise from the isolated species when captured in a matrix cage. Some extremely significant considerations in terms of matrix host/analyte molecular ratio are necessary to provide conditions relatively free of intermolecular interactions. Experimentation has indicated that intermolecular interactions appear in the characteristics of the IR spectra of organic compounds when the ratio of matrix gas to analyte is less than 5000:1 [23]. Calculations have shown that with larger species a higher ratio is required. For example, with a species having a diameter three times that of the matrix atom, a ratio of matrix gas to analyte of 10,000:1 is required to ensure 99% isolation.

This calculation provides valuable information on the practical limits of the GC/MI/FT-IR technique. In other words, with molecules containing multiple numbers of carbons and protons as analytes from the GC, it is unlikely that any of the compounds may be found in a truly isolated state. Thus, to some degree, intermolecular interactions will be present in the matrix. The degree to which these interactions can be defined and the consequences of the interactions can be derived in the spectral resolution and sensitivity of the technique.

C. IR Spectra of Matrix-Isolated Organic Compounds and Applications to Natural Products

Having provided a description of the matrix cage, the general characteristics of the MI/FT-IR spectra of organic compounds, as well as a number of selected types

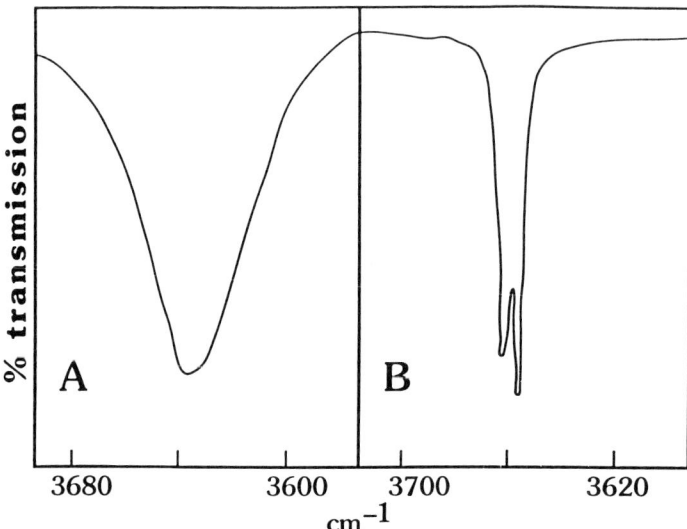

Fig. 11 Monomer OH stretching absorption of C_2H_5OH in (A) CCl_4 solution, and (B) Ar matrix. (From A. J. Barnes and W. J. Orville-Thomas, *Vibrational Spectroscopy: Modern Trends,* Elsevier, Amsterdam, 1977.)

of organic compounds common to natural products, will be discussed. One of the greatest advantages of the MI technique is the prominent sharpening of absorptions compared with those of other sample phases [24] (Fig. 11). This sharpening is due to four main factors: (1) the reduction of intermolecular interactions, (2) the absence of molecular rotation, (3) low temperature, and (4) well-defined uniform trapping sites. As a result of these factors, sensitivity is enhanced. Coupling this enhanced sensitivity with the capability to collect spectra for extended periods makes the GC/MI/FT-IR technique very sensitive, and detection limits of nanograms are achieved routinely.

A commercially important subset of natural products are those associated with aroma and flavor. Important contributors to this group of products are essential oils. Much work has been done on the characterization of these complex oils, and some of the more recent work has employed GC/FT-IR light pipe technologies [9,25–28].

Alcohols, aldehydes, ketones, esters, alkanes, and lactones are classes of compounds that occur frequently in natural products [29]. These functional groups can be part of an aliphatic, terpene, or aromatic structure. It has been well established that sensory threshold concentrations for components of essential oils vary by a factor of 10^6. This overwhelming difference is due primarily to the specificity of the olfactory receptors. In some cases the structure of compounds producing the same aroma may be very different, while in other cases, small changes in stereochemistries can alter the olfactory perception of the compounds. Thus determinations of structure for these components have significant commercial implications.

D. Matrix Shifts

During the course of investigations into the characteristics of the MI/IR spectra of selected types of organic compounds, it was found that the absorption maxima for a wide variety of organic compounds were susceptible to the type of matrix host employed (Fig. 12 and Table 1). Three types of phases were used: (1) Ar, (2) Xe, and (3) no matrix gas, bare gold disk (BARE). Shifts in band maxima as large as 20 cm^{-1} were noted between compounds matrix isolated in Ar versus the same compound matrix isolated in Xe at the same matrix gas/analyte ratio [30,31]. For example, differences were found in the MI/FT-IR spectra of a series of alkanes, esters, lactones, lactams, phenols, alcohols, amides, alkenes, and ketones when their spectra were compared in two matrices, Ar and Xe, as well in the absence of any matrix gas, BARE. The impact of these matrices on the characteristics of the IR spectra was compared with the impact observed when the spectra were

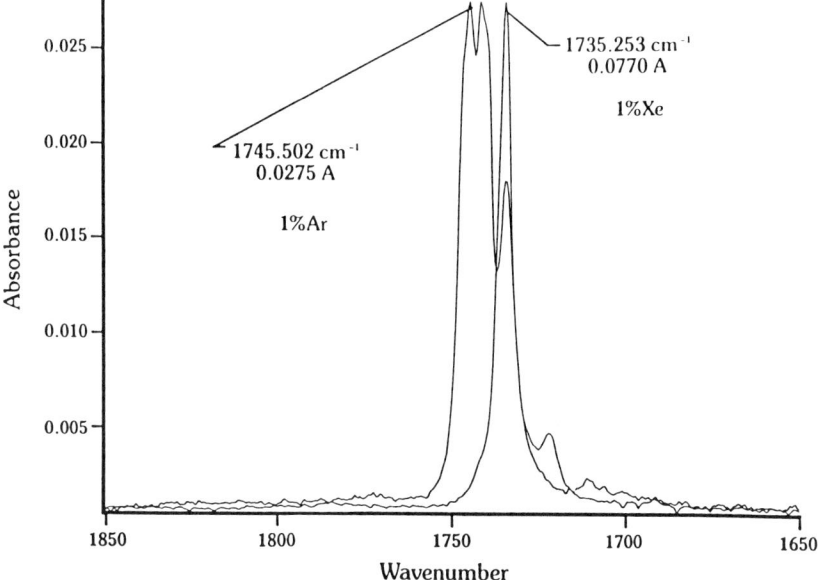

Fig. 12 MI/FT-IR spectra, carbonyl region, of butanol 5 ng each, in 1% Ar and 1% Xe matrices. (From Ref. 30.)

Table 1 Infrared Data on Selected Esters (cm^{-1})

Compound	Carbonyl absorption band(s)[a]					
	MIXe[b]	MIAr[c]	BARE[d]	SS[e]	BARE[f]	VP[g]
Methyldecanoate	1750	1754	1734	1738	1727	1760
	1730	1736			1734	
Methyldodecanoate	1749	1754	1739	1744	1727	1759
	1730	1736			1739	
Butylhexanoate	1744	1748	1734		1724	1755
					1738	
Butylheptanoate	1743	1748	1734	1738	1739	1755
			1720		1721	
Butyloctanoate	1743	1748	1736	1738	1722	1755
			1720		1734	
Butyldecanoate	1743	1748	1735	1738	1720	1751
			1721		1735	
Butyldodecanoate	1743	1748	1735		1720	1751
			1722			

[a]Most intense band listed first.
[b]Matrix isolation with Xe.
[c]Matrix isolation with Ar.
[d]30 ng on bare disk.
[e]Data from Refs. 48 and 49.
[f]10 ng on bare disk.
[g]Data from Refs. 50 and 51.

gathered in the vapor phase (VP) as well as in the condensed phase/solid state (SS). For the majority of compound classes, the major absorption bands fell between higher values for the vapor phase and lower values for the condensed phase when either Ar or Xe were used as matrix gas. The major absorption bands found in the Xe matrix were usually at lower energy than were comparable bands in the Ar matrix. In most cases, the values of absorptions for compounds deposited on the bare disk were lower than those found in the Ar matrix. Table 1 lists the carbonyl absorption values found for a series of alkyl esters as a function of sample phase. It was clear that the VP values were highest and the BARE disk values were lowest, while carbonyl absorptions of the compounds captured in the Xe matrix were lower than those in the Ar phase. These results documented that preconceptions of noble gases as inert hosts in the examination of FT-IR spectra at low temperatures were not valid.

As just discussed, vibrational bands of molecules trapped in matrices exhibit matrix shifts. These shifts can be considered somewhat like those which occur in room-temperature-solution IR spectra. Some of the same factors—electrostatic, inductive, dispersive, and specific compound interactions—are common to both phases. Since the molecule of interest is captured in a rigid "cage," the possibility of repulsive forces as factors in final band positions must be considered in the matrix case. Therefore, it was to be expected that the noble gas phase would exert its own effect on the IR spectra of organic compounds. However, the magnitude of the shifts between two inert noble gases, Ar and Xe, was not expected. The key to this effect probably lies with the characteristics of Ar and Xe. In a face-centered cube structure, the interstitial site diameters for Ar and Xe are very different, 1.56 and 1.80 Å, respectively. In addition, the atomic radii of Ar and Xe are also different, 3.75 and 4.34 Å, respectively. Thus the characteristics of the matrix site will be different for Ar and Xe, which in turn affects the energy of the absorption.

E. MI/FT-IR Spectra of Selected Compound Types

The characteristics of the MI/FT-IR spectra of numerous types of compounds captured in selected matrices after GC separation have appeared in the literature [32–47], and a brief description of the findings, with emphasis on natural product chemistries, is given here. In general, at 10 ng on column and below, very sharp absorption bands were produced independent of compound type. The full width at half height (FWHH) values averaged around 5 cm^{-1}. This narrow FWHH value was sustained as long as the matrix gas/analyte ratio was $\geq$ 10,000:1.

Alkenes
Terpenes comprise the bulk of this class of compounds, having aroma implications in essential oils. The unique features of terpenes rests in part with the geometry of the carbon–carbon double bond. MI/FT-IR spectra of selected *cis*-alkenes yielded narrow, weak absorption bands around 1665 cm^{-1}. The *trans*-alkenes yielded narrow, medium absorption bands around 970 cm^{-1}. It was interesting to find that the values for the C=C bond stretch and C—H bending were higher in the MI case than in the VP case (Table 2). This was not the general trend found for the major absorptions of a large range of organic compounds with different functional groups attached, as will be seen below. Additional resolution of the MI/FT-IR spectra of alkenes allowed for the discrimination of the actual position of the double bond within a given alkene (Fig. 13). The fingerprint region of each alkene was unique.

Carbonyl Compounds
One of the more diagnostic regions of the IR spectrum is that associated with the carbonyl group, approximately 1840 and 1600 cm^{-1}. Carbonyl compounds in the form of acyclic terpenes, cyclic terpenes, aromatics, lactones, and aliphatics comprise a very important subset of natural product components [29]. For

Table 2 Infrared Data on Selected *cis* and *trans* Alkenes (cm^{-1})

Compound	C=C stretch				C—H bend			
	MIAr[a]	MIXe[b]	VP[c]	SS[d]	MIAr	MIXe	VP	SS
cis-2-Heptene	1667	1665	1655	—	N.O.[e]	N.O.	N.O.	—
trans-2-Heptene	N.O.	N.O.	N.O.	1675	969	967	965	965
cis-2-Octene	1667	1665	1660	1655	N.O.	N.O.	N.O.	N.O.
trans-2-Octene	N.O.	N.O.	—	1675	968	966	—	964
cis-2-Nonene	1666	1664	1655	1655	N.O.	N.O.	N.O.	N.O.
trans-2-Nonene	N.O.	N.O.	N.O.	N.O.	972	969	965	965
cis-2-Decene	1667	1664	—	—	N.O.	N.O.	—	—
trans-2-Decene	N.O.	N.O.	—	—	968	967	—	—
cis-3-Decene	1663	1661	—	—	N.O.	N.O.	—	—
trans-3-Decene	N.O.	N.O.	—	—	969	967	—	—
cis-4-Decene	1663	1660	—	—	N.O.	N.O.	—	—
trans-4-Decene	N.O.	N.O.	—	—	972	970	965	—
cis-5-Decene	1661	1659	—	—	N.O.	N.O.	—	—
trans-5-Decene	N.O.	N.O.	1635	1675	970	969	965	967

[a]Matrix isolation with Ar.
[b]Matrix isolation with Xe.
[c]Data from Refs. 50 and 51.
[d]Data from Refs. 48 and 49.
[e]N.O., not observed.

example, aliphatic aldehydes are used, singly or in combination, in nearly all perfumes. One of the most potent fragrance and flavoring substances is an unsaturated aldehyde, 2-*trans*-6-*cis*-nonadienal. Aliphatic esters are used as flavor and fragrance compounds because of their taste and specific odor notes. Thus characterization of the MI spectra of the carbonyl class of compounds is necessary if the technique is to be employed successfully in structure elucidations.

For the majority of carbonyl-containing compounds examined under MI conditions, the energy of the carbonyl band was found between a high value set by the VP result and a low value set by the SS result. For example, the carbonyl absorption band for acetic acid in the MI, VP, and SS cases appeared at 1776, 1789, and 1714 cm^{-1}, respectively. Similar trends were evident for aldehydes, ketones, amides, and esters.

The position of the carbonyl absorption band in MI spectra was found to be a leading indicator of the type of compound present. The ranges of carbonyl absorptions for selected types of carbonyl-containing compounds are shown in Tables 3 to 5. The carbonyl absorptions for amides were found at the lowest

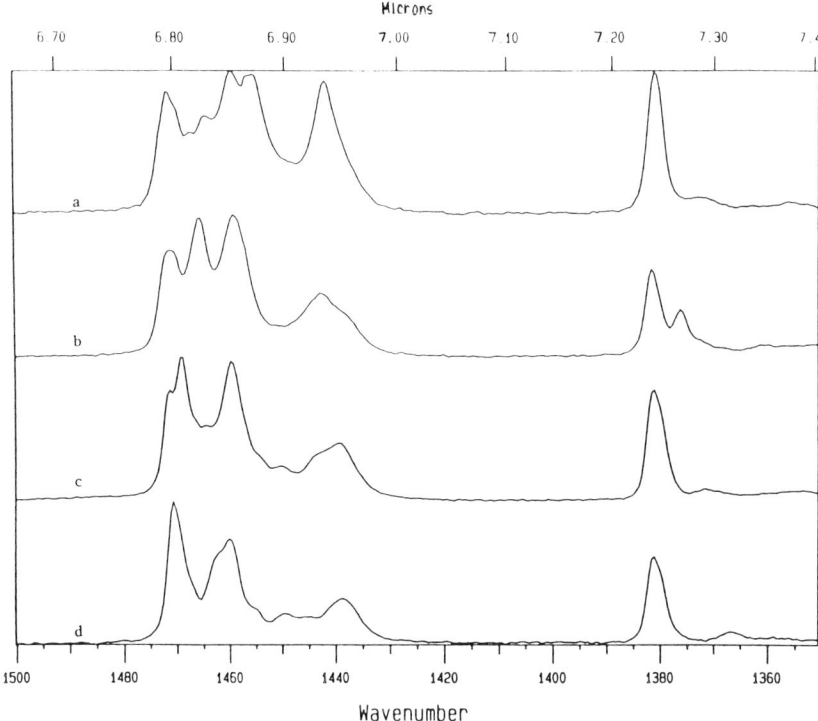

Fig. 13 MI/FT-IR spectra, fingerprint region of (a) *trans*-2-decene, (b) *trans*-3-decene, (c) *trans*-4-decene, and (d) *trans*-5-decene, 50 ng each. [From W. M. Coleman III and Bert M. Gordon, *Appl. Spectrom.*, *43:* 307 (1989).]

energy range, while the carbonyl absorptions for acids were found at the higher energies. Similar trends were found in the VP and SS spectra of these types of carbonyl-containing compounds.

In a number of cases, the MI/FT-IR spectra of carbonyl-containing organic compounds such as ketones, aldehydes, esters, and acids yielded multiple carbonyl absorption bands even though only one carbonyl group was present in the molecule. In some cases, nearest-neighbor interactions (aggregation) were the cause of the splitting due to matrix gas/analyte ratios lower than 10,000:1. In other cases, isolation of distinct conformers produced the multiple absorptions in halogen-substituted ketones [33,46] (Figs. 14 and 15). Acetophenone possesses a split carbonyl absorption, but both α-chloromethylacetophenone and α-dichloromethylacetophenone have two distinct carbonyl absorptions, indicating the pres-

Table 3 Infrared Data on Selected Ketones (cm^{-1})

Compound	Carbonyl absorption band(s)				
	MI(Xe)[a]	MI(Ar)[b]	BARE[c]	SS[d]	VP[e]
Cyclobutanone	1790	1796	1785	1783	1816
Cyclopentanone	1750[f]	1759	1751	1746	1765
	1750	1751	1733	1730	
	1729	1734			
Cyclohexanone	1719	1729	1709	1714	1732
	1725	1714	1719		
	1790	1723			
Cycloheptanone	1707	1714	1700	1702	1721
		1705			
Cyclooctanone	1711	1716	1697	1701	1720
	1704	1707			
		1760			
Cyclodecanone	1709	1713	1702	1702	1719
	1719	1723			
2-Pentanone	1724	1730	1717	1717	1730
2-Hexanone	1726	1730	1717	1717	1732
2-Heptanone	1726	1731	1715	1718	1731
	1717 sh[g]				
2-Octanone	1726	1731	1714	1717	1730
2-Nonanone	1726	1731	1714	1719	1732
3-Nonanone	1723	1728	1715	1717	1730
4-Nonanone	1723	1728	1714	1715	—
5-Nonanone	1723	1722	1714	1715	1728

[a] Matrix isolation with Xe.
[b] Matrix isolation with Ar.
[c] Bare gold disk at 10 K.
[d] Solid state; data from Refs. 48 and 49.
[e] Vapor phase; data from Refs. 50 and 51.
[f] Most intense band listed first.
[g] sh, Shoulder.

ence of discrete conformers. These splittings were observed at matrix gas/analyte ratios ≥ 10,000:1. Thus MI/FT-IR spectra should always be checked for concentration dependence to ensure that bands are correctly assigned. The compounds under examination should also be surveyed for the possibility of conformers.

The size and flexibility of the ring with a series of selected lactones was shown to influence the number of carbonyl absorptions in their MI/FT-IR spectra. For example, the MI/FT-IR spectrum of 2(5*H*)-furanone contained two carbonyl

Table 4 Infrared Data on Selected Lactones and Lactams (cm^{-1})

Compound	Carbonyl absorption band(s)				
	MI(Xe)[a]	MI(Ar)[b]	DISK[c]	SS[d]	VP[e]
δ-Valerolactone	1758[f]	1764	1733	1732	1785
	1751	1756	1722		
		1778			
γ-Valerolactone	1793	1804	1777	1773	1811
	1797	1800	1783		
	1786	1790	1767 sh[g]		
		1811	1742		
γ-Butyrolactone	1799	1804	1750	1771	1821
	1780	1801	1761		
	1774	1785	1771		
	1808	1778			
		1812			
β-Butyrolactone	1842	1842	1831	1823	1865
	1862	1851	1840		
		1843	1815		
		1873			
		1829			
2(5H)-Furanone	1789	1794	1748	—	—
	1780	1797	1779		
2-Pyrrolidinone	1731	1736	1694	1690	1759
	1721	1741	1710		
		1720	1732		
		1749			
1-Methyl-2-pyrrolidinone	1711	1717	1689	1689	1730
	1724	1720	1679 sh		
	1728	1727			
5-Methyl-2-pyrrolidinone	1723	1744	1700	—	—
	1732	1731	1685		
	1735	1754			
	1704	1712			
	1748				
δ-Valerolactam	1687	1698	1669	1690	1715
	1685	1644			
ε-Caprolactam	1687	1693	1668	1658	1711
	1702	1707	1637		
Oxindole	1754	1760	1706	1699	1772
	1748	1765	1697		
	1745	1750	1731		
Coumarin	1756	1760	1711	1704	1775
	1745	1748	1756	1755 sh	
	1771	1745			

Table 4 Continued

Compound	Carbonyl absorption band(s)				
	MI(Xe)[a]	MI(Ar)[b]	DISK[c]	SS[d]	VP[e]
Phthalide	1787	1793	1752	1748	1810
	1778	1809	1767		
	1802	1804	1789		
	1806	1781			
	1797				
2-Coumaranone	1824	1829		1811	1838
Pantolactone	1872	1883		1761	1815
	1879	1875		1740 sh	
	1841	1847			
	1902	1902			
3-Hydroxy-4,4,4-trichlorobutyric	1796	1803	1784	—	—
acid, β-lactone	1792	1818	1764		
	1817	1786			
	1764				

[a]Matrix isolation with Xe.
[b]Matrix isolation with Ar.
[c]Bare gold disk at 10 K.
[d]Solid state; data from Refs. 48 and 49.
[e]Vapor phase; data from Refs. 50 and 51.
[f]Most intense band listed first.
[g]sh, Shoulder.

absorptions. Conversely, the MI/FT-IR spectrum of γ-butyrolactone contained five absorptions and δ-valerolactone contained three absorptions in the carbonyl region. In a similar fashion, as the size of the ring in cycloketone structures increased, so did the complexity of the carbonyl absorption (Fig. 16).

Hydroxyls

Several types of hydroxyl-containing compounds are common to essential oils and play important roles in aroma and flavor formulations [29]. Saturated primary alcohols occur widely in natural products. Their odor is relatively weak and thus their use as components in fragrance compositions is limited. However, the alcohols serve as very important starting materials for the production of aldehydes and esters. Aliphatic acids are found in many essential oils and foodstuffs. They contribute to aromas but are less important as fragrances of flavor substances. Phenols, as another class of hydroxyl-containing compounds in essential oils and natural products, are used as fragrances and flavor compounds. An excellent

Table 5 Infrared Data on Selected Amides (cm^{-1})

Compound	Amide I band					N—H stretch				
	MI(Xe)[a]	MI(Ar)[b]	BARE[c]	VP[d]	SS[e]	MI(Xe)	MI(Ar)	BARE	VP	SS
N-Methylacetamide	1699 1687 1678	1707[f] 1697 sh[g] 1688 sh 1684 sh	1662 1641 1686	1730 1721 1711	1658	3481	3494	3281 3115	3490	3297 3093
N-Ethylacetamide	1696 1676	1701 1681	1663 1642 1683	1715	1655	3468 3464	3481	3284 3099	3480	3290
N-Propylacetamide	1697 1677	1701 1680	1661 1640	—	—	3463	3483	3280 3099	—	—
N-Isopropylacetamine	1693 1674	1699 1689 sh 1678	1662 1641 1679 1691	1713	—	3458	3464	3279 3091	3462	
N-Butylacetamide	1695 1675	1701 1680	1659 1638	1715	1645	3465	3485	3277 3090	3478	3270
N-Isobutylacetamide	1698 1676	1703 1695 sh 1682	1661 1641	—	—	3471	3481	3283 3098	—	
N-Amylacetamide	1695 1675	1701 1681	1659 1635	—	—	3463	3484	3276 3099	—	
Dimethylacetamide	1671 1663 1649	1676 1681 sh	1646 1663	1690	1646					
Diethylacetamide	1661 1665	1673 1666	1640	1682	1642					

83

Table 5 Continued

Compound	Amide I band					N—H stretch				
	MI(Xe)[a]	MI(Ar)[b]	BARE[c]	VP[d]	SS[e]	MI(Xe)	MI(Ar)	BARE	VP	SS
Dipropylacetamide	1661	1668	1644	—	—					
		1652								
Dibutylacetamide	1661	1669	1645	1678	1640					
		1650 sh								
Benzamide	1701	1708	1676	1727	1661	3536	3554	Broad	3550	3368
	1683	1689		1721		3421	3437		3436	3165
				1716						
Propionamide	1718	1726	1688	1745	1661	3539	3555	(3551)	3550	3363
	1706	1716 sh	1714	1725	1620 sh	3420	3435	(3432)	3430	3190
	1696	1699 sh						(3332)		
								(3200)		
2-Chloropropionamide	1708	1717	1691	—	1670	3535	3542	(3541)	—	3345
	1692	1700	1707			3417	3428	(3422)		3170
								(3320)		
								(3180)		
3-Chloropropionamide	1708	1722	1687	1740	1682	3535	3552	Broad	3550	3335
	1715	1702	1717	1725 sh	1635	3408	3433		3440	3154
	1721									
	1728									

[a]Matrix isolation with Xe.
[b]Matrix isolation with Ar.
[c]Bare gold disk at 10 K.
[d]Vapor phase; data from Refs. 50 and 51.
[e]Solid state; data from Refs. 48 and 49.
[f]Most intense band listed first.
[g]sh, Shoulder.

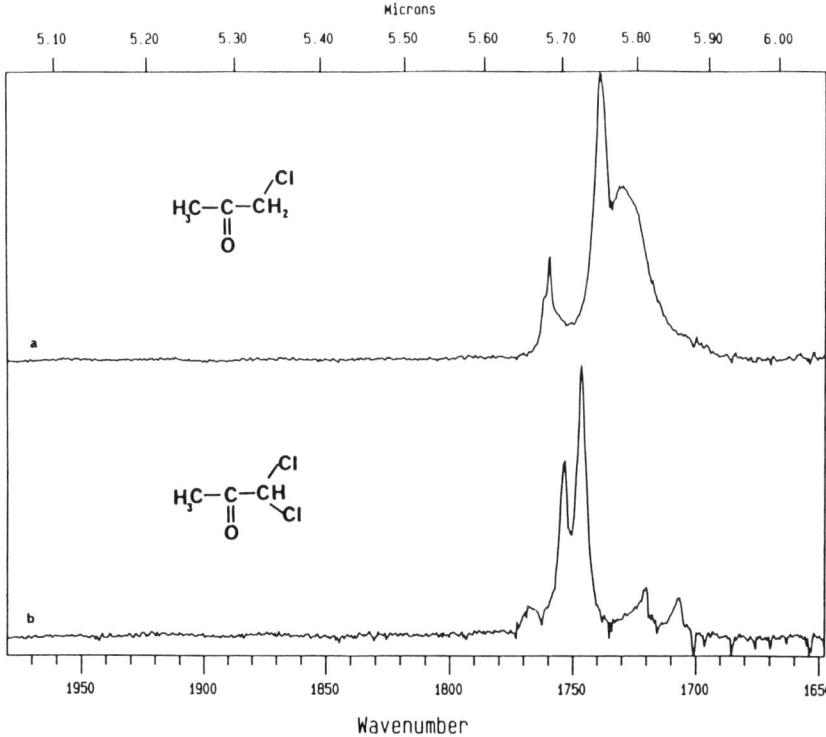

Fig. 14 MI/FT-IR spectra, carbonyl regions of (a) chloroacetone and (b) 1,1-dichloroacetone, 10 ng each. (From Ref. 46.)

example is thymol, from thyme and origanum oils, used in men's fragrances and in mouth care products.

The characteristics of the IR spectra of hydroxyl-containing compounds have long been recognized as depending on the nature of any intra- and intermolecular hydrogen bonding between the hydroxyl groups. Intermolecularly hydrogen-bonded OH stretching frequencies in the solid state and solution have been observed as broad bands between 3700 and 3200 cm^{-1}, with full width at half height (FWHH) values greater than 100 cm^{-1}. With the emergence of VP/IR techniques, the ability to observe sharp nonintermolecular bonded OH stretching bands around 3650 cm^{-1} was realized (Table 6), with FWHH values $\leq$ 20 cm^{-1}. Since these absorptions were well spaced from other bands, they became a significant diagnostic tool.

In the early stages of the development of MI/FT-IR technology, phenol was often used as a gauge for true matrix isolation. The appearance of one sharp band,

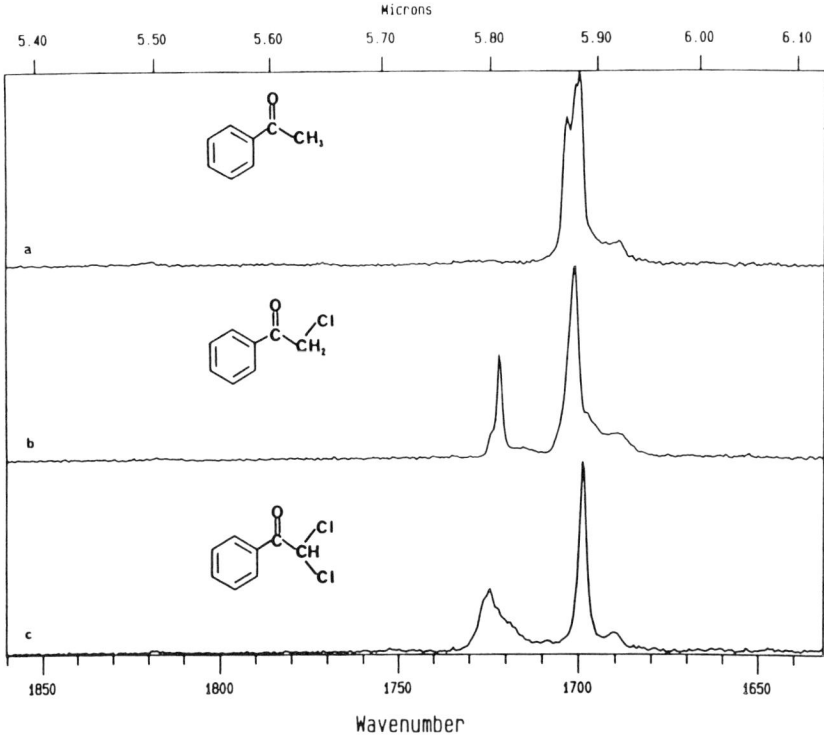

Fig. 15 MI/FT-IR spectra, carbonyl regions of (a) acetophenone, (b) 1-chloroacetophenone, and (c) 1,1-dichloroacetophenone, 10 ng each. (From Ref. 46.)

FWHH ≤ 10 cm^{-1}, around 3635 cm^{-1} with no accompanying broad absorptions was a clear indication that the matrix gas/analyte ratio was sufficient to inhibit intermolecular interactions. The position of the OH stretching bands for selected phenols in selected phases appears in Table 6. Regardless of substituent, the values in the MI case fall within a very narrow window. No significant change occurred even though meaningful changes were made to influence steric and inductive properties. These effects also seemed to have little effect on the VP values, but the range for the SS values was relatively wide, indicating significant effects due to the presence of intermolecular interactions. In the case of phenol, if a significant number of the molecules are not truly matrix isolated, the MI/FT-IR spectrum will reveal the presence of two types of OH groups, one intermolecularly hydrogen bonded, 3373 cm^{-1} (broad), and one matrix isolated, 3635 cm^{-1} (sharp). For the phenol compounds the OH stretching bands in the MI case were

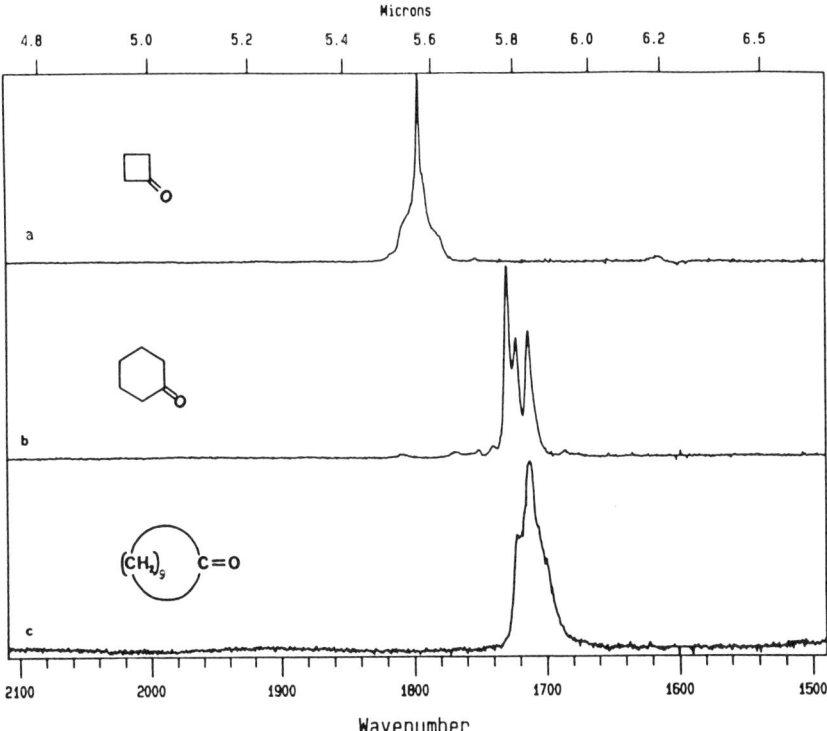

Fig. 16 MI/FT-IR spectra, carbonyl regions of (a) cyclobutanone, (b) cyclohexanone, and (c) cyclodecanone, 10 ng each. (From Ref. 46.)

found between high values set by the VP results and low values set by the SS results.

The positions of the OH stretching absorptions in the MI cases offer some of the best evidence for dramatic effects due to the inert matrix host noble gas. For the case of phenol, the OH absorption occurs at 3612 cm^{-1} in Xe and 3635 cm^{-1} in Ar. Similar differences were noted for linear alcohols (Table 6). It was interesting to note that the OH absorptions in Xe were lower in energy than those in Ar.

Table 7 contains data on the position of the nonintermolecular bonded stretching absorption of the OH group in linear carboxylic acids. These absorptions were very sharp, having FWHH values of ≤ 10 cm^{-1}. The MI and VP examples indicated a very narrow range of ± 1 cm^{-1}. Such was not the case, as expected, with the SS values. The OH bands were located between the extreme values set by the VP and SS data.

Table 6 Infrared Data on Selected Phenols and Alcohols

	O—H stretch				
Compound	MIXe[a]	MIAr[b]	BARE[c]	SS[d]	VP[e]
Phenol	3612	3635	3622	3650	3373
			3357		
o-Cresol	3614	3635	3625	3655	3448
			3350		
m-Cresol	3612	3636	3622	3650	3333
			3350		
p-Cresol	3615	3636	3627	3655	3364
			3350		
2,6-Dimethylphenol	3617	3634		3655	3574
1-Hexanol	3633	3660	3643	3670	3334
			3390		
1-Heptanol	3640	3660	3647	3670	3323
			3390		
1-Octanol	3631	3660	3640	3675	3329
			3400		
Furfuryl alcohol	3619	3643			3329

[a]Matrix isolation with Xe.
[b]Matrix isolation with Ar.
[c]Bare gold disk at 10 K.
[d]Data from Refs. 48 and 49.
[e]Data from Refs. 50 and 51.

Table 7 Comparative IR Group Frequency Data for Selected Aliphatic Carboxylic Acids (cm^{-1})

	C=O			O—H		
Acid	MI[a]	VP[b]	SS[c]	MI	VP	SS
Acetic	1777	1789	1715	3551	3580	3275
Pentanoic	1771	1780	1711	3551	3579	—
Hexanoic	1768	1778	1711	3552	3579	3191
2-Ethylhexanoic	1763	1770	1706	3549	3577	3170

[a]Matrix isolated with Ar.
[b]Data from Refs. 50 and 51.
[c]Data from Refs. 48 and 49.

Table 8 Oxygen–Hydrogen Stretching Positions for Selected Alcohols (cm^{-1})

Compound	MI[a]	VP[b]	SS[c]
1-Pentanol	3660	3678	3324
2-Pentanol	3642	3659	3351
3-Pentanol	3647	3662	3355
2-Methyl-2-butanol	3623	3644	3386
2-Methyl-1-butanol	3662	(3675[d])	3331
1-Octanol	3660	3675	3329
3-Methyl-1-butanol	3662	3670	3329
Furfuryl alcohol	3643	—	3329

[a] Matrix isolation with Ar.
[b] Data from Refs. 50 and 51.
[c] Data from Refs. 48 and 49.
[d] Value quoted measured for 2-ethyl-1-butanol.

A shift in the position of the nonbonded OH stretch to lower energy with the change from primary to secondary to tertiary carbon substitution has been established for VP spectra of aliphatic alcohols and can be seen in Table 8 for MI spectra. The overall range of the shift is approximately 40 cm^{-1} for both phases, decreasing in energy for primary to tertiary.

Thus the MI/FT-IR spectra of the OH-containing compounds were characterized by the presence of sharp absorptions in the region between 3600 and 3300 cm^{-1}. Yet within this broad range very narrow regions were found for each type of OH-containing compound. Consequently, with the absence of intermolecular hydrogen bonding, the OH absorption in MI experiments assumes a strong role, comparable to that of the carbonyl group in assignments of structure. Deviation from the matrix gas/analyte ratio of 10,000:1 produced broad absorption bands reminiscent of those found in condensed-phase IR spectra.

Alkanes

Saturated aliphatic hydrocarbons with straight as well as branched chains occur abundantly in natural products such as fruits. They have only a limited impact on odor and taste. The MI/FT-IR spectra of these compounds were unique in terms of the relationship of the energies of the major bands with VP and SS phases. Unlike the major bands discussed previously, the major bands of the alkanes were not found to be intermediate between high values set by the VP value and low values set by the SS phase (Table 9). For example, in the case of octane, the methyl C—H stretch was found at 2966 cm^{-1} in the MI case and at 2964 cm^{-1} in the VP case. No clear explanation for this behavior is currently apparent.

Table 9 Infrared Data on Selected Hydrocarbons (cm^{-1})

Compound	MIXe[a]	MIAr[b]	BARE[c]	SS[d]	VP[e]	MIXe	MIAr	BARE	SS	VP
Octane	2955	2966	2963	2962	2964	2927	2932	2931	2923	2930
	2870	2881	2878	2870	2880	2856	2863	2860	2850	2865
	1467	1471	1741	1467	1464	1464	1459	1461	1450	1464
	1377	1381	1381	1378	1384					
Nonane	2955	2965	2962	2960	2962	2927	2932	2930	2924	2930
	2870	2879	2876	2872	—	2856	2862	2860	2851	2862
	1467	1471	1470	1467	1453	1455	1459	1461	1450	1463
	1377	1381	1381	1378	1381					
Decane	2855	2966	2960	2960	2970	2927	2933	2828	2919	2938
	2870	2880	2873	2870	—	2856	2862	2858	2850	2862
	1467	1471	1469	1467	1468	1455	1459	1460	1450	1468
	1377	1381	1379	1378	1385					
Undecane	2965	2965	2959	—	2965	2951	2932	2927	—	2930
	28701	2879	2872	—	—	2856	2861	2854	—	2862
	467	1471	1469	—	1465	1455	1459	1460	—	1465
	1377	1381	1379	—	1385					
Dodecane	—	2965	2960	2962	2965	—	2933	2927	2926	2935
	—	2880	2876	2872	2880	—	2862	2857	2853	2862
	—	1471	1469	1465	1465	—	1459	1460	1450	1465
	—	1381	1380	1375	1382	—				

[a] Matrix isolation with Xe.
[b] Matrix isolation with Ar.
[c] Bare gold disk at 10 K.
[d] Data from Refs. 48 and 49.
[e] Data from Refs. 50 and 51.

F. IR Spectral Libraries

Due to the uniqueness of the MI/FT-IR spectra of the compounds described above, it was necessary to compile a library of MI spectra. When compiled, one overall trend appeared with the major absorption bands. In the overwhelming number of cases, the energies of the band maxima for MI compounds were unique and were intermediate between a higher value found in the vapor phase (light pipe) and a lower value found in the solid state (see above). Selected exceptions, alkanes and alkenes, have been noted previously. This shifting of the absorption maximum was the most obvious matrix effect, based on the fact that the vibrational levels of the analyte will be perturbed by the matrix, and thus the vibrational frequencies will be shifted from the gas-phase and solid-state values [33].

An evaluation of a set of search routines for the major components of four well-characterized essential oils indicated that the resolution of the search routine determined its effectiveness. For example, when using an acquisition with 1-cm^{-1} resolution, no meaningful information could be obtained from the search routines, principally because the spectra were gathered at 1 cm^{-1}. If, on the other hand, an acquisition of 32 cm^{-1} was used on the same spectra, acceptable matches were obtained for one of the search routines provided in the software. Such uniqueness of MI spectra has been reported in other works [13,35,37,40], thereby confirming the necessity for libraries based on spectra collected under MI conditions. It therefore seems quite clear that the uniqueness of the MI spectra from a library search perspective is due to (1) the natural narrow line width and (2) matrix-induced absorption maximum shifts.

G. Matrix Versus Light Pipe

A general comparison of the features of MI/FT-IR spectra with those obtained in the vapor phase have appeared in the literature [17]. The features of the two techniques were compared in terms of limit of detection, cost, run time, specificity, and spectral search compatibility. One of the attractive features of the light pipe compared to the matrix isolation system was cost. One of the attractive features of the MI technique was the much lower limit of detection, approximately 1 ng, compared with 10 ng for the VP case. Another feature was the natural very narrow line width of the MI spectra compared with the VP case. For example, this attribute led to the characterization and specification of substitution isomers of dichlorobiphenyl.

In terms of natural products, differences in sensory responses to compounds are often related to more subtle changes in geometry than just positional isomerization. For example, pinane, an important starting material in technical processes in the flavor and fragrance industry [29], can exist as a cis or trans isomer. This cis or trans isomer label does not refer to the configuration around a double bond but to the position of the methyl group relative to the bridge carbons (Fig. 17). For example, it is possible that the methyl group could be on the same side of the ring as the bridge, yielding the cis isomer, or on the opposite side of the bridge carbons, yielding the trans isomer.

Such subtle differences in structure have traditionally not yielded measurable differences in IR spectra. Somewhat unexpectedly, Fig. 18 reveals that spectra from the MI technique could distinguish between the two isomers in the fingerprint region of the IR spectrum. On the other hand, Fig. 19 reveals that spectra from the VP technique do not provide the required spectral resolution and thus are not capable of distinguishing between the isomers. The high temperatures (i.e., 250°C) employed in the VP case resulted in absorption bands with broader line widths, and consequently, differentiation between the two isomers is eliminated.

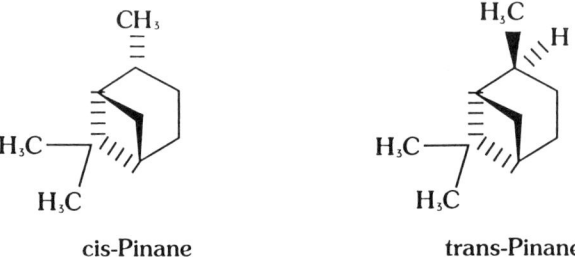

Fig. 17 Structures of *cis*-pinane and *trans*-pinane.

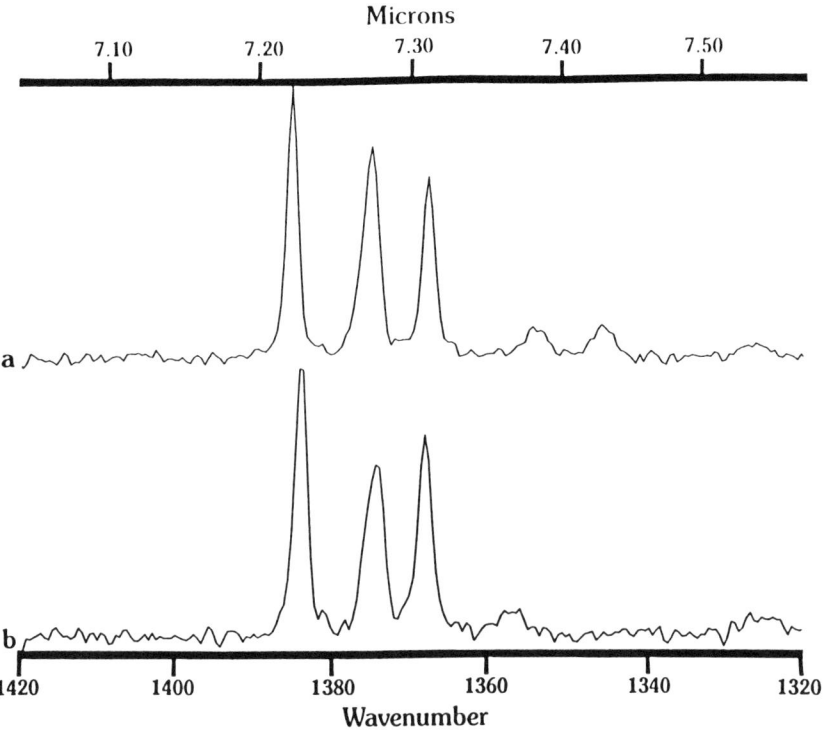

Fig. 18 MI/FT-IR spectra, methyl carbon–hydrogen bending region of (a) *cis*-pinane and (b) *trans*-pinane, 10 ng each. (From Ref. 31.)

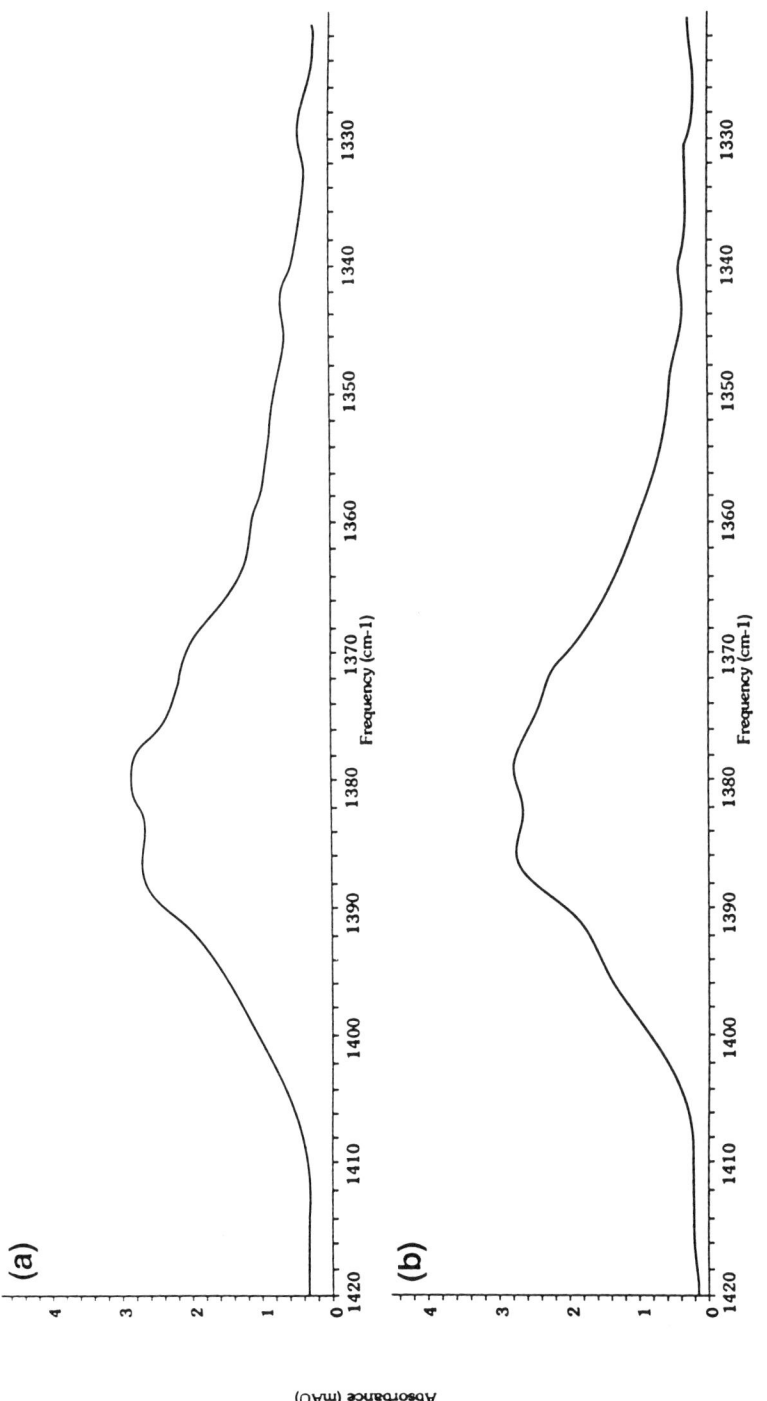

Fig. 19 VP/FT-IR spectra, methyl carbon–hydrogen bending region of (a) *cis*-pinane and (b) *trans*-pinane, 100 ng each.

IV. APPLICATIONS

Essential oils represent one of the more complex mixtures of volatile and semivolatile organic compounds, due to origins associated with natural products. Not only are multiple types of functional compounds present, but the ranges of concentrations can be large [29]. The current tendencies are for the development of methodologies capable of trace component identification. Thus the isolation and structure elucidation of these components requires high-resolution chromatographic separations coupled with advanced spectroscopic techniques.

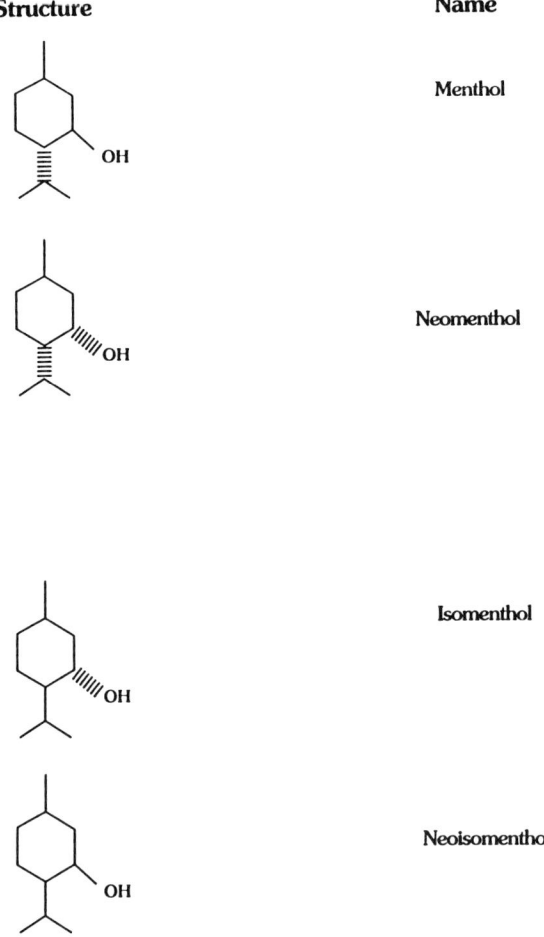

Fig. 20 Structures for selected cyclic menthol terpenes.

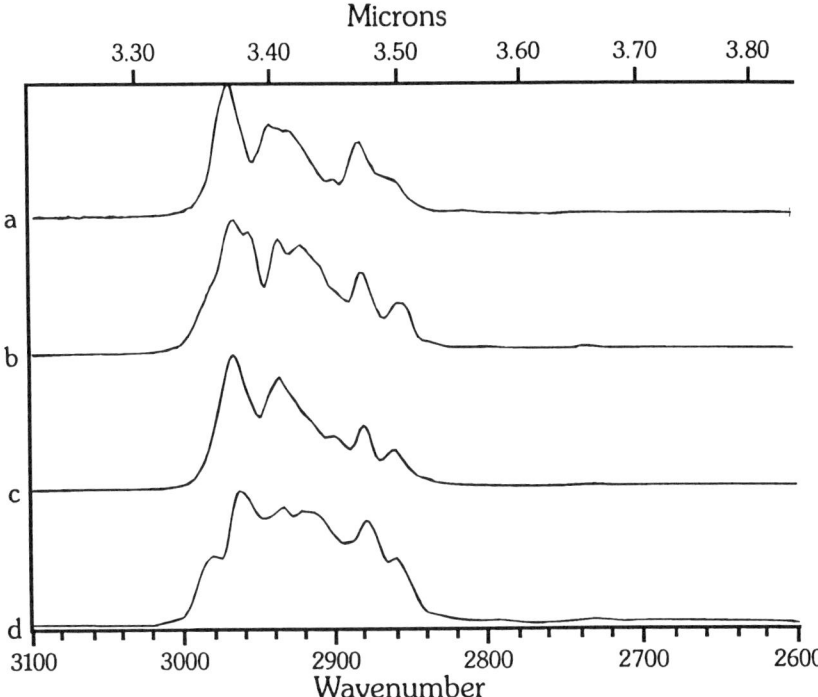

Fig. 21 MI/FT-IR spectra, carbon–hydrogen stretching region of (a) menthol, (b) neomenthol, (c) isomenthol, and (d) neoisomenthol, 25 ng each. [From W. M. Coleman III and Bert M. Gordon, *Appl. Spectrom., 43:* 303 (1989).]

A prime example of this separations/spectroscopic technique rests with the GC/MI/FT-IR approach. For example, (-)menthol is a major component of peppermint oil and has the characteristic cooling effect. The other isomers—neomenthol, isomenthol, and neoisomenthol—are structurally very similar (Fig. 20), but do not possess the cooling effect. Thus it is important to know the distribution of these components to assess the purity and effectiveness of the mixture. GC/MI/FT-IR has been employed successfully to characterize the differences in the IR spectra of these pure components (Fig. 21). The carbon–hydrogen stretching region of the MI spectra of these compounds revealed each to be distinct, even though the compounds differ in structure only in terms of the relative position of the hydroxyl groups. In like fashion, *p*-menthane can exist in two isomers based on the relative positions of the methyl and isopropyl groups. Mass spectrometry (Fig. 22) was not capable of differentiating between the isomers, yet the fingerprint regions of the MI/FT-IR spectra of the two isomers

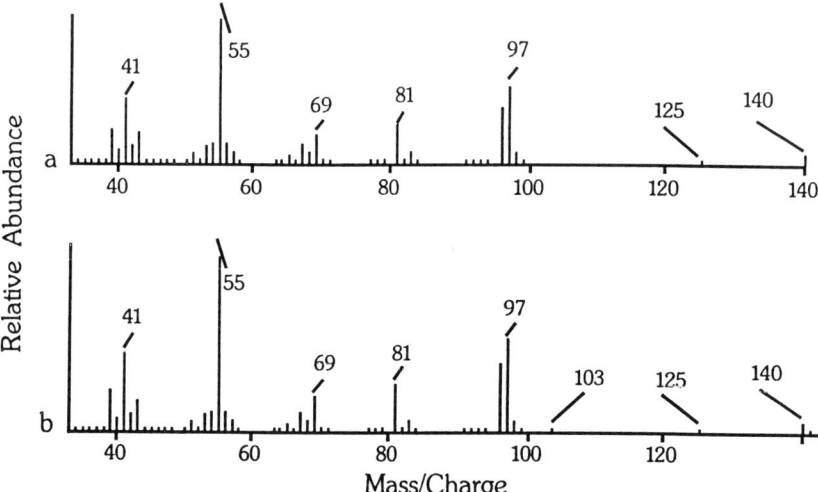

Fig. 22 Mass spectra of the two isomers of *p*-menthane. [From W. M. Coleman III and Bert M. Gordon, *Appl. Spectrom., 43:* 302 (1989).]

were unique (Fig. 23). With bicyclic natural product aliphatic systems such as pinane, which is similar to the cyclic systems of menthane, remarkably subtle differences in the MI/FT-IR spectra of the pure compounds were obtained (Fig. 18). The differences were noted in the relative intensities of the absorptions in the carbon–hydrogen bending region of the spectra. The real impact of MI/FT-IR spectra of natural products is fully realized when the natural products have been separated into their component parts by high-resolution GC. Several examples of the use of this approach are discussed next.

Soybean oil is the largest source of vegetable oil in the world [43]. However, because of its high content of unsaturated fatty acids, soybean oil must be hydrogenated to prevent flavor deterioration. Thus it is important to determine the presence and structure of any unsaturated species to ensure a stable product. GC/MI/FT-IR has been applied successfully to this problem, wherein identification of minor C_{18} triene and conjugated diene fatty acid methyl esters was accomplished. Examples of the differences in the MI/FT-RI spectra of the isomers are shown in Fig. 24.

Essential oils are and have been used successfully for aroma, flavor, and fragrance purposes since the ninth century. Stereochemical conformations and configurations have always played an important role in sensory responses to the oils. Figure 25 illustrates the flame ionization detector (FID) response for a coriander-based essential oil. Component c represents a very small component of

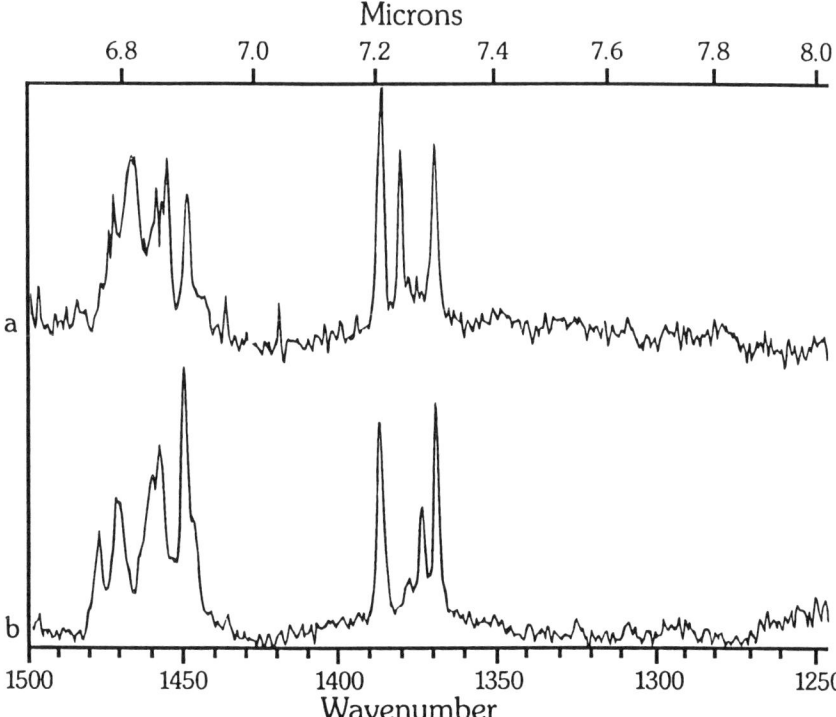

Fig. 23 MI/FT-IR spectra of the isomers of *p*-menthane, 2 ng each. [From W. M. Coleman III and Bert M. Gordon, *Appl. Spectrom., 43:* 302 (1989).]

the overall mix, yet through application of the GC/MI/FT-IR technique, an excellent IR spectrum was obtained (Fig. 25). Expansion of the carbon–hydrogen stretching region of the spectrum revealed several well-resolved absorptions (Fig. 26). Comparison via library search routines of matrix-isolated spectra indicated that the component was geraniol. A closely related compound, nerol, has physical properties similar to those of geraniol, and differs only in the trans versus cis conformation of the double bond. In terms of sensory aspects, nerol is preferred over geraniol for sweetness. Thus absolute identification was imperative. Even though both compounds have very similar hydrocarbon backbone structures, the MI/FT-IR spectra of the pure components was very different (Fig. 27), and identification of the component as geraniol and not nerol was possible.

In selected instances, the complexity of natural product extracts demands that multidimensional gas chromatography (MDGC) be employed to achieve acceptable resolution of components [16]. Such an application was demonstrated in a

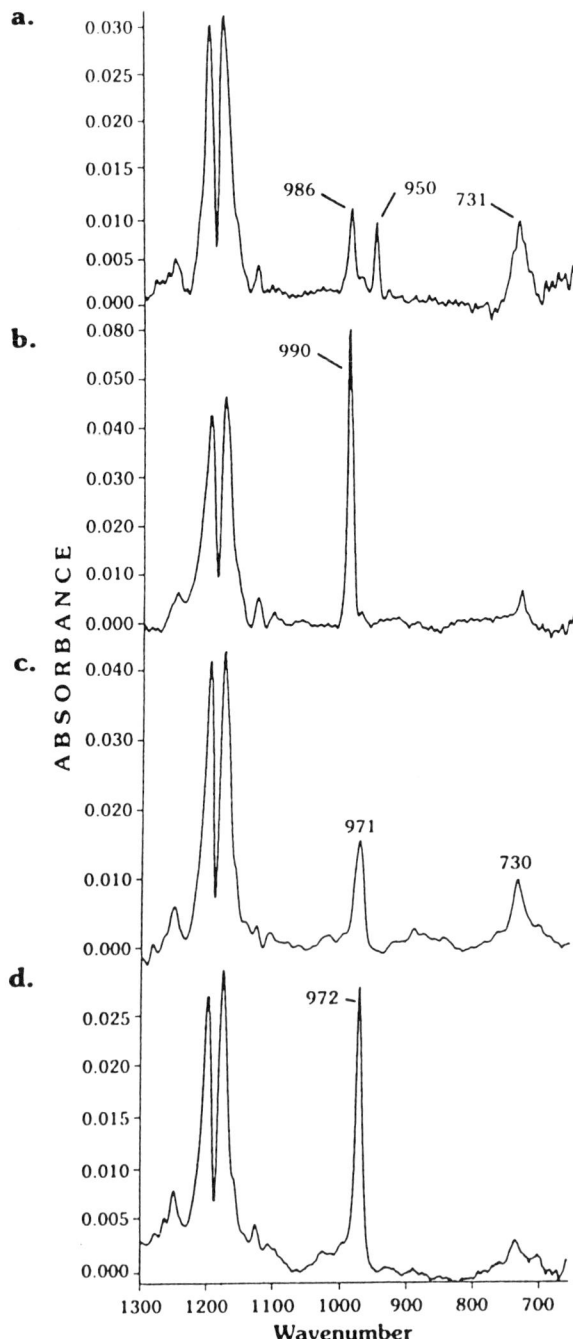

Fig. 24 Expanded IR spectra range showing the out-of-plane deformation bands for conjugated (a) *cis-trans* and (b) *trans* 18:2 dienes. Spectra for methylene-interrupted (c) *cis-trans* and (d) *trans-trans* 18:2 dienes shown for comparison. (From Ref. 43.)

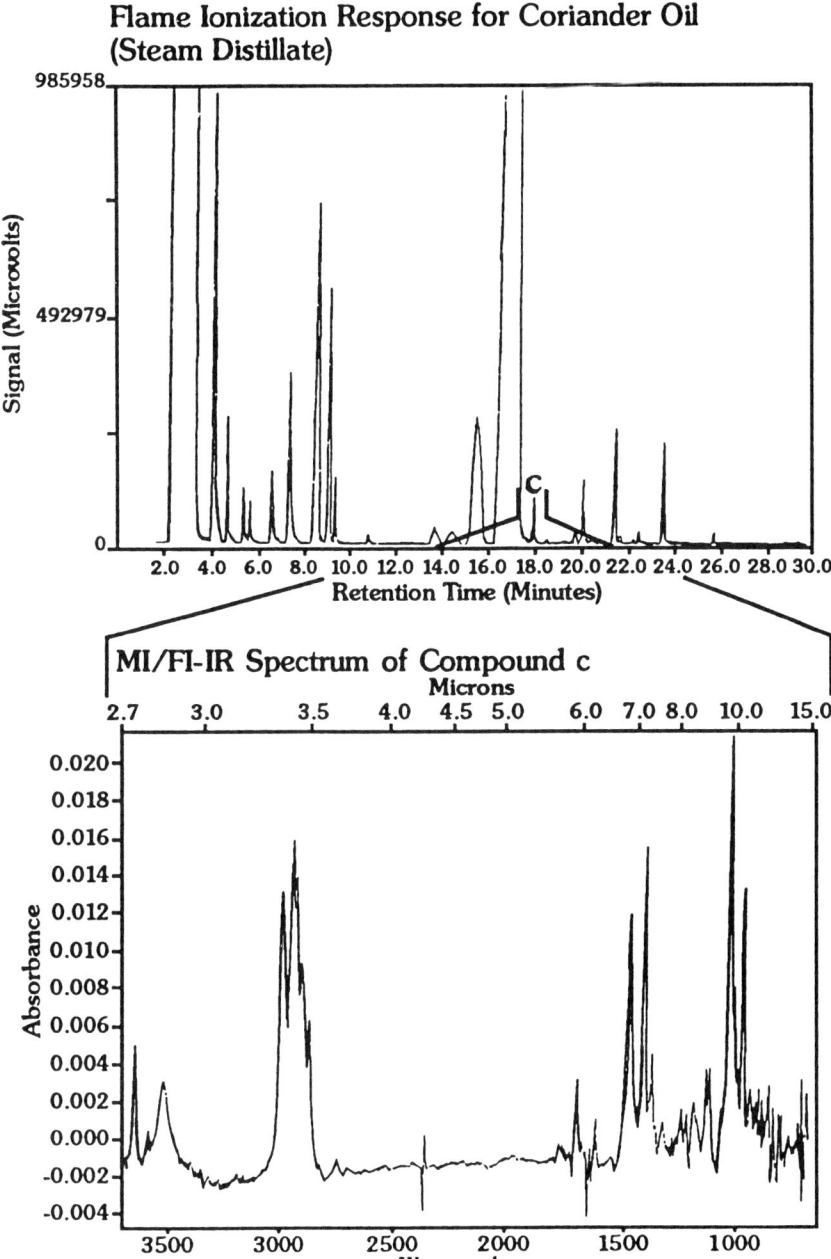

Fig. 25 Analysis of coriander oil: top, flame ionization response; bottom, MI/FT-IR spectrum of compound c. (From Ref. 45.)

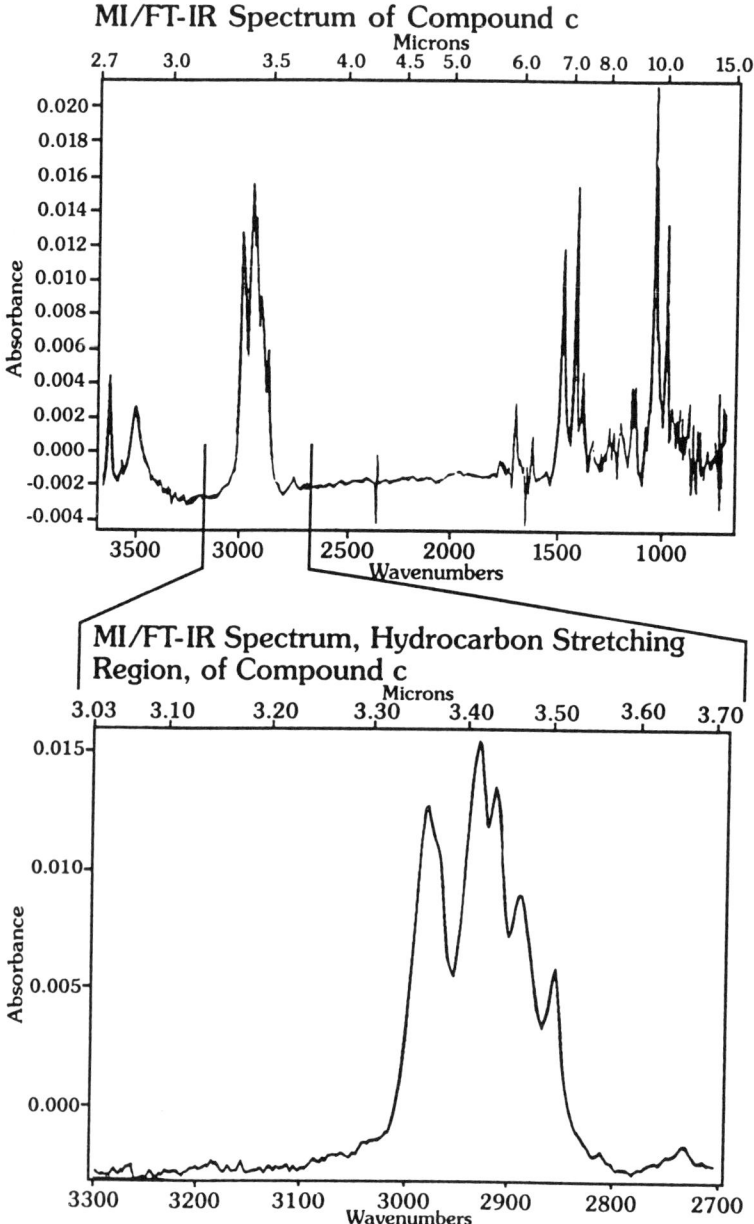

Fig. 26 MI/FT-IR spectra of compound c from Figure 25: top, full spectrum; bottom, hydrocarbon stretching region. (From Ref. 45.)

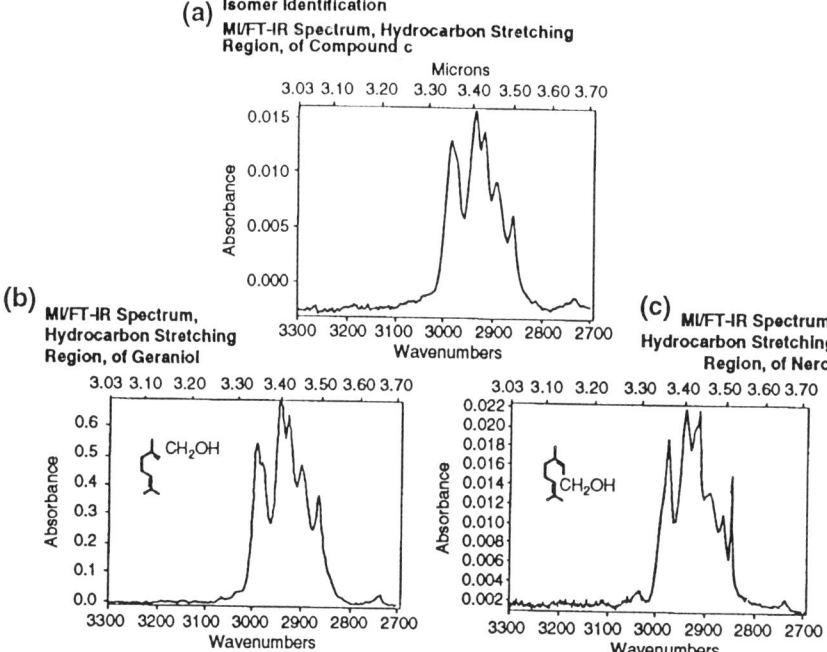

Fig. 27 Isomer identification via hydrocarbon stretching region: (a) compound c from Figure 23, (b) geraniol, (c) nerol.

study on the components of a tobacco essential oil. The complexity of the essential oil is shown in Fig. 28. The regions denoted by the up arrows represent fractions that were sequentially transferred (heartcut) to the analytical column. For example, heartcut 8, represented by retention times of 14.5 to 16.0 min, is shown in Fig. 29. MI/FT-IR spectra gathered on the components of this mixture were valuable for the identification of components when mass spectral information could not provide adequate structural information. For example, employing selected wavelength-reconstructed chromatograms, it was possible to segregate the carbonyl-containing compounds within the heartcuts. The carbonyl absorption of selected compounds under MI conditions had been shown to be a constructive piece of information in structure elucidation. Thus, by wedding information from MS with that from MI/FT-IR, coupled with enhanced separation of MDGC, positive identification of the majority of the components was possible. For example, the MS and MI/FT-IR information on a selected peak in Fig. 29 is presented in Fig. 30. The MS information gathered on this component was not sufficient to provide for definitive identification. Yet when combined with

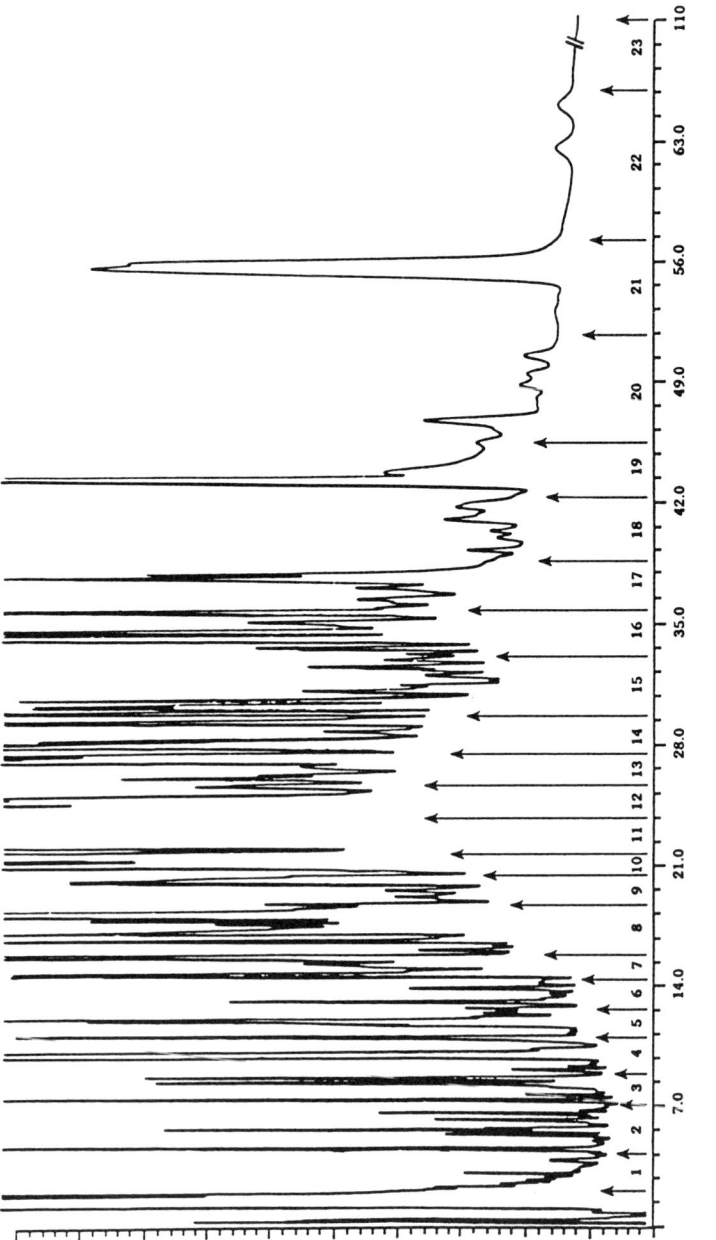

Fig. 28 Uncut precolumn chromatogram showing points at which the heartcuts were made. (From Ref. 16.)

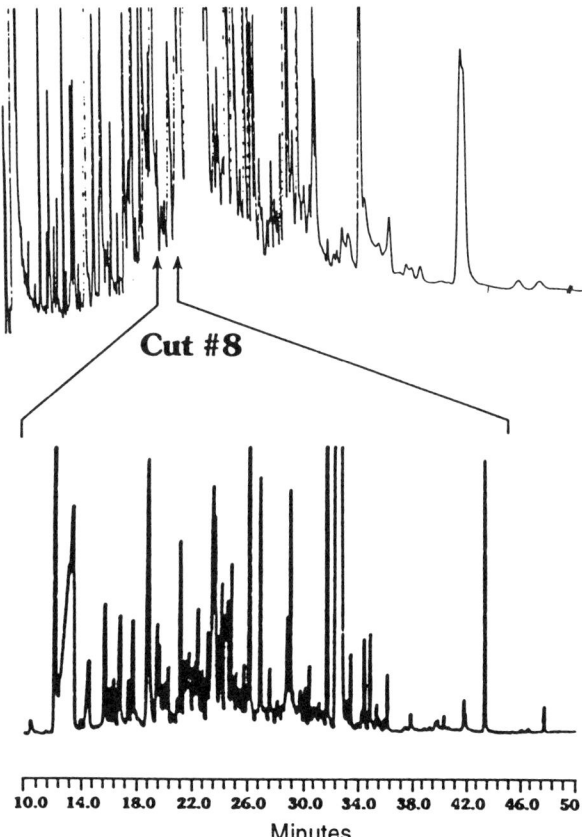

Fig. 29 FID trace of precolumn chromatogram and analytical column for heartcut 8 from a tobacco essential oil.

the MI/FT-IR information (i.e., a cyclic carbonyl compound), identification of the component became possible (Fig. 31).

V. SUMMARY

Information concerning the characteristics of the MI/FT-IR spectra of natural product components has been presented. These IR spectra were obtained on the components after their separation by gas chromatography. The special requirements of the hyphenated technique GC/MI/FT-IR were discussed, with particular emphasis on the nature of the interface. The unique features of MI/FT-IR spectra

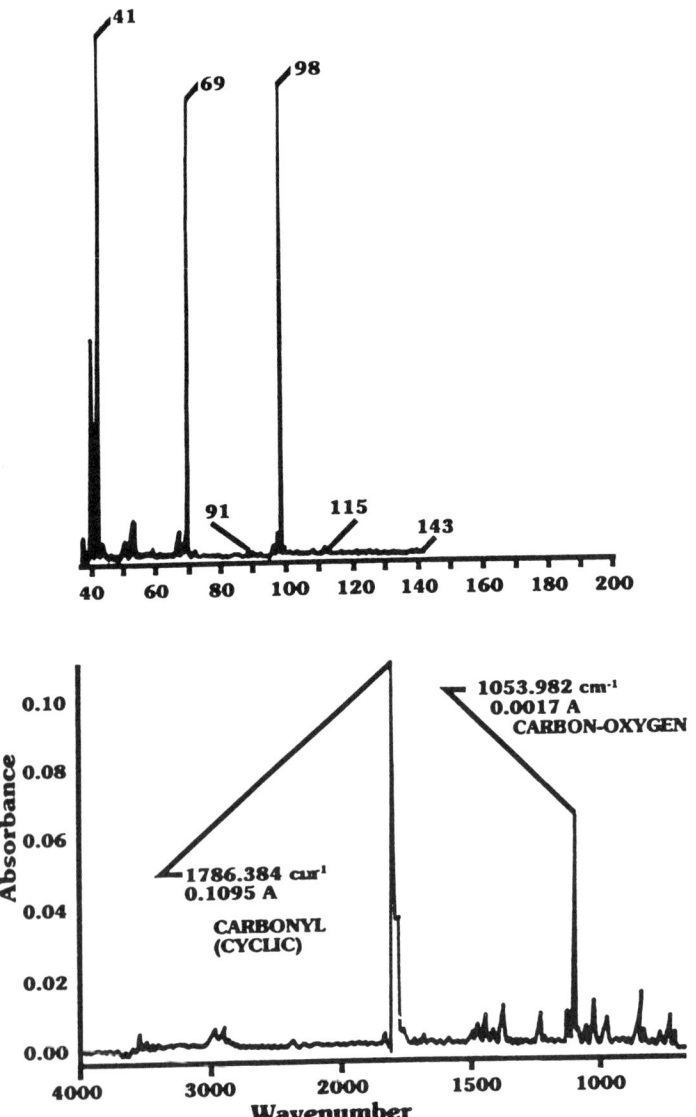

Fig. 30 Mass spectrum and MI/FT-IR spectrum of selected component from heart-cut 8.

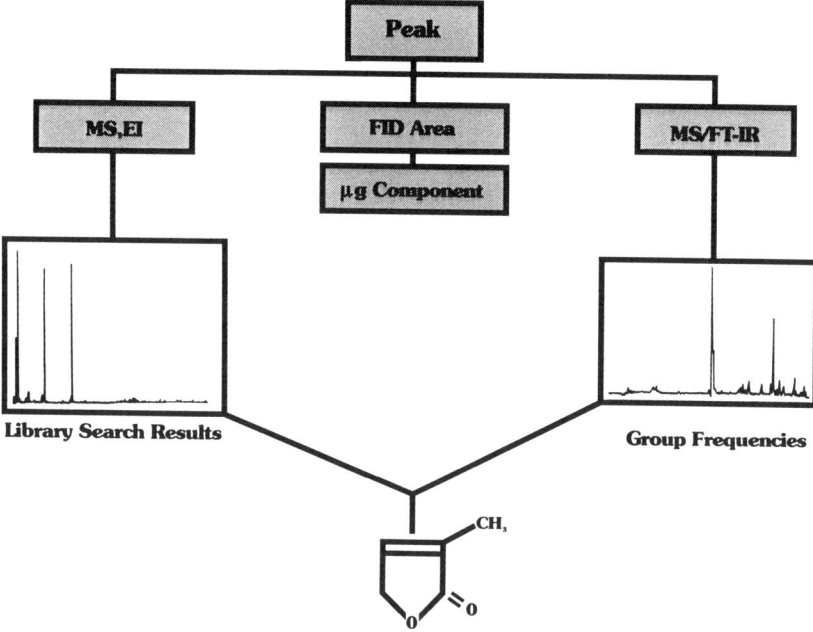

Fig. 31 Schematic representation of relationship between mass spectral and MI/FT-IR information for compound identification.

were discussed and contrasted/compared with the features of the FT-IR spectra obtained in other phases, such as vapor and solid states. An unprecedented observation relating to the influence of the inert nobel gas matrix host on absorption band positions was made. Previous assumptions that the nobel gas host were inert were shown not to be valid. Shifts in absorption maxima on the order of 20 cm^{-1} were observed depending on the selection of Ar or Xe as matrix host. Differences in atomic radii and interstitial site diameters for the two gases were advanced as potential causes for the observed shifts. Due to the low-nanogram-sensitivity capability of the GC/MI/FT-IR technique, examples of identifications of minor components of complex natural essential oils were cited. Subtle differences such as positional isomerization were detectable via MI techniques, while VP approaches were unable to provide isomer differentiation.

REFERENCES

1. R. M. Smith, *Supercritical Fluid Chromatography,* Royal Society of Chromatography, London, 1988.
2. T. A. Baillie, *Int. J. Mass Spectrom. Ion Process., 118/119:* 289 (1992).
3. X. Huang, J. A. Luckey, M. J. Gordon, and R. N. Zare, *Anal. Chem., 61:* 766 (1989).
4. H. J. Cortes, *Multidimensional Chromatography: Techniques and Applications,* Marcel Dekker, New York, 1990.
5. T. Hirschfeld, *Anal. Chem., 52:* 297A (1980).
6. P. R. Griffiths, J. A. de Haseth, and L. V. Azarraga, *Anal. Chem., 55:* 1361A (1983).
7. P. R. Griffiths and J. A. de Haseth, *Fourier Transform Infrared Spectrometry,* Wiley, New York, 1986.
8. S. F. Johnston, *Fourier Transform Infrared: A Constantly Evolving Technology,* Ellis Horwood, New York, 1991.
9. M. J. D. Low and S. K. Freeman, *Anal. Chem., 39:* 194 (1967).
10. R. Fuoco, K. H. Shafer, and P. R. Griffiths, *Anal. Chem., 58:* 3249 (1986).
11. G. T. Reedy, S. Bourne, and P. T. Cunningham, *Anal. Chem., 51:* 1535 (1979).
12. S. Bourne, G. T. Reedy, P. Coffey, and D. Mattson, *Am. Lab., 16*(6): 90 (1984).
13. J. F. Schneider, G. T. Reedy, and D. G. Ettinger, *J. Chromatogr. Sci., 23:* 49 (1985).
14. G. T. Reedy, D. G. Ettinger, and J. F. Schneider, *Anal. Chem., 57:* 1602 (1985).
15. J. F. Elder, Jr., B. M. Gordon, and M. S. Uhrig, *J. Chromatogr. Sci., 24:* 26 (1986).
16. B. M. Gordon, M. S. Uhrig, M. F. Borgerding, H. L. Chung, W. M. Coleman III, J. F. Elder, Jr., J. A. Giles, D. S. Moore, C. E. Rix, and E. L. White, *J. Chromatogr. Sci., 26:* 174 (1988).
17. B. M. Gordon, C. E. Rix, and M. F. Borgerding, *J. Chromatogr. Sci., 23:* 1 (1985).
18. E. Whittle, D. A. Dows, and G. C. Pimentel, *J. Chem. Phys., 22:* 1943 (1954).
19. D. W. Ball, Z. Kalafi, H. Akya, L. Fredin, R. H. Haugh, and J. L. Margrave, *A Bibliography of Matrix Isolation Spectroscopy,* Rice University Press, Houston, Texas, 1988, pp. 1954–1985.
20. B. J. Van Der Veken, Chapter 22 in *Matrix Isolation Spectroscopy,* A. J. Barnes, W. J. Orville-Thomas, A. Müller, and R. Gaufrès, Eds., NATO Advanced Study Institute, Series C, D. Reidel, Dordrecht, The Netherlands, 1981, p. 517.

21. A. J. Barnes, Chapter 2 in *Matrix Isolation Spectroscopy,* NATO Advanced Study Institute, Series C, D. Reidel, Dordrecht, The Netherlands, 1981, p. 13.
22. R. N. Perutz, Chapter 6 in *Matrix Isolation,* M. C. R. Symons, Ed., *Annual Reports on the Progress of Chemistry,* Vol. 82, Section C: Physical Chemistry, Royal Society of Chemistry, London, 1985, p. 157.
23. S. Cradock and A. J. Hinchcliffe, *Matrix Isolation,* Cambridge University Press, Cambridge, 1975.
24. W. J. Orville-Thomas, Chapter 1 in *Matrix Isolation Spectroscopy,* A. J. Barnes, W. J. Orville-Thomas, A. Müller, and R. Gaufrès, Eds., NATO Advanced Study Institute, Series C, D. Reidel, Dordrecht, The Netherlands, 1981, p. 1.
25. S. V. Compton and P. Stout, *Liq. Chromatogr. Gas Chromatogr., 8:* 920 (1990).
26. L. Jirovetz, G. Buchbauer, and W. Jager, *Biomed. Chromatogr., 492:* 109 (1989).
27. S. V. Compton and P. Stout, *Am. Lab., 10:* 38 (1992).
28. B. Lacroix, J. B. Huvenne, and M. Deveaux, *J. Chromatogr., 492:* 109 (1989).
29. K. Bauer and D. Garbe, *Common Fragrances and Flavor Materials,* VCH Verlagsgesellshaft, Weinheim, Germany, 1985.
30. W. M. Coleman III and B. M. Gordon, *Appl. Spectrosc., 41:* 1431 (1987).
31. W. M. Coleman III and B. M. Gordon, *Appl. Spectrosc., 43:* 1008 (1989).
32. W. M. Coleman III and B. M. Gordon, *Appl. Spectrosc., 43:* 1424 (1989). Preceding papers in this series appear as Parts I through XVI and describe the characteristics of the MI/FTIR spectra and selected types of organic compounds.
33. A. J. Barnes, W. J. Orville-Thomas, A. Müller, and R. Gaufrès, Eds., *Matrix Isolation Spectroscopy,* NATO Advanced Study Institute, Series C, D. Reidel, Dordrecht, The Netherlands, 1981, p. 531.
34. C. J. Wurrey, S. Bourne, and R. D. Kleopfer, *Anal. Chem., 58:* 483 (1986).
35. T. T. Holloway, B. J. Fairless, C. E. Freidline, H. E. Kimball, R. D. Kleopfer, C. J. Wurrey, L. A. Jønooby, and H. G. Palmer, *Appl. Spectrosc., 42:* 359 (1988).
36. D. F. Gurka, J. M. Brasch, R. J. Barnes, C. J. Riggle, and S. Bourne, *Appl. Spectrosc., 40:* 978 (1986).
37. S. Jagannathan, J. R. Cooper, and C. L. Wilkins, *Appl. Spectrosc., 43:* 781 (1989).
38. D. M. Hembre, A. A. Garrison, R. A. Crocombe, R. A. Yokley, E. L. Wehry, and G. Mamantov, *Anal. Chem., 53:* 1783 (1981).
39. E. L. Wehry and G. Mamantov, *Prog. Anal. Spectrosc., 10:* 507 (1987).
40. M. L. Rogers and R. L. White, *Appl. Spectrosc., 41:* 1052 (1987).

41. G. Mamantov, A. A. Garrison, and E. L. Wehry, *Appl. Spectrosc., 36:* 339 (1982).
42. R. L. White, *Chromatography/Fourier Transform Infrared Spectroscopy and Its Applications,* Marcel Dekker, New York, 1990.
43. M. M. Møssaba, R. E. McDonald, D. J. Armstrong, and S. W. Page, *J. Chromatogr. Sci., 29:* 324 (1991).
44. E. R. Baumeister, L. Zhang, and C. L. Wilkins, *J. Chromatogr. Sci., 29:* 331 (1991).
45. W. M. Coleman III and B. M. Gordon, *J. Chromatogr. Sci., 29:* 371 (1991).
46. W. M. Coleman III and B. M. Gordon, *Appl. Spectrosc., 42:* 304 (1988).
47. J. F. Schneider, J. D. Demirgian, and J. C. Stickler, *J. Chromatogr. Sci., 24:* 330 (1986).
48. C. J. Pouchert, *The Aldrich Library of FT-IR Spectra,* Vols. 1 and 2, Aldrich Chemical Co., Milwaukee, Wis., 1985.
49. *Sadtler Standard Infrared Grating Spectra Library,* Sadtler Research Laboratories, Philadelphia, 1985.
50. R. A. Nyquist, *The Interpretation of Vapor Phase Infrared Spectra Group Frequency Data,* Vol. 1, Sadtler Research Laboratories, Philadelphia, 1984.
51. *Sadtler Standard Infrared Vapor Phase Spectra Library,* Sadtler Research Laboratories, Philadelphia, 1985.

3
Statistical Theories of Peak Overlap in Chromatography

Joe M. Davis *Southern Illinois University at Carbondale, Carbondale, Illinois*

I. INTRODUCTION 110
II. STATISTICAL MODELING OF OVERLAP IN ONE-DIMENSIONAL SEPARATIONS 111
   A. Theories 111
   B. Basis for Assumption of Randomness 130
   C. Testing of 1-D Overlap Theories by Computer Simulation 131
   D. Application and Testing of 1-D Overlap Theories by Experiment 140
   E. Resolution Factors for 1-D Models of Overlap 151
   F. Amplitude Distribution of SCPs in 1-D Separations 156
   G. Determination Limits 156
III. STATISTICAL MODELING OF OVERLAP IN MULTIDIMENSIONAL SEPARATIONS 158
   A. Theories 158
   B. Testing of 2-D and n-D Overlap Theories by Computer Simulation 164
   C. Application of n-D Overlap Theory to 2-D Experimental Data 166
   D. Number n of Dimensions Necessary to Resolve m Components 166
IV. CONCLUSIONS 169

I. INTRODUCTION

The evolution of chromatography over the last 40 or so years has provided the scientist with increasingly sophisticated means to separate large numbers of compounds. The extent of this evolution is apparent by a comparison of chromatograms of the 1950s with those of today. One has a tendency to smile indulgently at this early work, in which the resolution of perhaps 20 to 30 peaks constituted a major accomplishment. Today, the separation of 10 times as many peaks on a single capillary column is routine (even mundane), and far more peaks can be separated by multidimensional means and by recycling.

Despite these accomplishments, at times we must wonder if the next generation will look back on our present separations with the same indulgence that we accord to chromatograms of the 1950s. Because of efforts by several groups over the last decade, one is fairly confident that the next generation will regard with indulgence our state-of-the-art chromatograms of complex mixtures. Indeed, a growing segment of *today's* scientists regards such chromatograms with indulgence. This attitude can be attributed to the development over the last decade of several probabilistic models by which the likelihood that pure chemical species form fused or overlapped peaks in chromatograms is calculated. For complex mixtures, these models predict that the number of components resolved as chemically pure peaks is low, that most peaks are multiplets comprised of two or more chemical species, and that the total resolution of mixture components by chromatography—at least as it is practiced today—is a virtually impossible task. These predictions cannot be dismissed offhandedly, because they are supported by extensive analyses of both digitally simulated and experimental chromatograms.

This chapter is a review of the literature on various models of overlap in chromatography by statistical means, the computer simulations and experiments by which these models were characterized and tested, and other statistical attributes of chemical separations. It is divided into two major sections: Section II addresses statistical concerns in a single dimension of space, such as capillary chromatography or electrophoresis. Section III addresses statistical concerns in multidimensional spaces, such as in two-dimensional thin-layer chromatography. Each section is divided further into several subsections.

To avoid unnecessary complication, a single variable is used to represent a physical attribute throughout the review. For example, peak capacity is represented here by n_c, although this attribute also has been represented by n and Nc. Otherwise, the reviewer has tried to be as faithful to researchers' original notation as possible without duplicating variables. Of course, the reviewer alone is responsible for any errors of interpretation.

II. STATISTICAL MODELING OF OVERLAP IN ONE-DIMENSIONAL SEPARATIONS

A. Theories

Several theories on the subject of overlap in one-dimensional (1-D) separations have been published in the last decade. This subject is still an active area of research, as is exemplified here by the discussion of two theories that have not yet been published. The theories are reviewed in chronological order.

Early Work

An early work that is often overlooked today is by Klein and Tyler [1], who in 1965 investigated the identification of compounds by mobility, elution volume, or retention value. The authors postulated that two compounds could be identified if they were separated in a chromatogram by an interval greater than δM, but could not be identified if they were separated by an interval less than or equal to δM. Simple reasoning would suggest that k compounds could be identified in an interval of length $k\ \delta M$.

The authors pointed out that this outcome is unlikely, because the number of compounds in any interval δM fluctuates. The authors then defined a density ρ, where ρ is the fractional occupancy of the interval δM. If the fluctuation is random, the probability $p(k, \delta M)$ of finding k compounds in δM is governed by Poisson statistics:

$$p(k, \delta M) = \frac{e^{-\rho}\rho^k}{k!} \tag{1}$$

From Eq. (1), Klein and Tyler then calculated the probability y that interval δM does not contain two or more compounds as the complement of the probability that it contains either no or one compound:

$$p(0, \delta M) + p(1, \delta M) = (1 + \rho)e^{-\rho} = 1 - y \tag{2}$$

Figure 1 is a graph of ρ versus y computed from Eq. (2). It is clear that low probabilities y are obtained only for small densities ρ. Specifically, if $y = 0.05$, then $\rho \approx 0.362$, and only 36 or so compounds can be identified in a region which, in principal, could hold 100 compounds.

It is fascinating today to review these predictions, which preceded similar ones by almost 20 years. Klein and Tyler perhaps are the founding fathers of statistical theories of overlap. Regrettably, their work was largely ignored.

Combinatorial Analysis

In 1982, Rosenthal [2] reported that mass spectra generated and interpreted by an in-house gas chromatograph/mass spectrometer/computer (GC/MS/COMP) often

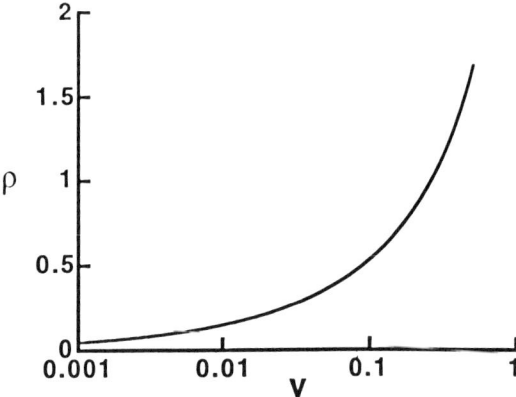

Fig. 1 Fractional occupancy ρ versus probability y that interval δM does not contain two or more compounds. (From Ref. 1.)

were erroneously identified or associated with low correlation factors when two or more compounds eluted within the 2-s interval required to complete one mass-spectral scan. The compounds eluting within this interval could not be deconvolved by a computer algorithm and, in effect, were overlapped.

Rosenthal gauged the severity of this coelution by combinatorial analysis. Each 2-s interval was identified with a hypothetical cell into which randomly distributed compounds could be put. A single compound in a cell was interpreted as a singlet peak, two compounds in a cell were interpreted as a doublet peak, and so on. By assuming that overlapping peaks were only doublets or triplets, Rosenthal calculated the probability P that a chromatogram contained s singlets, d doublets, and tr triplets as

$$P = \frac{c!m!}{s!d!tr!z!(2!)^d(3!)^{tr}c^m} \tag{3}$$

where m is the number of components, c the total number of cells, and z the number of cells in which no component is found.

Rosenthal then calculated the number of mass-spectral scans required to ensure a 50% probability ($P = 0.5$) that each compound would be a singlet. Figure 2 is a graph of the number of scans per chromatogram versus m needed to meet this specification. It is clear that the number of scans required to ensure that $P = 0.5$ increases substantially with increasing m. Rosenthal noted that for a chromatogram of 1-h duration and a 2-s scan time, only 45 compounds could be resolved when $P = 0.5$. For a 200-component mixture, the scan number would

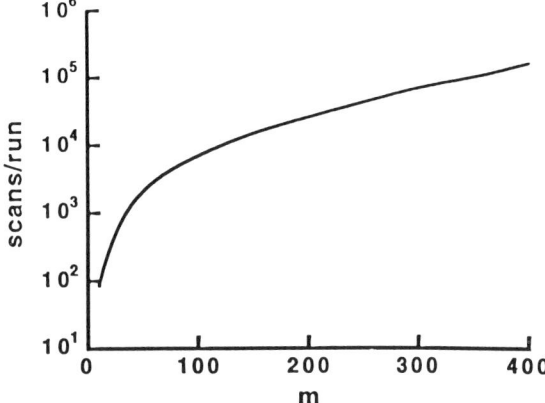

Fig. 2 Number of mass-spectral scans per chromatogram required to ensure m components are resolved with 50% probability versus m. (From Ref. 2.)

have to exceed 20,000 when $P = 0.5$, and the scan time would have to be less than 200 ms for a 1-h chromatogram. Rosenthal noted that this capability was beyond the realm of the then-current technology.

Variation of Overlap Probability with Peak Area

In 1983, Nagels et al. [3] determined the area distribution of observed peaks in chromatograms of phenolic compounds extracted from plant leaves and related this distribution to the likelihood that observed peaks overlapped. The authors distinguished clearly between component peaks or, alternatively, single-component peaks (SCPs), which are generated by pure compounds, and observed peaks, which consist of one or more SCPs. Sixty-two extracts were partially separated by high-performance liquid chromatography (HPLC) with a peak capacity n_c of 47 and were detected by ultraviolet (UV) absorption. The authors determined the frequency distribution of SCP areas by iteratively calculating computer-simulated chromatograms until the frequency distribution of observed-peak areas in the simulations agreed with that determined experimentally. In the simulations the authors assumed that SCPs were randomly distributed Gaussians with a constant standard deviation, whose value was determined by the elution window and the peak capacity.

The authors then generated 90,000 computer-simulated chromatograms from this frequency distribution and identified the various SCPs comprising observed peaks. The ratio AL, equal to the area of the largest SCP in an observed peak divided by the observed-peak area, was calculated for each observed peak. From these ratios the probability $PL_k(q)$ that the fractional area of the most abundant

SCP in any observed peak equaled or exceeded k ($0 \leq k \leq 1$) was calculated as

$$PL_k(q) = \frac{N_k(q)}{N(q)} \tag{4}$$

where q is a discrete class (or range) of relative observed-peak areas, $N(q)$ the total number of observed peaks in class q, and $N_k(q)$ the number of observed peaks in class q, for which AL exceeds k.

Figure 3a is a graph of $PL_k(q)$ versus q for the three k values 0.50, 0.90, and 0.95 when $n_c = 47$. The likelihood that an observed peak is an SCP is largest when the peak is either very small or is very large. In contrast, observed peaks of intermediate size, with 1 to 10% of the relative peak area, are likely to be

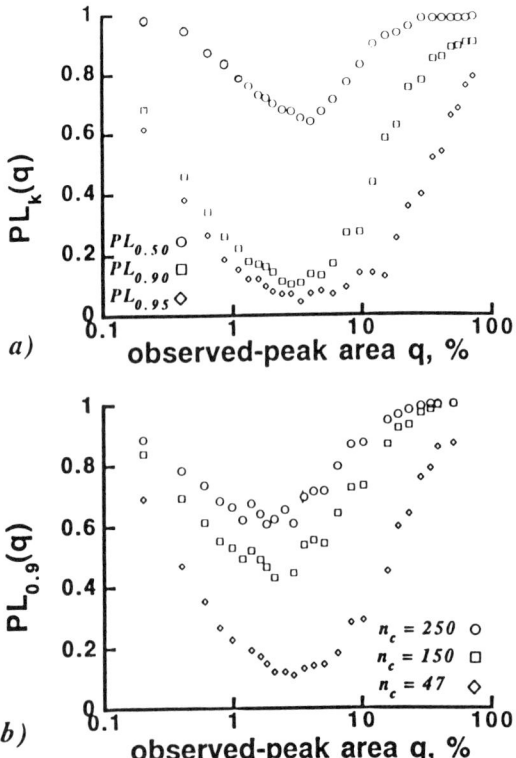

Fig. 3 (a) Graph of probability $PL_k(q)$ that the fractional area of the most abundant SCP in an observed peak equals or exceeds $k = 0.50$, 0.90, and 0.95 of the observed-peak area versus observed-peak area q. (b) Graph of $PL_{0.9}(q)$ versus observed-peak area q for different capacities n_c. (From Ref. 3.)

overlapping SCPs. Figure 3b is a graph of $PL_{0.9}(q)$ for three different values of n_c. The overall trends are identical to those in Fig. 3a, although larger values of n_c somewhat reduce coincidence.

Poisson Model

A shortcoming of Rosenthal's model was its discreteness. This issue was addressed in 1983 by Davis and Giddings [4], who assumed that SCPs within an interval X of a chromatogram were randomly distributed and could be represented by points located at their maxima or centers of gravity. They also assumed that adjacent SCPs could be resolved as separate peaks if the interval x_0 between them equaled or exceeded $4\sigma R_s^*$, where σ is the average standard deviation of adjacent SCPs and R_s^* is the resolution between them.

The assumption of randomness invoked Poisson statistics. The intervals between adjacent points in a Poisson process are distributed in accordance with probability density $P(x)$:

$$P(x) = \lambda e^{-\lambda x} \tag{5}$$

where $\lambda = \overline{m}/X$, $\overline{m}$ is the mean or expected number of components in X, and x is the interval between adjacent SCPs. The probability $P(x \geq x_0)$ that adjacent SCPs are resolved is

$$P(x \geq x_0) = \lambda \int_{x_0}^{\infty} e^{-\lambda x} dx = e^{-\lambda x_0} \equiv e^{-\alpha} \tag{6}$$

whereas the probability $P(x < x_0)$ that they are not resolved is

$$P(x < x_0) = 1 - e^{-\alpha} \tag{7}$$

The dimensionless parameter

$$\alpha = \lambda x_0 = \frac{\overline{m} x_0}{X} = \frac{\overline{m}}{n_c} \tag{8}$$

is the saturation of the chromatogram and equals the parameter p introduced by Klein and Tyler. The ratio X/x_0 is the peak capacity n_c, the maximum number of SCPs of width x_0 that can be resolved in interval X.

The expected number P_k of peaks containing k SCPs was calculated from Eqs. (6) and (7) as

$$P_k = \overline{m} e^{-2\alpha}(1 - e^{-\alpha})^{k-1} \tag{9}$$

The expected total number p of peaks was calculated as the sum of all P_k values:

$$p = \sum_{k=1}^{\infty} P_k = \overline{m} e^{-\overline{m}/n_c} = \overline{m} e^{-\alpha} \tag{10}$$

Parameter p is the average number of peaks in a statistically large number of chromatograms of saturation α. These chromatograms contain a variable integral number m of SCPs; the average number is $\bar{m}$. Typically, both p and $\bar{m}$ are nonintegral.

Figure 4 is a dimensionless graph of κ versus saturation α, where $\kappa = p/n_c$ or s/n_c and s is the expected number of singlet peaks. The straight line is the result expected if every SCP were resolved as a singlet peak. At low α, this ideal is nearly achieved. As α increases, however, the fraction of n_c utilized in resolving peaks and singlets becomes very small; the maximum fractions are $e^{-1} \approx 0.368$ for peaks and $e^{-1}/2 \approx 0.184$ for singlets.

The authors noted that any value of p is associated with two values of $\bar{m}$, except when $\bar{m} = n_c$. They suggested that $\bar{m}$ could be determined by varying the peak capacity n_c experimentally and graphing Eq. (10) in the form

$$\ln p = \ln \bar{m} - \frac{\bar{m}}{n_c} \tag{11}$$

From a graph of $\ln p$ versus n_c^{-1}, the number $\bar{m}$ of SCPs could be calculated from the slope ($-\bar{m}$) and intercept ($\ln \bar{m}$) of a line fit to the data (n_c^{-1}, $\ln p$).

Analogy Between Overlap and Depolymerization

In 1985, Martin and Guiochon [5] proposed an analogy between the chromatography of complex mixtures and the random depolymerization of a polymer. Just as a complex mixture containing m components is introduced to a chromatograph as one complex peak that ultimately separates into a distribution of singlets, dou-

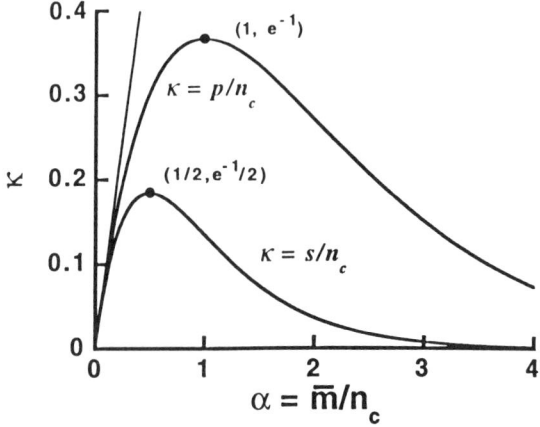

Fig. 4 Graph of $\kappa = p/n_c$ and s/n_c versus saturation $\alpha = \bar{m}/n_c$. Straight line corresponds to $p = \bar{m}$. (From Ref. 4.)

blets, triplets, and so on, so a single polymer comprised of m monomers at the beginning of a depolymerization ultimately degrades into a distribution of monomers, dimers, trimers, and so on.

For a random polymerization (e.g., a condensation polymerization), the mole fraction p_k of polymer containing k monomers is given by the Flory most probable distribution

$$p_k = (1 - \beta)^2 \beta^{k-1} \quad (12)$$

where β, the De Donder extent of reaction, is the fraction of functional groups that have reacted. The complement, $1 - \beta$, is the fraction of functional groups that have not reacted. Since both monomers and polymers have only one functional group, quantity $1 - \beta$ also equals the ratio of the total number p of polymers to the initial number m of monomers:

$$\frac{p}{m} = 1 - \beta \quad (13)$$

The same equations apply during random depolymerization. By analogy, Eq. (13) also equals the ratio of the number p of chromatographic peaks to the number m of resolvable components. By defining the extent of separation, γ, as $1 - \beta$,

$$\frac{p}{m} = \gamma \quad (14)$$

the authors reexpressed Eq. (12) as

$$p_k = \gamma^2 (1 - \gamma)^{k-1} \quad (15)$$

where p_k now is interpreted as the probability of forming a peak containing k components.

If m is identified with the expected number $\bar{m}$ of components in the Poisson model, then $\gamma = p/\bar{m} = e^{-\alpha}$ and Eq. (15) becomes identical to Eq. (9), divided by $\bar{m}$. The extent of separation, γ, is a quantitative criterion of merit for the goodness of separation in a complex chromatogram.

The analogy was extended further by identifying the polydispersity of a polymer with the polydispersity of peak areas in a chromatogram. If $\mu = 1 + \beta$ is the polydispersity index for a condensation polymerization, then

$$m = \frac{p}{\gamma} = \frac{p}{2 - \mu} \quad (16)$$

where $1 < \mu < 2$. By assuming that the polydispersity index of peak areas was equal to μ, the authors argued that m could be estimated from Eq. (16) by equating μ to

$$\mu = p \sum_{i=1}^{p} r_i^2 \tag{17}$$

where

$$r_i = \frac{a_i}{\sum_{k=1}^{p} a_k} \tag{18}$$

is the relative area associated with the ith peak of area a_i. The peak capacity n_c for SCPs of equal width was calculated as

$$n_c = -\frac{p}{(2-\mu)\ln(2-\mu)} \tag{19}$$

Alternative Means to Estimate $\overline{m}$ by Poisson Model

In 1985, Davis and Giddings [6] observed that the Poisson model could be used to estimate the expected number $\overline{m}$ of mixture components from a single chromatogram instead of multiple chromatograms. The authors expressed Eq. (11) as

$$\ln p' = \ln \overline{m} - \frac{\overline{m} x_0'}{X} \tag{20}$$

where p' is the number of intervals between adjacent maxima exceeding the arbitrarily chosen interval, x_0'. By selecting several values of x_0', several values of p' can be determined from a single chromatogram, and the data set $(x_0'/X, \ln p')$ can be fit to Eq. (20) to calculate $\overline{m}$. This procedure is designated the single-chromatogram method for determining $\overline{m}$ and is distinguished from the multiple-chromatogram method based on Eq. (11). Two estimates of $\overline{m}$ are calculable: one from the slope of Eq. (20), which is denoted m_{sl}, and one from the intercept of Eq. (20), which is denoted m_{in}.

The basis of this method is that X/x_0' is equivalent to an effective peak capacity, which is varied by selecting different values of resolution, R_s^*, instead of σ. In this context, a "peak" is a group of maxima in which all intervals between adjacent maxima in the group are less than x_0', and the number of such groups is the number p' of peaks. For intervals $x_0' < 2\sigma$ or so, however, p' does not vary with x_0', because adjacent SCPs are overlapped. These data must not be fit to Eq. (20).

Variance Associated with Poisson Model: Weighting Factors

In 1985, Davis and Giddings [6] determined by semiempirical means the variance $\sigma_{p'}^2$ associated with the number p' of peaks expected for capacity X/x_0', where x_0' is any arbitrary interval. From $\sigma_{p'}^2$, weighting factors for the coordinates $(x_0'/X, \ln p')$

in graphs of $\ln p'$ versus x_0'/X (or $\ln p$ versus n_c^{-1}), and uncertainties in m_{sl} and m_{in}, can be calculated.

Two independent sources of variation contribute to $\sigma_{p'}^2$. The first is the fluctuation of m in interval X due to Poisson statistics; the second is the fluctuation of intervals between adjacent SCPs, which occurs even for fixed m. The variance σ_A^2 associated with the first source was determined from propagation of errors to be

$$\sigma_A^2 = \overline{m}(1 - \alpha')^2 e^{-2\alpha'} \tag{21}$$

The variance σ_B^2 associated with the second source was determined empirically from computer simulations and was expressed as the log-normal function

$$\sigma_B^2 = \overline{m}\frac{0.0988}{\alpha'}\exp\left[-\frac{\{\ln(\alpha'/1.168)\}^2}{2.341}\right] \equiv \overline{m}f(\alpha') \tag{22}$$

where the coefficients were determined by least-squares regression. The variance $\sigma_{p'}^2$ thus was determined as

$$\sigma_{p'}^2 = \overline{m}[(1 - \alpha')^2 e^{-2\alpha'} + f(\alpha')] \tag{23}$$

The weight w of any datum in a graph of $\ln p'$ versus x_0'/X (or $\ln p$ versus n_c^{-1}) is thus

$$w = \frac{p'^2}{\sigma_{p'}^2} \tag{24}$$

where p' equals Eq. (10) with $n_c = X/x_0'$. Because w depends on $\overline{m}$, which is unknown, iterative procedures were suggested to determine both w and $\overline{m}$.

Probability Distribution of Observed Peaks and SCPs

Martin et al. [7] observed in 1986 that the likelihood of overlap decreases near the extremities of a chromatogram because no SCPs lie beyond the extremities. This decrease is negligible for large peak capacities but is significant when small capacities are considered. In a model that addressed this issue, the authors represented m SCPs by points located at their maxima or centers of gravity, defined an retention interval X by the span between the first and last SCP in the interval and assumed that the remaining $m - 2$ SCPs were distributed randomly within X. The authors further assumed that adjacent SCPs would be resolved if the span between them equaled or exceeded x_0. The probability $P_{m,n_c}(p)$ that p peaks are separated is the probability that $p - 1$ intervals between adjacent points are greater than or equal to x_0, which is

$$P_{m,n_c}(p) = \frac{(m-1)!}{(p-1)!} \sum_{i=0}^{\inf(m-p,n_c-p)} \frac{(-1)^i}{(m-p-i)!\,i!}\left(1 - \frac{p-1+i}{n_c-1}\right)^{m-2} \tag{25}$$

where inf (i, j) is the smaller of the two values i and j, and n_c is the peak capacity:

$$n_c = 1 + \frac{X}{x_0} \tag{26}$$

Note that here p represents an integral number of peaks in a single chromatogram and thus differs from the p of Davis and Giddings.

The average number $\bar{p}$ of peaks and the variance σ_p^2 of the probability distribution of p were calculated from Eq. (25) as

$$\begin{aligned}\bar{p} &= 1 + (m-1)[1 - (n_c - 1)^{-1}]^{m-2} \\ \sigma_p^2 &= (\bar{p} - 1)\{1 + (m-2)[1 - (n_c - 2)^{-1}]^{m-2} - (\bar{p} - 1)\}\end{aligned} \tag{27}$$

The expression for $\bar{p}$ is essentially equivalent to Eq. (10) when $m \gg 1$ and $n_c \gg 1$, but also describes the average number of peaks when these inequalities do not apply. The expression for σ_p^2 reported here differs from that originally published; the correction was noted elsewhere [8]. This expression is essentially equivalent to Eq. (22) when $m \gg 1$ and $n_c \gg 1$, but also describes the variance of p when these inequalities do not apply.

Figure 5a and b are histograms of $P_{m,n_c}(p)$ versus p for $m = 20$ SCPs when $n_c = 10$ and 20. The histograms are highly symmetrical, and the dashed curve is the normalized Gaussian

$$P_{m,n_c}(p) \approx (\sqrt{2\pi}\,\sigma_p)^{-1} \exp\left[-\frac{(p - \bar{p})^2}{2\sigma_p^2}\right] \tag{28}$$

with $\bar{p}$ and σ_p^2 defined by Eq. (27). Although the distribution of p is relatively wide, it is clear that the likelihood of achieving total resolution, with $p = m$, is very small.

The likelihood of total resolution can be calculated from Eq. (25) by equating p to m (only the $i = 0$ term need be considered):

$$P_{m,n_c}(m) = \left(1 - \frac{m-1}{n_c - 1}\right)^{m-2} \tag{29}$$

Figure 5c is a graph of $P_{m,n_c}(m)$ versus n_c for various numbers m of SCPs. The likelihood that all components are resolved is vanishingly small unless $m \ll n_c$. For example, the probability of resolving 30 SCPs is only 50% when $n_c = 1200$.

The minimum capacity $n_{c_{min}}$ necessary to separate all components with probability $P_{m,n_c}(m)$ can be calculated from Eq. (29) and is

$$n_{c_{min}} = 1 + \frac{m-1}{1 - [P_{m,n_c}(m)]^{1/(m-2)}} \tag{30}$$

For large m, this result simplifies to $-m^2/\ln[P_{m,n_c}(m)]$.

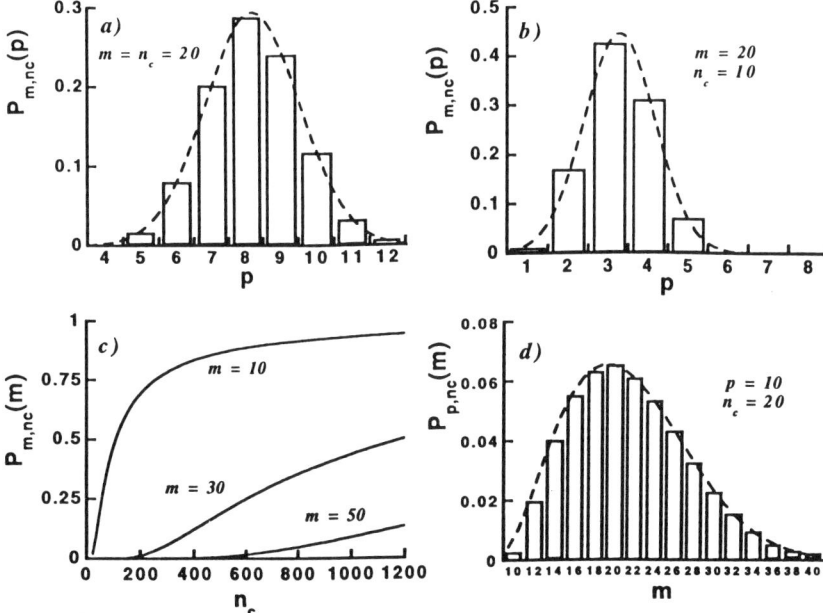

Fig. 5 (a) Probability $P_{m,n_c}(p)$ that $m = 20$ components produce p peaks versus p for $n_c = 20$. Dashed curve is a graph of Eq. (28). (b) As in (a), but $n_c = 10$. (c) Probability $P_{m,n_c}(m)$ that m components are resolved versus n_c for various m values. (d) Probability $P_{p,n_c}(m)$ that $p = 10$ peaks originate from m components versus m for $n_c = 20$. Dashed curve is calculated from Eq. (28). (From Ref. 7.)

Equation (25) describes the probability of observing p peaks when m SCPs are present. Given that p peaks are observed, the probability $P_{p,n_c}(m)$ that m SCPs are present can be calculated from Eq. (25) and Bayes' theorem. If the a priori probability that m SCPs are present in the chromatogram is assumed to be the same for all values of $m \geq p$, then

$$P_{p,n_c}(m) = \frac{P_{m,n_c}(p)}{\sum_{m'=p}^{\infty} P_{m',n_c}(p)} \tag{31}$$

Figure 5d is a graph of $P_{p,n_c}(m)$ versus m for $p = 10$ observed peaks when the peak capacity $n_c = 20$. The dashed curve is calculated from Eq. (28) with m now as a variable. Although the most probable value for m lies between 19 and 20, a small but real chance exists that m lies anywhere between 10 and 36 or 37. For appropriate values of p and n_c, Eq. (31) can be bimodal.

Statistical Basis for Eluant Optimization

In 1986, Herman et al. [9] developed criteria for the selection of optimal binary and ternary eluants for isocratic liquid chromatography. An equation developed by Schoenmakers et al. [10] was used to predict the capacity factors k' of 32 solutes for various mobile-phase volume fractions of water, methanol (MeOH), and tetrahydrofuran (THF). The smallest peak capacity capable of separating from 3 to 15 solutes with $R_s^* \geq 1$ was computed for each of 300 random selections of solutes from this 32-solute set. The peak capacity n_c was computed as

$$n_c = 1 + \frac{N^{1/2}}{4R_s^*} \ln \frac{1 + k_l'}{1 + k_f'} \tag{32}$$

where k_f' and k_l' are the capacity factors of the least and most retained solutes in a selection and N is the number of theoretical plates. These capacities were grouped according to polarity range PR, which specified the largest acceptable k' value. From these computations, the cumulative distribution function for the probability of total resolution ($R_s^* \geq 1$) was determined for various m values as a function of n_c. In this study, $N = 10,000$.

Figure 6a and b are graphs of the probability of resolution for optimal binary eluants versus n_c for various numbers m of components and the two PR values, 5 and 35. The required capacity n_c increases significantly with m but only modestly with PR. From these computations, the authors concluded that for even exceptional isocratic n_c values (e.g., 60), mixtures of intermediate polarity could contain no more than nine components if one desired at least a 90% probability of resolution. Similar results were found for ternary eluants composed of water, MeOH, and THF, except that slightly smaller n_c values were adequate for resolution. The authors noted that a shortcoming of the study was the lack of sufficient data for the solvent, acetonitrile. The use of these distribution functions in initiating searches for optimal eluant conditions is discussed here and elsewhere [11].

Based on these computations, the authors concluded that capacities smaller than those required for random SCPs would suffice to resolve mixtures if optimized eluants were used. Figure 6c is a graph of the capacity n_c sufficient to resolve all m components with 50% probability versus m for various optimized eluants. The ideal curve corresponds to $m = n_c$, and the trial-and-error curve is a graph of Eq. (29) derived by Martin et al., with $P_{m,n_c}(m) = 0.5$.

The authors also demonstrated that relatively small capacities are sufficient for the resolution of a fraction of mixture components. Figure 6d is a graph of the capacity n_c required for resolving a subset NI of m components with 90% probability, where $NI < m$, versus NI for various m values and PR = 10. These predictions were made for a ternary eluant. The authors observed that the required n_c rises rapidly with increasing NI but approaches a constant value when $NI/m >$

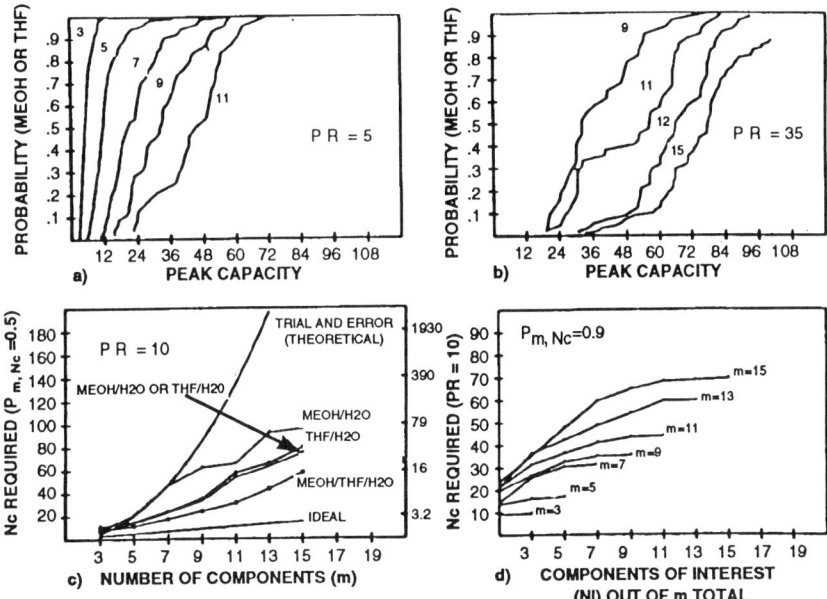

Fig. 6 Cumulative probability of resolving m components by optimized binary gradients versus n_c for polarity ranges PR of (a) 5 and (b) 35. Number adjacent to each distribution is m. (c) Capacity n_c required to ensure 50% probability of resolution versus m for PR = 10. Ideal curve corresponds to $m = n_c$; trial-and-error curve is a graph of Eq. (29), with $P_{m,n_c}(m) = 0.5$. Central curves correspond to optimized gradients. (d) Capacity n_c required to resolve with 90% probability NI of m components versus NI for PR = 10. (From Ref. 9.)

0.6. For the resolution of only a few mixture components, eluant optimization can reduce analysis time significantly.

Fourier Analysis

A shortcoming common to theories in which SCP centers are represented by points is the neglect of SCP amplitudes. At low saturation, this neglect is permissible (see below); at high saturations, these theories fail, because of this neglect. A new perspective was introduced in 1990 by Felinger et al. [12], who interpreted chromatograms as a linear combination $\mathbf{Y}(t)$ of individual stochastic processes

$$\mathbf{Y}(t) = \sum_{i=1}^{m} h_i u(t - m_i) \tag{33}$$

where t is time, m_i a shift constant, and h_i a scalar for the ith event $u(t)$. In this application, $\mathbf{Y}(t)$ is the observed peak height at time t, $u(t)$ a mathematical function for an SCP (e.g., a Gaussian), h_i the height of the ith SCP, and m_i its retention time. Regardless of the specific nature of h_i and $u(t)$, the Fourier transform, or power spectrum, of Eq. (33) is proportional to the power spectrum of $u(t)$. In other words, the power spectrum of the entire chromatogram contains information about $u(t)$, h_i, and m_i. The information about h_i, in turn, can be interpreted to provide an estimate of the number m of components.

The power spectrum, $F(\omega)$, of an infinitely long chromatogram can be computed either by the time averaging or ensemble averaging of $\mathbf{Y}(t)$, provided that the statistical properties of the chromatogram do not change with time. The time average is computed from the autocovariance function $C(t)$:

$$C(t) = \lim_{T \to \infty} \int_{-T/2}^{T/2} [Y(t') - \hat{Y}][Y(t + t') - \hat{Y}] \, dt' \tag{34}$$

where $\hat{Y}$ is the mean value of Y (i.e., the mean height of observed peaks). The function $C(t)$, when normalized to its value at $t = 0$, is the autocorrelation function. The power spectrum $F(\omega)$ is calculated from $C(t)$ by application of the Wiener–Khinchin theorem:

$$F(\omega) = 4 \int_0^\infty C(t) \cos(\omega t) \, dt \qquad C(t) = \frac{1}{2\pi} \int_0^\infty F(\omega) \cos(\omega t) \, d\omega \tag{35}$$

The function $F(\omega)$ is determined by ensemble averaging as

$$F(\omega) = \lim_{T \to \infty} \frac{2 E[|Z_T^{(k)}(\omega)|^2]}{T} \tag{36}$$

where E denotes the expectation value of the bracketed argument and $Z_T^{(k)}(\omega)$ is the Fourier transform of the kth replica of $\mathbf{Y}(t)$.

Equations (35) and (36) are both useful; Eq. (35) facilitates calculation of $F(\omega)$ from experimental or simulated chromatograms, and Eq. (36) facilitates derivation of models for $F(\omega)$.

Constant SCP Widths. In 1990, Felinger et al. [12] postulated that the heights and retention times of SCPs in a chromatogram were uncorrelated, that the probability density between adjacent SCPs was identical throughout the chromatogram, that SCPs could be described by the same function (e.g., a Gaussian) throughout the chromatogram, and that the standard deviations of all SCPs were constant. Following a lengthy and mathematically elegant analysis, the authors determined that

$$F(\omega) = \frac{2 a_h^2}{T} |g(\omega)|^2 \left[\frac{\sigma_h^2}{a_h^2} + 1 + 2 \operatorname{Re} \frac{\theta(\omega)}{1 - \theta(\omega)} + \frac{\delta(\omega)}{T} \right] \tag{37}$$

where a_h is the mean SCP height, σ_h the standard deviation of SCP heights, $g(\omega)$ the Fourier transform of the mathematical function for SCPs at the origin, $\theta(\omega)$ a function that depends on the probability density between adjacent SCPs, T the average interval between adjacent SCPs, and $\delta(\omega)$ the Dirac function. The last term in Eq. (37) drops out if a chromatogram is centered about its mean value when calculating $C(t)$. The ratio σ_h^2/a_h^2 depends on the SCP amplitude distribution. For an exponentially convoluted Gaussian SCP of standard deviation σ and time constant τ, the function $|g(\omega)|^2$ is

$$|g(\omega)|^2 = 2\pi\sigma^2 \frac{e^{-(\omega\sigma)^2}}{1 + (\omega\tau)^2} \tag{38}$$

The appropriate expression for a Gaussian SCP is obtained simply by equating τ to zero.

Table 1 reports the functions, $2\,\text{Re}[\theta(\omega)/(1 - \theta(\omega))]$, for various probability densities $f(t)$ between adjacent SCPs. This function is zero for randomly distributed SCPs, for which $f(t)$ is a negative exponential. This work was the first study in which non-Poissonian densities between adjacent SCPs were considered.

A connection between $F(\omega)$ and component number m was established by the identity

$$A_T = (2\pi)^{1/2}\sigma m a_h \tag{39}$$

where A_T is the total area under the chromatogram. If $X = mT$ is the span of the chromatogram in time, then $F(\omega)$ can be expressed as

$$F(\omega) = \frac{2A_T^2}{mX} \frac{e^{-(\omega\sigma)^2}}{1 + (\omega\tau)^2} \left[\frac{\sigma_h^2}{a_h^2} + 1 + 2\,\text{Re}\,\frac{\theta(\omega)}{1 - \theta(\omega)} + \frac{\delta(\omega)}{T} \right] \tag{40}$$

Table 1 Functions, $2\,\text{Re}[\theta(\omega)/(1 - \theta(\omega))]$, for Probability Densities $f(t)$ Associated with Constant-Width Fourier Model

Probability density	$f(t)$	$2\,\text{Re}[\theta(\omega)/(1 - \theta(\omega))]$
Exponential	$\exp(-t/T)/T;\ 0 \leq t$	0
Uniform	$1/b;\ 0 \leq t \leq b$	$2\dfrac{b\omega \sin(b\omega) + 2\cos(b\omega) - 2}{(b\omega)^2 - 2b\omega \sin(b\omega) - 2\cos(b\omega) + 2}$
Gamma	$t^{p-1}e^{-t}/\Gamma(p);\ 0 \leq t$	$2\,\text{Re}\left[\sum_{k=1}^{p} \binom{p}{k}(-i\omega)^k\right]^{-1}$
Delta	$\delta(t - T)$	$\delta(\omega - 2n\pi/T) - 1$

Source: Ref. 12.

Experimentally determined $F(\omega)$ values can be fit to Eq. (40) by nonlinear least squares to estimate the statistical parameters, m, σ, and τ. Alternatively, experimental autocorrelation functions $C(t)$ can be fit to theoretical functions calculated from $F(\omega)$ and Eq. (35).

Variable SCP Widths. The theory was extended in 1991 by Felinger et al. [13], who derived expressions of $F(\omega)$ for variable SCP widths. The authors observed that experimental control of both SCP density and width is difficult and that the latter is more difficult to control. In response to this lack of control, three expressions for $F(\omega)$ were derived, in which the variation of SCP widths was addressed. In the first two cases, widths were postulated to be independent of elution volume and were distributed either uniformly between two extremes or normally. In the third case, widths were postulated to increase linearly with elution volume. In all cases, the SCP shape was Gaussian and elution times were Poissonian.

Unlike for the constant-width Fourier model, the observed-peak height is not related simply to component number m. However, observed-peak area is so related. For the first two cases, a derivation based on the ensemble approach gave the result

$$F(\omega) = \frac{2A_T^2}{mX}\left[\left(\frac{\sigma_a^2}{a_a^2} + 1\right)E\{|g(\omega,\sigma)|^2\} + 2|E\{g(\omega,\sigma)\}|^2 \operatorname{Re}\frac{\theta(\omega)}{1-\theta(\omega)}\right] \quad (41)$$

where a_a and σ_a are the mean and standard deviation of SCP areas, $g(\omega,\sigma) = \exp[-(\omega\sigma)^2/2]$ is the zero-shifted Fourier transform of the Gaussian SCP shape, and the functions $E\{|g(\omega,\sigma)|^2\}$ and $|E\{g(\omega,\sigma)\}|^2$ depend on the distribution of SCP widths. Table 2 reports these functions for constant widths, uniformly distributed widths, and normally distributed widths.

The third case, in which SCP widths increase linearly with retention volume V_r, is of interest because of the relationship between σ and plate number N in isocratic chromatography:

$$\sigma = \frac{V_r}{N^{1/2}} \quad (42)$$

Here σ increases linearly with time. For SCPs with Poissonian retention times and uncorrelated areas and widths, the authors concluded from the ensemble approach that

$$F(\omega) = \frac{2A_T^2}{mX}\left[\left(\frac{\sigma_a^2}{a_a^2} + 1\right)E\{|g(\omega,\sigma)|^2\}\right] \quad (43)$$

where $E\{|g(\omega,\sigma)|^2\}$ is expressed by the second row and column entry in Table 2. Because $2\operatorname{Re}(\theta(\omega)/[1-\theta(\omega)])$ is zero for Poissonian retention times, Eq. (43)

Table 2 Functions $E\{|g(\omega,\sigma)|^2\}$ and $|E\{g(\omega,\sigma)\}|^2$ for the Variable-Width Fourier Model

Distribution of SCP widths	$E\{|g(\omega,\sigma)|^2\}$	$|E\{g(\omega,\sigma)\}|^2$
Constant	$e^{-(\omega\sigma)^2}$	$e^{-(\omega\sigma)^2}$
Uniform ($\sigma_1 \leq \sigma \leq \sigma_2$)	$\dfrac{\sqrt{\pi}[\text{erf}(\omega\sigma_2) - \text{erf}(\omega\sigma_1)]}{2\omega(\sigma_2 - \sigma_1)}$	$\dfrac{\sqrt{\pi}[\text{erf}(\omega\sigma_2/\sqrt{2}) - \text{erf}(\omega\sigma_1/\sqrt{2})]^2}{2\omega^2(\sigma_2 - \sigma_1)^2}$
Normal (mean: σ_{av}; std dev: s_n)	$\dfrac{\exp\left[-\dfrac{(\sigma_{av}\omega)^2}{2(s_n\omega)^2 + 1}\right]}{\sqrt{2s_n\omega^2 + 1}}$	$\dfrac{\exp\left[-\dfrac{(\sigma_{av}\omega)^2}{(s_n\omega)^2 + 1}\right]}{(s_n\omega)^2 + 1}$

Source: Ref. 13.

identically equals Eq. (41) for uniformly distributed σ values. This result is not coincidental but reflects the "loss of memory" associated with the power spectrum.

Experimental $F(\omega)$ values can be fit to Eqs. (41) and (43) to estimate several statistical parameters, including the number m of components, the σ extremes, σ_1 and σ_2, for uniformly distributed widths, and the mean σ_{av} and standard deviation s_n for normally distributed widths (the latter four parameters are incorporated in expressions in Table 2).

Saturogram

In work presently under development, M. Z. El Fallah and M. Martin (personal communication, 1993) are using the discrimination factor, d_0, to predict the number of components in highly saturated chromatograms. The parameter d_0 is defined as the ratio of the valley depth between two observed peaks to the height of the smaller observed peak and varies between zero (complete fusion of observed peaks) and 1 (complete resolution). Figure 7 is a graph of two overlapping Gaussians for which d_0 is defined.

The number $k(d_0)$ of adjacent maxima pairs for which d_0 equals or exceeds some chosen value depends on d_0 and the SCP density, λ. Various measures of $k(d_0)$ can be determined from a complex chromatogram and expressed as a saturogram, which contains all the information about overlap in the chromatogram. The saturogram can then be interpreted in terms of results established by computer simulation, and estimates of m can be calculated. Preliminary results show that values of m can be determined to within a few percent from the saturograms of chromatograms having five times the saturation addressed by Davis and Giddings.

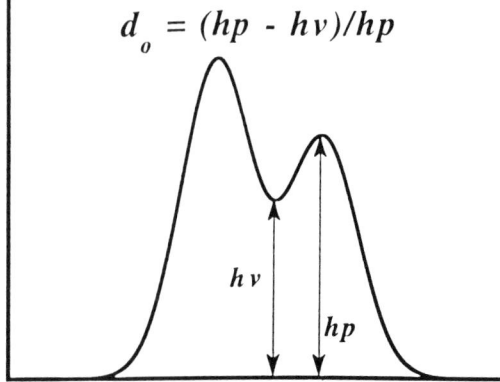

Fig. 7 Illustration of discrimination factor d_0. (From Ref. 31.)

Extended Poisson Model: Density Variations of SCPs

Davis [14] realized that the Poisson model could be modified to describe overlap in chromatograms containing SCPs with variable Poisson densities. The density of peaks, p/X, was calculated from Eq. (10) as

$$\frac{p}{X} = \frac{\overline{m}}{X} e^{-\overline{m}/n_c} = \frac{\overline{m}}{X} e^{-\overline{m}x_0/X} = \lambda e^{-\lambda x_0} \tag{44}$$

where $\lambda = \overline{m}/X$ is the constant SCP density. If this density instead varies along separation coordinate x, then $\lambda = \lambda(x)$. By taking the differential limit of the left-hand side of Eq. (44) (i.e., $p/X \rightarrow dp/dx$) and integrating, Davis expressed the number p of peaks as

$$p = \int_0^X \lambda(x) e^{-\lambda(x)x_0} dx \tag{45}$$

It is useful to define a dimensionless frequency $f(x)$ such that

$$\lambda(x) = \frac{\overline{m} f(x)}{X} = \lambda_P f(x) \tag{46}$$

where $\lambda_P = \overline{m}/X$ is the constant Poisson density and

$$\int_0^X f(x) \, dx = X \tag{47}$$

By introducing the variable $\zeta = x/X$, Eqs. (46) and (47) can be expressed as

$$\int_0^1 f(\zeta) \, d\zeta = 1$$

$$p = \overline{m} \int_0^1 f(\zeta) e^{-\overline{\alpha} f(\zeta)} d\zeta \tag{48}$$

where $\overline{\alpha} = \overline{m} x_0 / X$ is now the *average* saturation of the chromatogram; the product $f(\zeta)\overline{\alpha}$ is the *local* saturation at coordinate ζ.

Similar expressions can be derived for the number P_k of multiplet peaks by modifying Eq. (9):

$$P_k = \overline{m} \int_0^1 f(\zeta) e^{-2f(\zeta)\overline{\alpha}} (1 - e^{-f(\zeta)\overline{\alpha}})^{k-1} d\zeta \tag{49}$$

The SCP frequency $f(\zeta)$ typically will be unknown in an experimental chromatogram. It can be estimated, however, from the cumulative distribution function of maxima, which is then differentiated. This derivative is an approximation to $f(\zeta)$ and is denoted $f_a(\zeta)$. An estimate of $\overline{m}$ can then be calculated by minimizing the sum of squares SS of the following equation, which is based on the single-chromatogram method:

$$SS = \sum_{i=1}^{j} \left[p'_i - \overline{m} \int_0^1 f_a(\zeta) e^{-\overline{m} f_a(\zeta) x'_{0i} \zeta_i} d\zeta \right]^2 \tag{50}$$

Here $x'_{0\zeta}$ is x'_0/X and p' is the number of intervals between adjacent maxima exceeding $x'_{0\zeta}$. The subscript i represents the ith of j data points.

B. Basis for Assumption of Randomness

With the recent development of Felinger et al. [12], the statistical attributes of chromatograms containing other than randomly distributed SCPs can be calculated. For many years, however, theories addressed only the overlap of randomly distributed SCPs. In part, this consideration can be attributed to the relative simplicity with which such theories were derived. Beyond this simplicity, however, are several arguments that justify the legitimacy of the assumption of randomness. These arguments are now reviewed.

In 1983, Davis and Giddings [4] observed that the distribution of retention times for a mixture's components reflects the distribution of the components' standard-state free-energy differences, $\Delta\mu^0$, between the mobile and stationary phases. If differences in $\Delta\mu^0$ values for different compounds result from nonrecurring chemical, structural, or steric differences, the $\Delta\mu^0$ values will tend to be randomly distributed along the free-energy axis. This is especially true in high-resolution chromatograms, for which differences in $\Delta\mu^0$ for adjacent SCPs are much smaller than thermal energy RT.

In 1984, Herman et al. [15] constructed frequency distributions from published retardation factors R_f ($0.02 < R_f < 0.85$) of 163 steroids determined by silica gel TLC and seven different eluants. Because the R_f values were reported to only two significant figures, for each eluant the authors randomly selected only 30 of the steroids and computed the frequency distribution from them. This selection was then repeated 67 times, and the 67 resultant distributions were averaged. Thus seven averaged distributions were obtained (one for each eluant). For randomly distributed R_f values, the frequency distribution should be a negative exponential. The correlation coefficients resulting from fitting these averaged distributions to a negative exponential distribution ranged from -0.992 to -0.999.

In 1986, Martin et al. [7] interpreted the specific retention volumes V_g of 20 compounds on 65 stationary phases, as tabulated in McReynolds' handbook. The compounds were chosen selectively such that structural homologs were excluded, many functional groups were included, and all compounds had roughly the same number of carbon atoms. The authors then computed normalized cumulative distributions for both V_g and $\ln V_g$ to determine which of the two was more closely Poisson distributed. Their results suggested that $\ln V_g$ is more Poisson distributed than is V_g itself, and computer simulations supported this conclusion. This is consistent with the argument of Davis and Giddings, who suggested that components are randomly distributed along the $\Delta\mu^0$ axis.

In 1990, Felinger et al. [12] observed that the probability density for Poisson statistics should be viewed as the fundamental density for intervals between adjacent SCPs, because it is the result of the superposition of a large number of uncorrelated elementary subsequences. The distribution of intervals between adjacent SCPs in these sequences is immaterial. Complex mixtures are probably composed of many such sequences, each of which is formed by homologs or recursively correlated sets of compounds. The authors demonstrated this principle by superimposing four sets of periodically spaced coordinates with uncorrelated periods. The frequency distribution of the resultant superposition was described well by a negative exponential, except for small intervals.

C. Testing of 1-D Overlap Theories by Computer Simulation

The complexity of the chromatographic process has motivated some researchers to test their theories by interpreting computer-simulated chromatograms. The attractive features of simulations are that they are controllable by the researcher and that experimental artifacts are avoided. In fact, computer simulation is almost essential to this field of research; an equally thorough experimental program would require a substantially greater effort and would be fraught with uncontrolled phenomena.

Poisson Model

In 1984, Giddings et al. [16] generated 240 simulated chromatograms containing from 80 to 240 Gaussian SCPs, whose standard deviations σ were constant in each simulation but varied over a threefold range in different simulations. The SCP amplitudes were varied randomly over an 18-fold range and the numbers of peaks were determined by inspection.

Because the resolution R_s^* that resolves adjacent SCPs into maxima varies with the ratio of SCP amplitudes, the model was tested first by means largely independent of this ratio. In essence, all maxima contained between the departure of intensity from, and return of intensity to, the baseline were interpreted as one baseline-resolved peak, with $R_s^* = 1.5$. The close agreement between the expected and simulated results verified the predictions of theory.

To determine if an average value could be assigned to R_s^*, despite SCP amplitude variations, p was also identified with the number p_m of maxima. The authors then forced these numbers to agree with theory by fitting values of p_m to Eq. (11), with $\overline{m}$ approximated by the number m of SCPs, and determining optimal values of R_s^*. For chromatograms without noise, the best estimate of R_s^* was about 0.50. The results implied that the interval $x_0 = 4\sigma R_s^*$ required for resolution was about 2σ when p was identified with maxima.

Davis and Giddings [17] expanded this study in 1984 and determined by analysis of variance (ANOVA) the dependence of $\overline{m}$ and R_s^* values calculated from plots of $\ln p$ versus n_c^{-1} on various chromatographic attributes. The authors found that the identification of p with baseline-resolved maxima groups was associated with several difficulties, especially for tailing SCPs and highly saturated chromatograms. The authors also identified p with the number p_m of chromatographic maxima and showed that R_s^* values determined from noisy chromatograms were significantly smaller than 0.50, because p_m values in noisy saturated chromatograms were slightly greater than those in noiseless ones and p_m values in noisy unsaturated chromatograms were comparable to those in noiseless ones. The tailing of SCPs raised R_s^* to about 0.7 in the one case examined, because small maxima were obscured easily and large intervals x_0 were required. However, R_s^* remained very close to 0.50 for Gaussian SCPs, even when the amplitudes were varied randomly over a 180-fold range.

In 1984, Herman et al. [15] tested the Poisson model's ability to predict m by simulating chromatograms containing SCPs having several amplitude distributions. The model first was modified slightly; these modifications led to a reexpression of Eq. (11) as

$$\ln(p-1) = \ln(m-1) - \frac{m}{n_c - 1} \tag{51}$$

where n_c is the peak capacity.

Chromatograms containing randomly distributed Gaussian SCPs were simulated with one of three amplitude distributions: a numerical constant, a random number between 0 and 1, or a semiexponential value determined from the work of Nagels et al. [3]. The number p was identified with the average number of maxima in 20 simulated chromatograms, σ was equated to a constant, and X was adjusted as needed. The peak capacity n_c was calculated as

$$n_c = 1 + \frac{X}{x_0} \tag{52}$$

where $x_0 = 2.063\sigma$ (i.e., $R_s^* = 0.5158$). This resolution was the minimum interval between three adjacent SCPs of equal amplitude for which the authors could observe three distinct maxima.

Figure 8a to c shows graphs of $\ln(p-1)$ versus $(n_c - 1)^{-1}$ determined as outline above. The symbols represent the results of computer simulation; the dashed lines represent theory. In general, at low values of $m/(n_c - 1)$, the agreement between simulation and theory is good; at high values, the number p_m of maxima exceeds theory. The threshold value of $m/(n_c - 1)$ at which simulation and theory diverge is about 1.0 for SCPs with constant amplitudes and about 0.8 for SCPs with semiexponential amplitudes.

The number p_m of maxima is larger than expected for large values of $m/(n_c - 1)$, because the positions of maxima have little correlation with the positions of underlying SCPs. Figure 8d illustrates the summation of 20 SCPs of equal amplitude, all of which are separated from an adjacent neighbor by less than

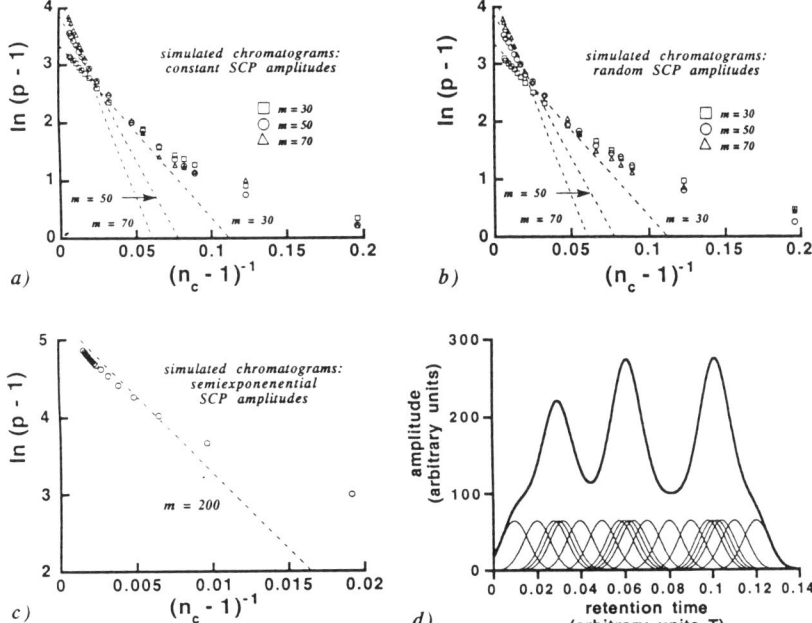

Fig. 8 Graphs of $\ln(p - 1)$ versus $(n_c - 1)^{-1}$ determined from simulated chromatograms containing SCPs with (a) constant, (b) random, and (c) semiexponentially distributed amplitudes. Dashed lines represent theory. (d) Response from overlap of 20 Gaussian SCPs of equal amplitude and standard deviation σ. Each SCP center is within 1.6σ of another. (From Ref. 15.)

1.6σ. The Poisson model predicts that only one peak, with a multiplicity of 20, should be formed. In fact, three maxima are observed.

The correct prediction by the Poisson model of the number of intervals between adjacent randomly spaced points (*not* SCPs) greater than a specified value was not verified until 1985 by Davis [18]. For each of four $\overline{m}$ values between 30 and 200, 200 values of m were selected by a Gaussian random-number generator, with $\sigma_m^2 = \overline{m}$. For each value, m uniform random numbers between 0 and 1 were generated and then ordered. The intervals between adjacent numbers greater than $x_0' = \alpha'/\overline{m}$ ($X = 1$) were interpreted as peaks. These simulations showed that the Poisson model correctly describes p' over a wide range of α' values, including those much greater than 0.5 to 1.0. Thus errors in statistical parameters determined from highly saturated chromatograms can be attributed to SCP amplitudes and not to serious errors in Eq. (10).

In 1985, Davis and Giddings [6] applied the single-chromatogram method to hundreds of computer-simulated chromatograms in which the saturation and the number, shape, and amplitude distribution of SCPs were widely varied. The saturation α was varied from 0.167 to 0.667 ($R_s^* = 0.5$); the SCP number was varied from 100 to 300; the SCP shape was varied from Gaussian to two kinds of exponentially convoluted Gaussians; and the amplitudes were varied from constant to exponential to random. Analyses of variance showed that errors in $\overline{m}$ determined by this method were most pronounced for SCPs that tail, for SCPs with wide amplitude ranges (e.g., exponential), and for chromatograms of high saturation. The authors concluded that, on average, $\overline{m}$ values accurate to 10% or so could be determined if the SCPs were Gaussian and α did not exceed 0.5 ($R_s^* = 0.5$). For tailing SCPs, the threshold α was somewhat lower (e.g., 0.333 to 0.5). Because α depends on $\overline{m}$, which is initially unknown, the authors suggested that it be approximated by a variation of an equation proposed by Herman et al. [15] or by Eq. (10).

Figure 9a and b are graphs of $\ln p'$ versus x_0'/X generated from simulated chromatograms containing Gaussian or exponentially convoluted Gaussian SCPs with 150-fold random amplitudes for the two saturations, $\alpha = 0.167$ and 0.500 ($R_s^* = 0.5$). The solid lines are the theoretically expected result. It is apparent that values of $\ln p'$ can be significantly greater than expected if the saturation is too large and SCPs tail significantly.

In 1986, Dondi et al. [19] applied the single-chromatogram method to experimental chromatograms of flavonoid and glycolic extracts and estimated that $\overline{m} \approx 50$. Because this method had not been tested by simulated chromatograms containing less than 100 SCPs, the authors simulated chromatograms containing 50 major SCPs with uniformly distributed heights between 1 and 10 units. From 100 to 500 minor SCPs, with uniformly distributed heights between 0 and 0.03 unit, then were added to mimic small components in the experimental chromatograms. In general, the $\overline{m}$ estimates determined from these simulations were correct, consistent, and affected only slightly by the minor SCPs.

Fourier Analysis

In 1990, Felinger et al. [20] tested the constant-width Fourier model by applying it to simulations containing randomly distributed Gaussian or exponentially convoluted Gaussian SCPs. Three procedures for calculating statistical parameters from the autocovariance function (ACVF) were tested. Two of these required digital smoothing of the ACVF prior to the computation of $F(\omega)$ by Fourier transform. The determination of the proper window for smoothing was crucial, and two window shapes were considered. The authors found that a smooth $F(\omega)$ could be obtained if the ACVF was truncated after about 4σ. Figure 10a is a graph of $F(\omega)$ versus $\omega\sigma$ computed from the total ACVF of a simulation and from the same ACVF truncated after $t > 4\sigma$. Figure 10b illustrates the goodness of fit of the theoretical $F(\omega)$ to the smoothed spectrum computed from a simulation.

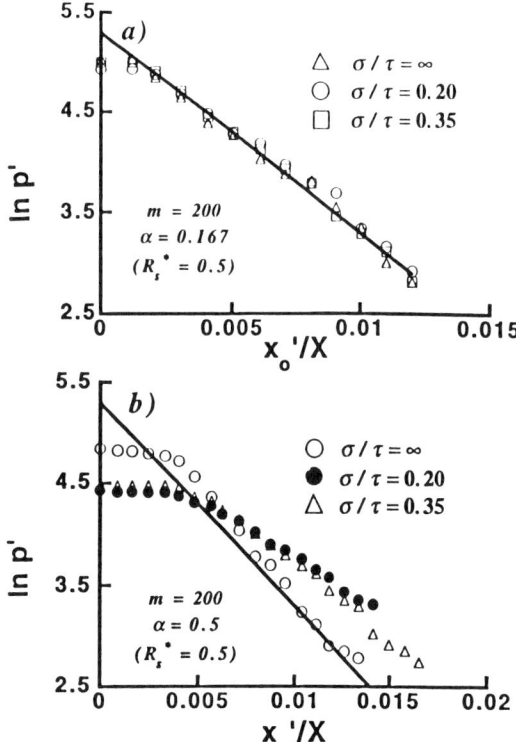

Fig. 9 Graphs of ln p' versus x_0'/X developed from simulated chromatograms containing 200 Gaussian ($\sigma/\tau = \infty$) or exponentially convoluted Gaussian ($\sigma/\tau = 0.20, 0.35$) SCPs having constant σ values and 150-fold random amplitudes. (a) $\alpha = 0.167$; (b) $\alpha = 0.500$. Solid lines represent theory. (From Ref. 6.)

To determine statistical parameters from simulated chromatograms, the authors approximated the relative standard deviation of SCP heights by the relative standard deviation of maxima heights. All three procedures gave essentially equivalent m estimates, and the first procedure was critiqued in detail. For symmetrical Gaussians, m estimates at low saturation (e.g., $\alpha < 0.667$) were comparable to those determined by the Poisson model; for asymmetrical Gaussians, however, they were superior. For simulations containing Gaussian SCPs with 150-fold random amplitudes, accurate m values were determined at high saturations (e.g., $\alpha = 1$ to 2; $R_s^* = 0.5$), and good estimates of σ and τ were calculated from simulations containing SCPs with 50- to 150-fold random amplitudes when $\alpha \leq 0.5$. Accurate m values could also be calculated from

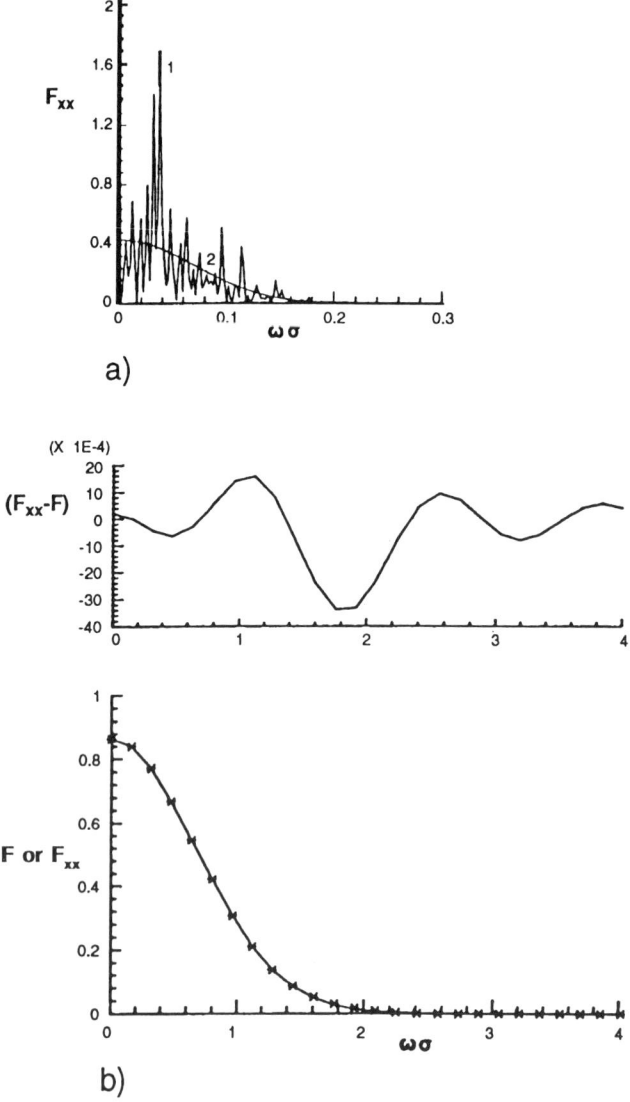

Fig. 10 (a) Power spectra F_{xx} computed from the (1) total ACVF of a simulated chromatogram and (2) ACVF truncated after $t = 4\sigma$ versus $\omega\sigma$. (b) Fit of theoretical $F(\omega)$ versus $\omega\sigma$ to smoothed power spectrum F_{xx} (represented by ×'s) computed from simulated chromatogram. Upper curve shows residuals. (From Ref. 20.)

chromatograms containing small numbers (e.g., 10) of SCPs and in the presence of easily filtered white noise. In the latter applications, a superiority to the Poisson model was demonstrated. Only fair m values were calculated, however, from modestly saturated chromatograms (e.g., $\alpha = 0.667$; $R_s^* = 0.5$) containing SCPs with exponential amplitudes. Here the relative standard deviation of maxima heights was less than the relative standard deviation of SCP heights.

In 1991, Felinger et al. [13] tested the variable-width Fourier model by applying it to simulated chromatograms containing 200 Gaussian SCPs with Poissonian retention times and standard deviations that principally varied either uniformly between two extremes, σ_1 and σ_2, or increased linearly throughout the chromatogram. Simulations containing SCPs with normally distributed widths were also addressed but not extensively, because the authors could not distinguish between $F(\omega)$ values for uniformly and normally distributed widths, as shown by Fig. 11a and b. The distribution of SCP areas was uniform over a 150-fold range, and α was varied between 0.333 and 1.0 ($R_s^* = 0.5$). The authors calculated $F(\omega)$ by using the first procedure in Ref. 20.

In their fittings, the authors approximated the relative standard deviation of SCP areas by the relative standard deviation of maxima (or band) areas. The area of any partially fused maximum was calculated by dropping a perpendicular at the valley to baseline.

Over the α range examined, estimates of m were accurate to within 10% or so; estimates of σ_{av} were usually accurate. The estimated ratio σ_2/σ_1, as calculated from σ_1 and σ_2 estimates, was somewhat biased for low α and uniformly distributed σ values. The origin of the bias is that information about σ_2/σ_1 is filtered from the ACVF.

The removal of information by smoothing actually can be advantageous. The authors showed that chromatograms containing SCPs with variable widths can be fit to the simpler Fourier model for constant SCP widths, if only the short-term correlations in $F(\omega)$ are used. Accurate estimates of both m and σ_{av} were so calculated.

In 1992, Felinger et al. [21] applied the constant-width Fourier model to calculate statistical parameters from simulations in which the intervals between adjacent SCPs were distributed in accordance with one of four probability densities: exponential (E), normal (N), uniform (U), and gamma (G). The motive of the study was to assess the applicability of the Fourier model to non-Poissonian chromatograms. In the simulations, α was varied from 0.333 to 0.667 ($R_s^* = 0.5$), and from 50 to 200 SCPs were distributed exponentially in intensity. The numbers p_m of maxima in simulations varied substantially for different probability densities, even at the same saturation. The theoretical $F(\omega)$ values for different densities were quite distinctive at low saturations but sometimes devoid of structure at high saturations, because long-distance correlations were minimal.

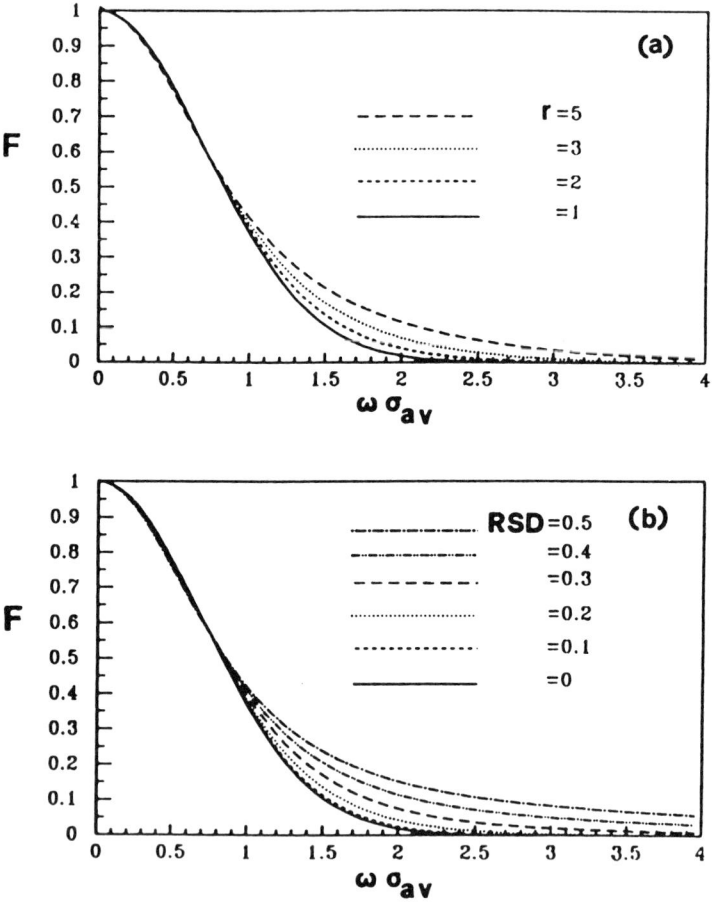

Fig. 11 Indistinguishability of spectra $F(\omega)$ versus $\omega\sigma_{av}$ for (a) uniformly distributed SCP widths ($r = \sigma_2/\sigma_1$) and (b) normally distributed widths (RSD = s_n/σ_{av}). (From Ref. 13.)

The statistical parameters m and σ, and various parameters of the four density functions (e.g., standard deviation of the normal density), were estimated from several sets of computer simulations. The ACVFs computed from simulations were fit to theoretical ACVFs for each of the four densities, with the relative standard deviation of SCP heights approximated by the relative standard deviation of maxima heights. Because the appropriate ACVF typically would not be known for a real chromatogram, the authors determined statistical parameters from the theoretical ACVF that best fit the data. By applying this criterion, the estimates m, σ, and usually (but not always) the density-function parameters were determined correctly, even when the best fit did not correspond to the correct probability density. This was a rather common outcome, which was attributed to the close resemblance of the densities for properly chosen density parameters and to random components in the ACVF. If data instead were fit to the proper ACVF, all statistical parameters were estimated accurately. Figure 12 shows graphs of ACVFs computed from simulations having different densities.

Extended Poisson Model

Computer simulations of chromatograms having linear, quadratic, sinusoidal, and exponential frequencies $f(\zeta)$ were generated by mapping uniform random numbers into cumulative distributions calculated from $f(\zeta)$ [14]. The numbers of peaks, singlets, doublets, and triplets in chromatograms containing SCPs with zero intensities (i.e., chromatograms consisting of normalized retention times having frequency $f(\zeta)$) agreed very well with Eqs. (48) and (49) over wide $\bar{\alpha}$ ranges. In simulations containing SCPs with exponential intensities, the number

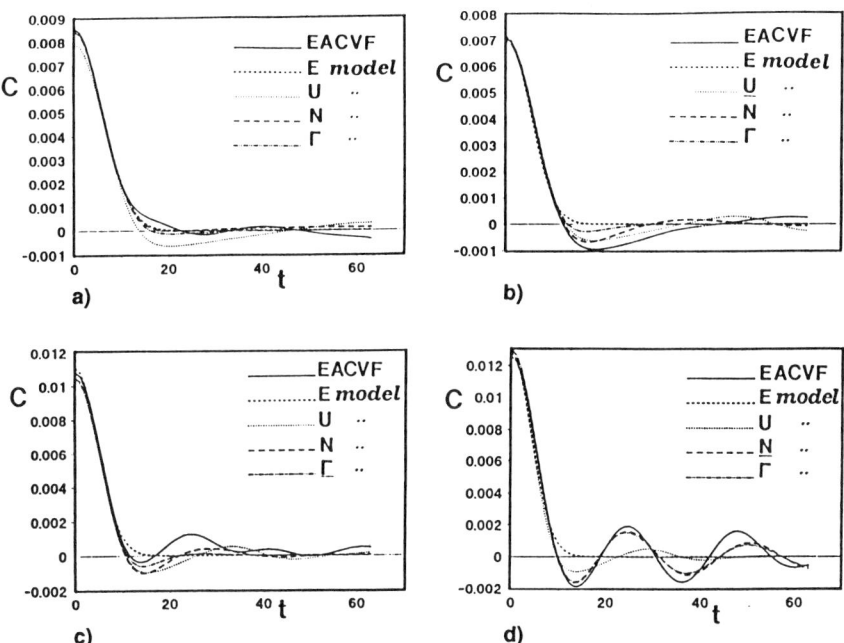

Fig. 12 ACVFs $C(t) \equiv C$ versus time t computed from simulated chromatograms ($\alpha = 0.333$) containing 200 SCPs with (a) exponential (E), (b) uniform (U), (c) gamma (G), and (d) normal (N) probability densities. Best fits of theoretical ACVFs for these four densities to simulation ACVFs are also graphed. (From Ref. 21.)

p_m of maxima varied with $\bar{\alpha}$ in a manner similar to that for Poissonian chromatograms (i.e., p_m was slightly less than predicted at low saturations and was greater than predicted at high saturations). Several approximations $f_a(\zeta)$ were calculated numerically as cubic splines or polynomial fits from simulated chromatograms containing SCPs with exponential intensities, for which $f(\zeta)$ was either linear or quadratic, $\bar{\alpha}$ was 0.25, local saturation $\bar{\alpha} f(\zeta)$ was less than 0.5, and m was either 250 or 625. On average, $\bar{m}$ estimates accurate to 3 to 4% were calculated from these $f_a(\zeta)$ values, the relative retention times of maxima, and Eq. (50).

D. Application and Testing of 1-D Overlap Theories by Experiment

The testing of overlap theories by computer simulation is only a screening mechanism, by which theories can be judged as promising. The application of these theories to experiment is almost a prerequisite for their acceptance by the scientific community.

Combinatory Analysis

In 1982, Rosenthal [2] tested Eq. (3) by analysis of a 46-component synthetic mixture by GC/MS/COMP. Mass-spectral scans were taken at 1.7-s intervals for approximately 30 min. Of the 35 maxima detected in the chromatogram, calculations based on Eq. (3) suggested that one or two might be doublets, which the deconvolution algorithm could not identify. In fact, careful inspection of the mass spectra confirmed that two doublets were present, in good agreement with theory.

Poisson Model

In 1984, Herman et al. [15] estimated $\bar{m}$ by the multiple-chromatogram method from gas chromatograms of the aromatic fraction of Emeraude oil and the petroleum distillate, fulgene. The peak capacity n_c was altered by proportional variation of the linear heating rate of, and the flow rate through, the capillary. Argon was used as the carrier gas, because small changes in flow rate substantially altered n_c.

Peak densities in eight regions of the aromatic-oil chromatograms were verified as constant to within 5%, and the standard deviations of SCPs, as approximated by the σ values of a four-component methylnapthalene mixture, were verified as constant to within 10%. For the fulgene sample, the number of maxima between undecane and dodecane having signal/noise ratios greater than 2 was identified with p, and the σ values of SCPs in this region were approximated by the average σ of SCPs from the methylnapthalene mixture. Capacities n_c were calculated from Eq. (52), with $x_0 = 2.10\sigma$ and interval X adjusted for SCPs at each end. The positive curvature in graphs of $\ln p$ versus n_c^{-1} indicated that many chromatograms were too saturated to interpret. From the two most efficient chromatograms, m_{sl} was estimated as 64.7 and m_{in} as 70.4, even though $p_m \leq 42$.

The aromatic-oil chromatograms were similarly analyzed; from the two most efficient chromatograms, m_{sl} was estimated as 138.6 and m_{in} as 137.2. Hence $\overline{m}$ was about 138, even though $p_m \leq 67$.

The authors noted that erroneous estimates of $\overline{m}$ could be calculated, if chromatograms were interpreted for which $m/(n_c - 1)$ was too large. This ratio can be estimated by dividing the number B of returns of signal to the baseline by the number p_m of maxima

$$\frac{B}{p_m} \approx \exp\left(\frac{-m}{n_c - 1}\right)^{1.857} \tag{53}$$

In 1985, Davis and Giddings [22] applied the single-chromatogram method to estimate $\overline{m}$, α, and the probability $P^{(1)}$ of singlet formation ($R_s^* = 1$) from three gas chromatograms provided by Lee and Karasek. Only regions of constant maxima density were interpreted statistically. The intervals X were determined by the retention-time differences between the first and last maxima in the interpreted regions. Graphs of $\ln p'$ versus x_0'/X were generated from five regions in the three chromatograms, and Table 3 reports the statistical parameters so computed. The standard deviations in m_{sl} and m_{in} were determined by assigning weighting factors [Eq. (24)]. Here α was estimated as $-\ln(p_m/m_{ave})$, where p_m is the number of maxima, and $m_{ave} = (m_{sl} + m_{in})/2$ ($m_{ave} \approx m_{in}$ for chromatogram IIC; simulations suggested this alternative assignment). It is apparent that m_{sl} and m_{in} agree closely except when α is considerably greater than 0.5. In particular, an internal consistency exists, in that the sum of $\overline{m}$ estimates determined from chromatograms IIA and B are equal to the $\overline{m}$ estimates determined from chromatogram A + B. Chromatogram C behaves as if it were Poissonian, even though homologous structure is present.

Table 3 Statistical Parameters m_{sl}, m_{in}, α, and $P^{(1)}$ Calculated from Chromatograms in Ref. 22

Chromatogram	Region	p_m	m_{sl}	m_{in}	α ($R_s^* = 0.5$)	$P^{(1)}$ ($R_s^* = 1.0$)
I		145	234 ± 8	267 ± 12	0.54	0.12
II	A	53	82 ± 9	79 ± 7	0.42	0.19
	B	52	74 ± 7	89 ± 8	0.45	0.16
	A + B	108[a]	165 ± 8	174 ± 10	0.45	0.16
			156 ± 11[b]	168 ± 11[b]		
	C	94	159 ± 8	208 ± 15	0.79[c]	0.04
III		78	127 ± 7	150 ± 10	0.57	0.10

Source: Ref. 22.
[a]Three maxima lay between regions A and B; hence $p_m = 53 + 52 + 3 = 108$.
[b]Sum of independent estimates in regions A and B.
[c]$m_{ave} = m_{in}$.

Figure 13 shows graphs of maxima density versus time and ln p' versus x_0'/X constructed from chromatogram I, which was developed from river-sediment extract. The upper chromatogram contains $p_m = 143$ maxima and is a computer simulation generated with statistical parameters estimated from the experimental chromatogram, for which $p_m = 145$. Here m was estimated as $m_{\mathrm{ave}} \approx 250$. No attempt was made to match SCP amplitudes in the simulation with those in the experimental chromatogram. The two chromatograms are very similar in appearance, as if one simply were a continuation of the other.

In 1986, Dondi et al. [19] applied the Poisson model to estimate the number of components in camomile flavonoid and glycolic extracts partially separated by gradient liquid chromatography. Estimates of $\overline{m}$ were made by applying the multiple-chromatogram method and a variation of the single-chromatogram method to both gradient-optimized and gradient-nonoptimized separations. Gradients were adjusted to maximize the number of maxima by using both simplex and Monte Carlo procedures. The baseline widths x_0 of three standard compounds determined n_c as X/x_0, with $R_s^* = 1$. For $R_s^* = 1$, p equaled the number of maxima groups containing adjacent maxima for which $R_s^* < 1$, as judged by comparisons of experimental maxima and valley heights to those expected for two Gaussian SCPs resolved by $R_s^* = 1$. A variation of the single-chromatogram method was also used in which R_s^* equaled both 0.5 and 1, with $p = p_m$ for the former R_s^*. The capacity was varied by changing flow rates and particle size and by working with coupled columns.

Figure 14a to c shows graphs of ln p versus n_c^{-1} generated from the chromatograms so developed. Both optimized and nonoptimized gradients and also coupled columns produced numbers p_m of maxima from glycolic extract that were consistent with the Poisson model, as shown by Fig. 14a. Figure 14b shows that the number p_m of maxima generated by camomile flavonoids followed the Poisson model for optimized gradients in C_{18} columns packed with either 5- or 10-μm particles. Figure 14c shows the internal consistency of the modified single-chromatogram method as applied to camomile flavonoids. In all cases, the number m of components estimated from the slope and intercept of these graphs was about 50 to 60.

The effect of chromatographic conditions on the predictions of the Poisson model first was addressed in 1988 by Coppi et al. [23], who gas-chromatographed a camomile extract on a 25-m OV-1 capillary and on 15- and 25-m lengths of Carbowax-20M capillaries. Three temperature programs were developed for each stationary phase to obtain Poissonian elution orders, which were verified by χ^2 tests. Graphs of ln p' versus x_0'/X were generated from retention times of maxima having relative areas greater than 0.01%. Table 4 reports estimates of m and other statistical parameters determined from these six chromatograms by the single-chromatogram method. It is clear that experimental conditions had little effect on m. The difference between m_{sl} and m_{in} determined from chromatogram V was

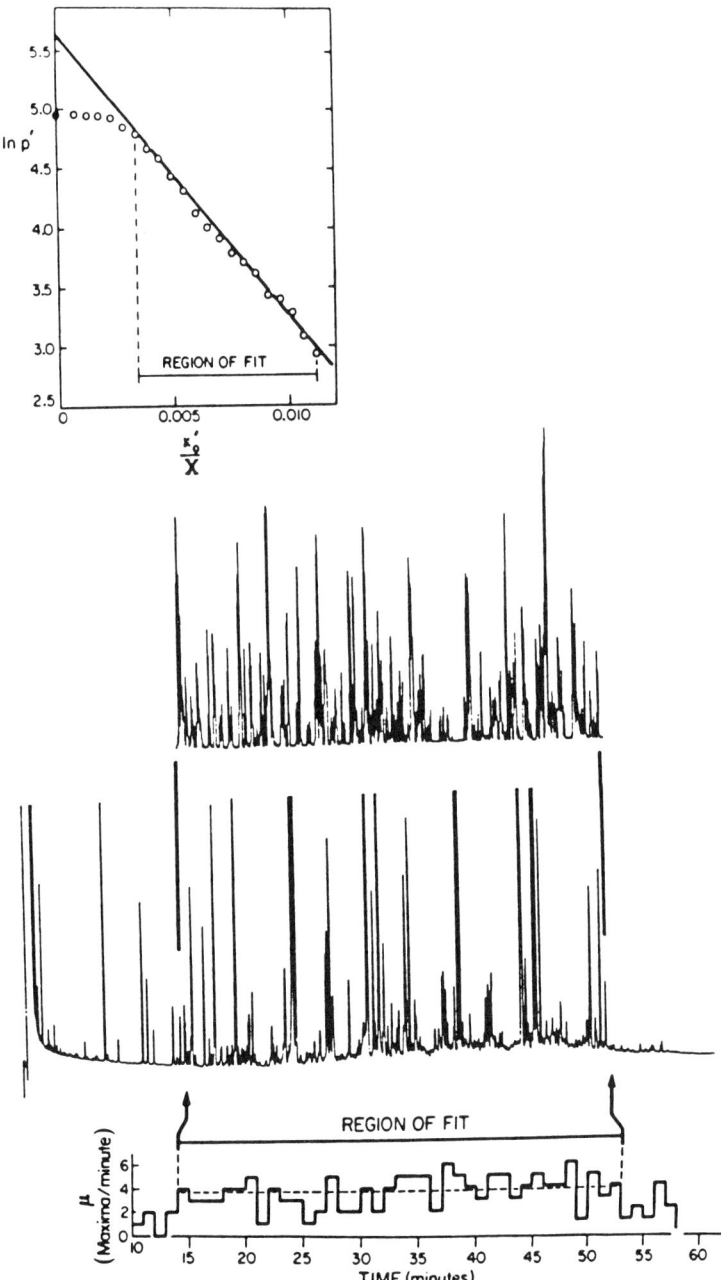

Fig. 13 Graph of maxima/minute versus time and ln p' versus x'_0/X determined from lower chromatogram of river-sediment extract. Upper chromatogram was simulated with statistical parameters estimated from experimental chromatogram. (From Ref. 22.)

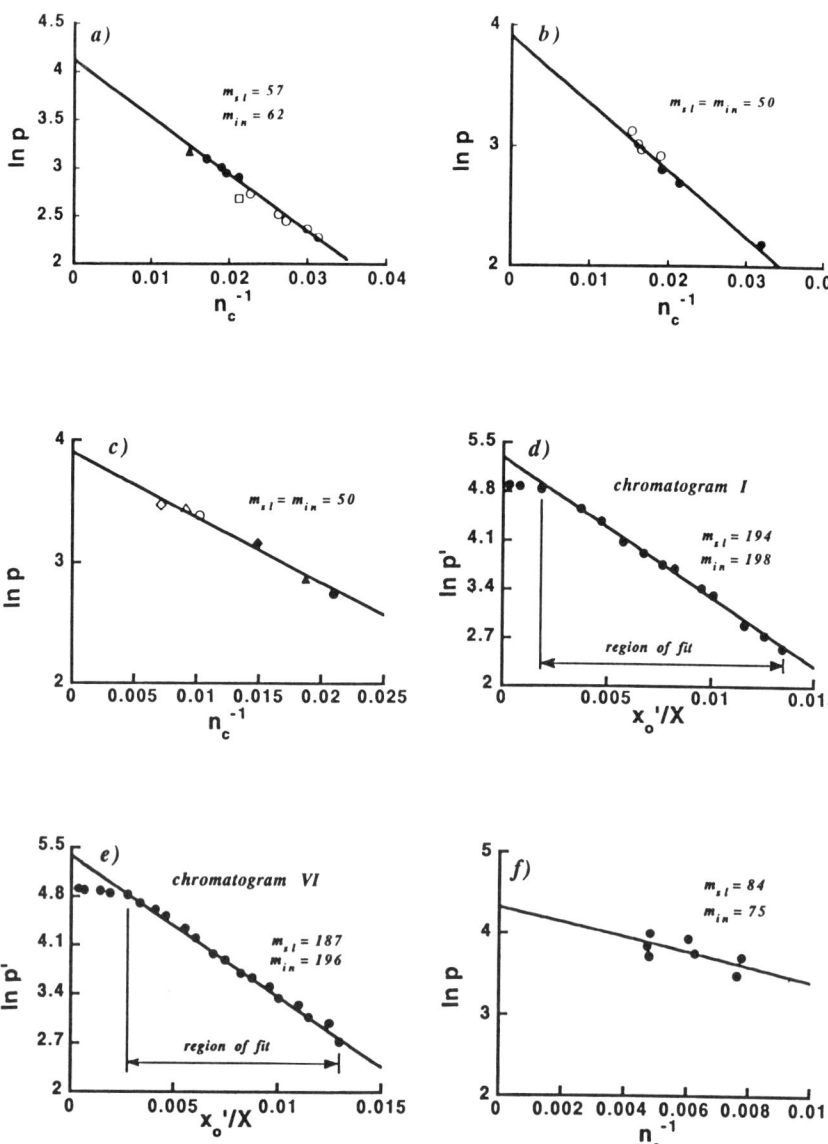

Fig. 14 Graphs of $\ln p$ versus n_c^{-1} and $\ln p'$ versus x_0'/X developed by Dondi et al. All solid lines represent least-squares fits. (a) For glycolic extract: ●, optimized gradient; ○, nonoptimized gradient; □, nonoptimized gradient that failed χ^2 test; ▲, coupled column. (b) For camomile flavonoid extract with optimized gradients: ●, 10-μm C_{18} column; ○, 5-μm C_{18} column. (c) For camomile flavonoid extract and modified single-chromatogram method: ○, ◇, and △, $p = p_m$ ($R_s^* = 0.5$); ●, ♦, and ▲, p defined by $R_s^* = 1$. (d) and (e) Graphs developed from chromatograms I and VI in Ref. 23. (f) Graph developed from *Swertia* herb extract. [(a)–(c) From Ref. 19, (d) and (e) from Ref. 23, (f) from Ref. 27.]

Table 4 Statistical Parameters m_{sl}, m_{in}, m_{ave}, α, $P^{(1)}$, and γ Calculated from Chromatograms in Ref. 23[a]

Chromatogram	p_m	m_{sl}	m_{in}	m_{ave}	α ($R_s^* = 0.5$)	$P^{(1)}$ ($R_s^* = 1.0$)	γ
I	132	194 ± 20	198 ± 25	196	0.30	0.20	0.67
II	165	216 ± 10	213 ± 18	214	0.26	0.30	0.77
III	129	192 ± 10	201 ± 17	196	0.42	0.20	0.66
IV	113	178 ± 10	181 ± 20	179	0.46	0.20	0.63
V	116	196 ± 10	235 ± 25	215	0.62	0.08	0.54
VI	137	187 ± 20	196 ± 42	191	0.33	0.30	0.72

Source: Ref. 23.
[a]Chromatograms I–III were developed on OV-1 stationary phase; chromatograms IV–VI were developed on Carbowax-20M stationary phase. Uncertainties represent 95% confidence levels.

attributed to a saturation greater than 0.5 and was eliminated by interpretation of chromatogram VI, which was developed on the longer Carbowax-20M capillary. Figure 14d and e shows graphs of ln p' versus x_0'/X determined from chromatograms I and VI.

In 1988, Davis [24] applied the single-chromatogram method to a chromatogram containing known numbers of singlet, doublet, and triplet peaks. The chromatogram was generated by Lee et al. [25] from a synthetic mixture of 113 polynuclear aromatic hydrocarbons, each of which had been chromatographed individually to determine its retention index. From these indices, Lee et al. identified each peak in the chromatogram as a singlet or multiplet. Davis photocopied the chromatogram, determined the relative positions of maxima with a digitizer pad, and generated a graph of ln p' versus x_0'/X from these positions. Only values of x_0'/X for which the digitization error was less than 10% were fit to Eq. (20). Table 5 reports the predicted and actual values of m, singlets s, doublets d, and triplets tr. The standard deviations of these parameters were calculated by weighting factors and propagation of errors. This close agreement between

Table 5 Actual and Predicted Numbers of SCPs, Singlets s, Doublets d, and Triplets tr in Chromatogram of Lee et al. [25][a]

Status	m_{sl}	m_{in}	m or m_{ave}	α	s	d	tr
Experiment	—	—	113	—	75	13	4
Prediction	106 ± 7	115 ± 9	110 ± 6	0.174 ± 0.051	77 ± 4	12 ± 3	2 ± 1

Source: Ref. 24.
[a]$p_m = 92$.

observation and theory was the first experimental verification of the Poisson model.

In 1990, Delinger and Davis [26] expanded this study by preparing a synthetic mixture containing 54 C_6 to C_{10} alkanes and alkenes, ethylbenzene, and cyclopentanone. The components' concentrations in this mixture were quasi-exponentially distributed. These 56 compounds first were gas-chromatographed as several simple mixtures of six or so components to determine the retention times of all SCPs. The maxima retention times in chromatograms of the 56-component mixture then were compared to these retention times to determine peak multiplicities. The Poissonian retention times were obtained on two gas capillaries by trial-and-error selections of various temperature programming rates. Once these rates were determined, different saturations were obtained by proportionally varying the heating rates and the flow rate of carrier gas.

Figure 15 shows graphs of the dimensionless ratios, p/n_c, s/n_c, d/n_c, and tr/n_c versus saturation α so determined. The solid curves are theory, and the symbols are experimental results. For the latter, α was calculated as $-\ln(p_m/m_{ave})$, where p_m is the number of maxima and m_{ave} is the weighted estimate of $\overline{m}$ determined as outlined in Ref. 24. Parameter n_c was calculated as $56/\alpha$ ($m = 56$). It is clear that experiment agreed well with theory, as long as α was less than 0.4 or so. At higher saturations, the amplitudes of SCPs could not be neglected. In fact, the number of singlets was somewhat lower, and the numbers of doublets and (especially) triplets was somewhat higher, than predicted.

The multiple-chromatogram method also was critiqued but found wanting, because σ varied substantially throughout chromatograms due to variation of the temperature-programming rate. Despite this, m_{in} was fairly accurate.

In 1990, Dondi et al. [27] applied the multiple-chromatogram method to chromatograms of *Swertia* herb extract partially separated by gradient liquid chromatography under a variety of conditions. These conditions included partial resolution of the extract on both 5- and 10-μm columns at flow rates of 1.0 and 2.0 mL/min and the use of three different organic modifiers (methanol, acetonitrile, and tetrahydrofuran). The numbers of maxima in all chromatograms were optimized either by a simplex algorithm or a Fibonacci search; the largest maxima number was 56 and the smallest was 34. All chromatograms were verified to have a constant maxima density. The capacity n_c was calculated at $R_s^* = 1.0$ from the mean baseline widths of three standards; the capacities at $R_s^* = 0.5$ were assumed to be twice the $R_s^* = 1.0$ values. Figure 14f is a graph of ln p versus n_c^{-1} ($R_s^* = 0.5$) determined from these optimized chromatograms, with $p = p_m$. The data are scattered, but the closeness of the values, $m_{sl} = 84$ and $m_{in} = 75$, suggested that $m \approx 80$ (similar scatter was obtained in computer simulations). These data indicated that m was independent of both packing size and organic modifier. The authors observed that this work reported the first graph of ln p versus n_c^{-1}, in

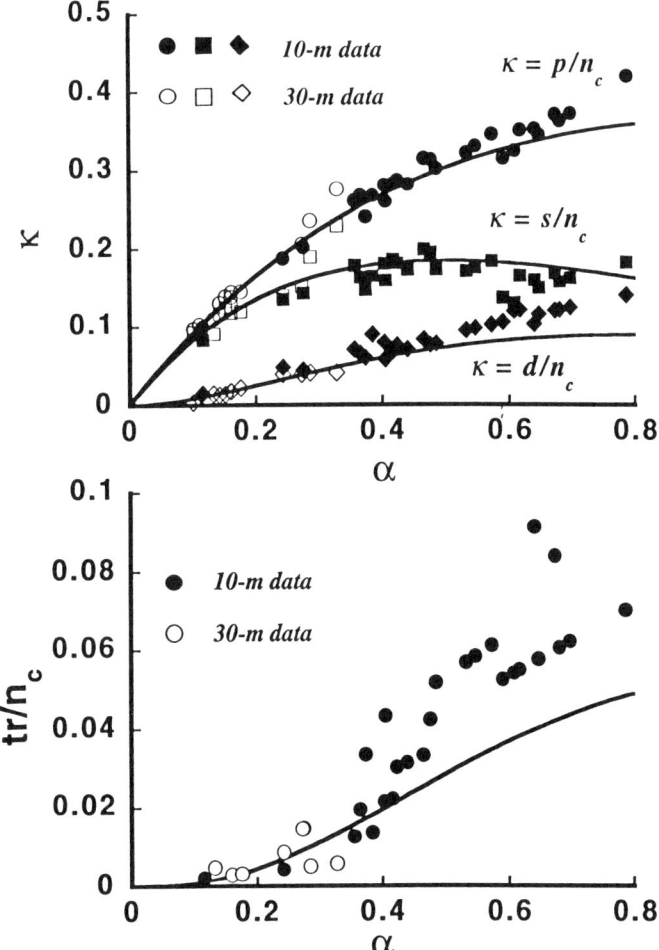

Fig. 15 Graphs of $\kappa = p/n_c$, s/n_c, and d/n_c and of tr/n_c versus α, as determined from synthetic 56-component mixture. Curves represent theory; filled and open symbols correspond to determinations from 10- and 30-m capillary lengths. (From Ref. 26.)

which data were deduced from chromatograms developed both on different columns and with different organic modifiers.

In 1991, Oros and Davis [28] expanded substantially the study initiated by Coppi et al. [23] and critiqued the effect of stationary phase on $\overline{m}$ values calculated by the single-chromatogram method from gas chromatograms. Chromatograms were generated from four complex mixtures of widely varying polarity on four capillaries of widely varying polarity. Of the 16 mixture/stationary-phase combinations, Poissonian retention times were obtained by trial and error for 12 combinations. For each combination, several estimates of $\overline{m}$ were made by proportionally varying the heating rate of, and the flow rate of carrier through, the capillary. The estimates so determined comprised a data set; for each mixture, different sets were compared by analysis of variance (ANOVA) at the 95% confidence level.

In brief, statistically equivalent $\overline{m}$ values were determined from chromatograms of lime oil developed on all four capillaries and from chromatograms of aliphatic hydrocarbons developed on the two capillaries that generated Poissonian retention times. Statistically equivalent m values were determined from chromatograms of the most polar mixture, peppermint oil, developed only on the three most polar capillaries. The $\overline{m}$ values determined from chromatograms of coal-tar extract were statistically different, but by only 10% or so, when chromatograms were developed on the two capillaries that generated Poissonian retention times. In general, either insignificant or small differences in $\overline{m}$ were found, as long as the mixture polarity was reasonably matched to the stationary-phase polarity.

Other work showed that chromatograms developed by simple linear temperature programs contained about the same or even fewer maxima than the Poissonian chromatograms and that $\overline{m}$ could vary with flow rate, if p_m varied with flow rate due to slope sensitivity.

D. Bowlin and J. M. Davis investigated the dependence of $\overline{m}$ values calculated by the single-chromatogram method on various settings of amplification, slope sensitivity, and threshold peak area for a particular GC electrometer and data processor (52). Unsurprisingly, the number p_m of maxima decreased as the amplification of the electrometer decreased and as the slope sensitivity and threshold peak area increased. It was found that the $\overline{m}$ estimates tracked the variation of p_m in a roughly proportional manner. These results demonstrated that $\overline{m}$ is not an absolute number but depends intimately on the number of maxima detected.

Analogy Between Overlap and Depolymerization

Martin and Guiochon [5] applied the depolymerization analogy to several experimental gas and two liquid chromatograms. Systematic variations in GC response factors were reduced by interpreting only fractions of chromatograms containing isomeric substances. Table 6 reports γ, m, and n_c values calculated from Eqs. (16) to (19) for three isomeric families of substituted aromatics in an Emeraude-oil fraction. The number m for the first two families almost equaled the number of possible isomers, whereas m for the third family was less than the number of possible isomers.

Similar calculations were reported for dodecane and tridecane fractions of Emeraude oil, for which the number of possible isomers substantially exceeded the estimated m values. The two fractions were then interpreted as one combined fraction; the m estimated from the combined fraction essentially equaled the sum of the two independent m values. This behavior also was observed for capacity n_c. The results are summarized in Table 6.

Other calculations were reported for the gas chromatogram of a Midway-oil fraction containing C_2 alkylbenzo[h]quinolines and liquid chromatograms of triglycerides in butterfat and cod liver oil. The nonlinearity of the light-scattering detector used to generate the liquid chromatograms was addressed. The oil fraction was estimated to contain three components, consistent with determinations by GC/MS and Shpol'skii spectroscopy. For the butterfat separation, the parameters $\gamma = 0.435$, $m \approx 83$, and $n_c = 99.4$ were determined from 36 maxima; for the cod-liver-oil separation, the parameters $\gamma = 0.298$, $m \approx 74$, and $n_c = 61.0$ were determined from 22 maxima.

Fourier Analysis

In 1993, Dondi et al. [29] interpreted gas chromatograms of a chamomile extract and a naphtha sample by Fourier analysis. Any periodic structure in these chromatograms was emphasized by scaling the time axis identically to that required to produce a constant temporal increment between C_9 and C_{14} homologs in a reference chromatogram. This scaling had no effect on the ACVF of the chamomile chromatogram, and the apparent correlations for $t > 3\sigma$ were judged from computer simulations of confidence intervals to be noise. Theoretical constant-width ACVFs for E, N, G, and U probability densities were fit to five

Table 6 Statistical Parameters γ, m, and n_c Estimated by Application of Depolymerization Analogy to Different Fractions of an Emeraude Crude Oil

Fraction	Family	p_m	γ	m	n_c
Substituted aromatics	Dimethylnaphthalene	7	0.671	10.4	26.4
	Trimethylnaphthalene	11	0.677	16.3	41.6
	3-Alkylphenanthrene	11	0.155	70.8	38.0
Isomeric alkanes	Dodecane	28	0.536	52.2	83.8
	Tridecane	22	0.411	53.6	60.2
	Dodecane + tridecane	50	0.476	105.0	141.5
				105.8[a]	144.0[a]

Source: Ref. 5.
[a] Sum of independent estimates for dodecane and tridecane families.

regions of this chromatogram. The ACVFs for the E, N, and G densities produced good fits, fairly consistent m values and σ values that agreed well with ones determined by fitting hydrocarbon SCPs in the reference chromatogram to Edgeworth–Cramér series. Overall, the chromatogram had a Poissonian-like character.

In contrast, the scaling had a pronounced effect on the naphtha chromatogram and revealed deterministic structure in the ACVF, which was attributable to a homologous series of components differing by one methylene unit. Figure 16 illustrates this structure as an enhancement of ACVF signal every 234 or so seconds. The ACVFs of 11 separate regions of this chromatogram indicated that SCP densities were gamma or normal and that the mean interval between SCPs varied throughout the chromatogram. As before, internally consistent m estimates were obtained from good fits, and σ values agreed fairly well (but not as well as before) with those determined from Edgeworth–Cramér fittings to the reference chromatogram. Despite the structure, overall the chromatogram had a pseudo-Poissonian character, principally because of the variability of the mean interval between SCPs.

Extended Poisson Model

Figure 17a is a gas chromatogram of lime oil, which was deliberately developed by a complex temperature program to produce a higher density of maxima in the

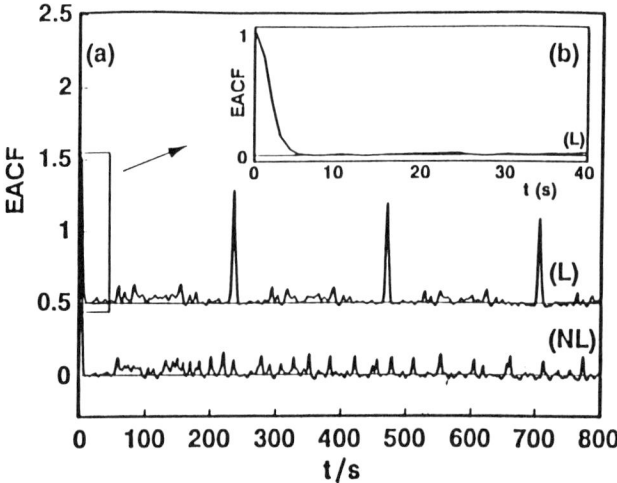

Fig. 16 Experimental autocovariance function EACF versus time t developed from naphtha chromatogram: (a) deterministic long-range correlations in the linearized (L) and nonlinearized (NL) chromatogram; (b) expansion of EACF at short times. (From Ref. 29.)

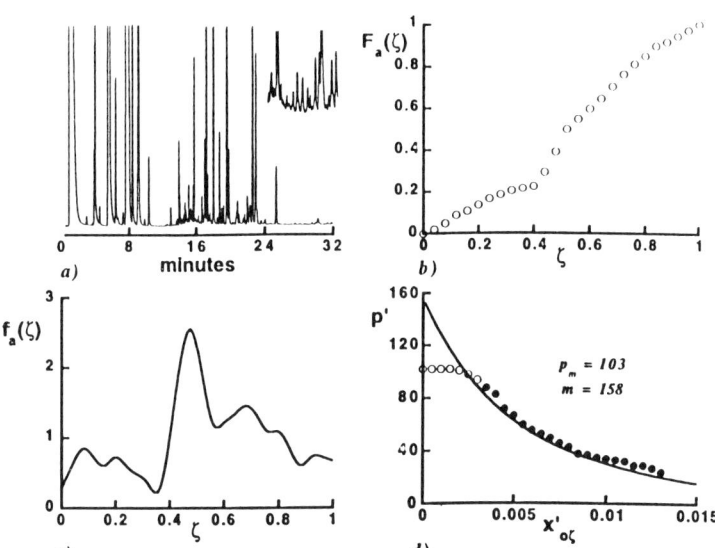

Fig. 17 (a) Lime-oil chromatogram containing high maxima density in chromatogram center. Insert is blowup of last 8 min of chromatogram; (b) cumulative distribution $F_a(\zeta)$ calculated from maxima retention times versus ζ; (c) approximate frequency $f_a(\zeta)$ calculated by numerically differentiating $F_a(\zeta)$ versus ζ; (d) graph of p' versus $x'_{0\zeta}$. Symbols represent $(x'_{0\zeta}, p')$ coordinates determined from retention times. Solid curve is least-squares fit to filled symbols; open symbols are excluded from fit. (From Ref. 14.)

chromatogram's center than elsewhere [14]. Figure 17b is the cumulative distribution $F_a(\zeta)$ calculated from the maxima retention times and represents the fraction of maxima, whose reduced retention times are less than ζ. Figure 17c is the approximation, $f_a(\zeta)$, calculated by numerically differentiating $F_a(\zeta)$. One observes that the highest value of $f_a(\zeta)$ coincides with the region of highest maxima density in the chromatogram. Figure 17d is a graph of p' versus $x'_{0\zeta}$ generated from the maxima retention times. The solid curve represents the least-squares fit of Eq. (50) to these data, with $\overline{m}$ determined as 158. The $\overline{m}$ estimate determined by application of the single-chromatogram method to three Poissonian chromatograms of the same mixture developed under identical conditions, except for the temperature program, is 165 ± 6. Thus the new estimate agrees with previous ones.

E. Resolution Factors for 1-D Models of Overlap

Computer simulations have shown that the optimal resolution factor R_s^* for the Poisson model is about 0.5, when SCPs are noiseless Gaussians, peaks are identified with maxima, and α is less than 0.5 to 1.0 (depending on one's criterion). As is well known, resolutions greater than 0.5 are required to resolve Gaussian SCPs of equal width but different amplitudes. Because chromatograms contain SCPs with variable amplitudes, one would anticipate that R_s^* should be larger than 0.5. This inconsistency has puzzled researchers and led to several studies.

Theory of Determination Limits

In 1987, Creten and Nagels [30] developed an analytical model for the probability that mixture components could be determined with errors less than a specified value. The probability depended on α and thus R_s^*. The authors repeated simulations similar to those of Giddings et al. [16], with SCP amplitudes governed by the quasi-exponential distribution of Nagels et al. [3]. Resolutions R_s^* were calculated in accordance with Eq. (10) and varied from 0.39 to 0.57. The authors judged these values to be unrealistic.

The authors noted that it is not possible to select an R_s^* that is independent of SCP intensities. For three SCP intensity distributions determined from plant extracts as detailed by Nagels et al. [3], from 46,000 to 126,500 simulations were generated to find "ideal" cases in which interval X was totally filled and the number p_m was maximized. This maximum number was interpreted as n_c, from which R_s^* was calculated as

$$R_s^* = \frac{X}{4\sigma n_c} \tag{54}$$

The resolutions so determined varied from 0.76 to 0.87, with a mean of about 0.8.

Discrimination Factor

In 1987, El Fallah and Martin [31] observed that R_s^* is insufficient to determine x_0, because it is independent of SCP amplitudes. To address this shortcoming, the authors introduced the discrimination factor d_0 for two overlapping maxima, as defined by Fig. 7. In general, the R_s^* required to maintain a constant d_0 between two maxima increases with the ratio of observed amplitudes, and the R_s^* required to maintain a constant ratio of observed amplitudes increases with increasing d_0.

The authors defined the effective peak capacity $n_{c_{\text{eff}}}$ as the largest number of maxima that could be accommodated within a fixed interval X while maintaining a constant d_0 between adjacent maxima. Because R_s^* depends on d_0, $n_{c_{\text{eff}}}$ depends on the amplitude distribution and elution order of SCPs. The authors examined the variation of $n_{c_{\text{eff}}}$ by computer simulations containing Gaussian SCPs with constant σ values and exponential intensities. Two extreme configurations of elution order maximized and minimized $n_{c_{\text{eff}}}$ (the minimum and maximum configurations, respectively). The average resolution $R_{s,\mathrm{av}}$ for these configurations was calculated as

$$R_{s,\mathrm{av}} = \frac{X}{4\sigma(n_{c_{\mathrm{eff}}} - 1)} \tag{55}$$

Quantity $R_{s,\mathrm{av}}$ varied slightly with the number m of SCPs in the simulations, as shown by the graph of $R_{s,\mathrm{av}}$ versus m^{-1} in Fig. 18a for $d_0 = 0.5$. For any m, the minimum configuration minimized $R_{s,\mathrm{av}}$ and the maximum configuration maximized $R_{s,\mathrm{av}}$. The intercepts correspond to $R_{s,\mathrm{av}}$ values extrapolated to an infinite number of components.

Figure 18b is a graph of $R_{s,\mathrm{av}}$ versus d_0 constructed from 10 and 20 SCPs arranged in the maximum and minimum configurations. Also graphed are the average $R_{s,\mathrm{av}}$ values calculated from 50 simulations containing 10 and 20 randomly distributed SCPs. Unsurprisingly, the $R_{s,\mathrm{av}}$ values for randomly distributed SCPs were intermediate between the extremes. Graphs similar to Fig. 18a then were constructed for various m and d_0 values and extrapolated to an infinite number of SCPs. Figure 18c is a graph of these extrapolated $R_{s,\mathrm{av}}(\infty)$ values versus d_0 for the maximum and minimum configurations. The mean $R_{s,\mathrm{av}}(\infty)$ at $d_0 = 0$, as determined from randomly distributed SCPs, was 0.71 ± 0.01. The authors interpreted this value as the appropriate value of R_s^* for the Poisson model. A closely related study was published in 1988 [32].

Super-resolution

In 1991, Schure [33] showed that the numerical deconvolution of computer-generated chromatograms by the constrained iterative relaxation method (CIRM) could be interpreted as a reduction of the average factor R_s^d necessary for resolution. The amplitude envelope $F(t)$ at time t of m randomly distributed Gaussian SCPs in a noise-free simulation is

$$F(t) = \sum_{j=1}^{m} \frac{B_j}{\sqrt{2\pi}\,\sigma_j} \exp\left[-\frac{(t - t_j')^2}{2\sigma_j^2}\right] \tag{56}$$

where B_j, σ_j, and t_j' are the detector response factor, standard deviation, and retention time of the jth SCP. Schure expressed this envelope by the convolution integral

$$F(t) = \int_{-\infty}^{\infty} G(t')\delta(t - t')\,dt' \tag{57}$$

where G is the broadening operator for a Gaussian SCP and the delta function $\delta(t - t')$ represents the delay time caused by retention. Because of constraints of noise and sampling rate, $\delta(t - t')$ is better represented in numerical deconvolution by another function, $W(t - t')$, of finite width.

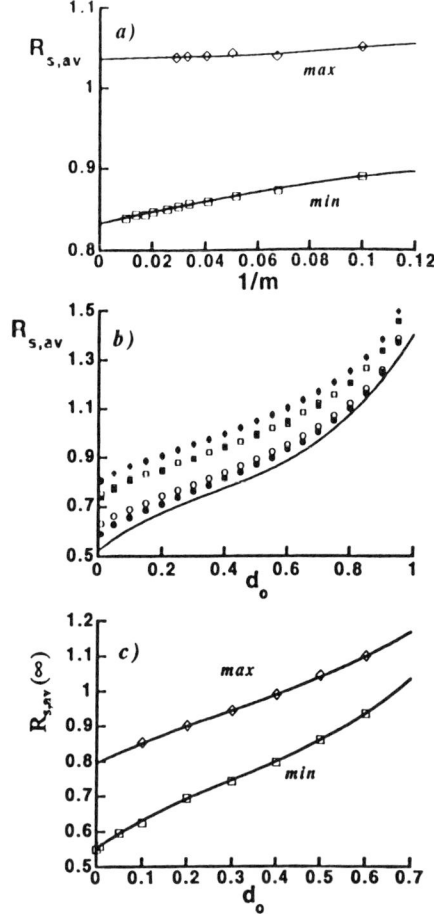

Fig. 18 (a) Graph of $R_{s,\mathrm{av}}$ versus m^{-1} for maximum and minimum configurations, $d_0 = 0.5$. (b) Graph of $R_{s,\mathrm{av}}$ versus d_0 for maximum ($\diamond\blacklozenge$) and minimum ($\bigcirc\bullet$) configurations and for randomly distributed SCPs ($\square\blacksquare$). For open symbols, $m = 10$; for filled ones, $m = 20$. Solid curve corresponds to resolution of two Gaussian SCPs with equal amplitudes. (c) Graph of $R_{s,\mathrm{av}}(\infty)$ versus d_0 for maximum and minimum configurations. (From Ref. 31.)

In accordance with CIRM, the value of W at discrete time t_i was computed iteratively from

$$W^{(k+1)}(t_i) = W^{(k)}(t_i) + r\{W^{(k)}(t_i)\}\{F(t_i) - [G*W^{(k)}]_{t_i}\} \tag{58}$$

where the superscript, k, denotes iteration number, the symbol, $*$, denotes convolution, and $r\{W^{(k)}(t_i)\}$ is the relaxation function.

Figure 19 is a graph of the minimum resolution R_s^d determined by CIRM necessary for the resolution of two Gaussians of different amplitudes versus height ratio. The curves are cubic-spline fits corresponding to different thresholds of resolution; the bottom curve corresponds to maxima resolution. An exponentially weighted average determined R_s^d as 0.352.

Schure then applied CIRM to computer-simulated chromatograms containing from 20 to 72 randomly distributed Gaussian SCPs with exponential amplitudes, constant σ values, and saturations α between 0.513 and 1.85 ($R_s^* = 0.8$; note $R_s^* \neq 0.5$!) to determine if this value of R_s^d, determined from only two SCPs, was consistent with an entire chromatogram, as interpreted by the Poisson model. The numbers of maxima before deconvolution, p_m, and after deconvolution, p_d, were counted by inspection of five chromatograms for each α. These results were ratioed, averaged, and then compared to the ratio of two expressions of Eq. (10) for different values of resolution:

$$\frac{p_d}{p_m} = \exp\left[\alpha\left(1 - \frac{R_s^d}{R_s^*}\right)\right] = \exp\left[\frac{4\sigma R_s^{*\,(1)}\overline{m}}{X}\left(1 - \frac{R_s^d}{R_s^{*\,(2)}}\right)\right] \quad (59)$$

where α is defined relative to $R_s^{*\,(1)}$. Superscripts are used in the final identity to distinguish the two supposedly equivalent resolution factors, R_s^*; Eq. (59) agrees most closely with simulations for $R_s^d = 0.352$ if $R_s^{*\,(1)} = 0.8$ and $R_s^{*\,(2)} = 0.5$. The former resolution is consistent with the work of Creten and Nagels [30]; the latter is consistent with the Poisson model [15–17]. The single R_s^* most consistent with Eq. (59) over the α range $0.794 \leq \alpha \leq 1.49$ ($R_s^* = 0.8$) and $R_s^d = 0.352$ is about

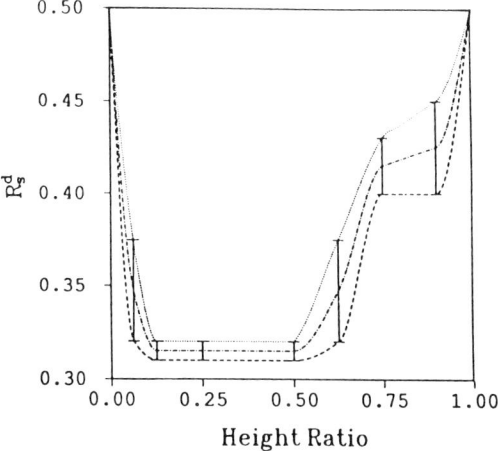

Fig. 19 Graph of super-resolution R_s^d versus height ratio, as computed from two overlapping Gaussian SCPs of equal widths but different heights. (From Ref. 33.)

0.6. Other estimates of R_s^* based on "shoulder" and "detectability" criteria were considered briefly.

F. Amplitude Distribution of SCPs in 1-D Separations

A common means for critiquing theories of overlap is the interpretation of computer-generated chromatograms. Among other attributes, a distribution of amplitudes must be assigned to SCPs. The most appropriate distribution appears to be exponential, as was shown by the following studies.

In 1983, Nagels et al. [3] determined by an iterative algorithm discussed previously the amplitude distribution of SCPs generated from plant-leaf extract and determined by HPLC with UV detection. The observed-peak areas were not described by either an exponential or gamma distribution, but the iteratively determined SCP areas were semiexponential. Figure 20a is a graph of the cumulative distribution of SCP areas versus the relative area of observed peaks. The circles represent the FDC data of Nagels et al.; the curve is a least-squares fit of an exponential cumulative distribution. The fit is not excellent but is fair. A related observation was made in 1985 by Nagels and Cretin [34], who showed that the observed-peak areas of plant-leaf extract, as determined by HPLC with electrochemical detection, were similar to those reported by Nagels et al. [3].

In 1986, Dondi et al. [19] determined the frequency distribution of observed-peak heights in a gradient LC chromatogram of camomile flavonoids. Figure 20b is a graph of the frequency of observed-peak heights versus observed-peak height. The solid curve is the least-squares fit of an exponential frequency to these data (the curve differs from that originally reported, which may not have been a fit). The fit is not excellent but is good.

In 1987, El Fallah and Martin [31] observed that a statistical theory of mixture composition predicted that for complex mixtures the relative number of mixture constituents having a mole fraction smaller than k increases exponentially with k. To the reviewer's knowledge, this theory was not published but was presented at a professional meeting [35].

G. Determination Limits

A consequence of overlap is that component quantitation is subject to error. Several probabilistic studies have assessed this error's magnitude. These studies are reviewed only briefly, not because they are unimportant but because they critique a consequence of overlap and not overlap itself.

In 1983, Nagels et al. [3] determined by computer simulations of plant-leaf chromatograms the probability $PC_e(q)$ that SCPs of relative area (or abundance) q were determined with an error smaller than e. Low probabilities were associated with small observed-peak areas, n_c values, and e values. The authors also observed that overlap, rather than detector characteristics, usually established determination limits for components of complex mixtures and defined the determination limit, DL_e^ω, as the minimum relative abundance necessary for a component's determination with probability ω, for which the relative error was less than e. From simulations, the authors determined that DL_e^ω decreased with increasing n_c and e.

Nagels and Creten [36] then introduced a new determination limit, DLS_e^ω, equal to DL_e^ω expressed as a relative mass amount for a mixture chosen randomly from a large, well-defined series of mixtures. In this study, the responses of 65 plant-leaf extracts partially separated by HPLC were determined at 34 UV

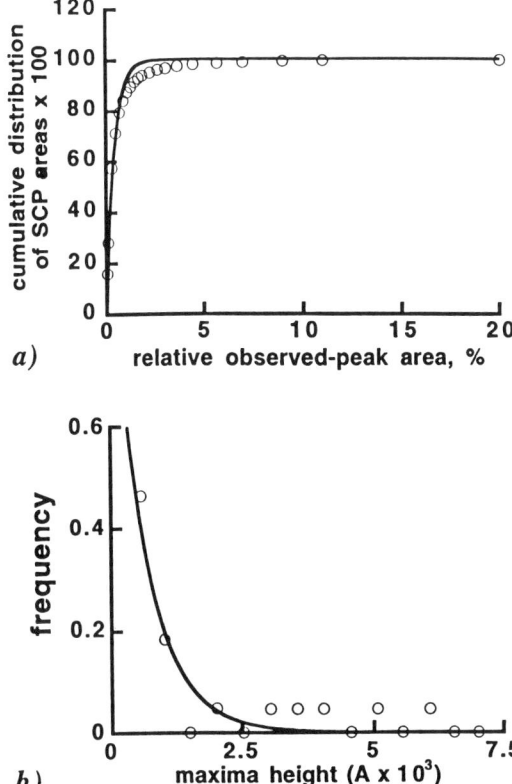

Fig. 20 (a) Cumulative distribution of SCP areas determined from plant-leaf extracts by Nagels et al. [3] versus relative observed-peak area. Solid curve is least-squares fit of exponential cumulative distribution. (b) Frequency of observed-peak heights versus observed-peak height. Solid curve is least-squares fit to a negative exponential. [(a) From Ref. 3, (b) from Ref. 19.]

wavelengths. Because responses differed at various wavelengths and SCPs were not necessarily random, the authors interpreted their results in terms of peak density, which was directly proportional to DL_e^ω, instead of n_c. From these responses and iterative computations, DLS_e^ω values for quercetin and caffeic acid were estimated over the interval 245 to 395 nm.

Nagels et al. [37] then established $DLS_{0.1}^{0.9}$ for 19 phenolic constituents of plant leaves. In this study, an empirical equation for estimating DLS_e^ω more simply than in Ref. 36 was verified and was also used in another study [38].

Creten and Nagels [30] then expressed the probability PC_e analytically as

$$PC_e = \vartheta^2 + (1 - \vartheta^2)P_e \tag{60}$$

where ϑ is the probability that no SCP falls in the interval $x_0 = 4\sigma R_s^*$ on either side of the center of the SCP to be determined and P_e is the probability that if overlap does occur, the relative error is less than e. Computer simulations established the variation of ϑ with interval X, and P_e was estimated from the cumulative distribution of observed-peak areas. The PC_e values so determined from Eq. (60) agreed reasonably well with ones computed by simulation [3], as did estimates of DL_e^ω.

El Fallah and Martin [39] computed by simulation the probability that determinations with errors smaller than specified values could be carried out. The SCPs were Gaussians in chromatograms of various saturation. In all cases, the probability decreased with saturation and decreasing error tolerance. For determination of the most abundant SCP of an observed peak, the probability was highest, when the observed-peak area was a small fraction of the chromatogram's area. For determination of a particular SCP in an observed peak, however, the probability was highest when the SCP area was a large fraction of the entire area. Surprisingly for the latter case, at high saturation and low error tolerance the probability leveled off or even slightly decreased with increasing relative SCP area.

III. STATISTICAL MODELING OF OVERLAP IN MULTIDIMENSIONAL SEPARATIONS

A. Theories

In the last 3 or so years, overlap theories have been derived or critiqued for separation spaces having more than one dimension. These efforts were motivated by the recognition that a single dimension of separation simply is not adequate for the resolution of complex mixtures. Because such spaces have large capacities, the probability of mixture resolution is much greater than in a single dimension. Nevertheless, some overlap occurs.

Early Work

A study often neglected today is by Connors [40], who in 1974 proposed that an n-dimensional TLC be partitioned into $M = (4\sigma)^{-n}$ cells. Each of m component peaks in this TLC has a constant standard deviation σ in each dimension and is represented by its n R_f values. If these are randomly distributed, the probability $p(k)$ that k components occupy any cell is

$$p(k) = \frac{e^{-\chi}\chi^k}{k!} \tag{61}$$

where $\chi = m/M$. The mean number N_k of cells containing k components is $Mp(k)$.

Poisson Model in Two Dimensions

A shortcoming of Connor's work is that components occupy discrete cells, much as in Rosenthal's work for 1-D separations. In 1991, Davis [41] represented single-component spots (SCSs) in a two-dimensional (2-D) separation by randomly distributed circles of diameter d_* and calculated the mean numbers of singlet spots s, doublet spots d, and triplets spots tr as

$$s = \overline{m}e^{-4\alpha} \qquad d = \overline{m}\frac{8\alpha^2 e^{-8\alpha}}{1 - e^{-4\alpha}}$$

$$tr \approx \frac{256}{5}\overline{m}\left\{\frac{\alpha^4 e^{-12\alpha}}{(1 - e^{-4\alpha})^2} + 2\frac{\alpha^6 e^{-12\alpha}}{[1 - (1 + 4\alpha)e^{-4\alpha}]^2}\right\} \tag{62}$$

where the saturation α is

$$\alpha = \frac{\overline{m}}{n_c} = \frac{\overline{m}\pi(\xi d_*)^2}{4A} \tag{63}$$

The spot capacity n_c is thus the ratio of the bed area A to the effective area, $A_0 = \pi(\xi d_*)^2/4$, of any SCS. The scalar, ξ, allows one to adjust this effective area, much as does the resolution factor R_s^* in 1-D separations.

This work showed that the spot capacity of a 2-D separation is utilized less effectively than the peak capacity of a 1-D separation. In other words, s/n_c is smaller at any α for a 2-D separation than for a 1-D separation, as shown in Fig. 21a. The reason for this behavior is simple. A 1-D SCP of width x_0 can be a singlet only if the centers of the nearest SCPs on both sides are no closer to its own center

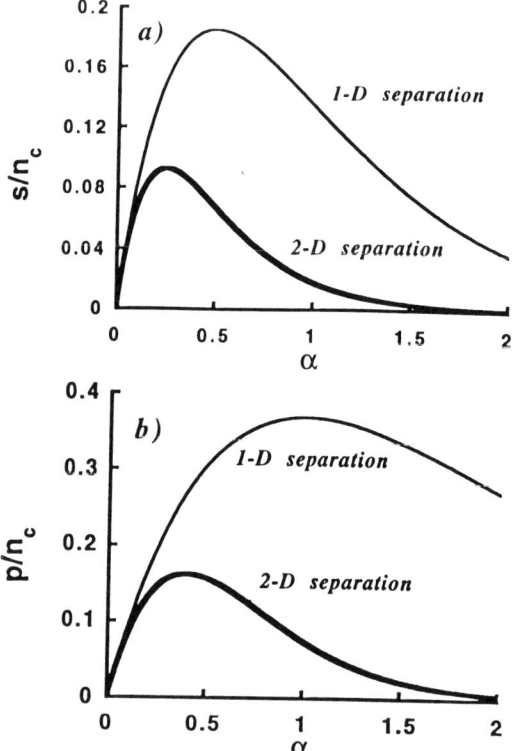

Fig. 21 (a) Graph of s/n_c versus α for 1- and 2-D separations; (b) graph of p/n_c versus α for 1- and 2-D separations. [(a) From Ref. 41, (b) from Ref. 44.]

than x_0, whereas a 2-D SCS of area A_0 can be a singlet only if the center of its nearest neighbor is excluded from a circle of area $4A_0$ centered about the SCS. Thus only two units of x_0 space must be sacrificed in 1-D separations, whereas four units of A_0 space must be sacrificed in 2-D separations.

Probability of Total Resolution in 2-D and 3-D Separations

In 1991, Martin [42] showed that the probabilities $P^{(1)} \equiv {_2}P_1$ and ${_3}P_1$ of forming singlets in 2-D and 3-D separation spaces containing m randomly distributed components are

$$ {_2}P_1 = \exp\left(-\pi \frac{m}{{_2}n_c}\right) \qquad {_3}P_1 = \exp\left(-\frac{4\pi}{3} \frac{m}{{_3}n_c}\right) \tag{64} $$

where ${_2}n_c$ and ${_3}n_c$ are the spot capacities of the 2-D and 3-D spaces, respectively. In these spaces, components are represented by circles (2-D) or spheres (3-D) of diameter d_*, and the capacities ${_2}n_c$ and ${_3}n_c$ are the products of the 1-D capacities of each dimension. Thus $_2n_c = X_1X_2/(d_*)^2 = A/(d_*)^2$ and $_3n_c = X_1X_2X_3/(d_*)^3 = V_3/(d_*)^3$, where X_k is the span of the kth dimension of separation and V_3 is the volume of the 3-D space. The first expression in Eq. (64) is identical to $s/\overline{m}$ in Eq. (62) for $\xi = 1$.

Equation (64) can be solved for the number m of components resolvable with probability $P^{(1)}$:

$$m = -\frac{n_c}{2\phi} \ln(P^{(1)}) \tag{65}$$

where $\phi = \pi/2$ for 2-D separations and $2\pi/3$ for 3-D separations. For well-designed separations, where $_2n_c$ may approach 10,000 and $_3n_c$ may approach 200,000, the numbers of components expected to be resolved with 50% probability in 2-D and 3-D spaces are 2200 and 33,000, respectively.

The probability of resolving all components is far less likely than resolving any one component. Martin argued that the probabilities $P^{(m)} \equiv {}_2P_m$ and $_3P_m$ of resolving all m components in 2-D and 3-D spaces are

$$_2P_m = \exp\left(-\frac{\pi m^2}{2 \, _2n_c}\right) \qquad _3P_m = \exp\left(-\frac{2\pi m^2}{3 \, _3n_c}\right) \tag{66}$$

when m, $_2n_c$, and $_3n_c$ are large and when $m/_2n_c$ and $m/_3n_c$ are small. The capacity required to resolve all m components with probability $P^{(m)}$ can be determined from these equations:

$$n_c = -\frac{\phi m^2}{\ln(P^{(m)})} \tag{67}$$

where ϕ is as defined immediately above.

Figure 22 is a graph of $_2P_m$ versus log m for the two capacities, $_2n_c = 10^4$ and 10^6. It is apparent that $_2P_m$ drops off rapidly with increasing m. A 50% or better probability of total resolution is obtained only if $m < 66$ for $_2n_c = 10^4$ and if $m < 664$ for $_2n_c = 10^6$.

Application of Theory of S. A. Roach to 2-D Separations

In 1992, Oros and Davis [44] critiqued a theory published in 1968 by Roach [45], which addressed the fusion of randomly distributed circles of constant diameter in a 2-D plane. The theory had been proposed as a model for the overlap of coal particulates on thermal precipitators. Also reported by Oros and Davis and summarized by Roach are equations derived in the late 1940s to address other problems of 2-D overlap (e.g., bacteria in culture dishes). These equations will not be discussed here because they are inaccurate except at very low saturations, as shown by upcoming graphs. One can consult the sources for further details [46–48].

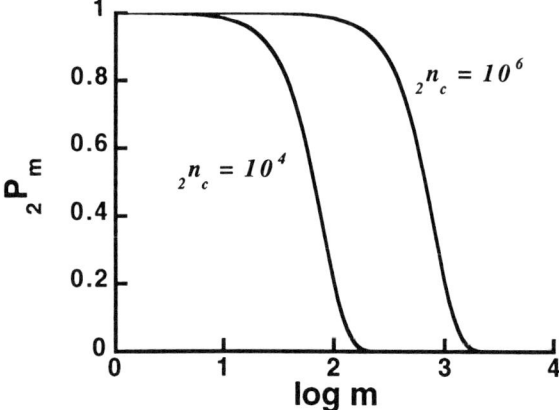

Fig. 22 Graph of probability $_2P_m$ of resolving m components in a 2-D separation versus log m for spot capacities $_2n_c = 10^4$ and 10^6. (From Ref. 43.)

In his seminal work, Roach argued that the expected number P_k of 2-D spots containing k SCSs was

$$P_k = \frac{\overline{m}e^{-4\alpha}(1 - e^{-4\alpha})^{k-1}}{k} \tag{68}$$

where α is defined by Eq. (63). The expected total number p of spots is the sum of all P_k values:

$$p = \sum_{k=1}^{\infty} P_k = \overline{m}\frac{4\alpha e^{-4\alpha}}{1 - e^{-4\alpha}} \tag{69}$$

Figure 21b is a graph of p/n_c versus α calculated from Eq. (69), with $\overline{m}/n_c$ expressed by α. The trends observed in Fig. 21a also are observed here; at any α, n_c is utilized less effectively in resolving 2-D spots than in resolving 1-D peaks.

Oros and Davis proposed a procedure analogous to the single-chromatogram method for estimating the number $\overline{m}$ of SCSs in a 2-D separation. In this procedure, a series of effective spot capacities n_c is defined by arbitrary values of ξd_*. Each n_c so defined determines a number p of spot groups. The coordinates, $(\xi d_*, p)$, can be fit to Eq. (69) by nonlinear least squares, with $\overline{m}$ as an unknown. As with its 1-D analog, the coordinates for small values of ξd_* must be excluded, because p is constant.

Theory for Overlap in n-*Dimensional Separations*

In 1993, Davis [49] extended the theory of Roach to describe overlap in separation spaces having n orthogonal dimensions ($n \geq 2$), in which n-dimensional spheres of diameter d_* were randomly distributed. The expected number P_k of n-dimensional spots containing k SCSs is [cf. Eq. (68)]

$$P_k = \frac{\overline{m} e^{-2^n \alpha}(1 - e^{-2^n \alpha})^{k-1}}{k} \tag{70}$$

and the expected number p of n-dimensional spots is [cf. Eq. (69)]

$$p = \sum_{k=1}^{\infty} P_k = \overline{m} 2^n \alpha \frac{e^{-2^n \alpha}}{1 - e^{-2^n \alpha}} \tag{71}$$

where $\overline{m}$ is a statistical approximation to the number m of SCSs and the saturation $\alpha = \overline{m}/n_c$ is

$$\alpha = \frac{\overline{m} \pi^{n/2} (\xi d_*)^n}{2^n V_n \Gamma(n/2 + 1)} \tag{72}$$

In Eq. (72), V_n is the volume of the n-dimensional space, Γ the gamma function, and ξ the scalar introduced by Davis [41].

If the n-dimensional space is an orthotope, then V_n equals the product of the n intervals X. Because the 1-D peak capacity n_{c_k} of the kth dimension is $X_k/\xi d_*$, Eq. (72) determines the n-dimensional spot capacity n_c as

$$n_c = \frac{2^n}{\pi^{n/2}} \Gamma\left(\frac{n}{2} + 1\right) \prod_{k=1}^{n} n_{c_k} \tag{73}$$

For $n = 3$ and $k = 1$, Eq. (70) is equivalent to the second expression of Martin in Eq. (64), if the distinction between n_c, Eq. (73), and $_3 n_c$ is taken into account and if the probability of singlet formation is interpreted as $P_1/\overline{m} = s/\overline{m}$.

Figure 23a is a graph of the fraction $p/\overline{m}$ of spots expected in 1-D, 2-D, 3-D, and 4-D separations, when the 1-D peak capacity of each dimension is 100, versus log $\overline{m}$. The phenomenal resolving power of high-dimensional spaces is apparent; in particular, $p/\overline{m}$ is 0.97 for $\overline{m} = 10^6$ and $n = 4$. Figure 23b is the equivalent graph when the 1-D capacity of subsequent dimensions is decreased by the factor 2 (i.e., $n_{c_2} = 50$, $n_{c_3} = 25$, and $n_{c_4} = 12.5$). This kind of behavior might be anticipated in a comprehensive separation, in which increasingly high linear velocities are required in subsequent dimensions. The loss of resolution is apparent, but resolution is still exceptional.

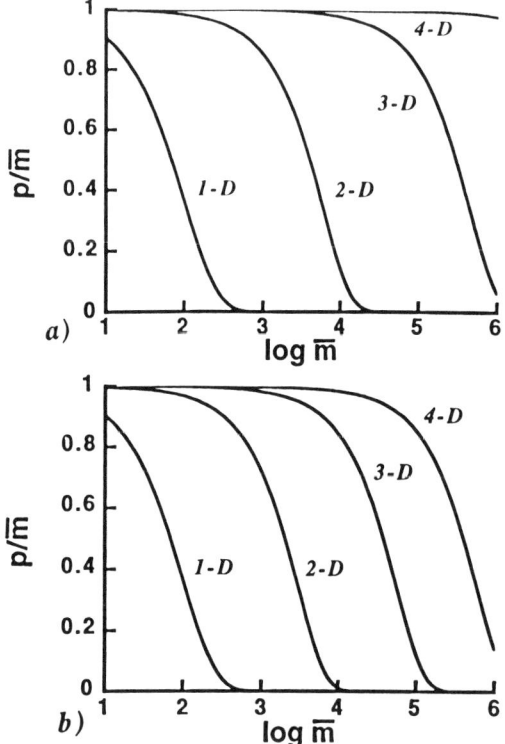

Fig. 23 Fraction $p/\bar{m}$ of components resolved in 1-D, 2-D, 3-D, and 4-D separations versus $\log \bar{m}$: (a) $n_{c_1} = n_{c_2} = n_{c_3} = n_{c_4} = 100$; (b) $n_{c_1} = 100$; $n_{c_2} = 50$; $n_{c_3} = 25$; $n_{c_4} = 12.5$. (From Ref. 49.)

B. Testing of 2-D and *n*-D Overlap Theories by Computer Simulation

Computer simulations of randomly distributed circles in a plane showed that the expression for s derived by Davis [41] in 1991 was acceptable over a wide saturation range but that expressions for d and tr were in error by more than 10% for α values greater than 0.20 or so. This discrepancy between simulation and theory indicated a basic problem with the theory.

In 1992, Oros and Davis [44] demonstrated the superiority of Roach's theory. Figure 24a and b are graphs of the expected number of doublets d and triplets tr predicted by Eq. (68) for $k = 2$ and 3. Also graphed are the predictions of Davis, Eq. (62), and some other predictions from the late 1940s [47,48]. The symbols represent the average number of spots, as defined by intersecting circles, computed by Davis [41] from 500 2-D simulations containing $m = 1000$ SCSs. It is evident that Roach's theory most accurately describes these results. Oros and Davis verified that Roach's theory is accurate for α values up to at least 0.65. The authors also verified the procedure for estimating $\overline{m}$ with very simple simulations, in which the coordinates of a spot were represented by the average coordinates of the circles comprising it.

The representation of SCSs in 2-D separations by circles is an extreme simplification, which was corrected in 1993 by Shi and Davis [50]. The authors assigned either constant, random, or exponential intensities to 100 or 200 double Gaussian SCSs having circular or elliptical contours in a series of computer simulations of 2-D separations. By forcing the numbers of maxima so determined

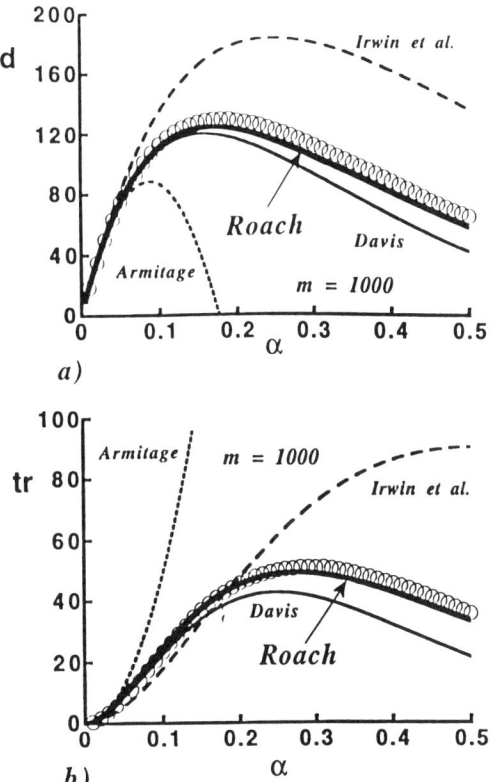

Fig. 24 Numbers of (a) doublets d and (b) triplets tr expected in 2-D separation containing $m = 1000$ SCSs versus α. Curves correspond to various theories; symbols represent averages of 500 computer simulations. (From Ref. 44.)

to agree with Roach's equation, the authors determined the 2-D spot capacity n_c for maxima over a wide range of saturation. The authors then defined the resolution R_s^* of two circular or elliptical double Gaussians and showed that

$$R_s^* = \frac{3\xi}{2} \tag{74}$$

Because n_c depends on ξ [see Eq. (63)], the authors were able to show that $R_s^* \approx 0.63$, as long as α is less than 0.13 or so ($R_s^* \approx 0.63$). This R_s^* is a bit larger than that for 1-D separations and also is physically more realistic.

The procedure proposed by Oros and Davis [44] to estimate $\overline{m}$ was verified by its application to these simulations, in which spot coordinates were identified with the coordinates of maxima. The errors in $\overline{m}$ values determined by this procedure were less than 10% or so for α values below 0.13 ($R_s^* \approx 0.63$). A second procedure for determining $\overline{m}$ was introduced and tested but was judged to be inferior to the original procedure.

In 1993, Davis [49] tested the extended theory of Roach by simple computer simulations with zero intensities. Figure 25 are graphs of the expected numbers of spots p, singlets s, doublets d, and triplets tr, divided by $\overline{m}$, expected in a three-dimensional (3-D) separation for various numbers $\overline{m}$. The solid curves represent theory, the symbols represent the average of 100 computer simulations, and the vertical error bars represent one standard deviation. The theory is followed modestly well, especially for large $\overline{m}$. Similar results were obtained for a 4-D separation, except that the agreement between simulation and theory was not as good. The discrepancy was attributed to edge effects, which cause n-dimensional spaces to be anisotropic near their boundaries.

C. Application of *n*-D Overlap Theory to 2-D Experimental Data

Table 7 is a comparison of N_k values predicted by Connors [40] in 1974 for four 2-D TLC systems. The predictions were made by identifying 4σ with 0.1 and dividing a square plane into $M = 100$ squares, graphing the components' R_f coordinates, and counting the coordinates' distributions among the squares. The R_f values were taken from the literature. The close agreement is suggestive that components in these 2-D TLCs were randomly distributed.

D. Number *n* of Dimensions Necessary to Resolve *m* Components

Martin et al. [7] and Martin [42] have demonstrated the improbability of resolving all mixture components by a single analysis. However, if enough orthogonal separations are made, each component can be isolated in at least one separation. Two theories for the number of separations required are reviewed here.

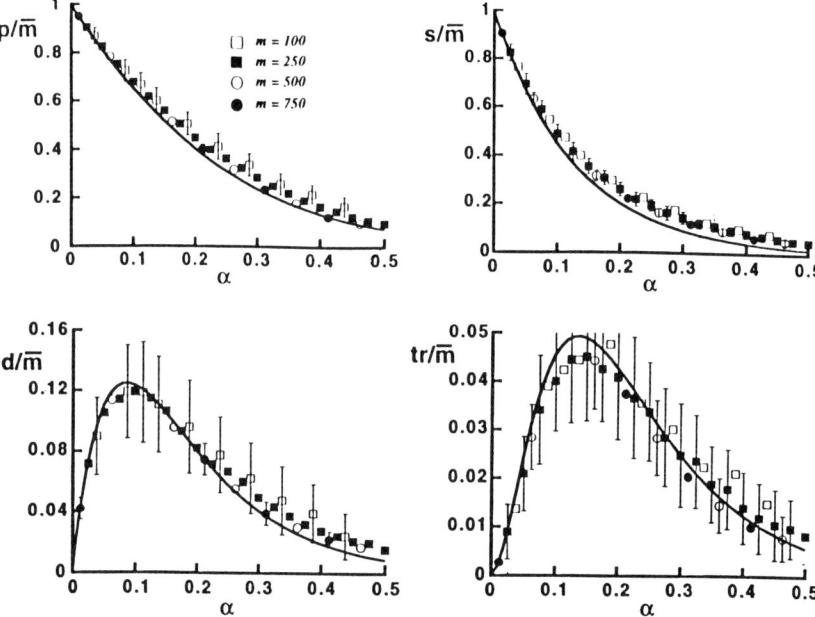

Fig. 25 Fractions of spots $p/\bar{m}$, singlets $s/\bar{m}$, doublets $d/\bar{m}$, and triplets $tr/\bar{m}$ expected in a 3-D separation for various numbers $\bar{m}$ of SCSs. Error bars represent one standard deviation. (From Ref. 49.)

Table 7 Comparison of Predicted Number N_k of Cells Containing k Components to Experimental Results for Four 2-D TLC Systems

	System 1 ($m = 26$)		System 2 ($m = 37$)		System 3 ($m = 16$)		System 4 ($m = 22$)	
k	N_k	Exp'l.	N_k	Exp'l.	N_k	Exp'l.	N_k	Exp'l.
0	77.11	81	69.07	71	85.21	87	80.25	84
1	20.05	13	25.56	23	13.63	10	17.66	13
2	2.61	5	4.73	4	1.09	3	1.94	1
3	0.23	1	0.58	2	0.06	0	0.14	2
4	0.01	0	0.05	0	0.00	0	0.00	0

Source: Ref. 40.

The first is by Connors [40], who in 1974 developed a theory similar to that of Klein and Tyler [1]. One envisions here an n-dimensional space, which is not of the sequential type discussed above but is simply comprised of n orthogonal 1-D separations, such as TLCs, of a mixture. Each of m components is represented by an n-dimensional coordinate comprised of its n R_f values (or retention times), and the space is divided into $M = (4\sigma)^{-n}$ cells, where σ is the standard deviation. If the R_f values of the components are distributed randomly, then the probability, $1 - y$, that a cell contains either no or one coordinate is

$$1 - y = e^{-\chi}(1 + \chi) \tag{75}$$

where $\chi = m/M = m(4\sigma)^n$. If one accepts that, on average, only one of the cells can contain more than one component, then $My = 1$ and Eq. (75) rearranges to

$$\frac{m - \chi}{m(1 + \chi)} = e^{-\chi} \tag{76}$$

Predictions based on this equation show that roughly four orthogonal separations are required to resolve about 100 compounds, when $4\sigma \approx 0.1$ (a component resolved in one separation need not be resolved in subsequent ones). Connors observed that Sunshine et al. [51] were able to resolve 113 out of 138 components by TLC with four solvent systems.

The second theory is by Martin et al. [7], who in 1986 observed that for n independent 1-D separations, the probability $Pb(n)$ that a mixture is resolved completely by at least one separation is the complement of the probability that resolution is incomplete for all n separations:

$$Pb(n) = 1 - \prod_{i=1}^{n} [1 - P_{m,n_c}(m)] \tag{77}$$

where $P_{m,n_c}(m)$ is expressed by Eq. (29). If the capacity n_c is identical in all dimensions, then

$$Pb(n) = 1 - \left[1 - \left(1 - \frac{m-1}{n_c - 1}\right)^{m-2}\right]^n \tag{78}$$

Equation (29) predicts that the probability of completely resolving a 10-component mixture by a single separation is only 20%, when $n_c = 50$. In contrast, if the mixture instead is subject to 10 independent separations ($n = 10$) for which $n_c = 50$ in each, the probability of complete resolution by at least one separation increases to 89%.

IV. CONCLUSIONS

The reviewer concludes with some personal observations, evaluations, and speculations. Perhaps the most important observation to make is that a remarkable consistency exists in this body of work. The predictions of various theories agree very well with one another, at least under the appropriate conditions for comparison.

Of the 1-D theories of overlap presently in use, the Fourier method of Felinger et al., probably is the most powerful. It is the most sophisticated theory to date, both mathematically and conceptually, and is capable of providing diagnostic information about low-saturation separations as well as many statistical parameters. In addition, the restrictions on its use are less severe than on other theories. The reviewer does reserve some judgment as to its application to high-saturation separations, however, since simulated chromatograms containing SCPs with exponential intensities (the most common type) have been interpreted for saturations only up to 0.667 ($R_s^* = 0.5$).

The Poisson-type theories (e.g., the works of Davis, Giddings, and Martin) are still very useful at low saturation, especially if theoretical simplicity and ease of analysis are considered. The Poisson theory of Davis and Giddings is still the only theory that provides estimates of multiplet numbers (e.g., doublets and triplets). Its major weaknesses are its limited range of saturation and its inapplicability to chromatograms with variable SCP densities. One is hopeful that the saturogram work of Martin will reduce the former limitation and that the extended Poisson model of Davis will reduce the latter.

The Poisson-type theory of Martin et al. is more detailed than that of Davis and Giddings and provides some unique insights but surprisingly has not been exploited experimentally. Certainly, it has a capability of describing overlap in simple chromatograms that the work of Davis and Giddings does not.

Future efforts may reflect a fusion of the predictions of the Fourier and Poisson models. For example, it is possible to adapt the theory developed by Davis and Giddings for the calculation of p, s, d, and so on, to any uncorrelated probability density between SCPs, including those now addressed only by the Fourier model.

The analogy between overlap and polymerization is highly insightful but has not been used, to the reviewer's knowledge, except in the paper that introduced it. Perhaps this lack of use can be attributed to insufficient characterization by simulation. A detailed examination could lead to fruitful results.

The R_s^* values suggested by Creten, Nagels, El Fallah, Martin, and Schure for the Poisson model are far more realistic than the empirically determined value, $R_s^* = 0.5$. For reasons not yet clear, the introduction of SCP amplitudes forces one to choose a less justifiable R_s^* to obtain agreement between theory and experiment or simulation.

The practical relevance of overlap theories for n-dimensional separations (with $n > 1$) awaits the test of time. The Poisson models may not prove very useful for spatial dimensions larger than two, because the likelihood of obtaining randomly distributed components in all dimensions of space is small. The reviewer hopes that the extended Poisson model can be applied to higher dimensions; preliminary efforts indicate that it can be.

A domain of multidimensional separations that has not attracted much attention is 2-D heartcutting. This is a common form of chromatography that should be considered as a fruitful area for future study. In addition, the possibility of extending n-dimensional theories of overlap to the interpretation of multivariate data other than overlapping mixture components should not be overlooked.

ACKNOWLEDGMENT

This work was supported by the National Science Foundation (CHE-9215908).

GLOSSARY OF SYMBOLS

A_0	effective area of 2-D SCS
A_T	total area of chromatogram
ACVF	autocovariance function
AL	ratio of area of largest SCP in observed peak to observed-peak area
a_a	mean SCP area
a_h	mean SCP height
a_i	area of ith peak
B	number of returns of signal to baseline
B_j	scaling factor for jth SCP
$C(t)$	autocovariance function
CIRM	constrained iterative relaxation method
c	number of cells
DL_e^ω	minimum relative abundance necessary to determination with probability ω and relative error less than e
DLS_e^ω	DL_e^ω expressed as relative mass for a randomly chosen mixture
d	number or expected number of doublets
d_0	discrimination factor
d_*	diameter of n-dimensional sphere
E	exponential density
E	averaging operator
$E\{|g(\omega,\sigma)|^2\}$	function defined in Table 2
$|E\{g(\omega,\sigma)\}|^2$	function defined in Table 2

$F(t)$	amplitude of chromatogram at time t
$F(\omega)$	power spectrum
$F_a(\zeta)$	cumulative distribution of maxima
$f(x)$	dimensionless variable Poisson frequency
$f(\alpha')$	normalized expression for variance of p' due to fluctuation of intervals between SCPs
$f(\zeta)$	dimensionless variable Poisson frequency in reduced coordinates
$f_a(\zeta)$	approximation to $f(\zeta)$
G	gamma density
G	broadening operator for Gaussian SCP
$g(\omega)$	Fourier transform of mathematical function for SCPs
$g(\omega,\sigma)$	Fourier transform of Gaussian SCP with variable standard deviation
h_i	scalar for ith event $u(t)$
k	generic number
$k(d_0)$	number of adjacent maxima pairs for which d_0 equals or exceeds some value
k'	capacity factor
k'_f	K' of least retained component of set
k'_l	K' of most retained component of set
M	number of cells in n-dimensional chromatogram
m	number of components
$\overline{m}$	expected number of components
m_{ave}	estimate of $\overline{m}$ calculated from m_{sl} and m_{in}
m_i	shift constant for ith event $u(t)$
m_{in}	estimate of $\overline{m}$ determined from intercept of Eq. (11) or (20)
m_{sl}	estimate of $\overline{m}$ determined from slope of Eq. (11) or (20)
N	normal density
N	number of theoretical plates
$N(q)$	number of observed peaks of reduced area q
N_k	average number of cells in n-dimensional TLC containing k components
$N_k(q)$	number of observed peaks of reduced area q, for which AL exceeds k
Nl	subset of m components
n	number of dimensions
n_c	peak or spot capacity
$_2n_c$	spot capacity for 2-D separations
$_3n_c$	spot capacity for 3-D separations
$n_{c_{\text{eff}}}$	peak capacity that maintains constant d_0
n_{c_k}	1-D peak capacity of kth dimension of n-dimensional separation

$n_{c_{min}}$	minimum capacity that resolves m components with given probability
n-D	n-dimensional
P	probability defined by Eq. (3)
$P(x)$	probability density defined by Eq. (5)
P_k	expected number of peaks or spots containing k components
$P_{m,n_c}(m)$	probability that m components are resolved
$P_{m,n_c}(p)$	probability that m components produce p peaks
$P_{p,n_c}(m)$	probability that p peaks are produced by m components
$_2P_1$	probability of singlet formation in 2-D space
$_3P_1$	probability of singlet formation in 3-D space
$_2P_m$	probability of resolving m spots in 2-D space
$_3P_m$	probability of resolving m spots in 3-D space
$P^{(1)}$	generic probability of resolving a singlet
$P^{(m)}$	generic probability of resolving m spots in 2-D and 3-D separations
$Pb(n)$	probability that mixture is resolved by at least one of n separations
PC_e	probability that an SCP is determined with a relative error smaller than e
$PL_k(q)$	probability that fractional area of most abundant SCP in observed peak equals or exceeds k
PR	polarity range
p	number or expected number of peaks
p_d	number of maxima after application of CIRM
p_k	probability that a peak contains k components; mole fraction of k-mer
p_m	number of maxima
$p(k)$	probability that k components occupy a cell
$p(k, \delta M)$	probability that k components occupy interval δM
p'	number of intervals between maxima exceeding x'_0
$\overline{p}$	expected number of peaks
q	classification of reduced area
R_f	retardation factor
R_s^*	resolution factor
$R_{s,av}$	resolution factor determined by $n_{c_{eff}}$
$R_{s,av}^\infty$	extrapolated to an infinite number m of components
R_s^d	super-resolution factor
r	σ_2/σ_1
r_i	relative area of ith maxima
$r\{W^{(k)}\}$	relaxation function for the kth iteration
SCP	single-component peak

SCS	single-component spot
SS	sum of squares
s	number or expected number of singlets
s_n	standard deviation of normally distributed σ's
T	mean interval between adjacent SCPs
t	time
t_i	discretized time t
t'_j	retention time of jth SCP
tr	number or expected number of triplets
U	uniform density
$u(t)$	mathematical function for SCP
V_g	specific retention volume
V_n	volume of n-dimensional space
V_r	retention volume
W	function of finite width that replaces δ in numerical deconvolution
w	weighting factor
X	extent of 1-D interval containing SCPs
x	coordinate of 1-D separation
x_0	minimum interval that resolves adjacent SCPs into peaks
x'_0	arbitrarily chosen number that defines effective peak capacity
$x'_{0\zeta}$	x'_0 in reduced coordinates
$\mathbf{Y}(t)$	linear combination of individual stochastic processes
$Y(t)$	observed-peak height at time t
$\hat{Y}$	mean height of observed peaks
y	probability that two or more compounds are not found in interval
$Z_T^{(k)}(\omega)$	Fourier transform of kth replica of $\mathbf{Y}(t)$
z	number of unoccupied cells

Greek Symbols

α	saturation
α'	saturation based on effective peak capacity
$\overline{\alpha}$	average saturation
β	De Donder extent of reaction
Γ	gamma function
γ	extent of separation
$\Delta\mu^\circ$	difference in free energy between mobile and stationary phases
$\delta(\omega);\ \delta(t-t')$	delta function
δM	interval of chromatogram
ζ	reduced separation coordinate
$\theta(\omega)$	function determined by probability density for SCPs

ϑ	probability that SCP does not fall in interval x_0
κ	dummy variable
λ	SCP density
λ_P	constant Poisson density
μ	polydispersity index for condensation polymerization
ξ	scaling factor in n-D separations
ρ	saturation parameter of Klein and Tyler
σ	SCP standard deviation
σ_a	standard deviation of SCP areas
σ_{av}	average standard deviation for normally or uniformly distributed widths
σ_h	standard deviation of SCP heights
σ_j	standard deviation of jth SCP
σ_m^2	variance in m
σ_p^2	variance in p
$\sigma_{p'}^2$	variance in p'
σ_1	smallest uniformly distributed σ
σ_2	largest uniformly distributed σ
τ	time constant of exponentially convoluted Gaussian
ϕ	coefficient in probability expressions for 2-D and 3-D separations
χ	density of n-dimensional TLC, as referenced to cell number
ω	frequency

REFERENCES

1. P. D. Klein and S. A. Tyler, *Anal. Chem., 37:* 1280 (1965).
2. D. Rosenthal, *Anal. Chem., 54:* 63 (1982).
3. L. J. Nagels, W. L. Creten, and P. M. Vanpeperstraete, *Anal. Chem., 55:* 216 (1983).
4. J. M. Davis and J. C. Giddings, *Anal. Chem., 55:* 418 (1983).
5. M. Martin and G. Guiochon, *Anal. Chem., 57:* 289 (1985).
6. J. M. Davis and J. C. Giddings, *Anal. Chem., 57:* 2168 (1985).
7. M. Martin, D. P. Herman, and G. Guiochon, *Anal. Chem., 58:* 2200 (1986).
8. M. Martin, D. P. Herman, and G. Guiochon, *Anal. Chem., 59:* 384 (1987).
9. D. P. Herman, H. A. H. Billiet, and L. deGalan, *Anal. Chem., 58:* 2999 (1986).
10. P. J. Schoenmakers, H. A. H. Billiet, and L. deGalan, *J. Chromatogr., 218:* 261 (1981).
11. L. de Galan, D. P. Herman, and H. A. H. Billiet, *Chromatographia, 24:* 108 (1987).
12. A. Felinger, L. Pasti, and F. Dondi, *Anal. Chem., 62:* 1846 (1990).
13. A. Felinger, L. Pasti, and F. Dondi, *Anal. Chem., 63:* 2627 (1991).

14. J. M. Davis, *Anal. Chem.*, in press.
15. D. P. Herman, M.-F. Gonnord, and G. Guiochon, *Anal. Chem., 56:* 995 (1984).
16. J. C. Giddings, J. M. Davis, and M. R. Schure, *Ultrahigh Resolution Chromatography*, ACS Symposium Series 250, Satinder Ahuja, Ed., American Chemical Society, Washington, D.C., 1984.
17. J. M. Davis and J. C. Giddings, *J. Chromatogr., 289:* 277 (1984).
18. J. M. Davis, Ph.D. dissertation, University of Utah, 1985.
19. F. Dondi, Y. D. Kahie, G. Lodi, M. Remelli, P. Reschiglian, and C. Bighi, *Anal. Chim. Acta, 191:* 261 (1986).
20. A. Felinger, L. Pasti, P. Reschiglian, and F. Dondi, *Anal. Chem., 62:* 1854 (1990).
21. A. Felinger, L. Pasti, and F. Dondi, *Anal. Chem., 64:* 2164 (1992).
22. J. M. Davis and J. C. Giddings, *Anal. Chem., 57:* 2178 (1985).
23. S. Coppi, A. Betti, and F. Dondi, *Anal. Chim. Acta, 212:* 165 (1988).
24. J. M. Davis, *J. Chromatogr., 449:* 41 (1988).
25. M. L. Lee, D. L. Vassilaros, C. M. White, and M. Novotny, *Anal. Chem., 51:* 768 (1979).
26. S. L. Delinger and J. M. Davis, *Anal. Chem., 62:* 436 (1990).
27. F. Dondi, T. Gianferrara, P. Reschiglian, M. C. Pietrogrande, C. Ebert, and P. Linda, *Anal. Chim. Acta, 485:* 631 (1990).
28. F. J. Oros and J. M. Davis, *J. Chromatogr., 550:* 135 (1991).
29. F. Dondi, A. Betti, L. Pasti, M. C. Pietrogrande, and A. Felinger, *Anal. Chem., 65:* 2209 (1993).
30. W. L. Creten and L. J. Nagels, *Anal. Chem., 59:* 822 (1987).
31. M. Z. El Fallah and M. Martin, *Chromatographia, 24:* 115 (1987).
32. M. Z. El Fallah and M. Martin, *Analusis, 16:* 241 (1988).
33. M. Schure, *J. Chromatogr., 550:* 51 (1991).
34. L. J. Nagels and W. L. Creten, *Anal. Chem., 57:* 2706 (1985).
35. M. Martin and G. Guiochon, *9th International Symposium on Column Liquid Chromatography*, Edinburgh, July 1–5, 1985.
36. L. J. Nagels and W. L. Creten, *Anal. Chim. Acta, 169:* 299 (1985).
37. L. J. Nagels, W. L. Creten, and L. van Haverbeke, *Anal. Chim. Acta, 173:* 185 (1985).
38. L. J. Nagels, W. L. Creten, and F. Parmentier, *Int. J. Environ. Anal. Chem., 25:* 173 (1986).
39. M. Z. El Fallah and M. Martin, *J. Chromatogr., 557:* 23 (1991).
40. K. A. Connors, *Anal. Chem., 46:* 53 (1974).
41. J. M. Davis, *Anal. Chem., 63:* 2141 (1991).
42. M. Martin, *Proceedings of the Congrès Mesucora '91*, Vol. 1, 1991, p. 3.
43. M. Martin, *Spectra 2000, 169*(Suppl.): 5 (1992).
44. F. J. Oros and J. M. Davis, *J. Chromatogr., 591:* 1 (1992).

45. S. A. Roach, *The Theory of Random Clumping,* Methuen, London, 1968.
46. R. B. Bourdillon, O. M. Lidwell, and J. E. Lovelock, *Studies in Air Hygiene,* Medical Research Council Special Report Series No. 262, H.M. Stationery Office, London, 1948.
47. J. O. Irwin, P. Armitage, and C. N. Davies, *Nature (London), 61:* 809 (1949).
48. P. Armitage, *Biometrika, 36:* 257 (1949).
49. J. M. Davis, *Anal. Chem., 65:* 2014 (1993).
50. W. Shi and J. M. Davis, *Anal. Chem., 65:* 482 (1993).
51. I. Sunshine, W. W. Fike, and H. Landesman, *J. Forensic Sci., 11:* 428 (1966).
52. D. Bowlin and J. M. Davis, *J. Chromatogr.,* submitted.

4
Capillary Electrophoresis of Carbohydrates

Ziad El Rassi *Oklahoma State University, Stillwater, Oklahoma*

I.	INTRODUCTION	177
II.	ELECTROPHORETIC SYSTEM	179
	A. Electrolyte Systems	179
	B. Capillary Column	190
	C. Detection Systems	197
III.	SEPARATION METHODOLOGIES AND APPLICATIONS	207
	A. Monosaccharides	207
	B. Polysaccharides	214
	C. Glycoproteins	224
	D. Glycosaminoglycans	238
	E. Glycolipids	242

I. INTRODUCTION

Since its early stages of development, electrophoresis has played important roles in the areas of carbohydrate chemistry and biochemistry. Paper electrophoresis was first employed for the analysis of carbohydrate samples [1–5], and later polyacrylamide slab gel electrophoresis was preferentially used in the separation of oligosaccharides [6,7] and in studies concerning glycoproteins [8–10] and proteoglycans [11–13]. With the advent of high-performance capillary electrophoresis (HPCE), we are now witnessing increasing applications of HPCE to the separation of complex carbohydrate samples. Certainly, HPCE represents an alternative to planar electrophoresis (i.e., paper or slab gel electrophoresis) in the analysis of glycoconjugate mixtures because of its high separation efficiencies, high speed, and precise instrumentation.

The intrinsic high resolving power of electrophoresis is particularly suitable for the separation of carbohydrates, which encompass a wide spectrum of compounds, many of which are isomers or slightly different from each other. The monosaccharides are divided into several classes, comprising the aldoses, ketoses, alditols, aldonic acids, uronic acids, and so on, each of which is further divided into several subclasses on the basis of the number of carbon atoms; and within each subclass a number of isomers exists, due to configurational differences. Further diversity in the monosaccharides results from naturally occurring modifications, such as deoxy, amnio, and acetylated derivatives. Oligo- and polysaccharides are composed of various combinations of monosaccharides forming linear or branched polymeric species, many of which may differ only in the position of attachment and anomeric configuration of the glycosidic linkages. Most oligosaccharides occur as side chains attached to polypeptides in glycoproteins and proteoglycans or to lipids in glycolipids. While the polypeptide chains of glycoproteins are produced under strict genetic control, their carbohydrate chains, in contrast, are enzymatically generated. Such processing enzymes are generally not available in sufficient quantities to yield uniform products, and consequently, any particular glycoprotein may have variable carbohydrate compositions at any of its glycosylation sites, a phenomenon known as microheterogeneity.

In electrophoresis, only ionic or ionizable species can differentially migrate under the influence of an applied electric field. With the exception of few naturally charged mono- and oligosaccharides, most carbohydrate molecules lack readily ionizable functions. Although neutral solutes can be separated by capillary electrophoresis by means of micellar electrokinetic capillary chromatography (MECC) [14], the MECC technique is largely inadequate for the separation of carbohydrates because of their hydrophilic nature. Unless they are conjugated to a relatively large hydrophobic moiety, carbohydrates show little or no partitioning in the hydrophobic micellar phases used in MECC. Furthermore, most carbohydrate species neither absorb nor fluoresce, a property that hinders their sensitive detection by modern analytical separation techniques.

To realize the full benefit of the high resolving power of HPCE in the separation of carbohydrates, several approaches have been introduced to overcome the various difficulties arising from the lack of readily ionizable functions and chromophores in the carbohydrate molecules. Polyhydroxy compounds such as carbohydrates can readily be converted to charged species by complex formation with other ions, such as borate and metal cations. At the reducing end of saccharides, suitable ultraviolet (UV) absorbing or fluorescent tags can be attached to the analyte molecules, a process called precolumn derivatization. Since HPCE is simply the instrumental version of planar electrophoresis employing detection systems that has been adapted from high-performance liquid chromatography (HPLC) with some modifications, it is not surprising to see that over a period of less than 6 years HPCE of carbohydrates has already made significant progress. Electrolyte systems, which were originally tested in traditional electrophoresis, as well as precolumn derivatization schemes, which afforded sensitive detection of carbohydrates separated by HPLC, have readily been adapted to carbohydrate HPCE. The current success of HPCE in the analysis of carbohydrates has also been in part the result of advanced capillary column technology and sophisticated detection techniques.

The aims of this chapter are (1) to review approaches and concepts that are most useful in the separation of carbohydrates by HPCE, (2) to shed some light on the dynamic behavior of carbohydrate species in the electrophoretic process, and (3) to discuss important applications.

II. ELECTROPHORETIC SYSTEM

A. Electrolyte Systems

As pointed out in the introduction, only some saccharides possess charged functional groups in their structures, which would allow their electromigration in an electric field. These saccharides are the aldonic acids, uronic acids, sialic acids, amino sugars (glucosamine, galactosamine), and compositional sulfated sugars of chondroitin, dermatan, keratan, and heparin. The separation of "neutral" carbohydrates by electrophoresis has often required the in situ conversion of these polyhydroxy compounds into charged species via complex formation with other ions. In planar electrophoresis, the most commonly used are borate complexes [1,15], but other inorganic oxyacids form anionic complexes with neutral saccharides (e.g., sodium germanate [16], stannate [3], arsenite [17], molybdate [18], wolframate [3], and metavanadate [19]). Also, carbohydrates can form cationic complexes with lead acetate [20] or cations of alkaline earth metals [21]. Furthermore, because of the ionization of the hydroxyl groups of the sugars at extremely high pH, sodium hydroxide was also useful for the electrophoresis of carbohydrates [20]. Recently, HPCE systems based mostly on sugar–borate complexes and to a lesser extent on sugar–metal cation complexes and high-pH sodium hydroxide solutions have been utilized.

The following sections are brief reviews of basic aspects of borate and metal ion complexes with sugars and self-dissociation of saccharides at highly alkaline pH. Since they have not found use in HPCE of sugars, complexation with metal-containing anions such as molybdate, tungstate, and so on, will not be discussed here, and the interested reader is advised to consult Refs. 3 and 22 for reviews on this topic.

Carbohydrate–Borate Complexes

Anionic complexes between polyhydroxy compounds and borate were first perceived by Böeseken in 1949 [23]. In these complexation reactions, it is the tetrahydroxyborate ion, $B(OH)_4^-$, rather than boric acid that undergoes complexation with the polyols [24]. This is because boric acid in aqueous media acts as a Lewis acid to form the tetrahedral anion $B(OH)_4^-$, and at alkaline pH (i.e., pH 8 to 10), where the complexation is most effective, the following equilibrium is very much shifted to the right:

$$B(OH)_3 + OH^- \rightleftharpoons B(OH)_4^- \qquad (1)$$
$$(B) \qquad\qquad\qquad (B^-)$$

In general, the borate complex formation can be described by the following equilibria [25,26]:

$$B(OH)_4^- + L \rightleftharpoons [BL^-]_{monodiol} + 2H_2O \qquad (2)$$

$$B(OH)_4^- + 2L \rightleftharpoons [BL_2^-]_{bisdiol\ or\ spirane} + 4H_2O \qquad (3)$$

where BL^- and BL_2^- are the mono- and diesters, respectively, L is the polyol, and $n = 0$ or 1. Equilibrium (2) is situated very much to the right, whereas (3) is dependent on the position of the hydroxyl groups in the polyol.

According to the equilibria above, polyols with adjacent hydroxyl groups could form cyclic borate esters with either five or six atom rings when $n = 0$ or 1, respectively (i.e., for 1,2-diols or 1,3-diols, respectively). The monocomplex, BL^-, is formed by the expulsion of two molecules of water, and the dicomplex, BL_2^-, by removal of four molecules of water from two molecules of polyol and a borate ion. The moncomplexes are less commonly called bidentate monoesters, whereas the dicomplexes are less often referred to as tetradentate diesters. In a few cases and depending on the structure of the polyol, tridentate monocomplexes (*1*) have been reported [27].

The complex formation described by reactions (2) and (3) is possible only if two hydroxyl groups in the polyol molecule are favorably situated. That is, the distance between the two hydroxyl groups in the polyol molecule should be of the same magnitude as the O—O distance in the borate ion. The O—O distances for

Tridentate monocomplex
(*1*)

trigonal and tetrahedral boron are 2.36 to 2.39 Å, respectively [28]. It thus appears that the borate ion would form complexes with those polyhydroxy compounds in which the oxygen atoms of at least two hydroxyl groups are separated by, or can easily approach each other to, a distance of approximately 2.4 Å. Such a distance is found in the *cis*-1,2-diols of five-membered ring compounds (O—O, 2.49 Å), which react more strongly with borate than do their trans isomers (O—O, 3.40 Å) [20]. On the basis of the O—O distance and conformational constraints, it is obvious that cyclic forms of sugars react less strongly with borate ions than do the open chains (acyclic) ones. For instance, for mannitol, which is a linear polyhydroxy compound, it has been reported that mannitol–borate complex is 700 times more stable than glucose–borate complex [29]. Also, with aqueous D-glucose solutions that contain smaller quantities of planar furanose and linear D-aldehydo forms, it has been reported that these two species are capable of forming stronger borate complexation than the pyranose form [29]. Reaction (4) shows the three different forms of glucose.

$$\begin{array}{c} \text{D-Glucose (linear aldehydo form)} \rightleftharpoons \text{a-D-Glucopyranose} \\ \rightleftharpoons \text{a-D-Glucofuranose} \end{array} \qquad (4)$$

For acyclic polyol compounds, the stabilities of borate esters of 1,3-diols have been found to lie in the same range as those of borate esters of 1,2-diols [26]. Whereas in the latter esters, *threo*-1,2-diols form more stable complexes than *erythro*-1,2-diols, in the former esters, *syn*-1,3-diols show higher relative stability than *anti*-1,3-diols [25]. This is again due to the fact that the O—O distances in threo and syn configurations are more favorably situated than in the erythro and anti configurations. The stability of both types of esters is enhanced on increasing the number of hydroxyl groups (see Fig. 1), and for more than two hydroxyl groups the borate esters with 1,2-diols generally are more stable than those of the 1,3-diols [25]. Statistically, the number of possibilities for a borate anion to bind to an alditol as a 1,2-diol increases with the number of hydroxyl groups in the molecule. The increase in the strength of complexation with increasing the number of hydroxyl groups in the polyol molecule can be explained as follows. According to Böeseken [23], for a simple aliphatic, acyclic diol (e.g., glycol, 1,2-propanediol), the position of the hydroxyl groups is not favorable, due to mutual repulsion of these groups, an impulse that they can obey due to the fact that they can rotate freely around the connecting axis of the carbon atoms to which the hydroxyl groups are bound. On the other hand, in the case of compounds with more than two adjacent hydroxyl groups, it can be expected that with the mutual repulsion of these groups it will be no longer possible for two of the groups to be 180° apart, as may be the case with the simple glycols, because a third adjacent hydroxyl group prevents this by its repulsion. With an increase in this number of adjacent groups two hydroxyl groups will be more and more favorably situated with reference to one another.

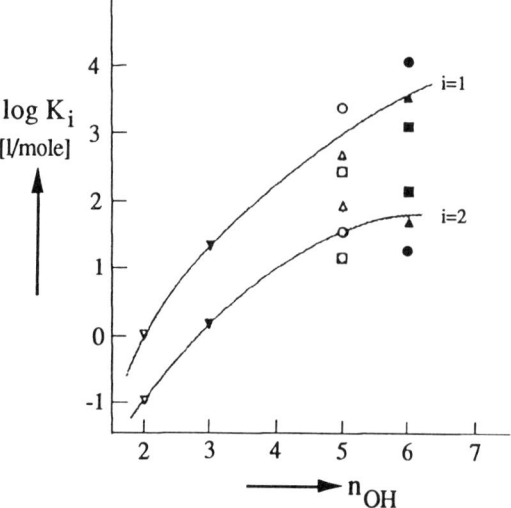

Fig. 1 log K_i ($i = 1$ for BL^-; $i = 2$ for BL_2^-) as a function of n_{OH} for borate complex formation of alditols, ▽, Glycol; ▼, glycerol; ○, xylitol; △, arabinitol; □, adonitol; ●, sorbitol; ▲, galactitol; ■, mannitol. (From Ref. 25.)

Furthermore, the complex formation is largely influenced by the presence of substituents in the polyol molecule as well as by their charges, locations, and anomeric linkages. Methylated sugars complex less than the parent unsubstituted sugars, and consequently, the electrophoretic mobilities of methylated sugars in zone electrophoresis are much lower [30]. The nature of the anomeric linkage of the substituent to the polyol molecule affects largely the complex formation. For instance, D-glucose with a methyl group at the C_1 position complexes with borate ions at the C_4 and C_6 hydroxyl groups of the glucopyranoside moiety, forming trans fused bicyclic systems [(2) and (3)] [1,31,32]. This complexation is much greater with the methyl-β- than the methyl-α-D-glucopyranoside, and consequently, the mobility of the α anomer is lower in zone electrophoresis with alkaline borate than that of the β anomer. This may be due to the fact that in the α anomer the glycosidic methoxyl group occupies an axial position and will interact strongly with the axial hydrogen atoms on C_3 and C_5, thus destabilizing the borate complex. This is not the case for the β anomer, where the substituent occupies an equatorial position, and consequently, is free from strong nonbonded interactions.

Methyl-α-D-glucopyranoside borate complex
(2)

Methyl-β-D-glucopyranoside borate complex
(3)

Furthermore, the location at which the substitution takes place may greatly reduce the complexation. In fact, while the mobility of 3-*O*-methyl-D-glucose in zone electrophoresis with alkaline borate decreased slightly with respect to that of the parent molecule, the mobility of 2-*O*-methyl- and 4-*O*-methyl-D-glucose became negligible. This suggest that the hydroxyl group at the C_2 and C_4 positions in the parent sugar are very important in complex formation with borate ions [1]. In fact, substituted 2,4-di-*O*-methyl-D-glucose showed little or no mobility with alkaline borate, while 3,4-di-*O*-methyl-D-glucose still exhibited a reasonable mobility in the same electrophoretic system. Another parameter that must be considered in corroborating mobilities in borate systems is the presence of charged substituents in the polyhydroxy molecule. Generally, a decrease of the stability of a borate ester is observed as a result of coulombic repulsion between a negatively charged substituent (e.g., COO^-) and BO_4^- moieties [25].

Mono- and dicomplex (or spirane) borate esters coexist in aqueous solutions [26,31], and their molar ratio is affected (1) by the relative concentration of borate ions and polyol molecules, (2) by the position of the hydroxyl groups in the polyol, and (3) by the presence of substituents. The six-membered ring dicomplexes or spiranes formed with methylglucopyranosides were reported to be hardly detectable by ^{11}B nuclear magnetic resonance (NMR) [32]. As discussed above, this finding may be attributed to the fact that the complex formation between borate and methylglucopyranoside is low. With free D-glucose, mono- and spirocyclic complexes coexist, the latter predominantly at high D-glucose/borate ratio [26] (see Fig. 2). Also, in the case of the polyols such as mannitol, mannitan, and fructose, which complex relatively easily with borate, spirane complexes with these sugars were detected when the excess borate was not too great [1,27]. The quantity of spiranes is directly connected, under otherwise identical conditions, with the position of two of the 1,2 (sometimes 1,3) hydroxyl groups. The less favorable the position of the hydroxyl groups, the smaller is the quantity of spiranes, and at large excess of borate/sugar ratio, the quantity of dicomplexes becomes negligible.

Under separation conditions normally used in electrophoresis, whereby 0.1 to 0.2 M borate are added to the running electrolyte and small plugs of 10^{-4} to 10^{-5} M sugar samples are introduced into the separation channel, anionic monocomplexes (i.e., BL^-) are likely to predominate and thus migrate differentially under the influence of an applied electric field. It should be noted that whether the injected sugar samples form BL^- or BL_2^- or both while migrating in a borate medium, all sugar molecules will be associated with a negative charge since mono- and dicomplex formations are dynamic. The magnitude of the charge will be influenced by the position of the equilibrium and therefore by the stability of the complex. According to equilibria 1, 2, and 3, at constant sugar concentration, the amount of complex increases with borate concentration according to the law of mass action and also with pH due to higher concentration of borate ions. According to a recent capillary zone electrophoresis (CZE) study by Honda et al. [33], at a given pH, resolution among various sugars increased with increasing

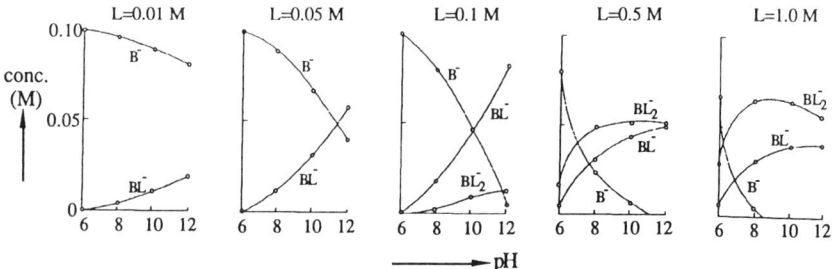

Fig. 2 D-Glucose (L) borate complex formation as a function of pH and [L]; 0.1 M boric acid, D$_2$O, 25°C. (From Ref. 26.)

borate concentration, and 0.2 M borate seems to be an adequate concentration while the system heating stays under control. Also, at constant borate concentration, Honda et al. [33] have noted that the complex formation as a function of pH varies among various carbohydrate species with an optimum in the pH range 10 to 11 for most monosaccharides, an observation that confirms earlier findings by Consden and Stanier in paper electrophoresis [34].

Another advantage of the use of alkaline borate buffers is that the carbohydrate–borate complexes show an enhanced absorption in the low-UV range which enables the detection of underivatized sugars in the nanogram quantities. This was demonstrated recently by Hoffstetter-Kuhn et al. [35] in the CZE of carbohydrates. It was also suggested that because an aqueous solution of a given sugar can coexist in different configurations (i.e., such as furanose and pyranose with their α and β conformers as well as both the open-chain form and its hydrated drivative), several sugar–borate complexes can be found rather than one definite complex. In fact, both glucose and xylose, which do not have any vicinal cis-hydroxyl groups in their pyranose forms, have formed negatively charged borate complexes and showed an increase in their UV absorption. This may suggest that the complex formation takes place in the open-chain forms of glucose and xylose.

It is clear from the overview above that under a given set of conditions, various sugars would undergo varying degrees of complexation, leading to differences in the electrophoretic mobilities of the complexed solutes and hence separation. Thus borate complexation is definitely an elegant approach for the high-selectivity separation of saccharides. In our laboratory we have exploited the various-sounding features of the sugar–borate complex to introduce novel micellar systems of adjustable surface charge density, and subsequently, tunable retention window for MECC of neutral and charged species. These micelles were based on the complex formation between alkylglucoside surfactants and borate ions at alkaline pH [36].

Highly Alkaline pH Electrolytes
The migration of carbohydrates during electrophoresis in sodium hydroxide solutions was first exploited by Frahn and Mills in 1959 [20]. Presumably, this migration is due to the ionization of the hydroxyl groups of saccharides at highly alkaline pH, yielding negatively charged species called alcoholates [37]. The ionization constants for carbohydrates are in the range of 10^{-12} to 10^{-14} (i.e., $pK_a = 12$ to 14). The pK_a values of some typical sugars are listed in Table 1. Electrophoretic migration measurements performed in 0.1 M sodium hydroxide solution [20] indicated that reducing sugars (e.g., glucose, galactose, mannose, etc.) are the most easily ionized, and that straight-chain alditols (e.g., glucitiol, mannitol) have on the average about the same acidity as cyclitols (e.g., inositols) and glycosides (e.g., methylglucopyranosides) of similar molecular weight and hydroxyl content. The higher acidity of reducing sugars is caused by the higher

lability of the hydrogen atom of the hemiacetal (anomeric) hydroxyl group, a condition that apparently stems from an electron-withdrawing polar effect (inductive effect) exerted on this group by the ring oxygen [37,38].

As can be seen in Table 1, the acidities of the aldoses are similar to one another except for 2-deoxy-D-ribose whose somewhat lower acidity may be explained by intramolecular hydrogen bonding. In D-ribose, the anionic oxygen at C_1 is capable of bonding with the adjacent hydroxyl group on C_2, whereas in the deoxy compound, such bonding cannot exist [37]. Another important feature of Table 1 is that the greater the number of hydroxyl groups, the greater the acidity. Glycerol has the same acidity as water; lactose and maltose, which are reducing disaccharides, seem to be somewhat more acidic than aldopentoses (arabinose, ribose, lyxose, and xylose). Further details on the ionization of carbohydrates at highly alkaline pH are documented in Ref. 37.

Recently, highly alkaline electrolyte solutions such as lithium, potassium, or sodium hydroxide at pH greater than 12 have been shown useful in the separation of underivatized saccharides by CZE [39–41]. Figure 3 shows the separation of six different saccharides by CZE using 20 mM (pH 12.3), 50 mM (pH 12.7), or 100 mM (pH 13) sodium hydroxide solutions as the running electrolytes [41]. As expected, the resolution among the various saccharides increased with pH due to

Table 1 Ionization Constants (Hydroxyl Group) of Carbohydrates in Water at 25°C

Compound	pK_a
D-Glucose	12.35
2-Deoxyglucose	12.52
D-Galactose	12.35
D-Mannose	12.08
D-Arabinose	12.43
D-Ribose	12.21
2-Deoxyribose	12.67
D-Lyxose	12.11
D-Xylose	12.29
Lactose	11.98
Maltose	11.94
Raffinose	12.74[a]
Sucrose	12.51
D-Fructose	12.03
D-Glucitol	13.57[a]
D-Mannitol	13.50[a]
Glycerol	14.40

Source: Ref. 37.
[a]Measured at 18°C.

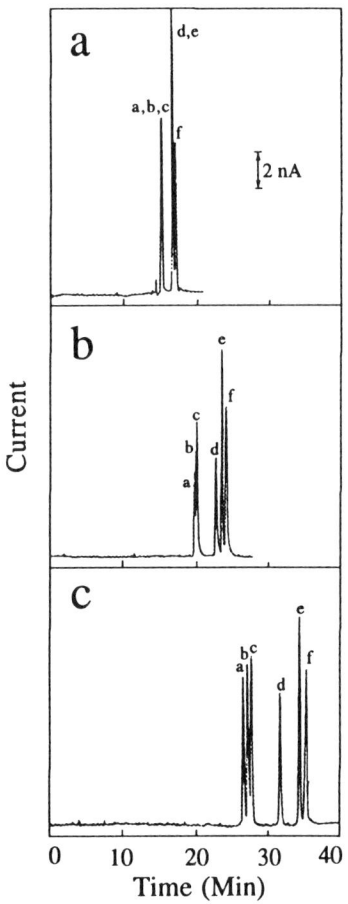

Fig. 3 Electropherograms of six saccharides in different concentration of NaOH: (a) 20 mM; (b) 50 mM; (c) 100 mM. Sugars: a, stachyose (terasaccharide); b, raffinose (trisaccharide); c, sucrose; d, lactose (both are disaccharides); e, galactose; f, glucose (both are monohexosaccharides). Concentrations of injected samples ranged between 80 and 150 μM. The fused silica capillary dimensions are 50 μm ID and 70 cm in length. The separation voltage is 10 kV. Injection is 10 s by gravity (10 cm height). The amperometric detection at constant potential is performed at 0.6 V (versus Ag/AgCl). (From Ref. 41.)

increasing ionization of the separated analytes. It should be noted that separations at extremely high pH can only be performed on naked fused silica capillaries since most coated and gel-filled fused silica capillaries will undergo hydrolytic degradation under such basic conditions. For further discussions, see Sections II.C and III.A.

Carbohydrate–Metal Cation Complexes

Complex formation between metal cations and carbohydrates has been known for a long time [22,42,43]. The hydroxyl groups of carbohydrates can coordinate to metal cations, thus forming sugar–metal complexes. In addition to paper electrophoresis [3,20,44,45], the metal complex formation has been exploited in thin-layer chromatography (TLC) [46] and in HPLC [47] of carbohydrates, whereby a ligand-exchange mechanism is considered to be involved in the separation of carbohydrates on TLC plates impregnated with metal ions or with HPLC metal-bearing cation-exchanger columns.

In general, strong complexing occurs between cations and a contiguous axial (a), equatorial (e), axial sequence of hydroxyl groups in carbohydrates, as was ascertained from the electrophoretic movement of compounds containing this sequence, and the immobility of many others lacking such an arrangement [48]. In addition to the a,e,a sequence, the rare 1,3,5-triaxial arrangement also favors coordination to cations. Similar to the a,e,a sequence is the *cis,cis*-1,2,3-triol grouping on a five-membered ring, whereby in a twist configuration, the three oxygen atoms are in a geometrical arrangement similar to the a,e,a arrangement [43]. The three oxygen atoms are then equidistant from the cation, thus favoring the complex formation with larger cations, while the smaller ones yield only weak complexes. Some compounds may not have the a,e,a sequence in their most stable conformations (e.g., β-D-ribopyranose), but they may acquire this sequence in less favored ones, and consequently, form weaker complexes with cations than compounds having the a,e,a sequence in their preponderant conformation (e.g., α-D-ribopyranose) [43,49]. Structures (*4*) and (*5*) show how β-D-ribopyranose can change from e,a,e to a,e,a configuration. If the difference in free energy between the noncomplexing and the complexing conformation is large, complex formation will not take place; an example is the case of *myo*-inositol.

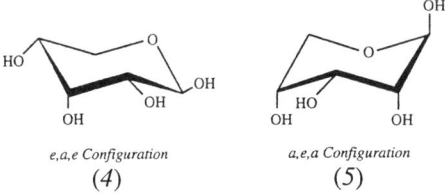

e,a,e Configuration　　　　*a,e,a Configuration*
(*4*)　　　　　　　　　　(*5*)

All of the alditols form complexes with cations showing good mobility in zone electrophoresis, but the extent of complexation varies considerably. For instance, the M_i values (defined as the electrophoretic mobility relative to *cis*-inositol in 0.2 M solution of calcium acetate in 0.2 M aqueous acetic acid) of hexitols range from 0.24 for D-iditol to 0.09 for allitol [44]. For these acyclic alditols, when three consecutive carbon atoms have the *threo-threo* configuration (*6*), the complex is most favored. An *threo-erythro* configuration (*7*) is less favored for complex formation, and an *erythro-erythro* arrangement (*8*) does not give rise to any noticeable complexation with cations. The more threo pairs of hydroxyl groups there are in the alditol, the stronger will be its complexes [43]. If one of the three hydroxyl groups is replaced by a methoxy group, complexing becomes weaker. If all three hydroxyl groups are methylated, complex formation becomes negligible [43]. Similarities can be drawn here to borate complex formation.

```
    |              |              |
  H—C—OH      HO—C—H         H—C—OH
    |              |              |
 HO—C—         H—C—OH         H—C—OH
    |              |              |
  H—C—OH      H—C—OH         H—C—OH
    |              |              |

 Threo-threo    Threo-erythro   Erythro-erythro
    (6)            (7)             (8)
```

As determined from electrophoretic data, even two consecutive hydroxyl groups can form a complex, although a very weak one, if they can approach each other sufficiently. Both *cis* and *trans* diols on six-membered rings will form weak complexes, but only *cis* diols on five-membered rings [50].

In general, cyclic monosaccharides yield tridentate complexes, whereby no more than three oxygen atoms can coordinate to one cation as shown in structure (9), where M^+ is the metal cation. In only a few cases involving disaccharides where complexation was reported to occur at more than three oxygen atoms, and tetra- and even pentadentate have been described [43]. Putting aside these exceptional cases, Angyal [43] arranged the effectiveness of complexation in the descending order as follows: 1,3,5-triaxial triol > a,e,a triol on a six-membered ring > *cis-cis* triol on a five-membered ring > acyclic *threo-threo* triol > acyclic *threo* pair adjacent to a primary hydroxyl group > acyclic *erythro-threo* triol > acyclic *erythro* pair adjacent to a primary hydroxyl group > acyclic *erythro-erythro* triol > *cis* diol on a five-membered ring > *cis* diol on a six-membered ring > *trans* diol on a six-membered ring. Generally, univalent cations all complex weakly, divalent ones complex better and trivalent ones form the strongest complexes with a given carbohydrate [20]. Also, the ionic radius of the cation is crucial, and the best radius for complex formation is 100 to 110 pm (e.g., Ca^{2+}, La^{3+}). Larger cations complex somewhat less strongly, whereas smaller cations complex weakly [43].

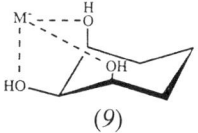

(9)

The rules above do not apply to amino sugars. The nitrogen atom in the sugar molecule has a great tendency to complex with transition metal ions, and if there is a free hydroxyl group in the vicinity, bidentate complexes will be formed that are much stronger than those of nitrogen free sugars. D-Glucosamine forms complexes with cations in the order $Cu^{2+} > Pb^{2+} > Zn^{2+} > Ni^{2+} > Cd^{2+}$, and Ca^{2+} and Mg^{2+} do not form complexes [51]. More details on metal–sugar complexes can be found in Refs. 42 and 43.

Recently, the use of electrolyte systems containing alkaline earth metals for the separation of neutral carbohydrates by CZE has been reported [52]. Separations in these media are based primarily on differences in the extent of complexation of the divalent metals with the carbohydrate solutes, and to a lesser extent on the bulk and shape of the molecule. Although these systems provided a different selectivity than that achieved with borate buffers, the resolution was in general inferior to that of borate buffers. Further discussions are provided in Sections III.A and III.B.

B. Capillary Column

HPCE of carbohydrates at alkaline pH (e.g., aqueous borate or sodium hydroxide solutions) has been performed on untreated fused silica capillaries since the majority of capillary coatings introduced so far would undergo hydrolytic degradation at pH > 8.5. Saccharides become negatively charged in alkaline borate or in sodium hydroxide solutions by forming anionic complexes or by self-dissociation, respectively. Under these conditions, the negatively charged sugar molecules are likely to be repelled by the negatively charged surface of untreated fused silica capillaries, a condition that prevents their surface adsorption, and as a result, high separation efficiencies can be obtained. The major drawback when using untreated fused silica capillaries, however, is that the naked silica surface can adsorb cationic contaminants that might exist in many "real-world" carbohydrate samples, a phenomenon that often leads to irreproducible separations.

On the other hand, ionic saccharides or those tagged with charged functions via precolumn derivatizations (see Section III.C) may be separated on capillaries having neutral, hydrophilic coatings at medium or low pH. In these cases, the rationale for using coated capillaries is to guarantee reproducible separations, and one should not be concerned with solute–wall interactions, since at relatively medium ionic strength (0.1 M) these species, whether native or derivatized, do not undergo any appreciable surface adsorption.

The situation becomes completely different when dealing with some glycoconjugates such as glycoproteins and glycopeptides. These species are multiply charged, and therefore surface adsorption becomes a major concern. In fact, due to the high surface/volume ratio of the capillary column used in HPCE, polyionic glycoproteins and glycopeptides have strong potential for electrostatic interactions with the charged surface of naked fused silica capillaries. This strong adsorption often gives rise to distorted peaks, poor resolution, low recovery of the separated analytes, and irreproducible separations, owing to the unpredictable changes in the magnitude of the electroosmotic flow (EOF). In view of this, untreated fused silica capillaries can be used in the separation of basic glycoproteins (also proteins) only under electrophoretic conditions whereby electrostatic interactions are minimized. These conditions include (1) the use of running electrolytes at high (> 9.0) or low pH (< 3.0) since at alkaline pH the protein and the silica surface are similarly charged (i.e., electrostatic repulsion predominates), and at low pH the silica surface is uncharged (i.e., minimum electrostatic interaction) [53–55], (2) the addition of relatively high salt concentrations to the background electrolytes so that the counterions can compete with the solutes for the available cation-exchange sites on the silica surface [56], and (3) inclusion of buffer additives in the running electrolyte that can interact with the analytes or the capillary surface, thereby reducing solute–wall interactions [57]. Although useful, these approaches are associated with various limitations, such as narrow pH range, high conductivity and low detection sensitivity, respectively. These facts severely restricts the applicability of untreated fused silica capillaries and thus do not provide a universal solution for the solute–wall interaction problem.

A more convenient approach to minimize solute–wall interactions has been the modification of the capillary inner wall to produce an inert nonadsorptive surface. Here, the coated capillaries, including their subsequent refinements, which have been applied to the HPCE of carbohydrate species, will be discussed in some detail, and then a brief description of other coatings will be provided. Thus the interested reader is advised to consult recent reviews [58–60] on capillary column technology since most the coatings introduced so far are potential candidates for use in HPCE of carbohydrates.

The first method for coating capillaries used in HPCE was reported by Hjertén [61], who modified glass or quartz tubes with non-cross-linked polyacrylamide. This led to the formation of a thin monomolecular layer, covalently attached to the surface via siloxane bonds, and the resulting coating proved useful in biopolymer separations. However, the polyacrylamide coating suffered from poor hydrolytic stability when operated at high pH. In an attempt to enhance the coating stability, Novotny and co-workers [62] attached the polyacrylamide coating to the silica wall by an Si—C bond as opposed to the siloxane linkage described by Hjertén. This procedure yielded improved stability and reproducibility in the pH range 2 to 10.5. However, this coating greatly reduced the electroosmotic flow and consequently, necessitated the use of extreme pH buffers to bring about the migration of various solutes by their own electrophoretic mobilities. In another effort to provide stable polyacrylamide coating, Huang et al. [63] immobilized the polyacrylamide via 7-*oct*-1-enyltrimethoxysilane in place of the most widely used 3-methylacryloxypropyltrialkoxysilane. This sialylating compound produced highly cross-linked siloxane sublayer, which afforded a more stable polyacrylamide open tubular capillary column.

In our laboratory [64–67] we have introduced a series of cross-linked polyether coatings that yielded stable and inert capillaries with various levels of

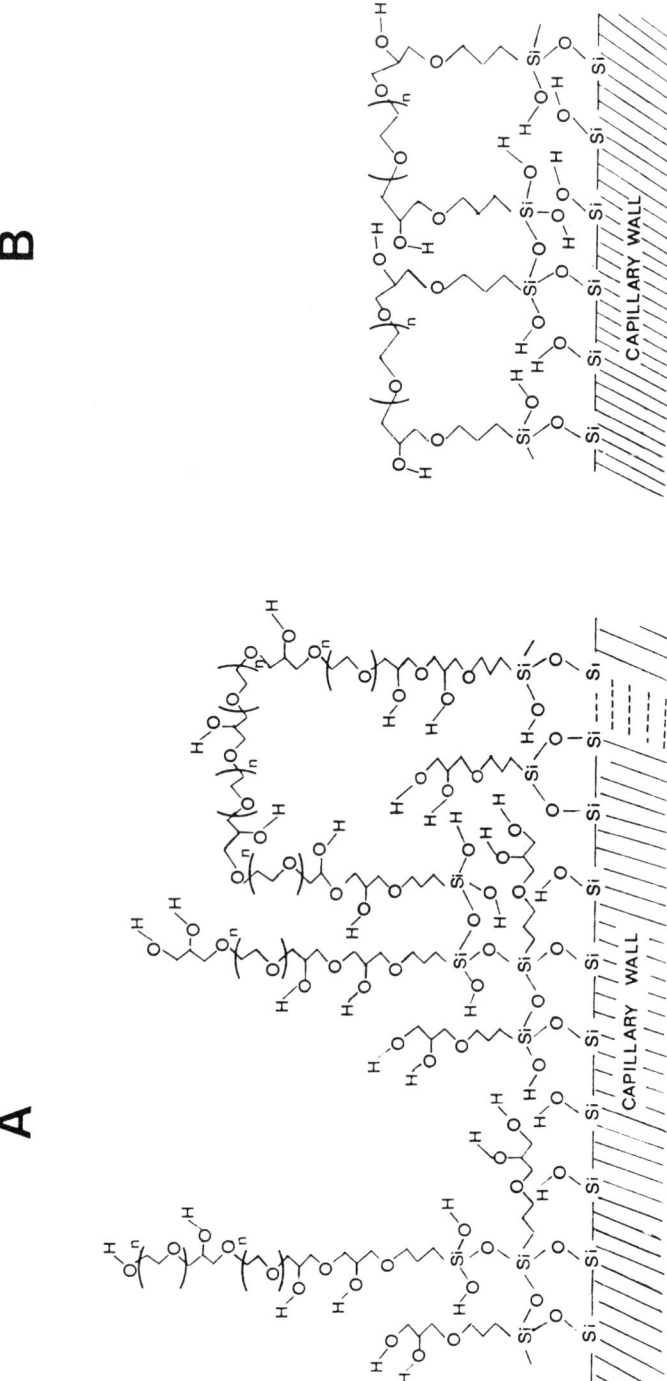

Fig. 4 Schematic illustration of the idealized structures of fuzzy (A) and interlocked (B) polyether coatings of fused silica capillaries. (From Ref. 64.)

electroosmotic flow (see Fig. 4). The polyether-coated capillary afforded high separation efficiencies for various biopolymers, including glycoproteins, glycopeptides, and various types of oligosaccharides. The coating procedures involved the covalent attachment of polyether chains of various length to the inner surface of fused silica capillaries via siloxane bonds, either directly using polyether chains with trimethoxysilane groups at both ends (interlocked polyether coatings; Fig. 4B), or by attaching a top polyether layer to a polysiloxane sublayer already anchored to the capillary inner surface (fuzzy polyether coatings; Fig. 4A). In contrast with other reported poly(ethylene glycol) (PEG) coating in which the PEG layer was not further reacted, the cross-linked polyether layer in the interlocked or fuzzy coatings produced chemically stable capillaries, and also provided the hydrophilicity and protection required for biopolymer separation.

Very recently, another capillary coating has been introduced by our laboratory [68], the idealized structure of which is shown in Fig. 5. This surface modification yielded fused silica capillaries with switchable EOF (anodal/cathodal). The capillary surface is a composite material consisting of unreacted silanol groups, a layer of positively charged quaternary ammonium functions, and a hydrophilic top layer of polyether chains. Because of the presence of positively and negatively charged groups (i.e., quaternary ammonium functions and unreacted silanol groups, respectively), the net charge of the capillary surface can be varied from positive to negative by changing the pH of the running electrolyte, thus enabling manipulation of the magnitude and direction of the electroosmotic flow from anodal to cathodal (Fig. 5b). Flow is described as anodal if the movement of the bulk solution is toward the anode. This EOF occurs when the net charge on the surface of the capillary is positive. Under this condition, the diffuse region of the electric double layer at electrolyte–capillary interface is populated by negatively charged counterions which, under the influence of an applied electric field, will migrate to the anode, dragging the solvent with them. Flow is described as cathodal if the EOF is in the direction of the cathode. The surface of the capillary is now negatively charged, and therefore the diffuse layer is rich in positively charged counterions. When an electric field is applied, these solvated counterions carry the solvent with them toward the cathode. Returning to Fig. 5, the long polyether chains were effective in shielding biomacromolecules from the charged inner surface of the capillary, thus minimizing electrostatic interaction of the solutes with both unreacted silanols and the quaternary ammonium groups that had been introduced. As a consequence, high separation efficiencies were achieved with proteins, nucleotides, and a series of acidic oligosaccharides.

A number of other coating procedures for HPCE have been developed in other laboratories. Typical coatings include poly(vinylpyrrolidone) [55], poly(ethylene glycol) [69–71], maltose [72], arylpentafluoro [73], polyethyleneimine (PEI) [74], hydrophobic C_{18} bonded capillary with adsorbed nonionic surfactant [75], poly(methylglutamate) coating [76], epoxy polymer coating [77],

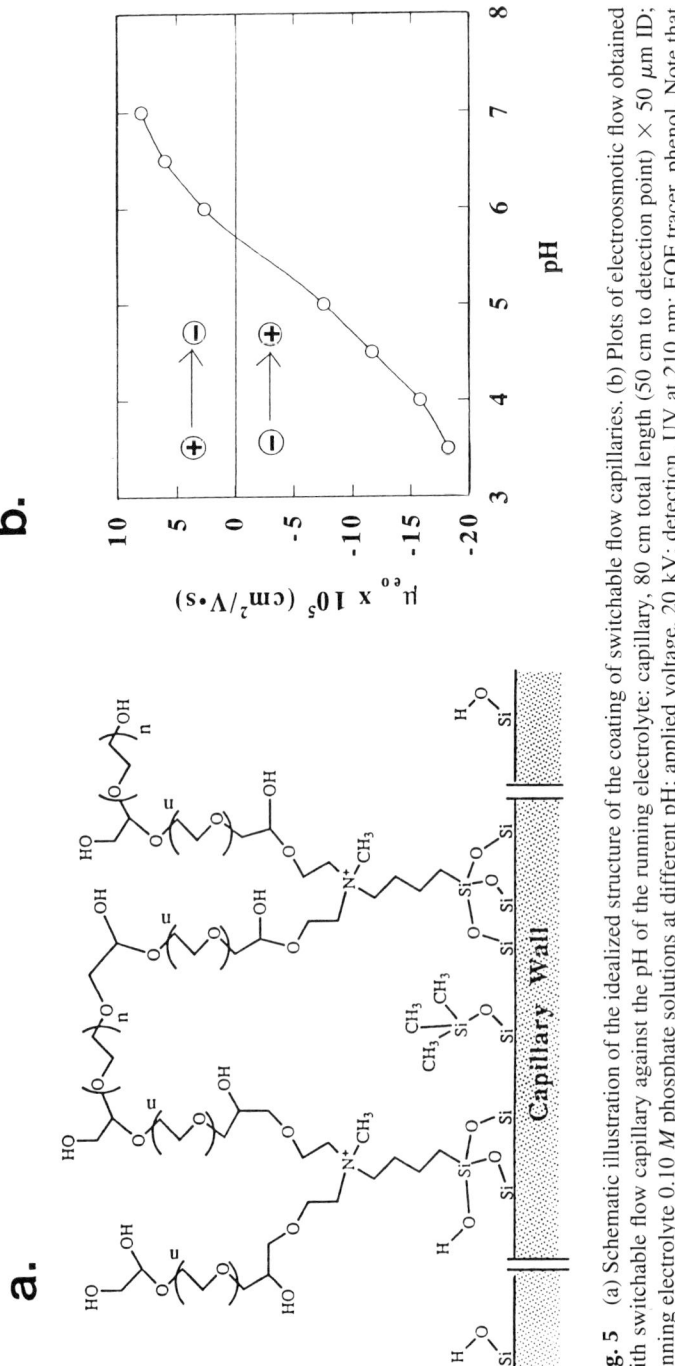

Fig. 5 (a) Schematic illustration of the idealized structure of the coating of switchable flow capillaries. (b) Plots of electroosmotic flow obtained with switchable flow capillary against the pH of the running electrolyte: capillary, 80 cm total length (50 cm to detection point) × 50 μm ID; running electrolyte 0.10 M phosphate solutions at different pH; applied voltage, 20 kV; detection, UV at 210 nm; EOF tracer, phenol. Note that ⊖→⊕ means anodal EOF, and ⊕→⊖ means cathodal EOF. (From Ref. 68.)

capillaries having surface-bound α-lactalbumin (an amphoteric protein) [78], poly(vinyl alcohol) [71], and cellulosic capillary coatings [79]. The PEI coating provided positively charged surface and was designed for the separation of basic proteins, which then would be repelled by the positively charged capillary wall, thus minimizing surface adsorption. The polyglutamate coating showed some limitations, including the potential for formation of anionic sites on the coating by hydrolysis of ester groups in the polymer and hydrophobic interaction of some proteins with the coating. The α-lactalbumin coating, which yielded capillaries with amphoteric walls, seems to be useful only for basic species. All other coatings were of the neutral type and produced acceptable results in terms of protein separations.

Finally, only a limited number of coated capillaries has become commercially available from different sources, and they are listed in Table 2. Referring to Table 2, the Micro-Coat reverses the EOF from cathodal to anodal and allows the separation of proteins (including glycoproteins) at pH values below their isoelectric points with minimum solute–wall interactions. The capillary coating from Bio-Rad has been designed to eliminate EOF and reduce solute adsorption on the capillary inner surface. The CElect capillaries were designed to provide reproducible EOF over a wide pH range. The CE-100-C18 capillary is normally used with

Table 2 Commercially Available Coated Capillaries

Source	Brand name	Nature of the coating
Applied Biosystems (Foster City, California)	Micro-Coat	Polymeric cationic coating normally applied to the capillary in an aqueous solution containing 2% ethylene glycol, forming a noncovalent coating on the capillary inner surface
Bio-Rad Laboratories (Hercules, California)	—	Linear hydrophilic polymer
Supelco, Inc. (Bellefonte, Pennsylvania)	CElect-P	Neutral hydrophilic phase
	CElect-H	Weakly hydrophobic C_1 phase
	CElect-H1	Moderately hydrophobic C_8 phase
	CElect-H2	Highly hydrophobic C_{18} phase
Isco, Inc. (Lincoln, Nebraska)	CE-100-18	Covalently attached octadecyl groups
	CE-200-glycerol	Covalently attached glycerol groups
	CE-300 sulfonic	Sulfonic acid groups linked to the capillary inner surface by an octyl linker

electrolyte containing nonionic surfactants [75] for the separation of oppositely charged species. While the CE-200-glycerol capillary is a hydrophilic coating and can be used for the separation of oppositely charged biopolymers, the CE-300 sulfonic coating provides a constant cathodal EOF and is suitable for the separation of negatively charged species.

Although CZE with open tubular capillaries provided adequate resolution for simple sugars, short oligosaccharides, glycopeptides, and glycoproteins, the high-resolution separation of some oligosaccharides of degree of polymerization greater than 20 necessitated the use of capillary gel electrophoresis (CGE) with polyacrylamide gel-filled capillaries [80,81].

The first method to prepare gel-filled capillaries with cross-linked polyacrylamide was introduced by Karger and Cohen [82,83]. In this method the capillary wall is first reacted with a silane bifunctional reagent such as 3-methylacryloxypropyltrialkoxysilane. Thereafter, a buffer solution containing the desired proportions of acrylamide monomers, N,N-methylenebisacrylamide, ammonium persulfate, and N,N,N',N'-tetramethylenediamine, is forced through the capillary tube and then allowed to polymerize inside the capillary for several hours while both ends of the capillary are immersed in the buffer solutions. The silane bifunctional coupling agent is to enable the covalent attachment of the gel matrix to the capillary wall, thus preventing its migration under high electric fields. As a refinement of this method, Bente and Myerson [84] used high pressures during the gelation step to prevent the formation of voids and the shrinkage of the gel matrix. Other methods for gel fabrication include the use of cobalt 60 radiation to initiate the formation of polyacrylamide gel [85] as well as the use of sequential polymerization techniques by isotachophoresis [86]. Additional information concerning these methods could be found in recently published books [87,88]. Gel-filled capillaries are commercially available from Applied Biosystems, Beckman Instruments, and J&W Scientific.

Due to the various problems encountered in CGE, such as bubble formation in the gel network and limited stability of gel-filled capillaries, size-based separations are increasingly carried out (1) in non-cross-linked gels polymerized in the capillary or introduced into the capillary by pressure after polymerization [89] and (2) in solutions of linear hydrophilic polymers [90]. With non-cross-linked gels, the same capillary column can be refilled with a new medium after several runs, thus providing a long lifetime for the capillary. Also, with polymer networks, the separations are more reproducible and easy to achieve with the same capillary column for an unlimited number of runs. The major drawback, however, is the lower separation efficiency than that for cross-linked gel media. These media are expected to find application in size-based separation of proteins, including glycoproteins. Polymer network sieving media are marketed by Beckman Instruments and Bio-Rad Laboratories.

C. Detection Systems

Precolumn Derivatization

Most carbohydrate species do not possess strong chromophores in their structures, a condition that does not allow their sensitive and direct detection in the UV. To overcome this difficulty, carbohydrates are generally tagged with a suitable chromophore or fluorophore. It is preferred that the tag also supply the charge necessary for electrophoresis over a relatively wide pH range. In precolumn derivatization reactions, the tagging should occur only at one reactive functional group of the analyte and should be complete, so that a single derivative is obtained in high yield. The polyhydroxy nature of carbohydrates is attractive as far as the attachment of a tag to the molecule is concerned. But derivatization based on the hydroxyl groups of the sugar molecule, which may differ in their relative reactivities, would lead to multiple tagging of the molecule, and consequently a distribution of derivatives rather than a single product would be obtained. To avoid this multiple derivatization, other functional groups on the sugar molecule must be considered. The carbonyl group in reducing sugars is by far the most widely used loci for tagging. In amino sugars and acidic carbohydrates, the amino group and the carboxylic function are also potential sites for derivatization, respectively.

To produce one kind of derivative in a given precolumn derivatization, the tag must also possess one single reactive site for attachment to the sugar molecule. Obviously, for UV and fluorescence detection, the tagging agent should exhibit a relatively high molar absorptivity and photoluminescence efficiency to ensure highly sensitive detection of the derivatized sugars. This requirement becomes very critical in HPCE, where the sample volume that can be injected is extremely small due the overall miniaturization of the separation system. While the choice of a tagging agent for UV detection is a relatively easy task, the list of compounds to choose from a potential fluorescing tag is rather limited. This list shrinks even further when considering the use of the very sensitive laser-induced fluorescence (LIF) detection instead of the lamp-based fluorescence detector. This is due to the inherent monochromatic nature (single wavelength) of laser sources, which dictates that the excitation maxima of the derivatives must match the laser line. As discussed below, when LIF is used as the detector, the most suitable tagging agent may not be commercially available and would require skilled chemists to make it. This makes LIF, in addition to other things, less practical than UV or regular fluorescence.

Exploiting the reactivity of the reducing end of saccharides, precolumn derivatizations via reductive amination for various carbohydrate species have been reported. Honda et al. [33] derivatized several monosaccharides with 2-aminopyridine (2-AP), and demonstrated the separation of the resulting pyridylamino derivatives in CZE as borate complexes. This precolumn derivatization was first exploited in paper electrophoresis [91] and then in HPLC of carbohydrates [92]. The 2-AP derivatives could be detected in the UV at 240 nm at the level of 10 pmol. It should be mentioned that the 2-AP sugar derivatives fluoresce at 400 nm when excited at 320 nm and thus could be monitored by on-column fluorometric detector [93]. Using reductive amination, our laboratory has demonstrated the utility of 6-aminoquinoline (6-AQ) [94], a strongly UV absorbing and also fluorescing tag, in the derivatization of oligosaccharides and subsequent separation of the 6-AQ derivatives in CZE. The 6-AQ derivatives showed maximum absorbance at about 270 nm, and the signal was eight times higher compared to that obtained with pyridylamino derivatives under otherwise identical electrophoretic conditions [94]. The quinolylamino derivatives could also be detected by fluorescence. In fact, 6-AQ possesses ideal fluorescence behavior in the sense that its excitation wavelength (355 nm) is far removed from its emission wavelength (550 nm), a condition that minimizes self-absorption. It is therefore expected that 6-AQ–sugar derivatives will afford low detection limits with a properly designed fluorescence detector. In schemes shown in reaction (5), N-acetylglucosamine (GlcNAc) was chosen as a typical saccharide to illustrate the precolumn derivatizations with 2-AP and 6-AQ [94].

(5)

Similar derivatization schemes via reductive amination were recently described for N-acetylchitooligosaccharides with a negatively charged fluorescing tag, the 7-amino-1,3-naphthalenedisulfonic acid (AGA) (10) [95]. The AGA–sugar derivatives were found to fluoresce at 420 nm when excited at 250 nm, with detection sensitivity on the order of 80 fmol [95].

7-Amino-1,3-naphthalenedisulfonic acid
(10)

Liu et al. [96] introduced a fluorogenic reagent, the 3-(4-carboxybenzoyl)-2-quinolinecarboxyaldehyde (CBQCA). This tagging agent is not commercially available, and its preparation has been described in Ref. 97. When tagging amino sugars with CBCQA, the derivatization is accomplished in the presence of cyanide ions, which catalyze the ring formation with CBQCA. In the case of neutral reducing sugar, the analyte is first reductively aminated with ammonia in the presence of sodium cyanoborohydride to produce 1-aminodeoxyalditol. This alditol is then reacted with CBQCA in the presence of cyanide. Reaction schemes (6) and (7) illustrate the CBQCA tagging of model sugars, glucosamine and glucose.

(6)

(7)

The CBQCA–sugar derivatives exhibited excitation maxima that closely matched the output wavelength of a helium–cadmium laser at 442 nm and that of the argon-ion laser at 457 nm [97,98]. The CBQCA–sugar derivatives could be detected at the attomole level. This is expected since laser beams can be focused onto small capillaries and provide a highly collimated beam, which allows efficient rejection of stray light during detection, thus leading to very low mass detection limits.

Purely UV-absorbing derivatives of saccharides were prepared by reductive amination using ethyl p-aminobenzoate (11) and p-aminobenzoic acid (12) [99] as the tagging agents. Ethyl p-aminobenzoate was first introduced by Sweely et al. for HPLC detection and the separation of various saccharides [100]. These UV-absorbing labels allowed the derivatizations of both aldoses and ketoses. This represents an advantage over 2-AP, which is suitable only for the derivatizations of aldoses and uronic acids. p-Amniobenzoic acid and ethyl p-aminobenzoate permitted the detection of fructose at mass detection limits as low as 0.3 and 0.14 pmol, respectively, and for aldoses the detection limits were as 15 and 7 fmol. The maximum absorbance wavelengths were 305 and 285 nm for ethyl p-aminobenzoate and p-aminobenzoic acid sugar derivatives, respectively [99].

Ethyl p-aminobenzoate
(11)

p-Aminobenzoic acid
(12)

Honda et al. [101] described the conversion of reducing carbohydrates to UV-absorbing species by tagging the analytes with 1-phenyl-3-methyl-5-pyrazolone (PMP) according to the scheme shown in reaction (8).

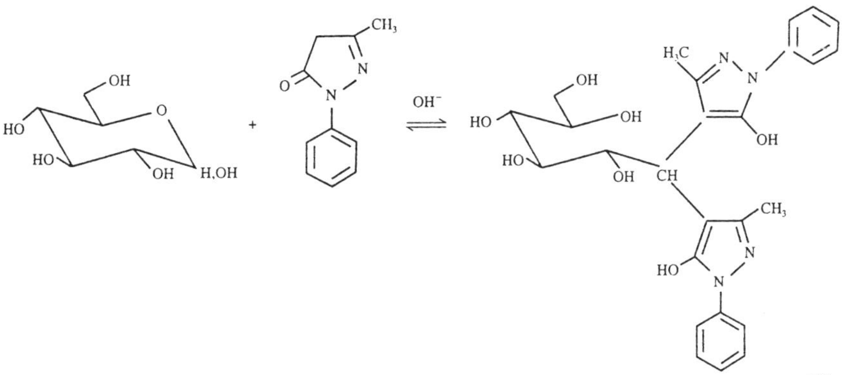

(8)

They originally devised this precolumn derivatization for the separation and detection of carbohydrates in HPLC [102]. The bis-PMP derivatives absorb strongly in the UV at 245 nm.

Indirect Detection

Indirect detection methods have been employed in the analytical separations (HPLC and HPCE) of species whose structures lack the necessary physical properties for direct detection and determination [103]. The key element for indirect detection is to maintain a large, continuous background signal at the detector by adding a detectable ionic species to the background electrolyte. When a nondetectable ion passes the detection window, there will be an increase or decrease in concentration of the detectable ion in the analyte zone (presumably via a displacement mechanism), resulting in either an increase or decrease in the background signal [103]. Thus the signal detected (i.e., peak) is derived from the detectable background ion rather than from the analyte itself. On this basis, almost all detection schemes can be made to function in the indirect mode without altering the instrumentation normally employed in direct detection. Furthermore, indirect detection is nondiscriminative (i.e., universal) and applies to a wide variety of compounds without the need of pre- or postcolumn derivatization techniques.

Thus far, indirect photometric [104], fluorometric [105,106], and to a lesser extent electrochemical detection schemes [107] have been introduced to HPCE. The topic of indirect detection has been reviewed by Yeung [108] and Yeung and Kuhr [103], and some of its fundamental aspects have been discussed by Bruin et al. [109].

In indirect detection, and according to Yeung and Kuhr [103], the attainable detection limit C_{lim} (in concentration units) is given by

$$C_{lim} = \frac{C_M}{DR \cdot TR}$$

where C_M represents the concentration of the detectable ion that generates the background signal, DR is the dynamic reverse, and TR is the transfer ratio. *Dynamic reverse* is defined as the ability to measure a small change on the top of a large signal, and is equal to the signal/noise ratio (*S/N*) of the background signal. *Transfer ratio* is defined as the number of molecules of the detectable ions displaced by each analyte molecule. It is desirable that the value of TR be close to unity [103]. In fact, the best detection sensitivity is achieved when the sample ions have an effective mobility close to that of the detectable ion in the background electrolyte [104]. It can be seen from the equation above that C_M should be as low as possible while generating a sufficient background signal, and the dynamic reverse should be as high as possible. However, the three parameters are not necessarily independent. For instance, decreasing C_M will increase TR, but at the expense of decreasing DR.

Garner and Yeung [106] were the first to apply the principle of indirect detection to the CZE of underivatized carbohydrates. Coumarin 343 was used as the background fluorescing ion for indirect laser-induced fluorescence. Coumarin was selected because it has a good solubility in aqueous solution and a high fluorescence yield with a molar absorptivity of 2×10^4 at 442 nm, which matches very well the 442-nm line of helium–cadmium (He-Cd) laser. Using this indirect LIF detection system, three simple sugars could be separated using 1 mM coumarin, pH 11.5, as the running electrolyte and a capillary of 18 μm ID. The high pH used in this study was to ionize the carbohydrates and render them amenable to electrophoresis as well as to indirect detection. In fact, the indirect detection of sugars requires a pH approaching 12 to have any appreciable fraction in the ionized form. But as the pH becomes very high, the concentration of hydroxide ions is no longer negligible relative to the concentration of the fluorophore. This resulted in a decrease of TR, and consequently in sensitivity. The optimum detection pH was found to be 11.65, but 11.5 was used for detection because the rate of degradation of fluorophore was decreased without any appreciable decrease of detection efficiency [108]. The sugars were injected at a concentration of 1 mM each, and the amount detected was about 640 fmol for each sugar.

Indirect photometric detection was also introduced to the detection of underivatized carbohydrates. Vorndran et al. [40] and Oefner et al. [99] have demonstrated the use of sorbic acid as both the background electrolyte and the chromophore in the indirect UV detection of several simple sugars. Sorbic acid was selected because (1) it has high molar absorptivity (ϵ = 27,800 at 256 nm); (2) it is compatible with the solvent system; (3) it carries a single charge, which ensures a good transfer ratio; (4) it does not interact with the capillary surface or with the analytes, and (5) it has an effective mobility that matches the ionic mobilities of carbohydrates. As with indirect LIF, charging of the sugars, either by complexation with borate or by dissociation at high pH, is a prerequisite, both for their separation based on differences in migration velocity and for their determination by means of charge displacement. As explained above, since the ionic strength of the running electrolyte should stay low, indirect detection precludes the use of borate ions due to the high concentration required (100 to 200 mM) for efficient complexation with the sugars. Also, at very high pH, greater than 11.5, which is required for the resolution of carbohydrates, the sensitivity of the indirect detection drops sharply. Because of these limitations, electrolyte systems that are permissible for indirect detection do not afford sufficient selectivity for weakly ionized carbohydrates, as shown in Fig. 6 (e.g., peaks 1 to 6).

Although the principle of indirect detection appears relatively simple and the method yields impressive limits of detection, several drawbacks can be pointed out. The major problem is the instability of the detection system, which yields drift or large disturbances of the baseline. In addition, an indirect detection system to be effective requires working at a low concentration of background electrolyte [103], a condition that results in lower efficiencies as a consequence of the stronger influence of concentration overload at higher sample concentrations and possible solute–wall interactions [109]. An additional obvious disadvantage is the limitation imposed by this detection mode on the selection of the composition and pH of the background electrolyte. In other words, there is not much room to manipulate selectivity and to optimize separations. Therefore, this mode of detection should be considered as the last resort whenever the solutes are not easy to derivatize or cannot be detected otherwise. As discussed above, the weak ionization of carbohydrates does not favor their high-resolution separation using indirect detection electrolyte systems. It should be noted, however, that indirect detection, which was first demonstrated in ion chromatography [110], is perhaps the best approach described so far for capillary electrophoresis separation of low-molecular-weight ionic species such as anions, cations, and organic acids [111,112], and more than 30 different ions can be separated in less than 3 min.

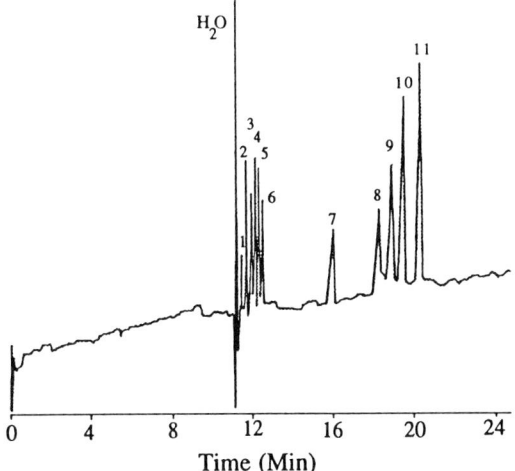

Fig. 6 Electropherogram of carbohydrates at sample concentrations in the range 0.95 to 2.66 mM. Electrolyte: 6 mM sorbate, pH 12.1; capillary: fused silica, 122 cm total length (100 cm to detection point) × 50 μm; applied voltage, 28 kV; temperature, 30°C. Zone identification: 1, raffinose; 2, 2-deoxy-D-ribose; 3, galactose; 4, glucose; 5, rhamnose; 6, mannose; 7, N-acetylneuraminic acid; 8, gluconic acid; 9, galacturonic acid; 10, glucuronic acid; 11, mannuronic acid. (From Ref. 40.)

This is because these ions are charged in their natural environment, and therefore the selection of pH and ionic strength of the running electrolyte may not be as critical for the outcome of separation as when dealing with weakly ionized solutes.

Direct Detection

Direct detection refers to sensing the separated species in their native forms by deriving the analytical signal directly from the analyte itself. Thus far, underivatized carbohydrates separated by HPCE have been detected by three different modes of direct detection. Low-wavelength UV has been the most widely used, and very recently, capillary-refractive index and capillary-amperometric detection have been introduced.

Low-wavelength UV detection is employed increasingly in HPCE of carbohydrates, especially for those compounds having carboxyl or other UV-absorbing groups. Very recently, even simple aldoses (mono- and disaccharides) have been detected at 195 nm [35]. Other carbohydrates that have higher UV molar absorptivities are more readily detected. These are the compositional disaccharides of heparin and heparan sulfate, which have been shown to detect well at 232 nm, reaching sensitivities at the level of 50 fmol of the disaccharides [113]. Also, using the same detection wavelength, HPCE of chondroitin sulfate [114,115] and dermatan sulfate disaccharides [115] was described. Hyaluronan oligosaccharides which detected poorly at 232 nm showed enhanced signal at 200 nm [114]. UV detection at 214 nm was also useful for monitoring colominic acid hydrolysate and heparin disaccharides [116] separated by HPCE.

Since many of the well-established HPLC detection systems have been adapted to HPCE, it is not surprising to see that the principle of amperometric detection is also finding its way to carbohydrate HPCE. Pulsed amperometric detection (PAD) at platinum and gold electrodes using strongly alkaline mobile phases is far more sensitive than refractive index and low-wavelength UV detection, permitting the determination of subnanomole quantities of carbohydrate species separated by HPLC [117]. This breakthrough in the detection of underivatized saccharides has recently prompted Zare and co-workers [41] to design an amperometric detection system for HPCE of carbohydrates which differed slightly from PAD in terms of the nature of the working electrode and signal processing. In PAD, a multistep potential waveform is applied to the working electrode, and the faradaic response is monitored at a given step potential of the applied waveform. Although this approach has solved the problem of electrode poisoning typically found with the oxidation of carbohydrates at gold or platinum electrode in the direct amperometric detection mode, the PAD detection system requires specialized pulse sequences, thus entailing expensive instrumentation. To keep the detection system simpler while providing comparable sensitivity to that of PAD, the amperometric detector introduced by Zare and co-workers operates at a constant potential of 0.6 V (versus Ag/AgCl) with an ultramicroelectrode consisting of a 25-μm metallic copper wire [41]. The running electrolyte was a strong alkaline solution (pH 13) to ensure the ionization of the native sugars, and allows their electrophoretic separations. The copper-wire microelectrode was found to resist the strong alkaline solution and showed no deterioration for hundreds of runs. This detection scheme permitted the realization of a linear range over three orders of magnitudes and a limit of detection for mono- and disaccharides in the femtomole range. A typical electropherogram is shown in Fig. 7. In contrast with indirect detection, here the pH can be increased to 13, thus permitting the separation of 15 saccharides (compare to Fig. 6). However, and as with the PAD design, the running liquid must not contain any electroactive species that would oxidize on the working electrode and yield a strong background signal. Thus the amperometric detection precludes the use of borate buffers, which are essential in attaining enough selectivity for the separation of a wide range of native and neutral saccharides. Nevertheless, the constant-potential amperometric detection permitted the determination of glucose and fructose in soft drinks [41]. Unfortunately, at the present time this type of detector is not available from commercial sources.

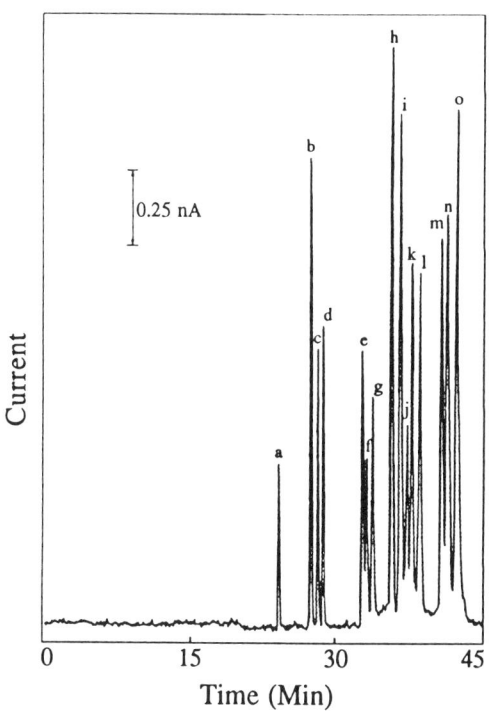

Fig. 7 CZE/amperometric detection at constant potential of a mixture containing 15 different saccharides at concentrations in the range 80 to 150 μM. Conditions: separation electrolyte, 100 mM NaOH (pH 13); fused silica capillary, 50 μm ID × 73 cm; separation voltage, 11 kV; injection 10 s by gravity (10 cm height). The amperometric detection at constant potential is performed at 0.6 V (versus Ag/AgCl). Saccharides: a, trehalose; b, stachyose; c, raffinose; d, sucrose; e, lactose; f, lactulose, g, cellobiose; h, galactose; i, glucose; j, rhamnose; k, mannose; l, fructose; m, xylose; n, talose; o, ribose. (From Ref. 41.)

Another direct detection mode for substances such as carbohydrates that neither absorb nor fluoresce is the refractive index (RI). Although not highly sensitive and its detection limit falls below that of low-wavelength UV, bulk RI detection has been used in many HPLC applications, including carbohydrates. The adaptation of this detection mode to micro-LC and HPCE originated from the original work of Dovichi and co-workers [118,119], which entailed the design of subnanoliter laser-based refractive index detectors for capillary formats. This laser-based RI detection was subsequently refined by other researchers for CZE applications [120–122]. The present detection method exploits the phenomenon of light interferences arising from side illumination of liquid-filled capillary tubes by laser light to measure changes in the RI (Δn). Interaction of the laser beam and the capillary produces a fan of scattered light, consisting of light and dark fringes, in the plane perpendicular to the capillary tube axis. As the migrating analytes reach the detection cell, the RI of the solution at this detection point changes, and consequently some of the fringes change their position. A small-area photodiode is located at the border of a fringe; a change in the position of the fringe results in a change in intensity monitored by the phohtodiode. This signal is then compared with that of a reference photodiode, and the solution's Δn value is measured [118].

The adaptation of RI detection to HPCE, however, has been accompanied by some difficulties arising primarily from the Joule heating encountered in HPCE. The Joule heating introduces changes in the refractive index of the medium, and consequently the thermal stability becomes the limiting factor in attaining relatively low detection limits. To overcome thermal fluctuations inside the capillary in HPCE, and to enhance the sensitivity of this mode of detection, Bruno et al.

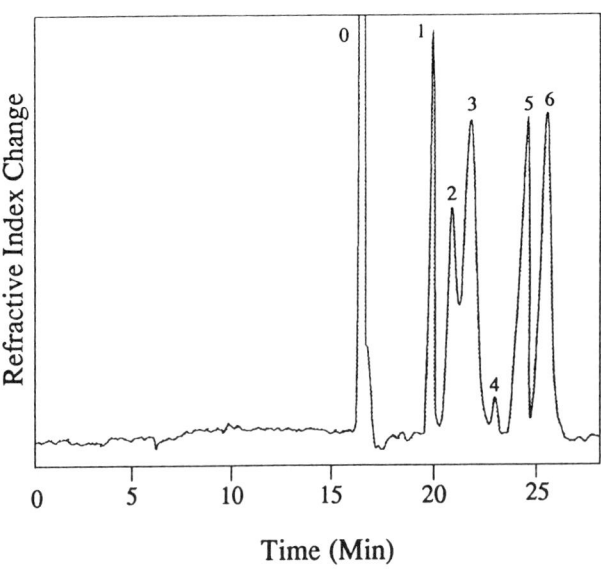

Fig. 8 Electropherogram of a mixture of five underivatized saccharides. The peaks are identified as buffer (0), sucrose (1), N-acetylglucosamine (2), cellobiose (3), impurity (4), N-acetylgalactosamine (5) and lactose (6). Conditions: each 1% except for sucrose, 0.5%; buffer 100 mM borate, pH 9.0; capillary length 70 cm, 55 cm to the detector, ID 50 μm; applied voltage, 14 kV; injection time, 7 s at 12 kV; thermocooler temperature, 27°C. (From Ref. 122.)

[122] described a RI detector with improved features. In their design they introduced the use of refractive index–matching fluids, position-sensitive photodiodes, and a thermoelectrically stabilized highly-symmetric RI cell design. These improvements in the overall detector performance were demonstrated in the CZE separation of synthetic underivatized saccharide mixture (Fig. 8). As can be seen in this figure, five different disaccharides could be detected using borate complex formation. Unfortunately, this type of detector is not yet commercially available.

III. SEPARATION METHODOLOGIES AND APPLICATIONS

The separation of carbohydrates, and in particular the glycan moieties of glycoproteins, glycolipids, and proteoglycans, is of primary importance in the biomedical and biotechnological areas. More than any other type of species, the separation of complex glycoconjugate mixtures requires separation techniques of high resolving power, such as HPCE. Although at its early stages of development, most HPCE applications involved proteins, peptides, and nucleic acid fragments and constituents. Over the last 5 years a number of HPCE methods have been developed to suit the separation of carbohydrate species. Our aim in this section is to review critically the HPCE methodologies that were developed for the separations of (1) monosaccharides, (2) disaccharides, (3) oligosaccharides, (4) glycoproteins and their glycan and glycopeptide fragments, (5) compositional sugars of glycosaminoglycans, and (6) glycolipids.

A. Monosaccharides

Monosaccharides constitute the basic units of carbohydrates. Sensitive and efficient HPCE methods for the separation and accurate quantitative determination of monosaccharides are of crucial importance for many areas of biochemistry, pharmacology, biotechnology, and food science. Figure 9 demonstrates the high-resolution separation, by capillary zone electrophoresis (CZE), of 12 monosaccharides derivatized with 2-AP. They were separated as anionic borate complexes in about 25 min [33]. Obviously, the migration velocity of each derivative was affected primarily by the extent of complexation with borate. In the case of aldopentoses, arabinose and ribose with *cis*-oriented hydroxyl groups at the C_3/C_4 positions were more retarded than lyxose and xylose with *trans*-disposed hydroxyl groups. The same behavior was also observed with aldohexoses [e.g., galactose (*cis*) and glucose (*trans*)] and hexuronic acids [e.g., galacturonic acid (*cis*) and glucuronic acid (*trans*)]. However, *N*-acetylhexosamines [e.g., *N*-acetylgalactosamine (*cis*) and *N*-acetylglucosamine (*trans*)], showed the reverse effect concerning the 3,4-disposition of hydroxyl groups. This was attributed to

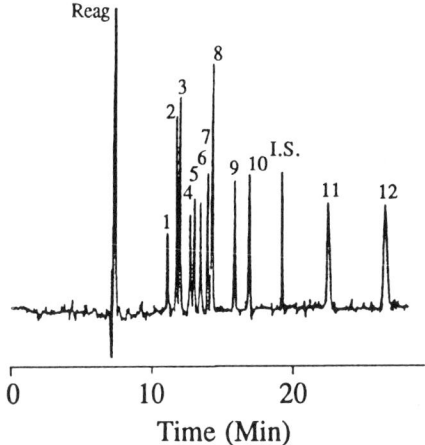

Fig. 9 Separation of N-2-pyridylglycamines derived from various reducing monosaccharides. Electrolyte, 200 mM borate buffer, pH 10.5; applied voltage, 15 kV; detection, UV at 240 nm. Peak assignment of parent saccharides: Reag, reagent (2-AP), 1, N-acetylgalactosamine; 2, lyxose; 3, rhamnose; 4, xylose; 5, ribose, 6, N-acetylglucosamine; 7, glucose; 8, arabinose; 9, fucose; 10, galactose; I.S. (internal standard), cinammic acid; 11, glucuronic acid; 12, galacturonic acid. (From Ref. 33.)

the contribution of the N-acetyl substituent at the C_2 position [33]. An important feature of this electrophoretic system was its usefulness in the quantitative determinations of the monosaccharide composition of various di- and oligosaccharides, such as lactose, melibiose, rutin, digitonin, and arabic gum, which were in good agreement with the theoretical values [33].

Using the borate complex formation approach, Bonn and co-workers [123] demonstrated the CZE separation of ethyl p-aminobenzoate derivatives of model monosaccharides. They reported virtually the same observations as those of Honda et al. [33] regarding the effect of pH and borate concentration on resolution of the derivatized saccharides. The resolution of selected pairs of saccharides increased monotonically with increasing borate concentration, and a maximum resolution was obtained in the pH range 10 to 11.

The high sensitivity of CZE in carbohydrate analysis was clearly demonstrated by Novotny and co-workers [96], who illustrated the electrophoretic separation of 13 CBQCA monosaccharide derivatives in about 22 min with separation efficiencies that ranged from 100,000 to 400,000 per meter (see Fig. 10). Besides tagging the sugar molecule with a fluorogenic group, the derivatization of monosaccharides with CBQCA provides each sugar derivative with an ionizable carboxylic group, thus allowing their electromigration. Therefore,

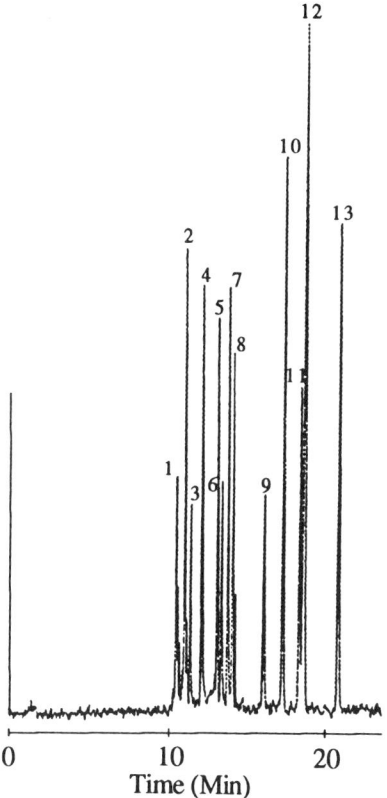

Fig. 10 Electrophoretic separation of a derivatized monosaccharide mixture. The derivatizing agent was CBQCA. Sample concentrations were 6.2 μM for glucosamine and galactosamine, 5.5 μM for galacturonic acid, and 4.4 μM for other sugars. Peak assignments: 1, D(+)-glucosamine; 2, D(+)-galactosamine; 3, D-erythrose; 4, D-ribose; 5, D-talose; 6, D-mannose; 7, D-glucose; 8, D-galactose; 9, impurity; 10, D-galacturonic acid; 11, D-glucuronic acid; 12, D-glucosaminic acid; 13, D-glucose-6-phosphate. Electrophoretic conditions were: buffer, 10 mM Na_2HPO_4/10 mM $Na_2B_4O_7 \cdot 10H_2O$, pH 9.40; capillary, 50 μm ID × 88 cm (58 cm effective length); applied voltage, 20 kV. (From Ref. 96.)

the inclusion of only 10 mM borate in the running electrolyte (pH 9.4) was enough to magnify small steric differences between closely related isomers and bring about their separation by CZE. Also, despite the fact that the introduction of a negatively charged group would weaken the complex formation with borate by virtue of coulombic repulsion (see Section II.A), the extent of complexation with the hydroxyl groups of CBQCA sugar derivatives was still enough to produce differential migration among the various derivatized monosaccharides. Thus the introduction of an ionizable group to a neutral sugar molecule through precolumn derivatization with a suitable tag is advantageous in CZE since a relatively low ionic strength borate buffer would be sufficient, a condition that would produce minimal Joule heating, and consequently high separation efficiency could be achieved.

Due to equal charge/mass ratios, the attachment of a charged group to closely related monosaccharide isomers through precolumn derivatization would only bring about group separation of the derivatives by direct CZE (i.e., in the absence of complex formation with suitable complexing ions) [124]. This is illustrated in Fig. 11a, where the positively charged 2-AP derivatives of the monosaccharide constituents of glycans were separated into groups using phosphate buffer containing small amounts of tetrabutylammonium bromide, pH 5.0. As shown in Fig. 11a, Gal and Man, differing only in the disposition of hydroxyl groups, emerged together as one peak separated from the single peak of GalNAc and GlcNAc, which have in their structures an additional acetyl group. This mixture was resolved when a 200 mM borate buffer, pH 10.5, was used as the running electrolye (Fig. 11b). At pH 10.5, the 2-AP sugar derivatives become deprotonated, reason for which a higher borate concentration than in the preceding case (i.e., CBQCA derivatives) is needed to complex them sufficiently with borate and in turn allow their differential electromigration.

The use of borate buffers was also extended to the analysis of reducing aldopentoses and aldohexoses as their 1-phenyl-3-methyl-5-pyrazolone (PMP) derivatives [101]. The PMP–sugar derivatives become negatively charged in aqueous basic solutions, due to the dissociation of the enolic hydroxyl groups. Under these conditions, and using 100 mM of sodium, potassium, or ammonium acetate as the running electrolyte, the PMP derivatives of isomeric aldopentoses or aldohexoses could not be separated because they have the same charge/mass ratio. In the presence of borate in the running electrolyte at alkaline pH, the PMP derivatives of aldohexoses or aldopentoses were readily converted to anionic borate complexes, and separated on the basis of the extent of complexation and the molecular size of the resulting complex. The separation efficiency was quite high, with a plate height of 4.1 μm for PMP–xylose. The mechanism of CZE separation of the PMP derivatives was the same as that of the 2-AP derivatives, although the optimum pH for borate complexation was shifted to 9.5 as opposed to 10.5 in the case of the 2-AP-derivatives, presumably due to the participation of the PMP substituent group in the complexation, perhaps through its hydroxyl groups (see Section II.C). An electropherogram depicting a typical separation of a mixture of six PMP–aldohexoses of the D-series is illustrated in Fig. 12. Also, the aldopentoses were well separated. But when a mixture of pentoses and hexoses is analyzed by CZE, the peaks of a few species of the pentose and hexose derivatives overlapped.

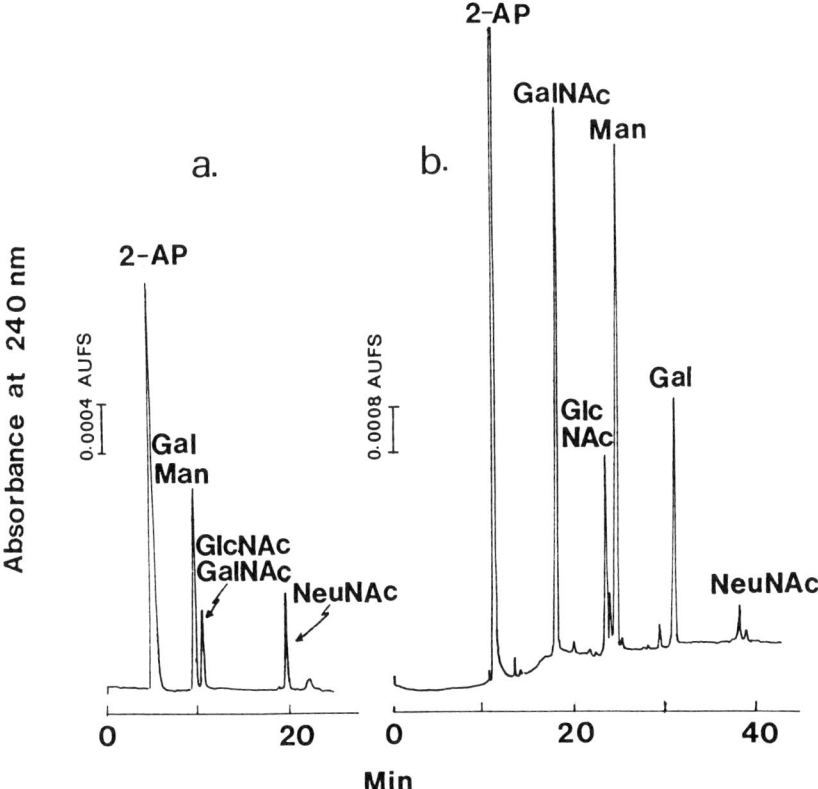

Fig. 11 Capillary zone electrophoresis of pyridylamino derivatives of standard monosaccharides. (a) Capillary, fused silica tube with polyether hydrophilic coating on the inner wall, 35 cm (to the detection point), 70 cm total length × 50 μm ID; electrolyte, 0.1 M sodium phosphate solution, pH 5.0, containing 50 mM tetrabutylammonium bromide; running voltage, 15 kV; current was about 55 μA; injection by electromigration for 8 s at 15 kV. (b) Capillary, uncoated fused silica tube, 50 cm (to the detection point), 80 cm total length × 50 μm ID; electrolyte, 0.2 M sodium borate, pH 10.5; running voltage, 18 kV; current was 75 μA; injection by electromigration for 1 s at 10 kV. (From Ref. 124.)

In another report, Honda et al. [52] demonstrated the complexation with divalent metal ions as an alternative to borate complexation to achieve CZE separation of monosaccharides. A mixture of five PMP-monosaccharides— arabinose, ribose, galactose, glucose, and mannose—initially coeluting using 100 mM sodium acetate as the background electrolyte, was fully resolved in an electrolyte system containing 20 mM calcium acetate. A typical elec-

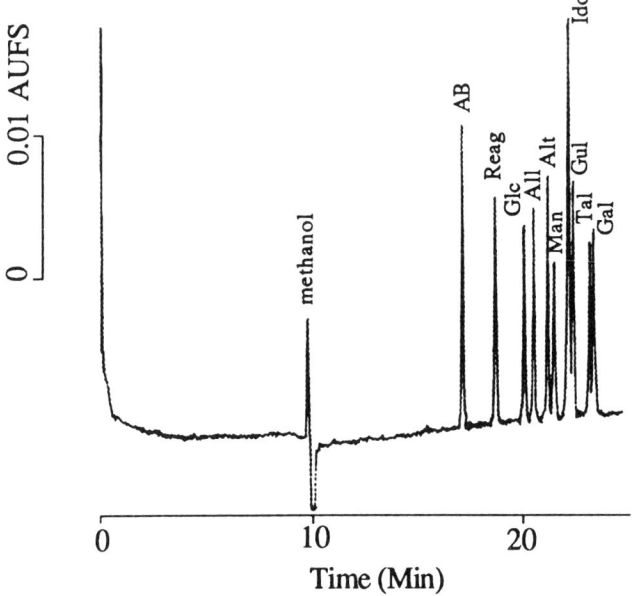

Fig. 12 Separation of PMP–aldohexoses by CZE. Capillary, fused silica tube, 63 cm (to the detection point), 78 cm total length × 50 μm ID; electrolyte, 0.2 M borate solution, pH 9.5; voltage, 15 kV; detection, UV at 245 nm. AB, amobarbital (internal standard); Reag, excess reagent (PMP); Glc, glucose; All, allose; Alt, altrose; Man, mannose; Ido, idose; Gul, gulose; Tal, talose; Gal, galactose. (From Ref. 101.)

tropherogram of four PMP–pentoses is shown in Fig. 13. The separation was presumably due to the relative ease of complexation of these derivatives with the metal ion. In fact, the PMP–pentoses eluted in the order of increasing complexing ability with the metal ion, which depends on the disposition of their hydroxyl groups (see Section II.A). Ribose–PMP with *erythro-erythro* disposed hydroxyl groups eluted first followed by lyxose–PMP (*erythro-threo*) and arabinose–PMP (*threo-erythro*) and then xylose–PMP (*threo-threo*). This interaction gives rise to a positive charge around the metal nucleus of the sugar–metal complexes, and consequently causes a relative reduction in the total negativity of the sugar derivatives, a condition that favors their migration toward the anode at a much slower rate than that of the unreacted reagent PMP (see Fig. 13). However, divalent metal ions have the tendency to adsorb electrostatically on the negatively charged fused silica wall, leading to a gradual inversion in the direction of the electroosmotic flow, from cathodal to anodal passing by zero EOF, as the capillary surface charge changes from negative to positive passing by neutral with

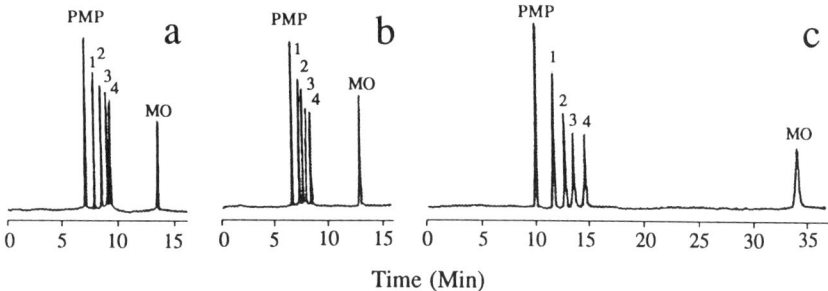

Fig. 13 Separation of pentose–PMPs in aqueous 100 mM solutions of calcium acetate (a), barium acetate (b), and strontium acetate (c). Capillary, fused silica (53 cm × 50 μm ID). Sample was introduced from the cathodic end of the capillary tube. Applied voltage, 10 kV; detection, UV at 245 nm. Peaks: PMP, excess reagent; 1, ribose–PMP; 2, lyxose–PMP; 3, arabinose–PMP; 4, xylose–PMP; MO, mesityl oxide (EOF neutral marker). (From Ref. 52.)

time [52]. With anodal flow, the electrophretic migration of the derivatives is in the same direction as the EOF, a condition that leads to rapid separation. The inversion of EOF is rather slow and can take up to 6 h to occur. To speed up the inversion and to ensure reproducible separations, the capillary column should be preconditioned before use with more concentrated metal ion solution than the one used as the separation electrolyte [52]. Other alkaline earth metal salts, including barium, strontium, and magnesium acetates, were also investigated [52]. While the elution order of selected PMP–aldopentoses stayed the same upon varying the nature of the metal ion in the electrolyte, as expected (see Section II.A) the electrophoretic mobility of the sugar–metal complexes was slightly higher with Ba^{2+} than with Sr^{2+} and Ca^{2+}, which is consistent with the fact that Ba^{2+} has a slightly larger ionic radius. Also, it seems that the binding of Ba^{2+} ions to the capillary surface is slightly stronger than that of Ca^{2+}, while the binding of Sr^{2+} is the weakest. This may explain why the anodal flow was higher in the presence of barium acetate electrolyte, and consequently the separation was faster (Fig. 13b). Overall, the separation efficiencies decreased in the following order: $Ba^{2+} > Ca^{2+} > Sr^{2+} \gg Mg^{2+}$ [52].

It is worth noting that the nature of the tagging agent used in precolumn derivatization can influence the migration pattern and resolution of the derivatized saccharides under otherwise identical conditions. For instance, while the ethyl p-aminobenzoate– [99] and PMP–sugar derivatives [101] behaved similarly to those derivatized with 2-aminopyridine in borate buffers, the p-aminobenzoic acid derivatives showed some differences [99]. This may be attributed to differences in the characteristic charges of the tags which would influence the extent of borate complexation of the derivatized sugars. p-Aminobenzoic acid differs from the other tags, among other things, by its carboxylic acid group. Another explanation has been provided in Ref. 99 to account for the differences in behavior among the various derivatives by postulating that while the sugar moiety in the p-aminobenzoic acid derivatives would assume an open-chain form that in the other two derivatives (i.e., ethyl p-aminobenzoate– and 2-AP–sugar derivatives) would be annular, a phenomenon that would affect the extent of complexation of borate ions, and consequently the migration and separation of the derivatized saccharides.

Underivatized monosaccharides were also analyzed by CZE using either alkaline borate buffers or high-pH electrolytes such as aqueous solutions of sodium hydroxide. The borate electrolyte systems worked well for direct UV detection at 195 nm [35] and RI detection [122], while the high-pH electrolyte systems allowed indirect detection of monosaccharides by fluorescence [39] or UV [40,99] and also direct detection by amperometry [41]. It is obvious that the migration pattern of derivatized saccharides would be different from that of underivatized saccharides in the presence of borate buffers due to the destabilizing effect of the tag on the sugar–borate complex (see Section II.A). As an example, the N-acetylhexosamines (i.e., GlcNAc and GalNAc) eluted in the order of GalNAc followed by GlcNAc [33] when derivatized with 2-AP and, as expected, in the reverse order when underivatized [35]. With the high-pH electrolyte systems, the migration order can be explained by the lability of the proton of the monosaccharide and by the charge/mass ratio. Using aqueous sodium hydroxide solution, pH 11.5, as the running electrolyte [39], sucrose (pK_a = 12.51) was detected first followed by glucose (pK_a = 12.35) and then fructose (pK_a = 12.03). This is the expected elution order when using positive polarity (anode to cathode). The stronger acid fructose is moving at a higher velocity upstream against the EOF, thus eluting last. Glucose, a weaker acid, eluted before fructose. Sucrose, a disaccharide, moves upstream against the flow at a lower velocity than glucose, due to its higher molecular weight, thus eluting first. The influence of acidity and size of the molecule is also shown in Fig. 6 [40]. They eluted in the order of increasing acidity: raffinose (a trisaccharide, pK_a = 12.74) < deoxyribose (pK_a = 12.52) < galactose (pK_a = 12.35) < glucose (pK_a = 12.35) < mannose (pK_a = 12.08).

B. Polysaccharides

Oligo- and polysaccharides can be classified as either homo- or heteropolymers, whether they consist of one or more than one type of monosaccharide units that alternate in a repetitive sequence. High-molecular-weight polysaccharides are usually analyzed through their degradation products (i.e., mono-, di-, or oligosaccharides). This depolymerization is usually achieved by chemical or enzymatic cleaving reactions. Only a few di-, tri-, and higher oligosaccharides occur free in nature, mostly in plants. In this section we attempt to outline the most useful HPCE separation methodologies for various oligosaccharides. For clarity, the oligo- and polysaccharides that constitute an integral part of glycoproteins and proteoglycans are treated separately in Sections III.C and III.D, respectively.

Disaccharides

The electrophoretic systems described in Section III.A for the separation of monosaccharides are equally applicable to the separation of disaccharides. In fact, most HPCE separation of disaccharides has utilized the principle of borate complexation as the separation media. With few exceptions, the mobility values of disaccharides tend to be lower than monosaccharides, due to the lower charge density on borate complexation.

While higher working temperatures in HPCE can lead in general to broader bandwidth and loss in separation efficiency, Hoffstetter-Kuhn et al. [35] have reported opposite observations in CZE of disaccharide–borate complexes. As shown in Fig. 14, by increasing the temperature of the capillary column and consequently of the background electrolyte, the separation efficiency of four common disaccharides improved dramatically. For instance, the baseline width of the lactose peak decreased from 140 s to 25 s as the temperature was increased from 20°C to 60°C. This was attributed to the increase in the reaction rate of the sugar–borate complex formation at higher temperatures. Besides the gain in efficiency, it was also noted that the sensitivity is significantly improved by raising the temperature. In fact, the carbonyl contents of the sugar solutions increase by a factor of 4 upon raising the temperature from 20°C to 60°C [35]. Furthermore, at elevated temperatures the electroosmotic flow velocity increased, thus providing a shorter separation time.

Thus far, the use of borate buffers seem to provide the best electrophoretic medium for the separation of disaccharides. In fact, when borate ions are substituted by divalent metal ions, incomplete CZE separations were observed [52]. Honda et al. [52] showed that a barium acetate buffer was unfavorable for the separation of cellobiose–melibiose and gentiobiose–lactose disaccharide pairs. Note that the electrophoretic mobilities of gentiobiose and lactose were 2.9 and 1.7×10^{-4} cm^2/V·s using 60 mM tetraborate, pH 9.3, as the background electrolyte [35]. The (1→6)-glycosidic linkage in gentiobiose seems to have favored more its ring opening as opposed to the (1→4) linkage in lactose. This would explain the higher mobility of lactose, since the borate complexation takes place preferentially in the open-chain form of the saccharide [35] (see Section II.A). CZE of disaccharides derived from glycosaminoglycans is discussed in Section III.D.

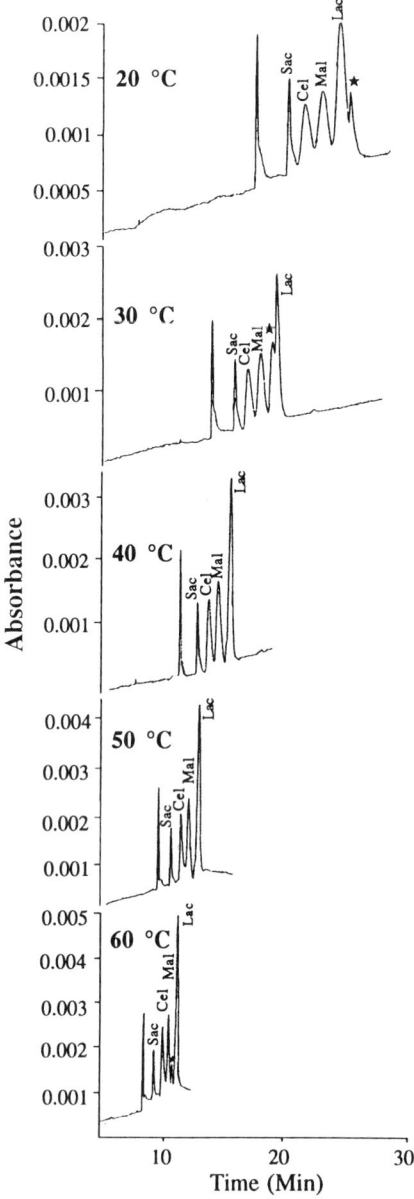

Fig. 14 Effect of temperature on CZE of underivatized disaccharides. Capillary, fused silica tube, 87 cm (to the detection point), 94 cm total length × 50 μm ID; electrolyte, 50 mM tetraborate solution, pH 9.3; running voltage, 20 kV. Sac, saccharose; Cel, cellobiose; Mal, maltose; Lac, lactose. (From Ref. 35.)

Linear Oligosaccharides

Several recent reports on CZE of oligosaccharides dealt with homologous series. Nashabeh and El Rassi [125] reported the separation of the pyridylamino derivatives of maltooligosaccharides having a degeee of polymerization (DP) from 4 to 7 using untreated fused silica capillaries. The positively charged sugar derivatives migrated ahead of the neutral marker (i.e., EOF marker) and were separated according to their size in the pH range 3.0 to 4.5 using 0.1 M phosphate solutions as the running electrolytes (Fig. 15). The inclusion of 50 M tetrabutylammonium bromide in the electrolyte solution decreased the EOF slightly, and consequently allowed the separation of the maltooligosaccharides at pH 5.0. However, as the pH approached the pK_a value of the derivatives (pK_a = 6.71), the homologus practically coeluted and moved virtually together with the bulk electroosmotic flow (EOF). In all cases, the resolution seems to decrease with an increased number of glucose units in the homologs. Similar observations were later reported

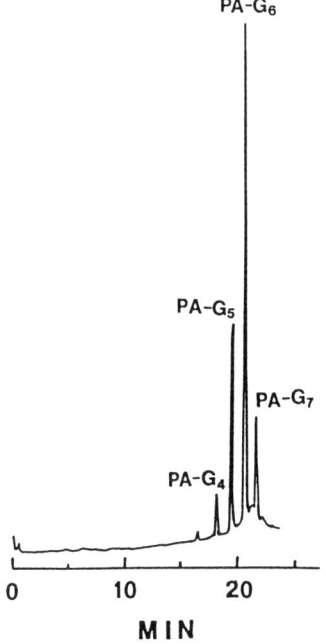

Fig. 15 Separation of pyridylamino derivatives of maltooligosaccharides. Capillary, fused silica tube, 50 cm (to the detection point) × 50 μm ID; electrolyte, 0.1 M phosphate, pH 4.0; voltage, 20 kV; injection, electromigration for 15 s at 18 kV; detection, UV 240 nm. Peaks: PA-G_4, PA-G_5, PA-G_6, and PAG$_7$ are the pyridylamino derivatives of maltotetra-, maltopenta-, maltohexa-, and maltoheptaose, respectively. (From Ref. 125.)

with the separation of standard CBQCA–maltooligodaccharide derivatives (DP 1 to 7) using a phosphate–borate buffer, pH 9.4 [97]. In this case, the negatively charged homologs migrating against the bulk flow, eluted in the order of decreasing size (i.e., maltoheptaose first and glucose last) due to the lower charge density of higher oligomers. Honda et al. [101] evaluated the CZE performance using borate buffers in the separation of PMP derivatives of homologous oligoglucans, such as α-(1→3)-linked (laminara-) oligoglucans, α-(1→6)-(isomalto-) oligoglucans, and β-(1→4)-(cello-) oligoglucans. All the homooligoglucans eluted in the order of decreasing size, and because of the unfavorable mass/charge ratio at high degrees of polymerization, the resolution between the homologs decreased as the number of recurring units increased. As expected, the rate of migration varied among series since the extent of their complexation with borate is largely influenced by the disposition of hydroxyl groups (i.e., by the type of interglycosidic linkage of the various oligosaccharides). In another report from the same laboratory [52], the PMP derivatives of a series of isomaltooligosaccharides were separated by CZE using an aqueous barium salt solution as the running electrolyte. The PMP isomaltooligosaccharides were separated from each other up to a DP value of 9, as opposed to a DP of 13 in the presence of borate, and the quality of the overall separation was not as good as that in borate buffer [101].

To examine the effect of the nature of the derivatizing agent on the spacing pattern between the migrating zones of homologs, a series of N-acetylchitooligosaccharides derivatized with either 2-AP or 6-AQ were separated by CZE [94] using the buffer system established for the maltooligosaccharides [125] as the running electrolyte, and a capillary having polyether interlocked coating (Fig. 16). As can be seen in Fig. 16b, plots of logarithmic electrophoretic mobility versus the number of GlcNAc in the homologous series were linear in the size range studied (i.e., up to a DP value of 6). Since 2-Ap and 6-AQ have similar characteristic charges, the slope of the straight lines, which is referred to as the N-acetylglucosaminyl group mobility decrement, δ_{GlcNAc}, was largely unaffected by the nature of the tag. This is to say that the spacing between two neighboring homologs is independent of the tagging agent provided that they carry similar charges.

To improve the resolution of homologous oligosaccharides with a higher degree of polymerization, two different approaches have been attempted. First, the use of coated capillaries having very low or virtually no electroosmotic flow can lead to a partial increase in the electrophoretic system resolution. Figure 17 depicts a typical electropherogram of the separation of pyridylamino derivatives of oligogalacturonide homologous series with a DP value in the range 1 to 18 on a coated capillary having a switchable (anodal/cathodal) EOF using 0.1 M phosphate solution as the running electrolyte [68]. The best separation of these oligosaccharides was achieved at pH 6.5, since at this pH the EOF is very low (for

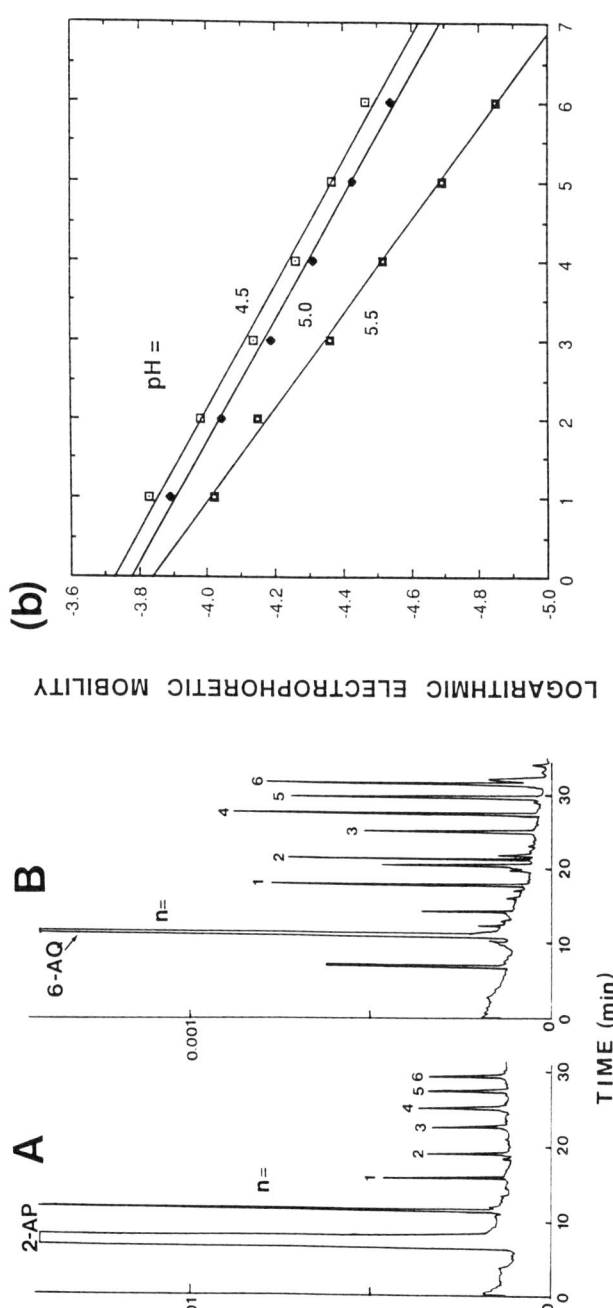

Fig. 16 (a) Electropherograms of pyridylamino (A) and quinolylamino (B) derivatives of N-acetylchitooligosaccharides. Capillary, fused silica tube with polyether interlocked coating on the inner walls, 50 cm (to the detection point), 80 cm total length × 50 μm ID; electrolyte, 0.1 M phosphate solution containing 50 mM tetrabutylammonium bromide, pH 5.0; voltage 18 kV. 2-AP = 2-Aminopyridine; 6-AQ = 6-aminoquinoline. (b) Plots of logarithmic mobility versus the number of GlcNAc residues in the 2-AP-GlcNAc$_n$ homologous series at various pH. Conditions are as in (a). (From Ref. 94.)

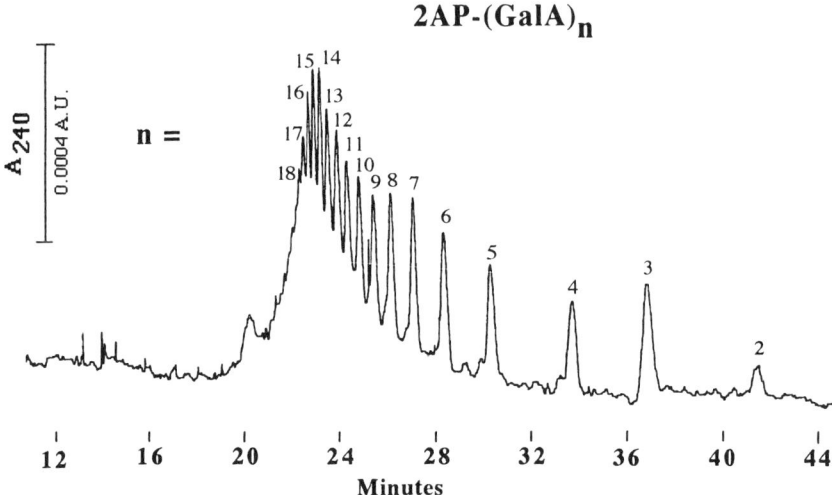

Fig. 17 Electropherogram of pyridylamino derivatives of oligogalacturonide homologous series obtained with switchable flow capillary at pH 6.0: running electrolyte, 0.10 M phosphate; hydrodynamic injection, 10 s, 20 cm differential height; applied voltage, -15 kV; detection, UV at 240 nm; capillary, 80 cm total length (50 cm to detection point) × 50 μm ID. (From Ref. 68.)

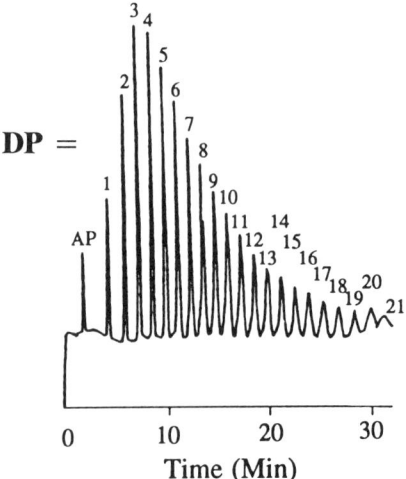

Fig. 18 Separation of pyridylamino derivatives of isomaltooligosaccharides by direct CZE. Capillary, fused silica coated with polyacrylamide, 25 μm ID, 20 cm in length; electrolyte 100 mM phosphate buffer pH 2.5; applied voltage, 8 kV; detection, UV at 240 nm. AP, 2-aminopyridine (excess reagent). (From Ref. 93.)

details on this coating, see Section II.B). Similarly, pyridylamino derivatives of isomaltooligosaccharides were completely separated from each other at least up to a DP value of 20 using fused silica capillaries in which the EOF was suppressed by chemically coating the capillary inner wall with linear polyacrylamide [93] (Fig. 18). In a second approach, Novotny and co-workers [80,81] suggested the use of polyacrylamide gel-filled capillaries with a high gel concentration for the separation of later-eluting polyionic oligosaccharides such as oligogalacturonides [82]. This is similar to the separation of oligonucleotides having the same charge/mass ratio using gel-filled capillaries. Separations in the gel media are based primarily on the sieving effect of the gel network. A comparison of the CZE separation of partially hydrolyzed dextrin 15 obtained with open tubular and gel-filled capillaries is shown in Fig. 19a and b, respectively. Note the improved resolution of the oligomers with a DP value 7 to 18 using the gel separation media. Although impressive separations can be achieved, the use of gel filled capillaries is accompanied with many difficulties such as poor gel-to-gel reproducibility, bubble formation under electrophoretic conditions and gel matrix collapse. A more convenient approach might be the use of coated capillaries with buffer containing polymeric species as sieving additives (see Section II.B).

Branched Oligosaccharides

Capillary electrophoresis was also applied to the separation of branched heterooligosaccharides derived form large xyloglucan polysaccharides (XGs) by enzymatic digestion [94]. Figure 20 illustrates the CZE separation of pyridylamino derivatives of xyloglucan oligosaccharides (2-PA-XG) obtained from cotton cell walls by cellulase digestion [96]. XGs possess a basic backbone identical to those of cellulose, a $(1\rightarrow4)$-β-linked D-glucan. Variations in XGs are caused by the differences in the nature and distribution of xylose, galactosyl–xylose, fucosyl–galactosyl–xylose, and in some cases arabinosyl–xylose side chains on the glucan backbone. Cellulase, a complex of enzymes, is able to digest the backbone of XGs after any glucosyl residue that does not subtend a side chain, thus liberating fragments of the polymer that reflect its branching patterns. The separation shown in Fig. 20 was performed on an interlocked polyether-coated capillary using 0.1 M phosphate containing 50 mM tetrabutylammonium bromide, pH 4.75. The peak numbering on the electropherogram (see Fig. 20) reflects the elution order obtained in reversed-phase chromatography (RPC) [126]. In CZE, the elution order was governed primarily by the number of sugar residues and the degree of branching, whereas in RPC the elution order was influenced primarily by the size of the oligosaccharide and the hydrophobic character of the sugar residues. For instance, fragment 4, which is smaller than fragment 3, was more retarded on the RPC column [126]. This may be attributed to the presence of fucosyl residue (i.e., 6-deoxygalactosyl) in structure 3, which is more hydrophobic than that of any other sugar residues in the molecule. The same reasoning can explain the elution order for fragments 6 and 7. In CZE, due to the fact that all 2-

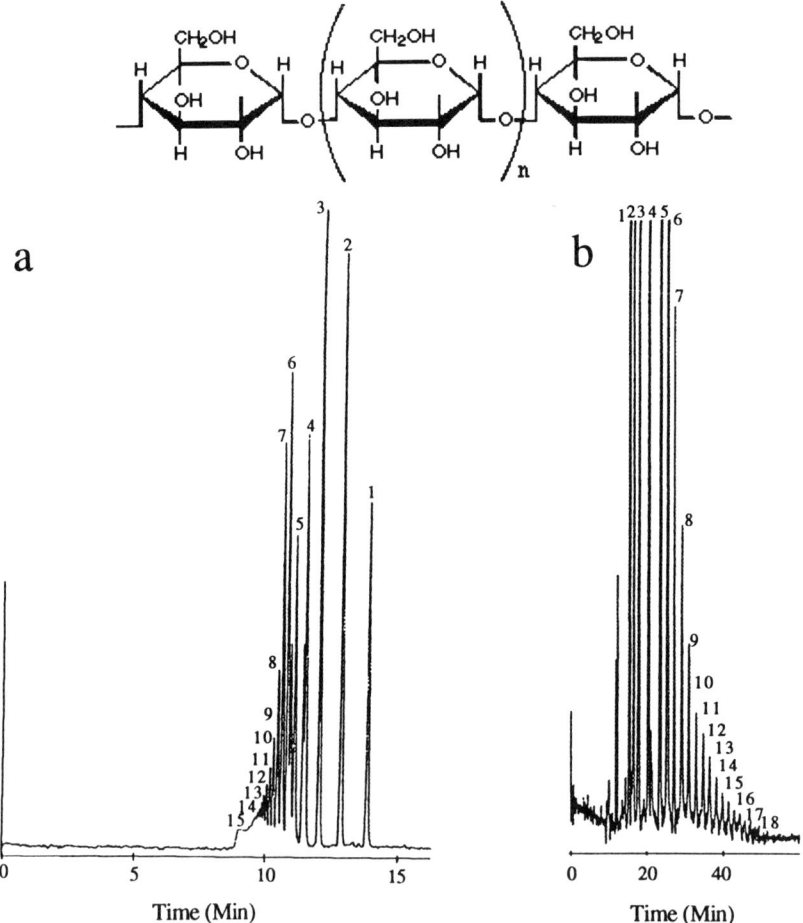

Fig. 19 Electrophoretic separation of CBQCA derivatives of a partially hydrolyzed polysaccharide (dextrin 15) in (a) open-tubular system and (b) polyacrylamide gel–filled column. (a) Capillary, 58 cm (to the detection point), 88 cm total length × 50 μm ID; electrolyte, 0.01 M disodium phosphate solution–0.01 M sodium teteaborate decahydrate, pH 9.4; running voltage, 20 kV. (b) Capillary, 19 cm (to the detection point), 26 cm total length × 50 μm ID; electrolyte, 0.1 M Tris–0.25 M borate–7 M urea, pH 8.33; applied voltage 269 V/cm. (From Ref. 81.)

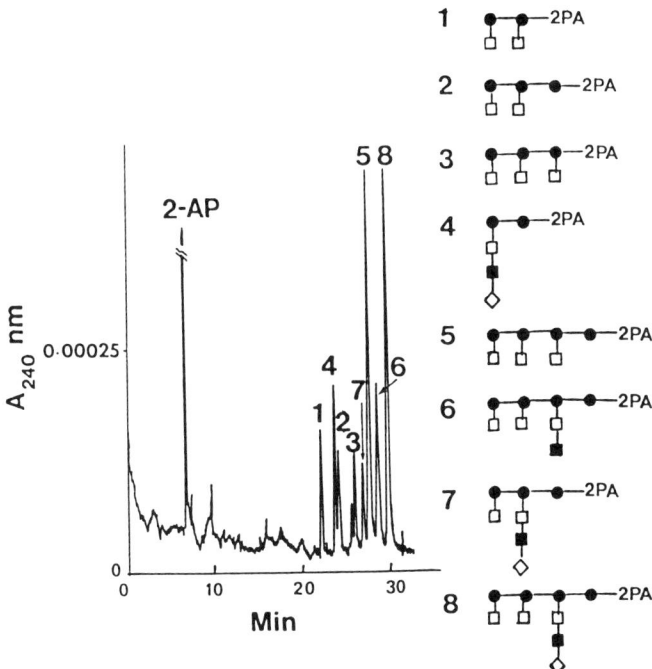

Fig. 20 Capillary zone electrophoresis mapping of pyridylamino derivatives of xyloglucan oligosaccharides from cotton cell walls. Capillary, fused silica tube with polyether interlocked coating on the inner walls, 50 cm (to the detection point), 80 cm total length × 50 μm ID; electrolyte, 0.1 M sodium phosphate solution containing 50 mM tetrabutylammonium bromide, pH 4.75; running voltage, 20 kV. Symbols: 2-AP, 2-aminopyridine; ●, glucose; □, xylose; ■, galactose; ◇ fucose. (From Ref. 94.)

PA-XG fragments possess the same charge, they migrated in the order of increasing size as the charge/mass ratio decreased. However, for the same number of residues but with slight differences in molecular weight, the less branched oligosaccharides eluted earlier than the more branched ones. In fact, structure 4, which has a slightly higher molecular weight than structure 2, eluted first. This may be explained by the fact that structure 2 is doubly branched as opposed to fragment 4, which is singly branched. The same behavior was observed for fragments 7 (doubly branched) and 5 (triply branched). Thus as the extent of oligosaccharide branching increased, the electrophoretic mobility decreased.

To interpret the electrophoretic behavior of the various 2-PA-XG fragments and to describe quantitatively the effects of the various sugar residues on their electrophoretic mobility, Nashabeh and El Rassi [94] have introduced a mobility indexing system for the branched xyloglucan oligosaccharides with respect to the linear pyridylamino derivatives of N-acetylchitooligosaccharides (2-PA-GlcNAc$_n$) homologous series, shown in Fig. 16a. The mobility index, MI, of the 2-PA-XG fragments were calculated using the following equation:

$$MI = 100n + 100 \left(\frac{\log \mu_s - \log \mu_{n+1}}{\log \mu_n - \log \mu_{n+1}} \right)$$

where μ_s is the electrophoretic mobility of the 2-PA-XG solute, and μ_n and μ_{n+1} are the electrophoretic mobilities for the two homologs with n and $n + 1$ repetitive units which eluted before and after the xyloglucan fragment, respectively. The difference between the mobility indices of an adjacent pair of 2-PA-XG fragments, ΔMI, referred to as the group mobility index decrement contributed by a sugar residue added to a parent oligosaccharide molecule, was estimated for selected 2-PA-XG fragments. The ΔMI values obtained for the adjacent pairs (1,2), (2,3), and (3,5), shown in Fig. 20, were almost identical: 52.5, 55.4, and 54.0, respectively. Thus the addition of a glucosyl residue to the linear core chain of the oligosaccharide showed a similar change in the mobility index decrement as the addition of a xylosyl residue at the glucose loci and behaved as one-half a GlcNAc residue in terms of its contribution to the electrophoretic mobility of the 2-PA-XG (see the structures in Fig. 20). However, the addition of a galactosyl residue to an already branched xylosyl residue exhibited less retardation [i.e., ΔMI of 38.4 for the xyloglucan pair (5,6)] than the addition of a glucosyl or xylosyl unit to the backbone of the xyloglucan oligosaccharide. The same observation was made about adding a fucosyl residue to a branched galactosyl residue. Thus, as the molecule becomes more branched, the addition of a sugar residue does impart a slightly less decrease in its mobility. This approach may prove valuable in correlating and predicting the effects of several parameters, such as the nature, position, and number of sugar residues, on the mobilities of complex carbohydrates. For clarity, the CZE of branched hetrooligosaccharides derived from glycoproteins is discussed in the following section.

C. Glycoproteins

Glycoproteins function as enzymes, transport proteins, receptors, hormones, and structural proteins. The carbohydrate (i.e., glycan) content of glycoproteins can vary from less than 1% to more than 60% by weight. Glycans perform many important biological roles, including direct or indirect influence on the activity of a glycoprotein, possible roles in clearance from circulation, targeting to a particular tissue, and influences on the solubility and stability of the protein [127]. Glycans are covalently attached to the polypeptide chains via serine or threonine (O-glycans) or via asparagine (N-glycans). Protein glycosylation can occur at two or more positions in the amino acid sequence, and the glycans at even a single position may be heterogeneous or may be missing from some molecules. This leads to populations of glycosylated variants of a single protein, usually referred to as glycoforms. However, the relative proportions of glycoforms are found to be reproducible, not random, and depend on (1) the environment in which the protein is glycosylated, such as the physiological state as well as the type of organism, tissue, and cell used in which the protein is made, (2) the manufacturing process, and (3) the isolation procedures. All these would affect the glycoform populations, thus affecting the function of a protein. In fact, work on tissue plasminogen activator [128] and uromodulin [129] has clearly demonstrated that the different glycoforms have differential activities. Thus the availability of high-resolution separation methods that allow the monitoring of the processes that affect the spectrum of glycoform populations of a given glycoprotein is of considerable importance. Although the full potentials of CZE has not been fully exploited in glycoform separations, the technique is projected to play an important role in this area. In this section we outline the various capillary electrophoretic systems employed in glycoprotein analysis. These approaches include (1) the resolution of glycoprotein's glycoforms, (2) glycopeptide mapping, and (3) oligosaccharide mapping.

Glycoforms

Kilàr and Hjertén [130] reported the separation of human transferrin glycoforms by HPCE. At least five components, corresponding to the di-, tri-, tetra-, penta-, and hexasialo transferrins, which differ from each other by one negative charge, were resolved by both CZE and capillary isoelectric focusing (CIF) due to charge differences in the carbohydrate moiety of the transferrin molecules. The capillary columns used in this study were coated with a layer of linear polyacrylamide on the inner wall to suppress EOF, and consequently, provide better resolution and sharper focusing of the closely related glycoforms by CZE and CIF, respectively (for details on the coating, see Section II.B). The authors also monitored the action pattern of neuraminidase, an exoglycosidase that liberates specifically the negatively charged sialic acids from the terminal nonreducing positions in glycans, on the electrophoretic behavior of the various isoforms. As shown in Fig. 21, the electrophoretic analyses of samples taken from the enzymatic digestion at various time intervals demonstrated the gradual removal of sialic acid as manifested by the changes in the relative proportions of the different isoforms with time. The electrophoretic pattern of the final product was completely different from the starting material and showed one main component, the asialotransferrin. Thus, in the case of transferrins, the major source of microheterogeneity seems to be the variation in the terminal sialic acid of the glycans. This microheterogeneity leads to broad or smeared bands in traditional electrophoresis. HPCE, with its high separation efficiencies, on the contrary, is very suitable for such difficult separations.

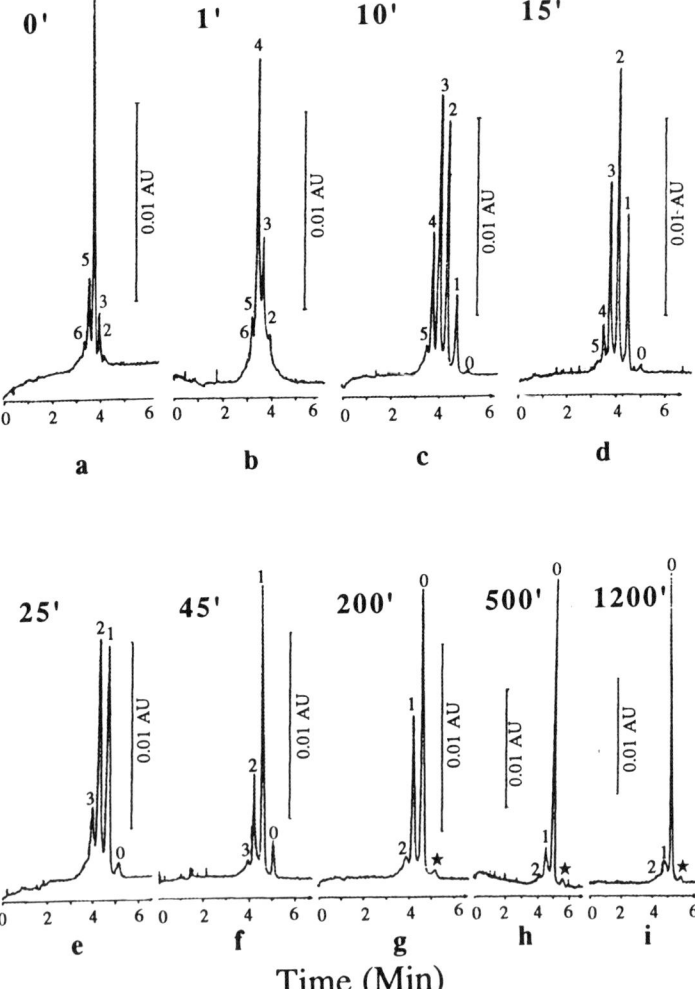

Fig. 21 CZE of iron-free transferrin following incubation with neuraminidase. Capillary, 18.5 cm × 50 μm ID; electrolyte, 18 mM Tris–18 mM boric acid–0.3 mM EDTA, pH 8.4; running voltage, 8 kV. The samples for electrophoresis were taken after various incubation times: (a) 0; (b) 1; (c) 10; (d) 15; (e) 25; (f) 45; (g) 200; (h) 500; (i) 1200 min. The proportions of the transferrin isoforms (asialo, mono-, di-, trisialo, etc., marked 0, 1, 2, 3, etc.) changed with time. The sample taken after 20 h still contained transferrin molecules having one and two sialic acids. The small peak (labeled with an asterisk) appeared after 50 to 80 min, but did not increase in size on prolonged incubation time (g–i). (From Ref. 130.)

The potentials of CZE in separating glycoforms were also demonstrated by Grossman and co-workers [131], who reported the separation of ribonuclease B (RNase B) isoforms using 20 mM CAPS [(cyclohexylamino)propanesulfonic acid], pH 11.0. The glycoforms of RNase B have the same polypeptide chain but differ in the extent of glycosylation on the asparagine 34 amino acid residue. It has been found that this single glycosylation site can accommodate at least five different high-mannose glycans. At the high pH values used in this study, the glycoforms of RNase B acquire higher negative charges as the carbohydrate moieties became ionized, a condition that permitted their separation. More recently, a more elaborate study on the same glycoprotein was reported by Rudd et al. [132]. In this work, the separation of the five different glycoforms of RNase B has been achieved through the formation of anionic borate complexes with the hydroxyl groups of the glycan moiety (Fig. 22). the relative proportions of the various glycoforms correlated with the relative proportions of the high-mannose glycan populations (i.e., Man$_9$-Man$_5$) determined by other more established

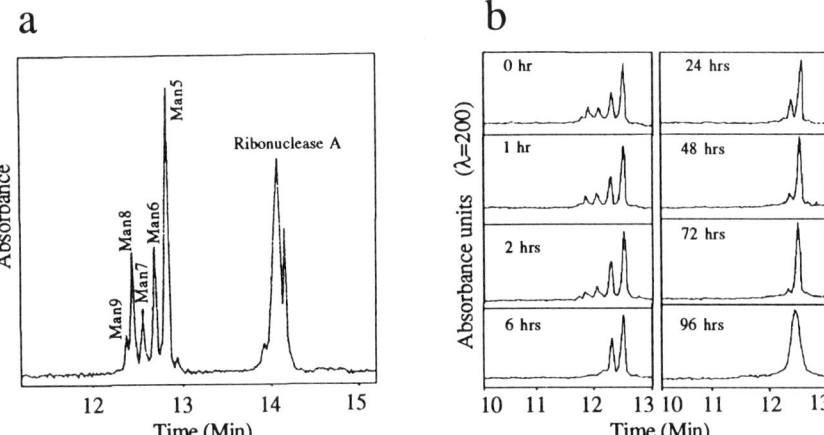

Fig. 22 (a) CZE profile of ribonuclease showing the nonglycosylated form of the protein, ribonuclease A, and the glycoforms of the same protein, collectively known as ribonuclease B. RNase A is a contaminant of RNase B as supplied by Sigma. RNase B remained unaffected during the digestion of the oligosaccharide component of RNase B with *A. saitoi* α(1-2) mannosidase. (b) CZE profile of RNase B showing the time course for the digestion of the glycoprotein with the exoglycosidase, *A. saitoi* α(1-2) mannosidase. Capillary, fused silica 72 cm × 75 μm ID; applied voltage, 1 kV for 1 min and 20 kV for 19 min, temperature, 30°C; detection, UV at 200 nm; injection 1.5 s; electrolyte, 20 mM phosphate containing 50 mM sodium dodecyl sulfate, 5 mM borate, pH 7.2. (From Ref. 132.)

analytical methods [e.g., mass spectrometry, high-performance anion-exchange chromatography (HPAEC), and size-exclusion chromatography on Bio-Gel P4]. To further substantiate the presence of the various glycoforms of RNase B, the time course for the digestion of the protein with *A. saitoi* α(1-2) mannosidase, an exoglycosidase that specifically cleaves the mannose from the nonreducing end, was monitored by CZE. Figure 22b shows that after 25 h the glycoform populations carrying Man_9–Man_6 structures were all reduced to a single population carrying Man_5. This once again confirms that HPCE offers a direct method for analyzing glycoforms at the protein level with high resolution and precision. Therefore, HPCE may become a useful tool, for instance, in studying the variation in glycoform populations from one cell type to another or in monitoring the production of a recombinant glycoprotein.

An additional glycoprotein, ovalbumin, was recently analyzed for its content of glycoforms by CZE [133]. Here the microheterogeneity of the glycoprotein is augmented by its phosphate content. Ovalbumin has one asparagine residue that can accommodate at least nine different carbohydrate structures of the high-mannose and hybrid-type N-glycans. There are also two potential phosphorylation sites at two serine residues, one at position 68 and the other at position 344. In this study, the various glycoforms were separated via borate complex formation with the hydroxyl groups of the carbohydrate moieties of the protein using untreated fused silica capillary. This glycoprotein is a strongly acidic species and therefore would not undergo adsorption onto the naked capillary surface when using alkaline borate. Puterscine (i.e., 1,4-butanediamine), a doubly charged cationic species, was added in small amounts (1 mM) to the borate buffer to slow the electroosmotic flow and bring about improved resolution of the ovalbumin glycoforms. Using these conditions, five major protein peaks were separated, indicating the presence of protein glycoforms (see Fig. 23). Upon dephosphorylation of the glycoprotein with calf intestinal alkaline phosphatase or potato acid phosphatase, the five peaks were still resolved but shifted in the position of a more rapid migration time, a behavior consistent with a loss of negative charge. Based on this observation, it was suggested that all ovalbumin glycoforms are phosphorylated to the same degree and that heterogeneity among ovalbumin isoforms resides solely in the carbohydrate structures. Also, the same electrophoretic system was shown to permit the separation of pepsin glycoforms.

Another important application of HPCE has been in the area of recombinant glycoproteins. Yim reported the fractionation of the human recombinant tissue plasminogen activator (rtPA) by CZE [134]. Tissue plasminogen activator is a fibrin-specific protein that has been approved for the treatment of myocardial infractions. The CZE analysis of two main glycosylation variants (types I and II) of the same glycoprotein showed different electrophoretic migration patterns. The study further elucidated the microheterogeneity of the glycoprotein, as was manifested by the partial resolution of almost 15 glycoforms in a protein that has

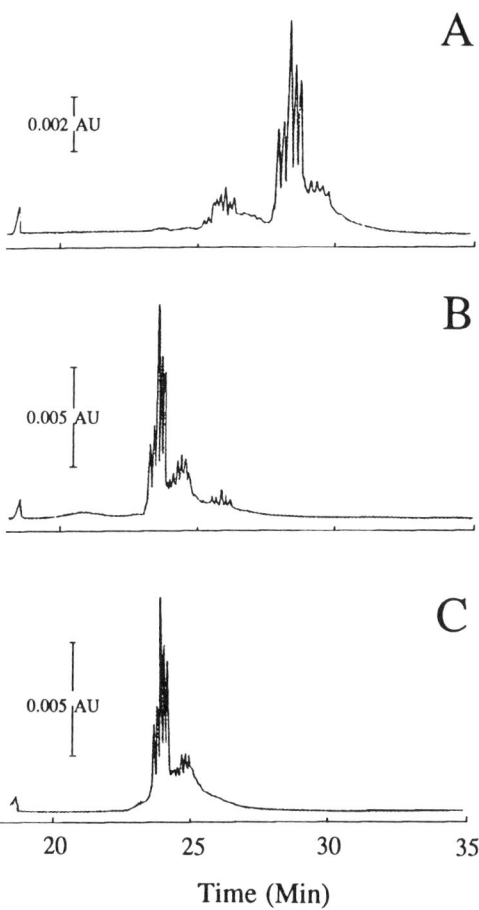

Fig. 23 Enzymatic dephosphorylation of ovalbumin: (A) untreated ovalbumin (intact); (B) ovalbumin treated with calf intestinal alkaline phosphatase; (C) ovalbumin incubated with potato acid phosphatase. Capillary, 87 cm × 50 μm; electrolyte, 100 mM borate containing 1 mM putrescine; applied voltage, 25 kV. (From Ref. 133.)

only four possible N-glycosylation sites. This report compared the CZE profile of an rtPA sample to that of a desialylated rtPA obtained through neuraminidase treatment, an approach similar to that introduced by Kilàr and Hjertén for human transferrin [130]. The desialylated rtPA exhibited a much simpler CZE profile, indicating that the glycoprotein microheterogeneity is mostly the result of various levels of sialylation. Along the same lines, Watson and Yao [135] extended the use of CZE to the separation of glycoforms of recombinant human granulocyte-colony-stimulating factor (rhGCSE) produced in Chinese hamster ovary cells. This glycoprotein contains only two O-linked carbohydrate moieties that differ only in having one or two sialic acid residues. Due to its relative simplicity compared to other more complex glycoproteins, the rhGCSE yielded two well-resolved and equally sized peaks using phosphate–borate buffer pH 8.0 and an untreated fused silica capillary. Under these conditions, the acidic protein was repelled from the negatively charged capillary wall, and no apparent solute adsorption was observed. The resolution of the two equally present glycoforms was further enhanced upon the addition of 2.5 mM of 1,4-diaminobutane to the phosphate–borate buffer. The 1,4-diaminobutane has apparently reduced the electroosmotic flow and further minimized solute–wall interaction. These effects allowed improved resolution, but prolonged the separation time. As expected, the two glycoforms migrated in the order of increasing numbers of sialic acids, since the separation was carried out in the positive polarity (i.e., anode to cathode). Of course, when the glycoforms were incubated with neuraminidase, a single peak was obtained that eluted earlier than either of the original two sialylated species.

The utility of CZE was also demonstrated in the separation of the glycoforms of recombinant human erythropoietin (r-HuEPO) [136], a glycoprotein hormone produced in the kidney of adult mammals that acts on bone marrow erythroid progenitor cells to promote their development into mature blood cells. The glycoprotein has a polypeptide backbone of 165 amino acids and contain two types of carbohydrates: three N-linked complex oligosaccharides at the asparagine positions 23, 38, and 83 and one O-linked oligosaccharide chain at serine position 126. The effects of the pH, buffer type, organic additives, and capillary preequilibration time on the resolution of the microheterogeneity of r-HuEPO were investigated. The main factors for improving the resolution were regulation of the electroosmotic flow of the running buffer and the reduction of solute–wall interactions. For instance, in the pH range 6.0 to 9.0, pH 6.0 was found to give the best resolution. At this pH, a combination of two effects, reduction in the EOF and increase in the differences in charge between the glycoforms, resulted in enhanced separations. Optimum resolution of the four glycoforms was obtained with a mixed acetate–phosphate buffer.

Glycopeptide Mapping

Besides its important role in the characterization of closely related glycoforms, CZE with its high resolving power and unique selectivity is well suited for the separation and characterization of peptide and glycopeptide fragments of glycoproteins. Figure 24 illustrates the CZE mapping of the tryptic peptide fragments of human α_1-acid glycoprotein (AGP) as well as the submapping of its glycosylated and nonglycosylated fragments [124]. Prior to CZE runs, the entire digest was first fractionated into peptide and glycopeptide fragments on a silica-bound concanavalin A (Con A) column, a lectin-affinity sorbent. Three pooled

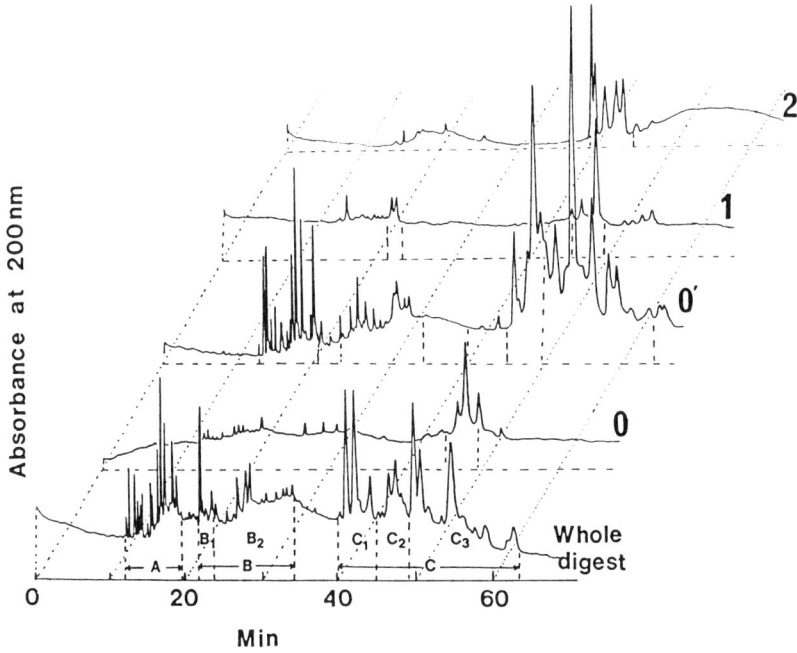

Fig. 24 Capillary zone electrophoresis tryptic mapping and submapping of human AGP. Capillary, fused silica tube with hydrophilic coating on the inner walls, 45 cm (to the detection point), 80 cm total length × 50 μm ID; electrolyte, 0.1 M phosphate solution, pH 5.0; running voltage, 22.5 kV; injection by electromigration for 4 s at 22.5 kV. Fraction 0, Con A nonreactive (excluded from the column); fraction 0', Con A nonreactive (unretained by the column); fraction 1, Con A slightly reactive (eluted with buffer); fraction 2, Con A strongly reactive (eluted with the haptenic sugar). (From Ref. 124.)

fractions were obtained, the first two being Con A nonreactive and Con A slightly reactive eluted with 20 mM phosphate pH 6.5 containing 0.1 M NaCl, while the third fraction interacted strongly with the Con A column and eluted with the heptanic sugar (i.e., methyl-α-D-mannopyranoside). The three fractions were then analyzed by CZE using a fused silica capillary with fuzzy polyether coating and 0.1 M phosphate, pH 5.0, as the running electrolyte. As seen in Fig. 24, the CZE mapping of the entire digest reveals the microheterogeneity of the glycoprotein as manifested by the excessive number of peaks for a protein of 181 amino acid residues with 20 trypsin cleavage sites (8 lysine and 12 arginine residues). In fact, by neglecting all sources of heterogeneity in the protein, the tryptic digest should result only into 12 peptide and 5 glycopeptide fragments, and three single amino acids (two lysine and one arginine). However, more than any other serum glycoproteins, AGP is a highly heterogeneous protein [137]. One of the unique aspects of the primary structure of polypeptide chain of pooled human AGP is its peculiar structural polymorphism. Substitutions were found at 21 of the 181 amino acids in the single polypeptide chain [138], which is responsible in part for multiple peptide and glycopeptide fragments in the tryptic digest. Another source of multiple fragments in the tryptic digest is the microheterogeneities of the oligosaccharide chains attached [137]. As shown in (*13*), there are five carbohydrate classes attached to AGP having different degrees of branching and sialylation. Classes A, B, and C are the bi-, tri-, and tetraantennary complex N-linked glycans, respectively, whereas BF and CF are the fucosylated B and C structures [138]. Two additional glycans exist [139]; one has two additional fucose linked to the GlcNAc residues marked with asterisks, and one has an outer chain prolonged by Galβ1-4GlcNAc at either of the Gal residues marked with an arrow. These different glycan structures are the major source of AGP microheterogeneity. In fact, the variation in the terminal sialic acid causes charge heterogeneity in the glycopeptide fragments cleaved at the same location by trypsin. The differences in the extent of glycosylation among a population of the protein molecules lead to fragments having the same peptide backbone but with or without carbohydrate chains, and the variation in the nature of the oligosaccharide chains at each glycosylation site yields several glycopeptides that have the same peptide backbone but differ in their oligosaccharide structures [124].

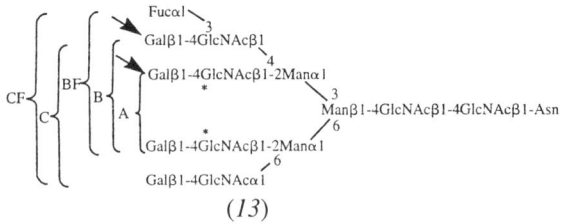

(*13*)

As can be seen in Fig. 24, the CZE submapping of Con A reactive peptides produced peaks that are missing from the submaps of all other collected fractions (i.e., 0, 0′, and 1) but whose components are found in the entire map (see area C_1, Fig. 24). This approach allows the monitoring of a group of peptides as well as the assessment of glycosylated fragments in the entire map. This methodology is expected to work also with other glycoproteins, and the CZE submapping of all the glycosylated tryptic fragments with different types of glycans may require the use of more than one lectin column in the prefractionation step [124].

Glycans

The potentials of CZE were also demonstrated in the analysis of branched oligosaccharides derived from glycoproteins. Using ovalbumin as a model protein, Honda et al. [93] demonstrated the separation of glycans by CZE with on-

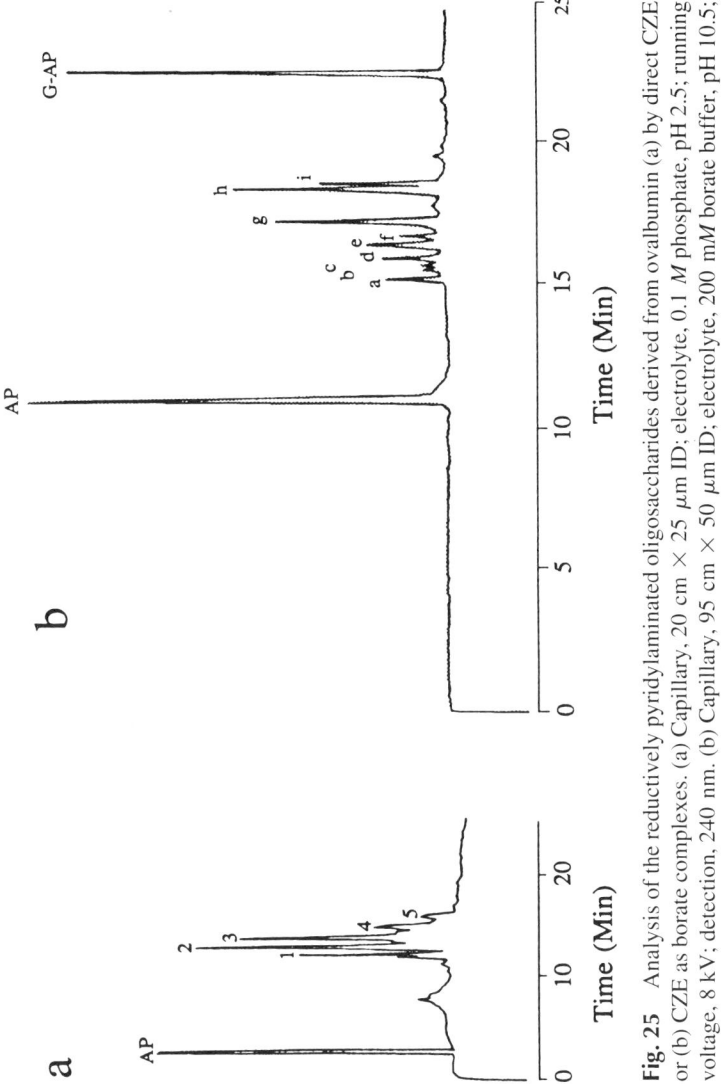

Fig. 25 Analysis of the reductively pyridylaminated oligosaccharides derived from ovalbumin (a) by direct CZE or (b) CZE as borate complexes. (a) Capillary, 20 cm × 25 μm ID; electrolyte, 0.1 M phosphate, pH 2.5; running voltage, 8 kV; detection, 240 nm. (b) Capillary, 95 cm × 50 μm ID; electrolyte, 200 mM borate buffer, pH 10.5; running voltage, 20 kV; fluorescence detection with excitation at 316 nm and emission at 395 nm. Peak assignments in (a): 1, heptasaccharide; 2, octasaccharides; 3, nonasaccharide; 4, decasaccharide; 5, undecasaccharide. Peak assignments in (b): refer to the illustration provided in the text; peaks b and c were not assigned. (From Ref. 93.)

column fluorometric detection. In this study, the oligosaccharides of ovalbumin were released with anhydrous hydrazine and tagged with 2-AP after re-N-acetylation. As shown in Fig. 25, the oligosaccharides were electrophoresed using two different electrolytes, an acidic phosphate buffer whereby the derivatized glycans are positively charged (cationic immonium ions) due to the protonation of the amino group of the tag (i.e., direct CZE) and an alkaline borate buffer that allows the in situ conversion of the derivatives to anionic borate complexes (i.e., indirect CZE). Both modes of separation demonstrated characteristic features based on their separation mechanisms. Direct CZE gave good separation of the oligosaccharide derivatives on the basis of molecular size (Fig. 25a), or in other words, the number of monosaccharide units, but could not resolve solutes having the same degree of polymerization. On the other hand, CZE as borate complexes separated the oligosaccharide derivatives based on structural differences in the outer monosaccharide residues (Fig. 25b). The greater the number of unsubstituted mannose units, the more retarded are the derivatives [93]. However, the borate system failed to resolve high-mannose oligosaccharides having the same number of outer mannose residues, such as structures h_1 and h_2 in (*14*). Similarly, hybrid-type oligosaccharides having the same number of peripheral mannose or galactose residues, but differing in the total number of monosaccharide units were not resolved [see structures g_1 and g_2 in (*14*)]. In both cases, however, the use of direct CZE with phosphate buffers gave satisfactory separations. In another report from the same laboratory [140], the authors extended this approach to an analysis of 32 different *N*-glycosidically linked asialooligosaccharides. The oligosaccharides were cleaved from various glycoproteins, including fetuin, human transferrin, immunoglobulin G, human α_1-acid glycoprotein (AGP), ribonuclease B (RNase B), and invertase, which can be classified into three major categories: complex, high-mannose, and hybrid glycans. Two-dimensional plots of the relative mobilities of the derivatives in the dual separation modes described above (i.e., phosphate and borate buffers) to reductively pyridylaminated glucose showed that the three types of oligosaccharides were distributed in separate domains. Furthermore, in each domain the plot moved toward the left as the degree of polymerization increased and upward as the number of peripheral mannose and galactose residues increased. It was suggested that such plots can be used as a convenient tool for the approximate identification of glycans.

Capillary electrophoresis can also be used to elucidate the differences in glycan structures of the same glycoprotein but from different sources [124]. In this regard, both human and bovine AGP oligosaccharidees were cleaved from their corresponding glycopeptide fragments by treating the tryptic digest of the proteins with peptide-*N*-glycosidase F (PNGase F), an endoglycosidase that cleaves all types of N-linked oligosaccharides between the asparagine and the carbohydrate units. The CZE mapping of the pyridylamino derivatives of the liberated oligosaccharides is portrayed in Fig. 26. Both electropherograms show

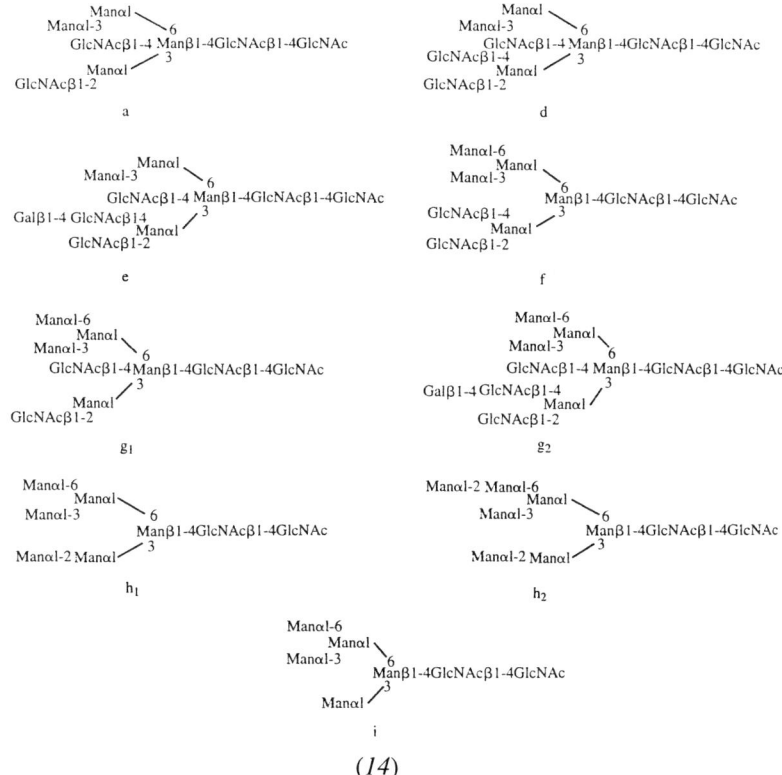

(*14*)

several well-defined peaks and a few minor peaks, eluting after the excess 2-aminopyridine. The glycans of AGP are known for their microheterogeneities, caused primarily by variation in their terminal sialic acid [see (*13*) for the glycan structures of AGP], a fact that may explain the presence of several peaks in the CZE maps. However, the map of bovine AGP glycans shows an elution pattern different from that of the oligosaccharides derived from the human glycoprotein. Both human and bovine AGP have been found to have the same sialic acid, galactose, and mannose content [141]. The major differences are such that 50% of the sialic acid in bovine AGP are *N*-glycolylneuraminic acid, and the fucose content is very low. These differences are accentuated by CZE mapping of both types of glycans (compare Fig. 26a and b). Based on these results, CZE can play an important role in the field of glycan separation and characterization.

More recently, Hermentin et al. [142] analyzed the reducing oligosaccharides released from AGP using both high-pH anion-exchange chromatography with pulsed amperometric detection (HPAEC-PAD) and CZE with UV detection at 190 nm. According to the authors, the CZE analysis proved to be 4000 times more sensitive than HPAEC-PAD. In fact, the carbonyl function of the N-acetyl

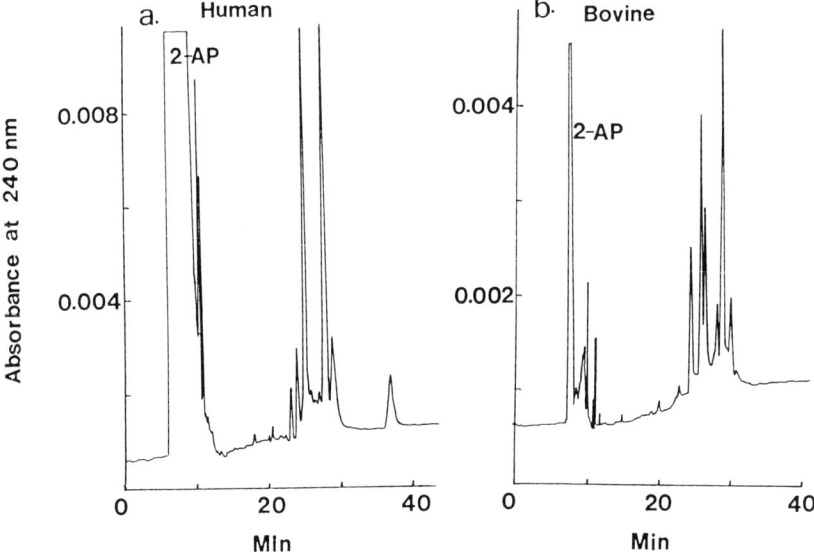

Fig. 26 Capillary zone electrophoresis mapping of pyridylamino derivatives of human (a) and bovine (b) AGP oligosaccharides. Capillary, fused silica tube with hydrophilic coating on the inner walls, 45 cm (to the detection point), 80 cm total length × 50 μm ID; electrolyte, 0.1 M phosphate solution, pH 5.0, containing 50 mM tetrabutylammonium bromide; running voltage, 18 kV; current, 80 μA; injection by electromigration for 2 s at 18 kV. (From Ref. 124.)

and carboxyl groups present in the molecules enabled their direct UV detection at concentrations in the femtomole region. This approach has the advantage of avoiding derivatization and sample cleanup processes. In this study, the authors also compared the mapping profiles of AGP glycans released by conventional hydrazinolysis or by digestion with peptide-N-glycosidase F. Hydrozinolysis proved best with practically no loss of N-acetylneuraminic acid, while PNGase F digestion resulted in partial desialylation of the liberated N-glycans in the presence of SDS [142].

High-mannose oligosaccharides released from RNase B by digesting the protein with PNGase F were also analyzed by CZE [94]. As shown in Fig. 27, the peak labeled with an arrow was identified as $(GlcNAc)_2$-Man_5 using an oligosaccharide standard. As pointed out earlier, the polypeptide chain of RNase B is known to have only one glycosylation site, which can accommodate five different high-mannose glycans. This may explain the presence of several peaks in the CZE map besides that of $(GlcNAc)_2$-Man_5 oligosaccharide, which according to Liang

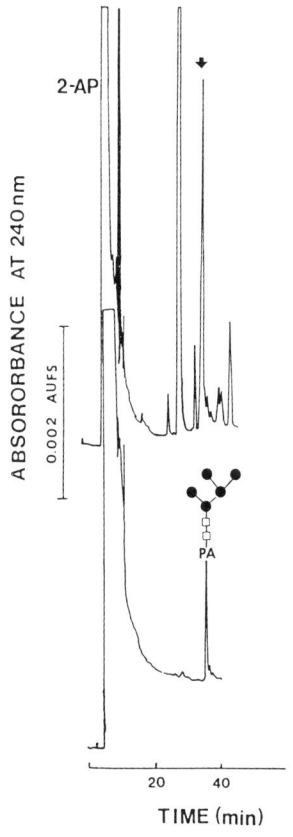

Fig. 27 Capillary zone electrophoresis mapping of 2-pyridylamino derivatives of high-mannose oligosaccharides cleaved from bovine ribonuclease B (top electropherogram), and of 2-PA-(GlcNAc)$_2$-Man$_5$ standard (lower electropherogram). Capillary, fused silica tube with polyether interlocked coating on the inner walls, 50 cm (to the detection point), 80 cm total length × 50 μm ID; electrolyte, 0.1 M phosphate solution containing 50 mM tetrabutylammonium bromide, pH 5.0; running voltage, 18 kV. 2-AP, 2-aminopyridine; □, GlcNAc; ●, mannose. (From Ref. 94.)

et al. [143], is the most predominant carbohydrate moiety of bovine RNase B. Additional CZE analysis of glycans was reported by Liu et al. [96]. N-Linked oligosaccharides from bovine fetuin were released through hydrazinolysis and then derivatized with CBQCA. With on-column LIF detection, this tag permitted the CZE of subpicogram amounts in a phosphate–borate buffer, pH 9.5. Four major peaks as well as few minor peaks were well resolved.

D. Glycosaminoglycans

Glycosaminoglycans (GAGs), also called mucopolysaccharides, are unbranched polysaccharides of alternating uronic acid and hexosamine residues. They occur naturally in the cartilage and other connective tissues, which are collectively called the ground substance. Proteins and GAGs in the ground substance aggregate covalently and noncovalently to form a variety of macromolecules of enormous molecular mass (up to tens of millions of daltons) known as proteoglycans or mucoproteins [144]. GAGs exhibit a variety of biological functions and can be altered in disease states [145]. In GAG, while the amnio group of the hexosamine residue is either N-acetylated or N-sulfated, the uronic acid may be either D-glucuronic acid or L-iduronic acid. Moreover, the repeating disaccharide units (i.e., uronic acid–glucosamine disaccharide) are O-sulfated to varying degrees at C_6 and/or C_4 of the various glucosamine residues and at C_2 of the uronic acid residues. Hyaluronic acid, chondroitin sulfates (chondroitin-4-sulfate and chondroitin-6-sulfate), dermatan sulfate, keratan sulfate, heparin, and heparan sulfate are the most common GAGs. Usually, GAGs occur together, and there is no ideal method for the separation of various GAGs. Therefore, prior to their analysis, the GAGs are depolymerized via various polysaccharide lyases (e.g., chondroitinase and hyaluronidase), a process that cleaves the hexoaminide linkages to give the disaccharide repeating units with the unsaturated uronic acid residue at the nonreducing termini. Using these approaches, several researchers have applied HPCE to the separation of the compositional saccharides of GAGs from various cartilage tissues.

Linhardt and co-workers [115] applied CZE to the separation and quantitative determination of the disaccharides derived from chondroitin sulfate, dermatan sulfate, and hyaluronic acid. Exhaustive treatment of these GAGs with polysaccharide lyases released nine different disaccharides bearing unsaturated uronic acids [see (15)] that can be detected by UV absorbance at 232 nm without prior derivatization (molar absorptivity, $\epsilon = 5000$ to $6000\ M^{-1}\ cm^{-1}$). These disaccharides, having a net charge from -1 to -4, were well resolved by CZE using a borate buffer pH 8.8, primarily on the basis of their net charge and secondly, on the basis of charge distribution. The use of phosphate buffer, in place of borate buffer, at pH 8.8 yielded an increase in the separation time and a decrease in resolution. The nonsulfated disaccharides 1 and 2 eluted first from the capillary followed by the monosulfated 3–5, the disulfated 6–8, and the trisulfated 9 disaccharides [for structures, see [15]. Despite the fact that the two nonsulfated disaccharides, structures 1 and 2, differed only by the chirality at C_4 in the hexosamine residue, these disaccharides were well resolved by CZE. In another report from the same laboratory [113], the authors investigated the electrophoretic behavior of eight commercial disaccharide standards derived from heparin, haparan sulfate, and derivatized heparins of the structure $\Delta UA2X(1\rightarrow 4)$-

[Structural diagram showing chondroitin sulfates, dermatan sulfate, and hyaluronic acid, with chondroitin lyase and hyaluronate lyase producing disaccharide products]

2. [Δdi-0S] where $X^2, X^4, X^6 = H$
3. [Δdi-6S] where $X^2, X^4 = H, X^6 = SO_3^-$
4. [Δdi-4S] where $X^2, X^6 = H, X^4 = SO_3^-$
5. [Δdi-UA2S] where $X^4, X^6 = H, X^2 = SO_3^-$
6. [Δdi-S_E] where $X^4, X^6 = SO_3^-, X^2 = H$
7. [Δdi-S_D] where $X^2, X^6 = SO_3^-, X^4 = H$
8. [Δdi-S_B] where $X^2, X^4 = SO_3^-, X^6 = H$
9. [Δdi-triS] where $X^2, X^4, X^6 = SO_3^-$

1. [Δdi-HA]

(15)

D-GlcNY6X (where ΔUA is 4-deoxy-α-L-threo-hex-4-enopyransyluronic acid, GlcN is 2-deoxy-2-aminoglucopyranose, S is sulfate, Ac is acetate, X may be S, and Y is S or Ac). Heparin and heparan sulfate are structurally similar GAGs differing primarily in their relative content of *N*-acetylglucosamine, O-sulfation, and glucuronic acid. Using heparinases (I, II, and III) as the degrading enzymes, heparin and heparan sulfate can be dpolymerized through an eliminative mechanism to yield eight different disaccharides of structure (16) [113]. Using a borate buffer pH 8.8, two of the standard heparin/heparan sulfate disaccharides, having an identical charge of -2, ΔUA2S(1→4)-D-GlcNAc (structure 3) and ΔUA(1→4)-D-GlcNS (structure 4) were not fully resolved. The resolution of these two saccharides could be improved by preparing borate buffer in deuterated water or eliminating boric acid. Surprisingly, baseline resolution was achieved in micellar solution of sodium dodecyl sulfate (SDS) in the absence of buffer. Since the two saccharides (structures 3 and 4) are charged and polar, it is unlikely that separation of these solutes was caused by differential partitioning in the SDS micelles. These electrophoretic systems were then applied to the determination of disaccharide composition of porcine mucosal heparin and that of bovine kidney heparan sulfate. Both GAGs were found to have an equimolar content of disaccharide ΔUA2S(1→4)-D-GlcNAc (structure 3) and ΔUA(1→4)-D-GlcNS (structure 4). These analysis required 15 ng of polysaccharides and a 20-min analysis time, and the results were comparable to those obtained by the more

1. $X^2 = X^6 = H, Y = Ac$
2. $X^2 = H, X^6 = SO_3^-, Y = Ac$
3. $X^2 = SO_3^-, X^6 = H, Y = Ac$
4. $X^2 = X^6 = H, H = SO_3^-$
5. $X^2 = X^6 = SO_3^-, Y = Ac$
6. $X^2 = H, X^6 = Y = SO_3^-$
7. $X^2 = SO_3^-, X^6 = H, Y = SO_3^-$
8. $X^2 = X^6 = Y = SO_3^-$

(16)

established strong anion-exchange HPLC, which requires a 40-μg sample and a 90-min analysis time [113].

The separations of chondroitin sulfate disaccharides and hyaluronic oligosaccharides were also reported by Carney and Osborne [114]. This work discussed the various factors affecting the separation of the GAG disaccharides, such as pH, borate concentration, buffer ionic strength, and the inclusion of sodium dodecyl sulfate (SDS) in the running electrolyte. Although the disaccharides were highly charged and too polar to partition with SDS micelles, the presence of SDS in the electrolyte improved the resolution of the electrophoretic system. The best conditions for optimum resolution of the disaccharides of chondroitin sulfate and hyaluronan are summarized in Fig. 28a. The structure of these disaccharides are, of course, the same as those shown above when discussing the work by Linhardt et al. [115]. Using these conditions, the disaccharides of beagle cartilage chondroitin sulfate were identified by CZE. Also, six oligosaccharides derived from hyaluronan by digestion with testicular hyaluronidase were fully resolved by CZE in about 15 min (see Fig. 28b). It should be noted that the use of a single lyase in the depolymerization step of hyaluronan usually yields a mixture of disaccharides and higher-oligosaccharide products.

Additional uses of CZE in the analysis of unsaturated disaccharides derived from GAGs by digestion with chondroitinase AC and ABC were reported by Honda et al. [146]. The conversion of these saccharides to their PMP derivatives improved the system sensitivity (see Fig. 29a). The electrophoretic system also proved suitable for the quantitative estimation of human urinary chondroitin sulfates (see Fig. 29b).

Damn et al. [116] reported CZE methods for the quality control of natural and synthetic heparin fragments. Because of the anti-blood-clotting activity of heparin, the production of natural and synthetic heparin fragments for pharmaceutical use relies on the availability of analytical procedures for the efficient characterization of intermediates and final products. In this study, using a low-pH electrolyte system (i.e., 0.2 M phosphate, pH 4.0) and controlling the capillary

Capillary Electrophoresis of Carbohydrates / 241

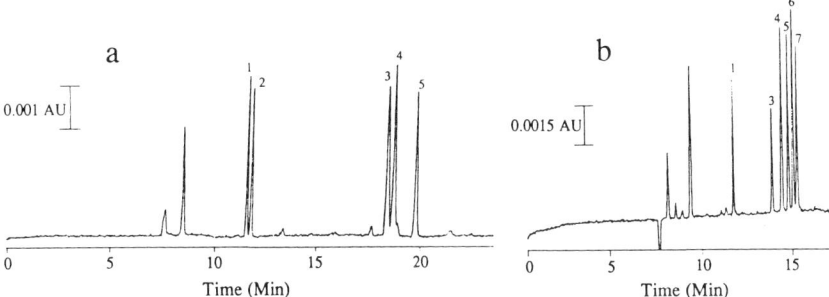

Fig. 28 (a) Electropherogram of standard unsaturated disaccharides of chondroitin sulfate and hyaluronan. Applied voltage, 15 kV; temperature, 40°C; electrolyte, 40 mM phosphate containing 40 mM SDS and 10 mM borate, pH 9.0; capillary, 72 cm × 50 μm ID; detection, UV at 232 nm. Peaks: 1, Δdi-HA; 2, Δi-0S; 3, Δdi-6S; 4, Δdi-4S; 5, Δdi-UA2S. For disaccharide structures, see (15) in the text. (b) Electropherogram of oligosaccharides derived from hyaluronan by digestion with testicular hyaluronidase. Conditions are as in (a) except that the column was monitored at 200 nm. Peak 1 is the unsaturated disaccharide of hyaluronan (Δdi-HA), and peaks 3, 4, 5, and 6 are the saturated hexa-, octa-, deca-, dodeca-, and tetradecasaccharides of hyaluronan, respectively. (From Ref. 114.)

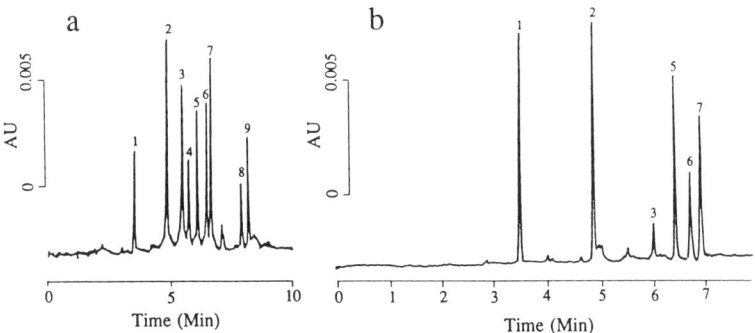

Fig. 29 (a) Electropherogram of chondroitin ABC-digested mixture of chondroitin sulfates A–E, chondroitin and hyaluronic acid by CZE after derivatization with PMP. Capillary, fused silica (51 cm × 50 μm ID); electrolyte, 100 mM borate buffer, pH 9.0; applied voltage 25 kV. Peaks: 1 came from the buffer for enzymatic digestion; 2, PMP (excess reagent); 3, PMP derivative of Δdi-0S; 4, PMP derivative of Δdi-HA; 5, sodium benzoate (internal standard); 6, PMP derivative of Δdi-4S; 7, PMP derivative of Δdi-6S; 8, PMP derivative of Δdi-S_D; 9, PMP derivative of Δdi-S_E. (b) Analysis of the PMP derivatives of unsaturated disaccharides derived from the GAG fraction of a urine sample digestion with chondroitinase ABC by CZE. Conditions and peak assignment are as in (a). (From Ref. 146.)

column temperature at 40°C allowed the separation of the nine most common heparin disaccharides, the mapping of the oligosaccharides derived from heparin after heparinase treatment, and the assessment of the quality of synthetic heparin pentasaccharide preparations. According to the study, it seems that at least for this type of molecule, CZE forms an alternative to HPAEC.

Liu et al. [80] demonstrated the advantages of using gel-filled capillaries with high gel concentration in the separation of enzymatically degraded hyaluronic acid from human umbilical cords. Here again, CGE afforded the high-resolution separation of hyaluronic acid-derived oligosaccharides tagged with CBQCA. The use of less concentrated gels gave poor results with the same oligosaccharide derivatives.

E. Glycolipids

Carbohydrates are also integral parts of some lipids, the glycolipids. Among glycolipids, gangliosides are the most complex group. They are sialic acid–containing glycosphingolipids found at high concentrations embedded in the plasma membranes of vertebrate cells and believed to provide recognition sites on the cell surface. As shown in (17), a ganglioside molecule is composed of a hydrophilic sialooligosaccharide chain and a hydrophobic moiety, ceramide, which consists of sphingosine and fatty acid. The complex carbohydrate head of gangliosides, which extends beyond the surface of cell membranes, acts as a specific receptor for certain pituitary glycoprotein hormones.

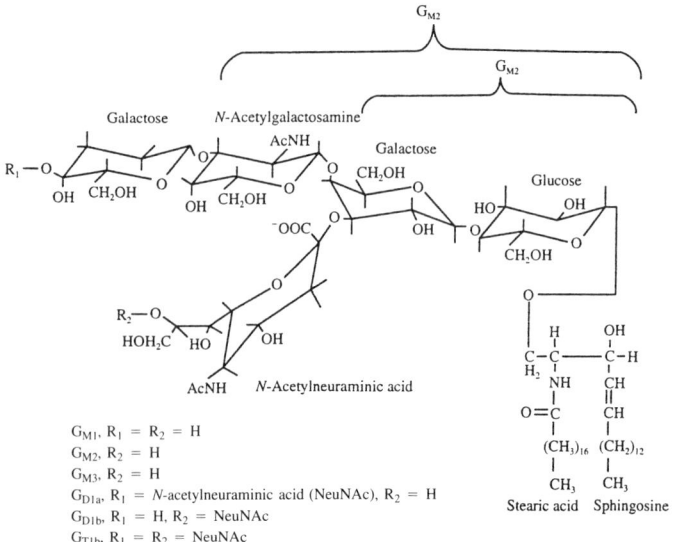

(17)

Growing interest in the biological functions of gangliosides has generated a need to develop improved methods for their isolation and analysis; the scope of HPCE has been extended recently by Liu and Chan [147] to the separation of these glycolipids. As anionic amphiphiles with similar sizes of hydrophilic head groups and hydrocarbon tails, gangliosides are known to exist as stable micelles in aqueous solutions, a phenomenon that often hinders their efficient separation as monomeric species. These researchers [147] demonstrated that CZE can separate some ganglioside micelles, and consequently permitted studies of the micellar properties of these amphiphilic species using untreated fused silica capillaries and on-column direct UV detection at 195 nm. The ganglioside micelles were analyzed successfully within 10 min with mass sensitivity on the order of 10^{-11} mol. As shown in Fig. 30a, baseline resolution of a mixture of three ganglioside micelles (i.e., G_{M1}, G_{D1b} and G_{T1b}) was achieved using 2.5 mM potassium phosphate, pH 7.40, as the running electrolyte. The separation was facilitated primarily by the varying content of the sialic acid residues in the ganglioside micelles, which imparted them with different electrophoretic mobilities. The observed migration velocities of the ganglioside micelles seem to be largely unaffected in the pH range 7.0 to 11.0, and decreased monotonically when the pH was varied from 6.0 to 4.0. This is because of the diminishing dissociation of the carboxyl group of the sialic acid residues as the pH was decreased [147]. Increasing the ionic strength of the running electrolyte decreased the migration and broadened the peaks of the gangliosides. This may be due to the fact that increasing the ionic strength would decrease the electrostatic repulsion between the ganglioside monomers, thus increasing the number of monomers in the aggregates, and consequently the size of the ganglioside micelles. This increase in the size of the micelles would result in decreasing their migration velocity. The authors also studied the time course of mixed micelle formation of gangliosides. As shown in Fig. 30b, both G_{D1b} and G_{T1b} were first separated into two distinct peaks after initial mixing. However, upon incubation at 37°C, complete fusion between both micellar peaks could be observed in less than 2.5 h. The fusion process was temperature dependent. In fact, at 50°C the formation of mixed micelles between G_{D1b} and G_{T1b} was complete within 30 min. In contrast, no fusion of the ganglioside peaks was observed at 0°C even after 75 h. The mixed micelle formation seems to be dependent on the sialic acid content of the individual gangliosides [147]. Whereas mixed micelle formation between G_{D1b} and G_{D1a} and that between G_{D1b} and G_{T1b} required 1.5 and 3.0 h, respectively, the aggregation of G_{M1} with polysialogangliosides (i.e., G_{D1b} and G_{T1b}) was 6- to 36-fold slower under otherwise identical incubation conditions. In addition, no fusion was observed between G_{M1} and G_{M2} after 2 days of incubation. Based on these observations it was suggested that polysialogangliosides (e.g., G_{D1a} and G_{T1b}) may have higher propensities than monosialoganglioside. Thus the high resolution, high speed, and quantitative aspects of CZE were clearly demonstrated in

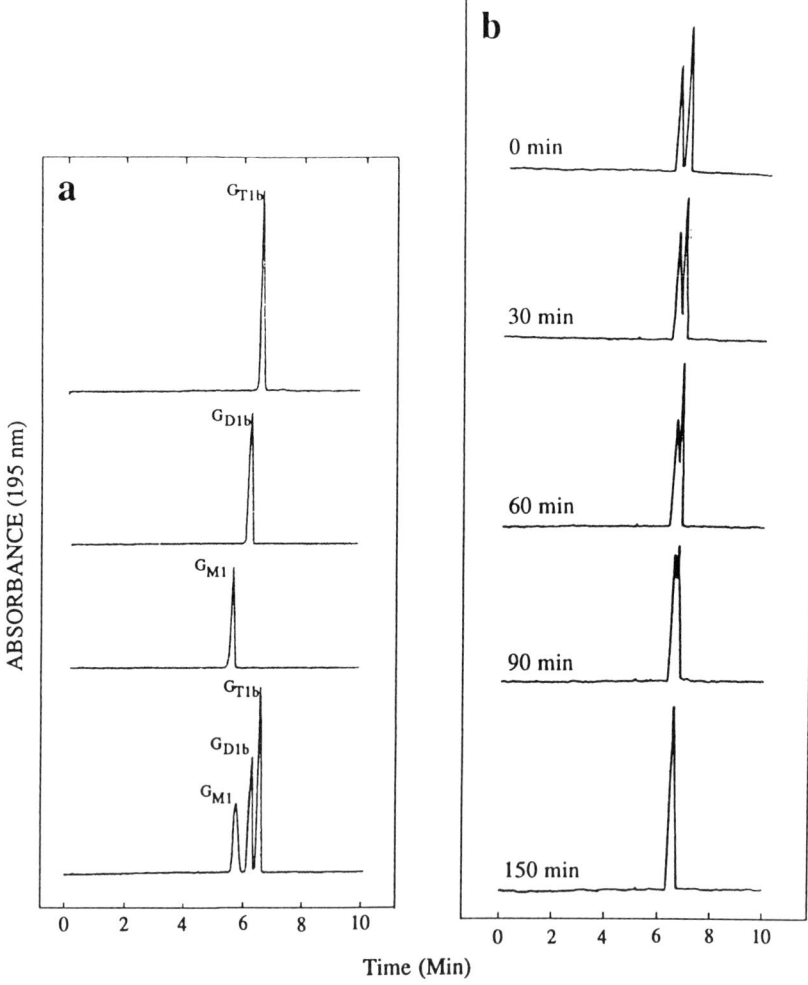

Fig. 30 (a) Capillary electrophoresis of a mixture of a mixture of G_{M1}, G_{D1b}, and G_{T1b}. Individual G_{M1}, G_{D1b}, and G_{T1b} micelles and a mixture of these three gangliosides shortly after mixing were analyzed by CE. The buffer was 2.5 mM potassium phosphate, pH 7.40. Detection was by UV at 195 nm. (b) Time course of mixed micelle formation between G_{D1b} and G_{T1b}. Equimolar concentrations of polysialogangliosides G_{D1b} and G_{T1b} (165 μM) in 2.5 mM potassium phosphate, pH 7.40, were mixed by vortexing and incubated in a water bath at 37°C. At time intervals, the electrophoretic patterns of the ganglioside mixtures were analyzed. Electrophoretic conditions are as in (a). (From Ref. 147.)

monitoring processes that may have important implications in the distribution and function of gangliosides in biological membranes.

ACKNOWLEDGMENTS

The author acknowledges previous financial support from the Oklahoma Center for the Advancement of Sciences and Technology, Oklahoma Health Research Program, grant HN9-004; from the University Center for Water Research; from the University Center for Energy Research; from the Agriculture Station, grant CSRS OKL 02109; and from the College of Arts and Sciences, Dean Incentive Program at Oklahoma State University. Also gratefully acknowledged is current financial support from Oklahoma Water Resources Research Institute/USGS Section 104, Project 08, and from the Cooperative State Research Service, U.S. Department of Agriculture, under Agreement 92-34214-7325.

REFERENCES

1. A. B. Foster, *Adv. Carbohydr. Chem., 12:* 81 (1957).
2. A. B. Foster, *Methods Carbohydr. Chem., 1:* 51 (1962).
3. H. Weigel, *Adv. Carbohydr. Chem., 18:* 61 (1963).
4. S. Hase, S. Hara, and Y. Matsushinua, *J. Biochem., 85:* 217 (1979).
5. M. C. Tarvis, D. R. Threlfall, and D. R. Friend, *J. Phytochem., 16:* 849 (1977).
6. P. Jackson, *Anal. Biochem., 196:* 238 (1991).
7. K.-B. Lee, A. Al-Hakim, D. Loganathan, and R. J. Linhardt, *Carbohydr. Res., 214:* 155 (1991).
8. R. P. Rago, D. Ramirez-soto, and R. D. Povetz, *Carbohydr. Res., 236:* 1 (1992).
9. R. J. Stack and M. T. Sullivan, *Glycobiology, 2:* 85 (1992).
10. O. P. Das and E. J. Henderson, *Anal. Biochem., 158:* 390 (1986).
11. M. K. Cowman, M. F. Slahetka, D. M. Hittner, J. Kim, M. Forino, and G. Gadelrab, *Biochem. J., 221:* 707 (1984).
12. J. E. Turnbull and J. T. Gallagher, *Biochem. J., 251:* 597 (1988).
13. K. G. Rice, M. K. Rottink, and R. J. Linhardt, *Biochem. J., 244:* 515 (1987).
14. S. Terabe, *Trends Anal. Chem., 8:* 129 (1989).
15. Z. Deyl, *J. Chromatogr. Libr., 18B:* 13 (1983).
16. P. J. Antikainen, *Acta Chem. Scand., 13:* 312 (1959).
17. G. L. Roy, A. L. Laferriere, and J. O. Edwards, *J. Inorg. Nucl. Chem., 4:* 106 (1957).
18. E. J. Borune, D. H. Hutson, and H. Weigel, *J. Chem. Soc.,* 4252 (1960).
19. F. Searle and H. Weigel, *Carbohydr. Res., 85:* 515 (1980).

20. J. L. Frahn and J. A. Mills, *Aust. J. Chem.*, *12:* 65 (1959).
21. J. A. Mills, *Biochem. Biophys. Res. Commun.*, *6:* 418 (1961–62).
22. J. A. Rendelman, *Adv. Carbohydr. Chem.*, *21:* 209 (1966).
23. J. Böeseken, *Adv. Carbohydr. Chem.*, *4:* 189 (1949).
24. J. G. Dawber and G. E. Hardy, *J. Chem. Soc. Faraday Trans.*, *80:* 2467 (1984).
25. V. Van Duin, J. A. Peters, A. P. G. Kieboom, and H. Van Bekkum, *Tetrahedron, 41:* 3411 (1985).
26. M. Makkee, A. P. G. Kieboom, and H. Van Bekkum, *Recl. Trav. Chim. Pays-Bas, 104:* 230 (1985).
27. C. F. Bell, R. D. Beauchamp, and E. L. Short, *Carbohydr. Res.*, *147:* 191 (1986).
28. C. C. Christ, J. R. Clark, and H. T. Evans, Jr., *Acta Crystallogr.*, *11:* 761 (1958).
29. H. B. Davis and C. J. B. Mott, *J. Chem. Soc. Faraday Trans. I, 76:* 1991 (1980).
30. A. B. Foster, *Chem. Ind.*, 828 (1952).
31. G. R. Kennedy and M. J. How, *Carbohydr. Res.*, *28:* 13 (1973).
32. O. Nobel and F. R. Taravel, *Carbohydr. Res.*, *184:* 236 (1988).
33. S. Honda, S. Iwase, A. Makino, and S. Fujiwara, *Anal. Biochem.*, *176:* 72 (1989).
34. R. Consden and W. M. Stanier, *Nature, 169:* 783 (1952).
35. S. Hoffstetter-Kuhn, A. Paulus, E. Gassmann, and H. H. Widmer, *Anal. Chem., 63:* 1541 (1991).
36. J. Cai and Z. El Rassi, *J. Chromatogr.*, *608:* 31 (1992).
37. J. A. Rendelman, Jr., *Adv. Chem. Ser.*, *117:* 51 (1971).
38. F. W. Baker, R. C. Parish, and L. M. Stock, *J. Am. Chem. Soc.*, *89:* 5677 (1967).
39. T. W. Garner and E. S. Yeung, *J. Chromatogr.*, *515:* 639 (1990).
40. A. E. Vorndran, P. J. Oefner, H. Scherz, and G. K. Bonn, *Chromatographia, 33:* 163 (1992).
41. L. A. Colón, R. Dadoo, and R. N. Zare, *Anal. Chem.*, *65:* 476 (1993).
42. S. J. Angyal, *Pure Appl. Chem.*, *35:* 131 (1973).
43. S. J. Angyal, *Adv. Carbohydr. Chem. Biochem.*, *47:* 1 (1989).
44. S. J. Angyal, D. Greeves, and J. A. Mills, *Aust. J. Chem.*, *27:* 1447 (1974).
45. S. J. Angyal, D. Greeves, L. Littlemore, and V. A. Pickles, *Aust. J. Chem., 29:* 1231 (1976).
46. J. Briggs, P. Fnich, C. Matulevicz, and H. Weigel, *Carbohydr. Res.*, *97:* 181 (1981).
47. K. B. Hicks, *Adv. Carbohydr. Chem. Biochem.*, *46:* 17 (1988).
48. S. J. Angyal and K. P. Davies, *Chem. Commun.*, 500 (1971).
49. R. E. Leukinski and J. Reuben, *J. Am. Chem. Soc.*, *98:* 3089 (1976).

50. S. J. Angyal and J. A. Mills, *Aust. J. Chem., 32:* 1993 (1979).
51. M. Miyazaki, S. Nishimura, A. Yoshida, and N. Okubo, *Chem. Pharm. Bull., 27:* 532 (1979).
52. S. Honda, K. Yamamoto, S. Suzuki, M. Ueda, and K. Kakehi, *J. Chromatogr., 558:* 327 (1991).
53. H. H. Lauer and D. McManigill, *Anal. Chem., 58:* 166 (1986).
54. Y. Walbroehl and J. W. Jorgenson, *J. Microcolumn,* Sept. 1, p. 41 (1989).
55. R. M. McCormick, *Anal. Chem., 60:* 2322 (1988).
56. J. S. Green and J. W. Jorgenson, *J. Chromatogr., 478:* 63 (1989).
57. A. Emmer, M. Jansson, and J. Roreaade, *J. Chromatogr., 547:* 544 (1991).
58. J. R. Mazzeo and I. S. Krull, *Biotechniques, 10:* 638 (1991).
59. T. Wehr, *Liq. Chromatogr. Gas Chromatogr., 11:* 14 (1993).
60. Z. El Rassi and W. Nashabeh, in *Capillary Electrophoresis Technology,* N. A. Guzman, Ed., Marcel Dekker, New York, 1993, pp. 383–434.
61. S. Hjertén, *J. Chromatogr., 347:* 191 (1985).
62. K. A. Cobb, V. Dolnik, and M. Novotny, *Anal. Chem., 62:* 2478 (1990).
63. M. Huang, W. P. Vorkink, and M. L. Lee, *J. Microcol. Sep., 4:* 233 (1992).
64. W. Nashabeh and Z. El Rassi, *J. Chromatogr., 559:* 367 (1991).
65. W. Nashabeh and Z. El Rassi, *J. High Resolut. Chromatogr., 15:* 289 (1992).
66. W. Nashabeh and Z. El Rassi, *J. Chromatogr., 632:* 157 (1993).
67. W. Nashabeh, J. T. Smith, and Z. El Rassi, *Electrophoresis, 14:* 407 (1993).
68. J. T. Smith and Z. El Rassi, *J. High Resolut. Chromatogr., 15:* 573 (1992).
69. G. J. M. Bruin, J. P. Chang, R. H. Kuhlman, K. Zegers, J. C. Kraak, and H. Poppe, *J. Chromatogr., 471:* 429 (1989).
70. T. Wang and R. A. Hartwick, *J. Chromatogr., 594:* 325 (1992).
71. M. Gilges, H. Husmann, M.-H. Kleemiss, S. R. Motsch, and G. Schomburg, *J. High Resolut. Chromatogr., 15:* 452 (1992).
72. G. J. M. Bruin, R. Huisden, J. C. Kraak, and H. Poppe, *J. Chromatogr., 480:* 339 (1989).
73. S. A. Swedberg, *Anal. Biochem., 185:* 51 (1990).
74. J. K. Towns and F. E. Regnier, *J. Chromatogr., 516:* 69 (1990).
75. J. K. Towns and F. E. Regnier, *Anal. Chem., 63:* 1126 (1991).
76. D. Bentrop, J. Kohr, and H. Eugelhardt, *Chromatographia, 32:* 171 (1991).
77. J. K. Towns, J. Bao, and F. E. Regnier, *J. Chromatogr., 599:* 227 (1992).
78. Y.-F. Maa, K. J. Hyver, and S. A. Swedberg, *J. High Resolut. Chromatogr., 14:* 65 (1991).
79. J. T. Smith and Z. El Rassi, *Electrophoresis, 14:* 396 (1993).

80. J. Liu, V. Donik, Y.-Z. Hsieh, and M. Novotny, *Anal. Chem., 64:* 1328 (1992).
81. J. Liu, O. Shirota, and M. Novotny, *J. Chromatogr., 559:* 223 (1991).
82. B. L. Karger and A. S. Cohen, U.S. patent 4,865,707, 1989.
83. B. L. Karger and A. S. Cohen, U.S. patent 4,865,706, 1989.
84. P. F. Bente and J. Myerson, U.S. patent 4,810,456, 1989.
85. J. A. Lux, H. F. Yin, and G. Schomburg, *J. High Resolut. Chromatogr., 13:* 436 (1990).
86. V. Dolnik, K. A. Cobb, and M. Novotny, *J. Microcol. Sep., 3:* 155 (1991).
87. R. S. Durbow, in *Capillary Electrophoresis: Theory and Practice,* P. D. Grossman and J. C. Colburn, Eds., Academic Press, New York, 1992, p. 133.
88. S. F. Y. Li, *J. Chromatogr. Libr., 52:* 173 (1992).
89. A. S. Cohen, S. Carson, A. Belenkii, and B. L. Karger, presented at the *4th International Symposium on HPCE,* Amsterdam, Feb. 9–13, 1992.
90. M. Zhu, D. L. Hansen, S. Burd, and F. Gannon, *J. Chromatogr., 480:* 311 (1989).
91. S. Hase, S. Hara, and Y. Matsushima, *J. Biochem., 85:* 217 (1979).
92. H. Takemato, S. Hase, and T. Ikenaka, *Anal. Biochem., 145:* 245 (1985).
93. S. Honda, A. Makino, S. Suzuki, and K. Kakehi, *Anal. Biochem., 191:* 228 (1990).
94. W. Nashabeh and Z. El Rassi, *J. Chromatogr., 600:* 279 (1992).
95. K.-B Lee, Y.-S. Kim, and R. J. Linhardt, *Electrophoresis, 12:* 636 (1991).
96. J. Liu, O. Shirota, D. Wiesler, and M. Novotny, *Proc. Natl. Acad. Sci. USA, 88:* 2302 (1991).
97. J. Liu, Y. H. Hsieh, D. Wiesler, and M. Novotny, *Anal. Chem., 63:* 408 (1991).
98. J. Liu, O. Shirata, and M. Novotny, *Anal. Chem., 63:* 413 (1991).
99. P. J. Oefner, A. E. Vorndan, E. Grill, C. Huber, and G. K. Bonn, *Chromatographia, 34:* 308 (1992).
100. W. T. Wang, M. C. Lee Doune, Jr., B. Ackerman, and C. C. Sweeley, *Anal. Biochem., 141:* 366 (1984).
101. S. Honda, S. Suzuki, A. Nose, K. Yamamoto, and K. Kakehi, *Carbohydr. Res., 215:* 193 (1991).
102. S. Honda, E. Akao, S. Suzuki, M. Okuda, K. Kakehi, and J. Nakamuro, *Anal. Biochem., 180:* 351 (1989).
103. E. S. Yeung and W. G. Kuhr, *Anal. Chem., 63:* 275A (1991).
104. F. Foret, S. Fanali, L. Ossicini, and P. Bocek, *J. Chromatogr., 470:* 299 (1989).
105. W. G. Kuhr and E. S. Yeung, *Anal. Chem., 60:* 2642 (1988).
106. T. W. Garner and E. S. Yeung, *J. Chromatogr., 515:* 639 (1990).
107. T. M. Olefirowiecz and A. G. Ewing, *J. Chromatogr., 499:* 713 (1990).

108. E. S. Yeung, *Acc. Chem. Res., 22:* 125 (1989).
109. G. J. M. Bruin, A. C. van Asten, X. Yu, and H. Poppe, *J. Chromatogr., 608:* 97 (1992).
110. H. Small and T. E. Miller, *Anal. Chem., 54:* 462 (1982).
111. W. R. Jones, P. Jandik, and R. Pfeifer, *Am. Lab.,* May, p. 40 (1991).
112. A. Weston, P. R. Brown, P. Jandik, A. L. Heckenberg, and W. R. Jones, *J. Chromatogr., 608:* 395 (1992).
113. S. A. Ampofo, H. M. Wang, and R. J. Linhardt, *Anal. Biochem., 199:* 249 (1991).
114. S. L. Carney and D. J. Osborne, *Anal. Biochem., 195:* 132 (1991).
115. A. Al-Hakim and R. J. Linhardt, *Anal. Biochem., 195:* 68 (1991).
116. J. B. L. Damn, G. T. O. Verklift, B. W. M. Vermeulen, G. F. Fluitsma, and G. W. K. van Dedem, *J. Chromatogr., 608:* 297 (1992).
117. D. C. Johnson and W. R. La Course, *Anal. Chem., 62:* 589A (1990).
118. D. J. Bornhop, T. G. Nolan, and N. J. Dovichi, *J. Chromatogr., 384:* 181 (1987).
119. D. J. Bornhop and N. J. Dovichi, *Anal. Chem., 59:* 1632 (1987).
120. C. Y. Chen, T. Demana, S. D. Huang, and M. D. M. Morris, *Anal. Chem., 61:* 1590 (1989).
121. J. Pawliszyn, *Anal. Chem., 60:* 2796 (1988).
122. A. E. Bruno, B. Krattiger, F. Maystre, and H. M. Widmer, *Anal. Chem., 63:* 2689 (1991).
123. A. E. Vorndran, E. Grill, C. Huber, P. J. Oefner, and G. K. Bonn, *Chromatographia, 34:* 109 (1992).
124. W. Nashabeh and Z. El Rassi, *J. Chromatogr., 536:* 31 (1991).
125. W. Nashabeh and Z. El Rassi, *J. Chromatogr., 514:* 57 (1990).
126. Z. El Rassi, D. Tedford, J. An, and A. Mort, *Carbohydr. Res., 215:* 25 (1991).
127. T. W. Rademacher, R. B. Parekh, and R. A. Dwek, *Annu. Rev. Biochem., 57:* 785 (1988).
128. R. B. Parekh, R. A. Dwek, P. M. Rudd, J. R. Thomas, and T. W. Rademacher, *Biochemistry, 28:* 7670 (1989).
129. R. S. Smagula, H. van Haleck, J. M. Decker, A. V. Muchmore, C. E. Moody, and A. P. Sherblom, *Glycoconjugate J., 7:* 609 (1990).
130. F. Kilàr and S. Hjertén, *J. Chromatogr., 480:* 351 (1989).
131. P. D. Grossman, J. C. Colburn, H. H. Lauer, R. G. Nielsen, R. M. Riggin, G. S. Sihampalam, and E. C. Rickard, *Anal. Chem., 61:* 1186 (1989).
132. P. M. Rudd, I. G. Scragg, E. Coghill, and R. A. Dwek, *Glycoconjugate J., 9:* 86 (1992).
133. J. P. Landers, R. P. Oda, B. J. Madden, and T. C. Spelsberg, *Anal. Biochem., 205:* 115 (1992).
134. K. W. Yim, *J. Chromatogr., 559:* 401 (1991).

135. E. Watson and F. Yao, *J. Chromatogr., 630:* 442 (1993).
136. A. D. Tran, S. Park, P. J. Lisi, O. T. Huynh, R. R. Ryall, and P. A. Lane, *J. Chromatogr., 542:* 459 (1991).
137. K. Schmid, J. P. Binette, L. Dorland, F. G. Valiegenhart, B. Fourmet, and J. Montreuil, *Biochim. Biophys. Acta, 581:* 356 (1979).
138. K. Schmid, in *The Plasma Proteins,* 2nd ed., F. W. Putnam, Ed., Vol. 1, Academic Press, New York, 1975, pp. 183–228.
139. H. Yoshima, A. Matsumoto, T. Mizuochi, T. Kawasaki, and A. Kobata, *J. Biol. Chem., 256:* 8476 (1981).
140. S. Suzuki, K. Kakehi, and S. Honda, *Anal. Biochem., 205:* 227 (1992).
141. R. Got, R. Bourvillon, and P. Cornillot, *Biochim. Biophys. Acta, 58:* 126 (1962).
142. P. Hetmentin, R. Witzel, R. Doenges, R. Bauer, H. Hampt, T. Patel, R. B. Paretch, and D. Brajel, *Anal. Biochem., 206:* 419 (1992).
143. C. J. Liang, K. Yamashita, and A. Kobata, *J. Biochem., 88:* 51 (1980).
144. D. Voet and J. G. Voet, *Biochemistry,* Wiley, New York, 1990, pp. 258–260.
145. R. Varma and R. S. Varma, in *Mucopolysaccharides and Glycosaminoglycans of Body Fluids in Health and Disease,* Walter de Gruyter, New York, 1983.
146. S. Honda, T. Ueno, and K. Kakehi, *J. Chromatogr., 608:* 289 (1992).
147. Y. Liu and K.-F. Chan, *Electrophoresis, 12:* 402 (1991).

5
Environmental Applications of Supercritical Fluid Chromatography

Leah J. Mulcahey, Christine L. Rankin, and Mary Ellen P. McNally
E. I. DuPont de Nemours & Co., Inc., Wilmington, Delaware

I.	INTRODUCTION	252
II.	POLYCHLORINATED BIPHENYLS	253
	A. Flame Ionization Detection	255
	B. UV Detection	255
	C. Mass Spectrometric Detection	258
III.	PESTICIDES AND HERBICIDES	259
	A. Flame Ionization Detection	259
	B. UV Detection	265
	C. Electron Capture Detection	269
	D. Thermionic Detection	270
	E. Photometric Detection	273
	F. Atomic Emission Detection	273
	G. Fourier Transform Infrared Detection	275
	H. Mass Spectrometric Detection	277
	I. Ion Mobility Detection	280
	J. SFE/SFC	280
IV.	PHENOLS	281
	A. UV Detection	282
	B. Flame Ionization Detection	286
	C. Miscellaneous Detection	287

V. POLYNUCLEAR AROMATIC HYDROCARBONS	288
A. UV Detection	288
B. Mass Spectrometric Detection	293
C. Miscellaneous Detection	295
D. Multidimensional Techniques	300
VI. CONCLUSIONS	301

I. INTRODUCTION

Supercritical fluid chromatography (SFC) has been labeled as a niche technique, amenable to compounds that have not been able to be analyzed either by gas or liquid chromatography (GC, LC). Defined limitations for this niche have been compound properties such as solubility or volatility: that is, those that do not readily volatilize at the temperatures used for GC or are not readily soluble in the common LC solvents. Detection has also been a significant contributing factor to its utilization. The most commonly used and reliable detector in LC has been UV (ultraviolet detection); however, nonvolatile analytes that do not contain a chromophore have suffered in routine analysis because a universal detector such as the flame ionization detector (FID) in GC has not been found. SFC offers the capability of interfacing with a wider variety of detectors, such as the FID, providing the potential for broadening the scope of routinely analyzed compounds.

In deference to the label "niche technique," supercritical fluid chromatography has shown widespread applicability to a variety of compound types, polarities, solubilities, and sizes [1–7]. It is able, even though not commonly utilized in this regard, to replace applications of both gas and liquid chromatography. As is common in business and market strategies, the product that arrives at the market first is generally the most successful—however, not always the most enduring. Even though SFC shows broad-based applicability, the substitution of SFC for LC and GC is practical, although not guaranteed. The future of supercritical fluid chromatography still depends on ingenuity both in salesmanship and science. There are many industrious people active in this regard.

The issues of efficiency, analysis time, limits of detection, and sample capacity for packed and capillary SFC columns have been argued extensively. As in all column chromatography techniques, each has its advantages, disadvantages, and more appropriate applications. In this chapter, references have been presented that have made comparisons of packed and capillary column applications without prejudice to either technology. In some cases, optimization of column choice or chromatographic conditions would have made one or the other separation choice more opportune. The work reviewed here can be thought of as a starting point for further analysis.

For ease of perusal, this chapter has been subdivided into compound class and then further subdivided by detector type. As can easily be recognized, some classes of compounds are detected predominantly with one specific detector, probably because of their amenability to this detection type. Where relevant, real sample matrices have been presented. However, as in all new technologies, test mixtures with a wide range of compounds have been analyzed; conditions for these analyses have also been included.

It has been hypothesized by several authors, and we currently affirm this hypothesis, that supercritical fluid chromatography conditions can be used as a starting point for supercritical fluid extraction (SFE) [8]. Matrix effects cannot always be predicted directly, but solubility information concerning the analyte of interest can be obtained. For this reason, some SFE conditions have been presented briefly to broaden the compound base for compounds that can be analyzed by supercritical fluid extraction, and therefore can be analyzed using SFC.

The selection of a technology to separate compounds is dictated by a variety of factors: instrument availability, number of samples, required accuracy and precision, desired method sensitivity, the amount of sample preparation needed, and the cost. Environmental applications can be conducted for a variety of reasons: population consideration because of residential atmospheric pollution, technical advances to monitor pollution control, assessing plant and animal effects due to agricultural systems, identification of substances which explain events or activities that have taken place in our environment, and preserving industrial work areas by detecting harmful components which could be injurious to personnel [9]. We have chosen to present the state of the art in environmental applications of supercritical fluid chromatography because of both our expertise and our interest. The examples presented in this chapter illustrate broadly that SFC is a technology that can perspicaciously separate compounds of environmental concern.

II. POLYCHLORINATED BIPHENYLS

Polychlorinated biphenyls (PCBs) have been manufactured and used since 1929 [10]. These widespread pollutants have been used as heat transfer fluuids, hydraulic fluids, and dielectric fluids. Improper disposal practices of PCBs have been the most significant environmental contamination cause. Generally, PCBs have been analyzed using gas chromatography coupled with electron capture detection [11,12], more infrequently by reversed-phase liquid chromatography [13]. Supercritical fluid chromatography has been reported, but not widely, for the analysis of PCBs. More work has been conducted in the supercritical fluid extraction of these compounds. Therefore, reported SFE conditions have been presented in tabular form (Table 1) as starting points in terms of density, pressure,

Table 1 Supercritical Fluid Extraction Condition[a] and References for the Extraction of Polychlorinated Biphenyls

PCB	Temperature (modifier)[b]	Pressure	Matrix	Refs.
Trichlorobiphenyl	Not given	Not given	River sediment	19
Aroclor 1260	75°C (5% acetone or methanol)	400 atm (5900 psi)	Contaminated soil	20
Aroclor 1260	Not given	Not given	Soil	21
Aroclor 1242, 1254, 1260	40–60°C (2% methanol)	100–200 atm (1500–3000 psi)	Certified EC-1 sediment	22
Aroclor 1254	45°C	300 atm (4400 psi)	River sediment	23
Aroclor 1254	40°C	100 atm (1500 psi)	Lab-spiked Cecil subsoil (wet and dry)	24,25
Unspecified PCB	50°C	125 atm (1800 psi)	River sediment	26
PCB-33, PCB-77, PCB-153	1.03 (reduced temperature)	0.86 (reduced pressure)	Spiked soil	27
Aroclor 1260	50°C	150	Fish tissue, milk, blood, soil	28–30

[a]These conditions can be considered as starting points for supercritical fluid chromatography.
[b]If used.

and temperature for the analysis of PCBs. Logically, no column choice can be presented but the matrix has been given and a general idea of polarity of matrix can aid in column selection.

A. Flame Ionization Detection

Lee and co-workers at Bringham Young University used an on-line two-dimensional supercritical fluid system (SFC/SFC) with a packed capillary coupled to an open-tubular column arrangement with two separate pumps to analyze Aroclor 1242 [14]. The instrumentation utilized a switching valve interfaced in-line with a cryogenic trap for refocusing of analyte fractions. As illustrated in the chromatograms of Fig. 1, the Aroclor was first separated according to polarity using a 60 cm × 250 µm ID fused silica column packed with 7-µm 300-Å aminosilane bonded silica. The second separation (Fig. 1B) of the starred peak of Fig. 1A shows the shape-selective separation achieved using a 10.5 m × 50 µm ID fused silica open-tubular column coated with a liquid-crystalline polysiloxane stationary phase (SB-Smectic-50 from Lee Scientific). This combination of columns is a feasible option for the PCB isomers, which frequently have similar volatilities and polarities but differ in their shapes.

B. UV Detection

Perchlorination of individual congeners of PCBs and PCTs (polychlorinated terphenyls) to fully chlorinated decachlorobiphenyl and tetradecachlorinated terphenyls is the typical quantitation method of PCBs. Onuska et al., at the National Water Research Institute in Ontario, evaluated perchlorination via quantitation of two resultant chromatograms from SFC using open-tubular and microbore columns [15]. The microbore C_{18} column used in these studies separates individual congeners in Aroclor mixtures on the basis of the relative extents of hydrophobic affinities. Ortho-substituted chlorine congeners that cause the two phenyl rings of biphenyl no longer to be coplanar tend to decrease the hydrophobicity of PCBs relative to chlorine atoms substituted in other positions. Therefore, when biphenyl rings have chlorine atoms in the ortho position they elute earlier on nonpolar phases. Detection was conducted at 208 nm with a detection limit of 1 µg/mL for PCBs with a 0.125-µL UV cell.

A negative temperature gradient coupled with density programming was used for the supercritical fluid separation of PCBs [16]. Separations were conducted on 1-mm-ID packed columns of Deltabond phenyl and C_{18}, 15 and 20 cm in length, respectively, as well as on 25-cm Spheri-5 cyanopropyl columns. Three supercritical fluid mobile phases were utilized: nitrous oxide, sulfur hexafluoride, and carbon dioxide. The UV detector, which was coupled to the end of the columns via a short piece of 25-µm-ID fused silica tubing, was set at wavelengths of 200 and 238 nm.

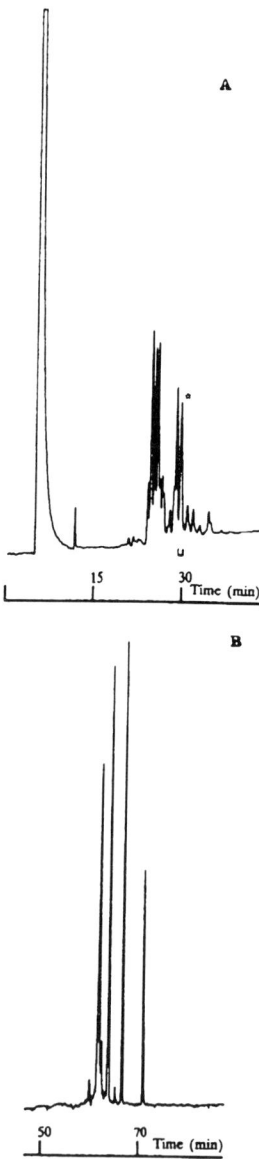

Fig. 1 Two-dimensional SFC chromatograms of a polychlorinated biphenyl (Arochlor 1242) sample. Conditions: (A) CO_2; 90°C; pressure program from 130 to 414 atm at 3 atm/min; asterisk, heartcut peak between 29 and 30 min; (B) CO_2; 110°C; pressure program from 70 to 120 atm at 20 atm/min, and then 120 to 414 atm at 5 atm/min, after an initial 2-min isobaric period. (From Ref. 14.)

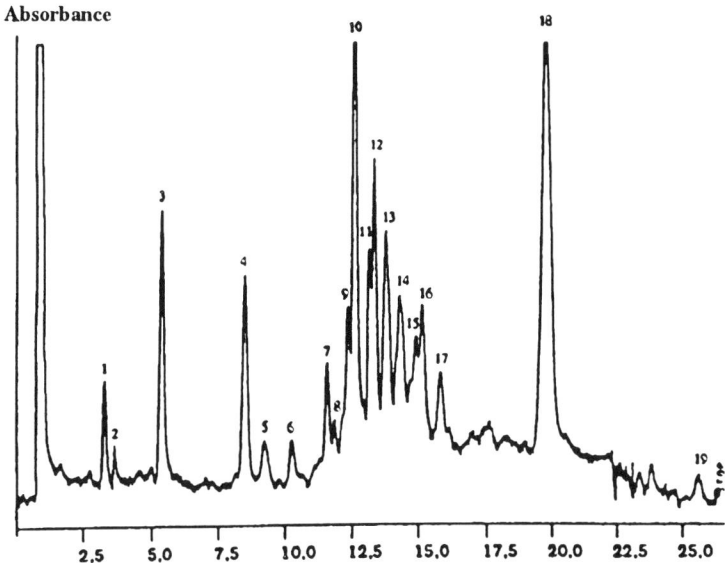

Fig. 2 Supercritical fluid chromatogram of sediment extract. 1, Acenaphthene; 2, fluorene; 3, hexachlorobenzene; 4, anthracene; 5, PCB 65,44; 6, PCB 71,37; 7, PCB 59; 8, unknown; 9, PCB 92; 10, PCB 101; 11, PCB 87,86; 12, PCB 110,115; 13, fluoranthene; 14, PCB 138; 15, PCB 128; 16, pyrene; 17, PCB 156; 18, chrysene; 19, OCDD. Operating conditions: 20 cm × 1 mm ID. Deltabond C_{18} column, supercritical CO_2 held at 12.6 MPa; two-step temperature programmed at 1.5°C/min from 70 to 60°C after a 4-min isotherm period and at 10°C/min from 60 to 35°C. UV detection at 200 nm. (From Ref. 16.)

The separation conditions ultimately used to separate PCBs and PAHs from a sediment extract, as illustrated in Fig. 2, were isoconfertic at 120 atm (1800 psi), with a two-step reverse-temperature program: isothermal for 5 min, 1.5°C/min from 70 to 60°C, and 10°C/min from 60 to 35°C. For the example in Fig. 2, a C_{18} column with carbon dioxide and a UV wavelength of 200 nm was used.

The same authors had previously published a more extensive study on the retention of PCBs in supercritical fluid chromatography [17]. In this report, the effects of column packing, mobile phase, temperature, and density were examined. The influence of density changes were investigated using decachlorobiphenyl and a cyanopropyl column with both nitrous oxide and carbon dioxide mobile phases and a C_{18} column with CO_2 as the only mobile phase. The results indicated that there was no significant difference between the two different mobile phases for a given stationary phase. However, with the C_{18} stationary phase, the column interaction with the PCB was stronger and higher densities were needed to affect elution. The main parameters that controlled elution of the

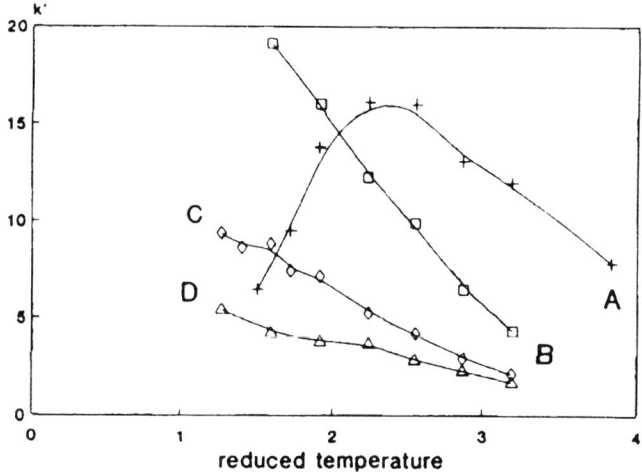

Fig. 3 Capacity factor versus reduced temperature on the cyanopropyl column at various densities: (A) 0.369; (B) 0.417; (C) 0.496; (D) 0.580 g/mL. (From Ref. 17.)

PCBs were determined to be the weight and size of the congeners; shape and electron configuration of the PCB analyte also influenced the retention. Substitution in the meta or para position or one substituent in the ortho position does not affect the free rotation of the molecule and it remains coplanar. Coplanar PCBs are retained more strongly than nonplanar congeners within a group of chloro homologs. An almost linear relationship between the capacity factor and the number of chlorine substituents for coplanar PCBs was observed. This was not true for the ortho-substituted PCBs. Figure 3 illustrates the effect of reduced temperature (experimental temperature, T_e, divided by critical temperature, T_c) at constant density. As illustrated, the capacity factor decreases with increasing reduced temperature above T_c. At a lower density (curve A of Fig. 3), there was first an increase followed by a decrease similar to the relationship found between retention and temperature at constant pressure. This peak maximum was found only in those cases where the density was less than the critical density.

C. Mass Spectrometric Detection

Direct supercritical fluid injection mass spectrometry (MS) utilized supercritical fluids for solvation and transfer of materials with a chemical ionization source [18]. One of the compounds examined in this investigation was PCB-like, 3,3-dichloro-4,4-diaminobiphenyl. Although no chromatographic separation was used up front, the advantage of using supercritical fluid introduction was shown as providing the potential to utilize the selective solvating power of supercritical fluids to develop rapid methods for direct extraction from complex substrates into the mass spectrometer. One hundred nanograms of this chloro-substituted biphenyl was injected into ammonia at 150°C and 400 atm (5900 psi).

III. PESTICIDES AND HERBICIDES

The analysis of pesticides and herbicides is often accomplished by HPLC with ultraviolet (UV) detection or by GC with some type of element specific detection (ECD, NPD). Both methods of analysis offer advantages and disadvantages. The lack of a universal detector for HPLC limits the scope of its applicability to agriculturally active compounds with UV chromophores, while GC is limited to thermally stable, volatile compounds. Many of the pesticides and herbicides that must be analyzed are thermally labile, or are not volatile, and therefore require derivitization for GC analysis. Specifically, supercritical fluid chromatography offers advantages for the analysis of pesticides and herbicides in that the available detectors span the range of traditional "GC-like" and "LC-like" detectors, and low operating temperatures may be used. The analysis of pesticides and herbicides by supercritical fluid chromatography is reviewed below. Figures 4 to 6 show the structures of the herbicides, insecticides, and fungicides discussed in this chapter, respectively.

A. Flame Ionization Detection

Wright and Smith [31] demonstrated the rapid analysis of carbamate pesticides with flame ionization detection. They used a short capillary column (0.9 m × 25 μm ID) coated with 5% phenyl polymethylsiloxane that had been cross-linked with azo-*tert*-butane. Pure carbon dioxide was used as the mobile phase, and pressure programming at rates of 100 atm/min was used. The compounds investigated were propoxur, chlorpropham, carbaryl, and phenmedipham. Baseline resolution of these compounds was achieved in less than 90 s. In a later publication [32], Wright and Smith again demonstrated the rapid analysis of carbamate pesticides. A mixture of six pesticides was separated on a 1.5 m × 25 μm 5% phenyl polymethylsiloxane column with pressure programming at 50 atm/min. The baseline resolution of propoxur, dicamba, carbaryl, 2,4-D, Silvex, and phenmedipham was achieved in approximately 120 s. This mixture was further expanded [33] with the addition of picloram and chloramben. Again, baseline resolution was obtained on a 1.5-m capillary column in approximately 120 s.

Richter [34] also reported the separation of four carbamate pesticides on a 15 m × 50 μm ID fused silica column with an SE-33 stationary phase (0.25 μm film thickness). Aldicarb, methomyl, diflubenzuron, and phenmedipham were baseline resolved in a little more than 20 min.

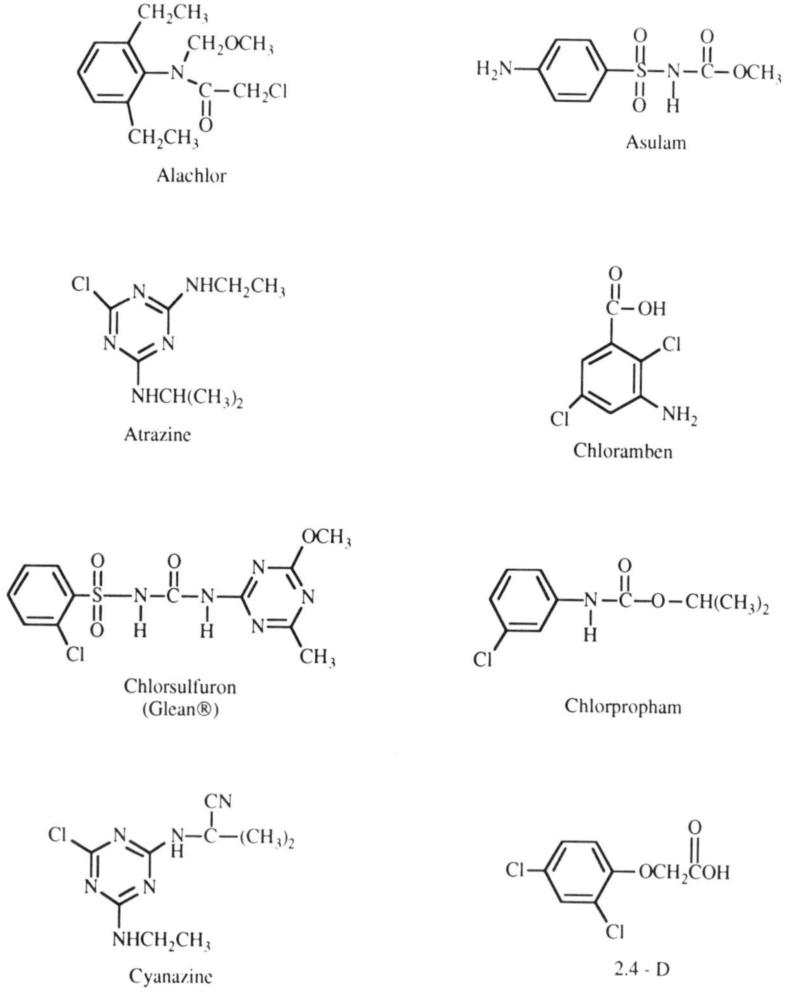

Fig. 4 Structures of herbicides.

Wheeler and McNally [35] reported the analysis of Karmex, Harmony, Glean, and Oust herbicides by capillary SFC with FID detection. Short capillary columns were used for this work (3 m × 100 μm ID). Relative standard deviations were determined for peak area, the linear range was determined, and the limit of detection was explored. Relative standard deviations for peak area were in the range 3 to 5% for the compounds of interest. The linear range was

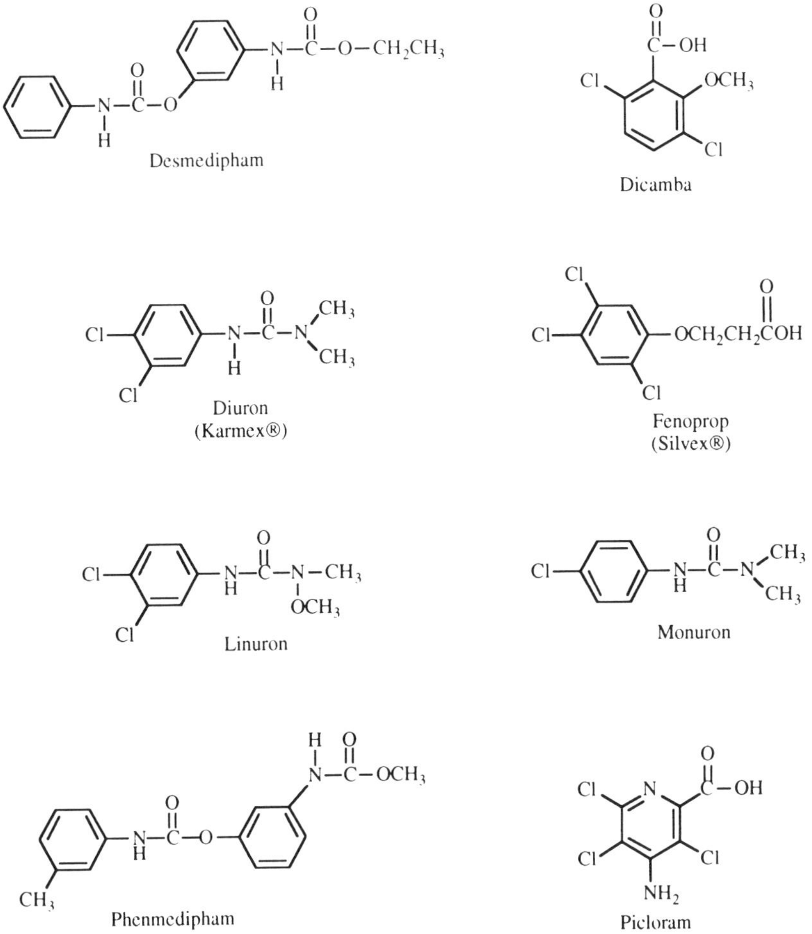

Fig. 4 Continued

found to be fairly compound dependent, and detection limits were reported in the range 20 to 80 µg/mL.

Novotny et al. [36] studied nonvolatile trace organics from water by microcolumn liquid chromatography (packed capillary) with UV detection and capillary SFC with FID detection. Isolation of the organic compounds from water was accomplished through the use of Sep-Pak C_{18} cartridges. Acceptable recoveries of 2,4,5-T, 2,4-D, *op,p'*-DDD, methoxychlor, atrazine, dioctylphthalate, pyrene, pentachlorophenol, carbazole, and hexachlorobenzene were demon-

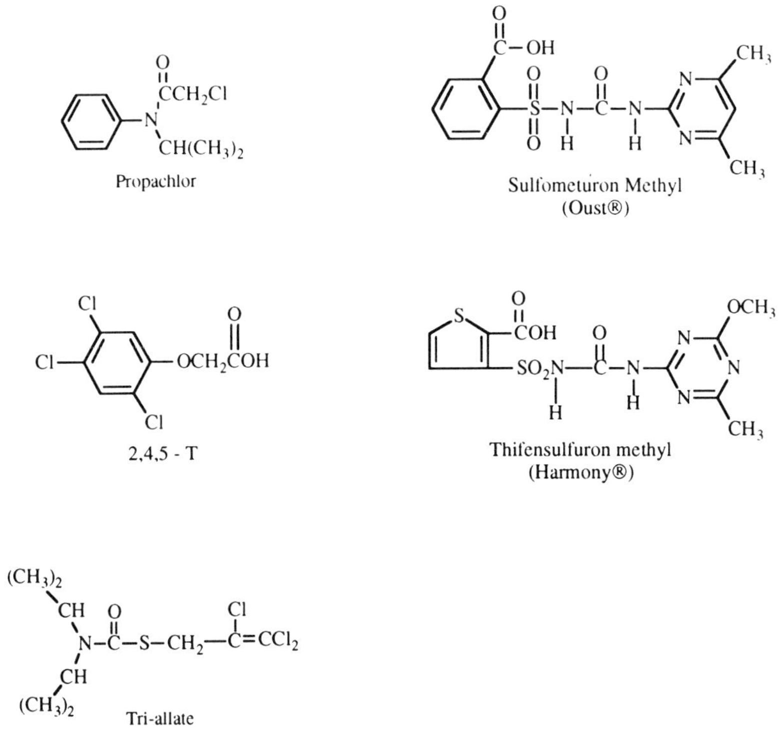

Fig. 4 Continued

strated for this isolation method by microcolumn liquid chromatography. However, under SFC conditions with flame ionization detection, the use of the Sep-Pak cartridges resulted in interfering peaks. Therefore, water samples for SFC analysis were preconcentrated through lyophilization. A lake and a river water sample were analyzed by capillary SFC (10 m × 50 μm, SE-30). The capillary SFC analyses required less time than the microcolumn LC separations (50 min as opposed to 10 h). However, the sensitivity of the SFC analysis was limited due to the small injection volume.

Lee et al. [37] investigated multidimensional SFC (SFC/SFC) for the analysis of carbofuran, its 3-keto, and 3-hydroxy metabolites extracted from the gullet of a bird. They demonstrated a flow-switching interface that allowed two 50-μm-ID capillary columns to be used in tandem. The use of solvent venting allowed for the injection of large volumes into the SFC/SFC system. For the analysis of

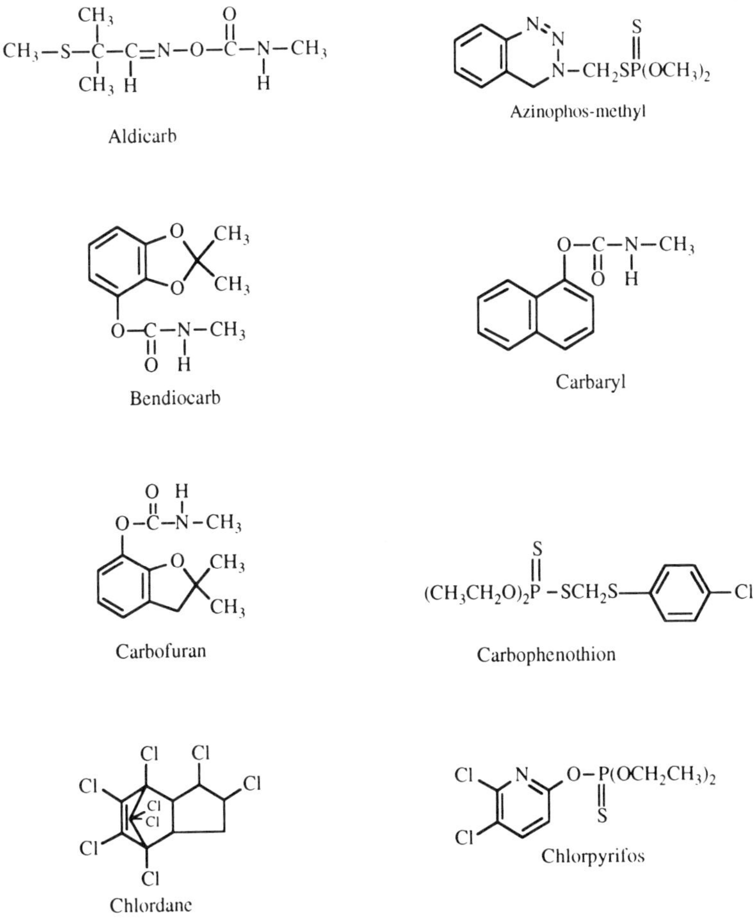

Fig. 5 Structures of insecticides.

carbofuran and its metabolites a 1.0-μL aliquot of bird extract was injected. A biphenyl column was used as the first column, followed by a glyme column. Using this instrumentation, carbofuran and its two metabolites were detected at levels in the range 1 to 10 ng.

Ashraf et al. [38] explored solvent-vented injection of 1.0-μL samples onto a 2 m × 110 μm ID retention gap connected through a venting valve to a 10 m × 50 μm ID fused silica capillary column. During the injection process the sample

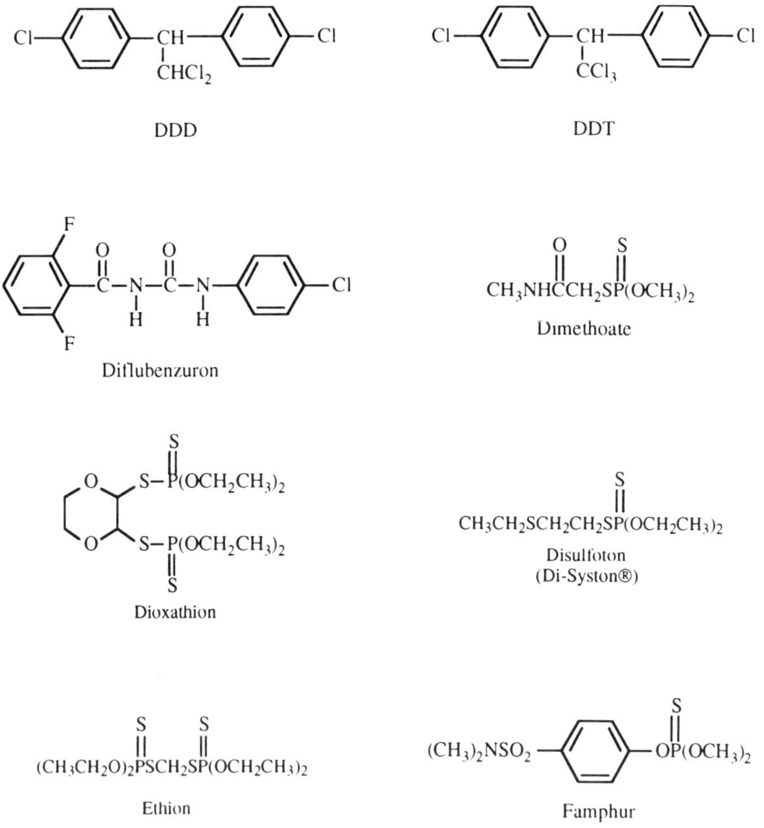

Fig. 5 Continued

loop was purged with nitrogen gas, and the valve was switched to the inject position for either a controlled amount of time to deliver 0.2 µL of sample, or for the full length of the chromatographic run to deliver 1.0 µL of sample. After injection the venting valve was positioned so that the flow of nitrogen passed through to vent, therefore concentrating the injected compounds on the precolumn. The venting valve was then switched, bringing the analytical column in-line so that chromatography could be performed. This injection technique was successful for up to 1.0-µL samples of 20 ppm solutions of the two triazine herbicides, atrazine and cyanazine. Two pyrethroids and one benzophenylurea compound were also injected using this technique. Detection of these analytes

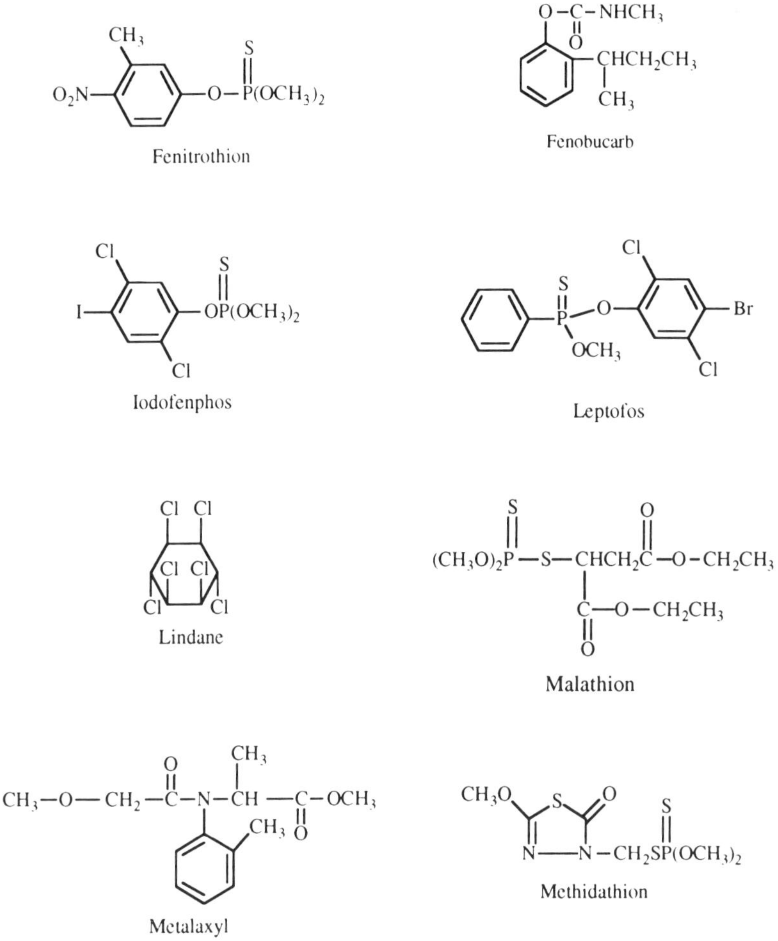

[Chemical structures of: Fenitrothion, Fenobucarb, Iodofenphos, Leptofos, Lindane, Malathion, Metalaxyl, Methidathion]

Fig. 5 Continued

was accomplished at concentrations of 5 ppm with good signal/noise ratio and good peak shape.

B. UV Detection

Games et al. [39] used analytical-scale packed columns with modified carbon dioxide and UV detection at 254 nm to analyze carbamate pesticides. The pesticides analyzed were chlorpropham, pirimcarb, methiocarb, carbaryl, phen-

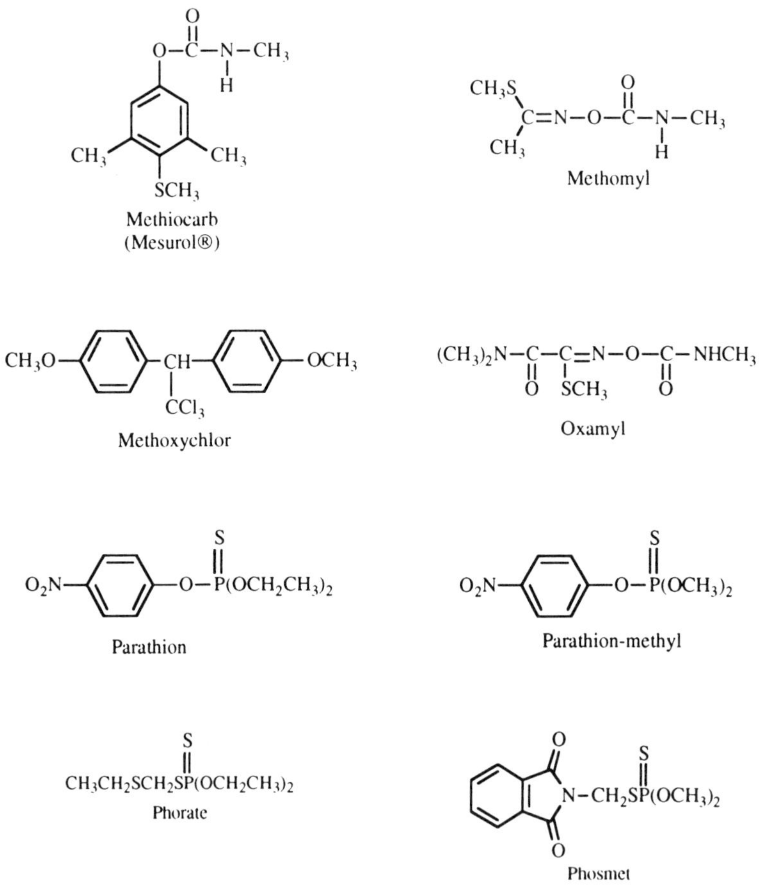

Fig. 5 Continued

medipham, and asulam. Resolution of the compounds of interest was achieved in approximately 5 min on a 100 × 4.6 mm ID LiChrosorb column. The mobile phase consisted of carbon dioxide with 12% methanol, and the flow was increased from 2 mL/min to 4 mL/min after the first 2 min.

Wheeler et al. [35] compared packed and capillary supercritical fluid chromatography with HPLC using representative herbicides and pesticides. Five herbicides were analyzed by HPLC, packed column SFC, and capillary column SFC. For both the capillary and packed column SFC experiments, UV detection was used. In the case of the packed column system, the UV detector used was the detector of the Hewlett-Packard 1082B supercritical fluid chromatograph. For the

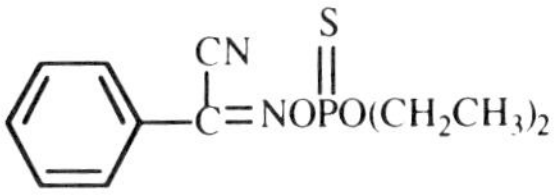

[Chemical structures shown: Phoxim, Pirimicarb, Propoxur, Tetrachlorvinphos, Thiodicarb (Larvin®)]

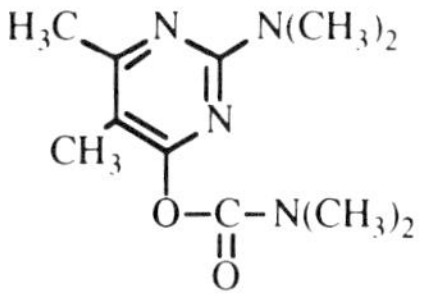

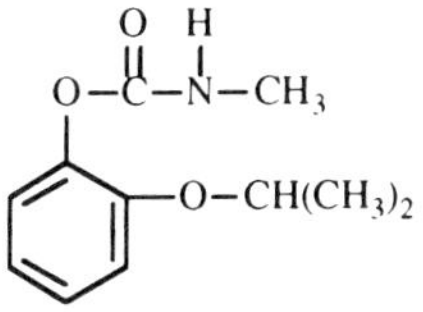

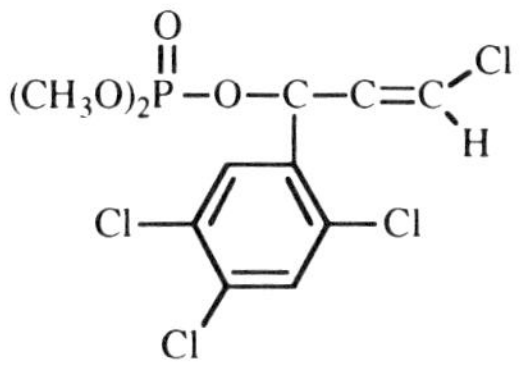

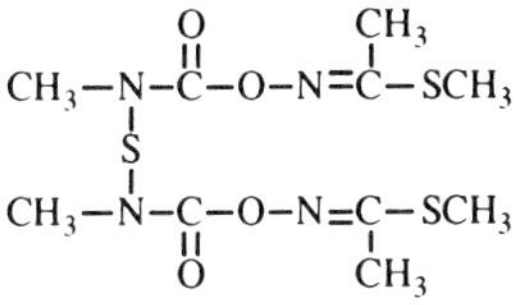

Fig. 5 Continued

capillary system, a Kratos 770 UV detector was modified to use fused silica tubing with the polyimide coating removed as the flow cell. Limits of detection, reproducibility, and linearity of response were compared. The compounds investigated were the moderately polar herbicides Oust, Glean, Karmex, Harmony, and Nustar. Oust, Glean, and Harmony are sulfonylurea herbicides. Karmex is a (phenylmethyl)urea and Nustar is a silicon fungicide. Their results indicated that faster analyses, lower detection limits, and greater injection to injection reproducibility were obtainable with packed column SFC. No appreciable difference in the linearity of response between the three techniques was observed.

France and Vorhees [40] investigated the use of a multichannel UV detector for capillary SFC analysis of the pesticides bendiocarb and carbaryl and the herbicides alachlor, diuron, and metalaxyl. A Hewlett-Packard 8452A photodiode array spectrophotometer was used to acquire UV spectra of the compounds of interest. The flow cell consisted of a fused silica capillary (0.32 mm ID) with the polyimide coating removed, resulting in a detector cell volume of approximately

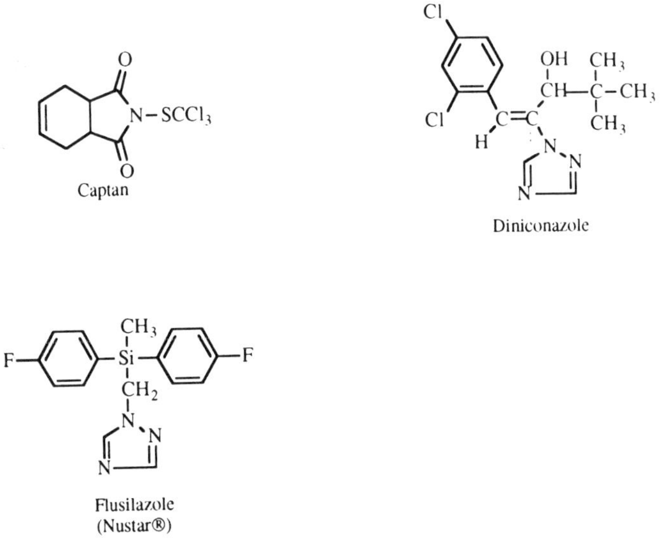

Fig. 6 Structures of fungicides.

710 nL. The chromatographic column was 12 m × 100 μm ID 5% phenyl methylpoly siloxane. Baseline resolution of the five compounds of interest was achieved in 31 min. Full spectra over the range 190 to 310 nm were collected at the time of maximum response for each compound. The limit of detection of bendiocarb ($S/N = 5$) was 3.8 ng.

McNally et al. [41] examined the retention behavior of a variety of compounds on packed SFC columns with modified carbon dioxide as the mobile phase. Compounds containing methyl, phenyl, nitro, amide, carboxamide, and chloro functional groups were examined to gain an understanding of retention characteristics in packed column SFC. The modifiers investigated in this study were methanol, ethanol, isopropanol, hexanol, and tetrahydrofuran at 2% w/v in carbon dioxide. The stationary phase used was silica, which was chosen to simulate polar matrices, such as soil and plant materials that are of interest in supercritical fluid extraction. Very specific interactions of the compounds of interest were observed which were dependent on the functional groups present, the polarity of the compound, the steric interactions between individual functional groups of the molecules, and the polarity of the modifier used.

Taylor et al. [42] investigated the behavior of triazine and triazole herbicides on packed column SFC. An analytical scale (25 cm × 4.6 mm ID) Deltabond CN column was used for this work. The flow of carbon dioxide was held constant, while the flow of methanol was increased during the chromatographic run. Baseline resolution of the eight compounds of interest was obtained in approximately 6 min. Over the course of those 6 min, the methanol concentration was increased from 2.4% to approximately 30%. The outlet pressure during the separation was 270 atm (4000 psi) and the oven temperature was maintained at 60°C. The effects of carbon dioxide flow rate, outlet pressure, and oven temperature on the separation were also explored.

Shah and Taylor [43] also investigated the separation of some ureas on an analytical-scale packed cyanopropyl column. Baseline resolution of six compounds of interest was achieved in approximately 8 min. Methanol concentration was held constant at 2%, and the flow rate was set at 3 mL/min. The compounds separated were dimethylcarbanilide, (dimethylphenyl)urea, (diphenylmethyl)urea, monuron, diuron, and carbanilide.

C. Electron Capture Detection

Kennedy and Wall [44] investigated the use of an electron capture detector (ECD) for the analysis of agrochemicals by capillary SFC. Separations of a triazole fungicide metabolite were obtained on a 5 m × 50 μm SB-methyl-100 column. A frit restrictor was used to provide back-pressure and was placed at the entrance to the cell. The operating temperature of the detector was 350°C. Makeup gas of 10% methane in argon was used. They found that the optimum conditions for the operation of the system were obtained with makeup gas at a flow rate of 15 mL/min, and the restrictor positioned approximately 3 cm from the column nut at the entrance to the detector. Operating under these conditions, pressure programming from 100 to 350 atm (1500 to 5100 psi) at 40 atm/min was performed without a substantial increase in background. They estimated that a minimum detection limit of approximately 35 pg ($S/N = 3$) could be achieved for their compound of interest.

Chang and Taylor [45] also evaluated the performance of the electron-capture detector with capillary supercritical fluid chromatography. They used capillary columns (6 m × 50 μm) and frit restrictors to provide back-pressure to the system. The effect of makeup gas flow rate (10% methane/argon) was investigated, and subsequent work was done with a makeup gas flow rate between 20 and 30 mL/min. They also determined that the temperature of the detector affected the sensitivity in a compound-dependent manner. Detection limits of 0.64 pg were reported for the pesticide triallate. Chlordane, Arochlor 1254, and Arochlor 2565 were also separated and detected using the ECD. A separation of seven thermally labile pesticides was accomplished in approximately 20 min, and is shown in Fig. 7. Attempts at the detection of captafol and captan by SFC/ECD were unsuccessful, although these compounds had been detected in SFC/FID

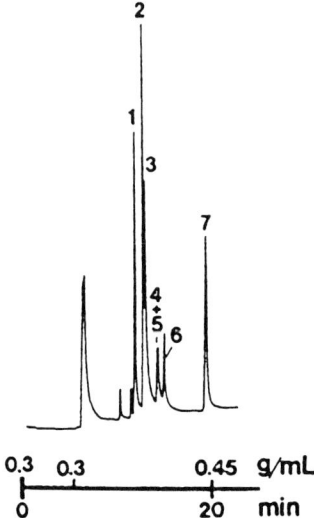

Fig. 7 Chromatogram of (1) metobromuron, (2) fenitrothion, (3) fenchlorphos, (4) chlorbromuron, (5) tetrachlorvinphos, (6) diuron, and (7) tetradifon on an SD-phenyl-5 with ECD detection. The mobile phase was density programmed from 0.3 to 0.4 g/mL at 0.01 g/mL per minute at 100°C. (From Ref. 45.)

experiments. It was postulated that these compounds could not be vaporized in the detector cavity of the ECD, due to their high melting points.

D. Thermionic Detection

Mathiasson et al. [46] investigated the use of capillary SFC with nitrous oxide as the mobile phase and a thermionic nitrogen–phosphorus detector for the analysis of polar nitrogen-containing compounds. Detector performance was optimized by systematic variation of the makeup gas flow rate (nitrogen), the hydrogen and air flow rates, and the bead current. Detector response with increasing nitrous oxide pressure was explored. A dependence of peak area on the system pressure was observed for some of the compounds of interest. The effect of the addition of methanol modifier to the nitrous oxide on the detector response was also explored. Concentrations of methanol above 0.8% resulted in a significant loss in signal from the detector. The test compounds for this work were free amines and their carbamate and amide derivatives.

Ashraf et al. [47] described the use of a "three-electrode" thermionic detector (TID) for the analysis of agrochemicals. Both the phosphorus- and nitrogen-selective modes of this detector were explored with capillary SFC as the means of sample introduction to the detector. The compounds of interest were two triazine herbicides, two pyrethroids, a benzophenylurea, and a chlorophenyl vinyl diethylphosphone. Solutions of these compounds spanned the concentration range of 2.5 ppb to 25 ppm. The capillary column used was a 10 m × 50 μm ID biphenyl methylpolysiloxane column. Optimization experiments were conducted to determine appropriate values for hydrogen, air, and nitrogen flow rates (makeup gas), and the position of the alkali source in relation to the flame tip. Significant baseline rise was observed during pressure programming. In the phosphorus mode, good sensitivity was demonstrated for the vinyl phosphone, with detection in the picogram range. Responses were obtained for nitrogen-containing compounds while the detector was operated in the phosphorus mode, but detection was only slightly improved over what could be obtained with flame ionization detection. In the nitrogen mode, the background current was observed to be much lower than in the phosphorus mode. However, the lower background current did not result in a less severe baseline increase with pressure programming. This detector was found to demonstrate enhanced sensitivity in the nitrogen mode, with detection limits in the range 0.6 to 60 pg for the compounds of interest.

Knowles and Richter [48] studied the effect of using modified carbon dioxide and nitrous oxide on the retention behavior of aldicarb, methomyl, diflubenzuron, and phenmedipham. Analyses were carried out on capillary columns (15 m × 50 μm ID) with a nitrogen–phosphorus detector (NPD). Similar separations of the four compounds were obtained when carbon dioxide and nitrous oxides were used as the mobile phase. Upon the addition of 1% THF as a modifier, changes in retention were observed, with the later eluting peaks (diflubenzuron and phenmedipham) showing larger changes in retention.

Brinkman et al. [49] investigated the analysis of organophosphate pesticides by packed capillary SFC with a thermionic detector for phosphorus-selective detection. The compounds of interest were malathion, phoxim, ethion, dimethoate, and azinphos-methyl in onion and tomato extracts. The column was 130 × 0.32 mm ID packed with C_{18} packing material (5 μm). Experiments were conducted to optimize the detector performance with regard to mobile-phase composition, hydrogen and air flow rates, as well as the distance between the jet and the bead. Attempts at separating the five compounds of interest with pure carbon dioxide were unsuccessful; therefore, the addition of methanol and 2-propanol as modifiers was explored. The addition of 1.5% methanol to the carbon dioxide modifier resulted in baseline resolution of the five compounds of interest in approximately 9 min. The addition of 3.5% 2-propanol also allowed for baseline resolution of the organophosphate pesticides; however, the resulting peak shape was poorer than the peak shape obtained with methanol modifier. Calibration plots were prepared for standards dissolved in acetone, and compared to plots obtained when the compounds of interest were spiked into tomato and onion

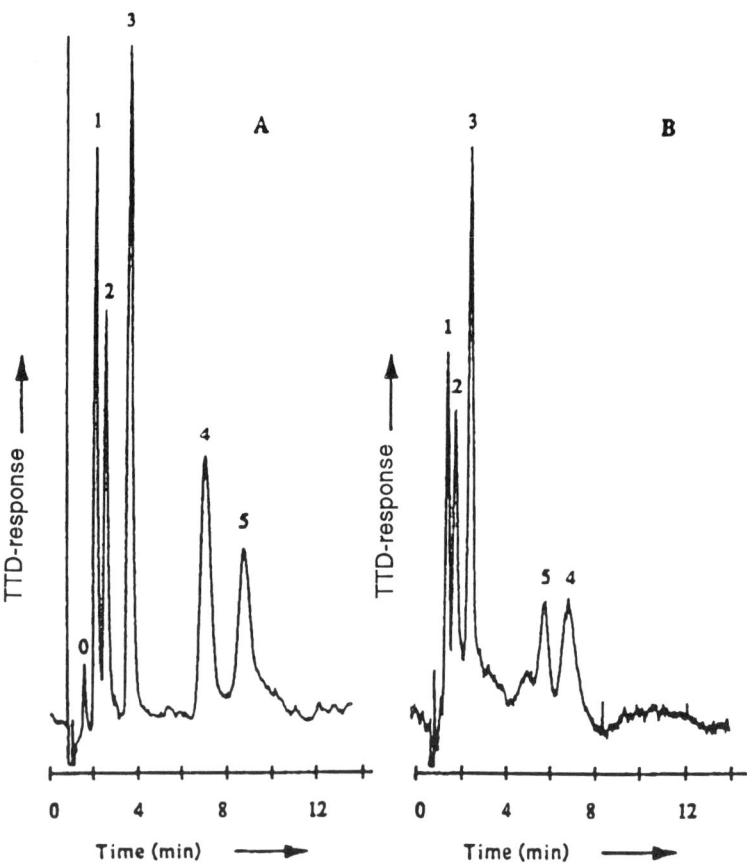

Fig. 8 Chromatogram of organophosphorus pesticides: (0) contamination in ethione; (1) malathion; (2) phoxim; (3) ethione; (4) dimethoate; (5) azinphos-methyl. Chromatogram A shows the separation of 1 ng per component in onion extract. Mobile phase: CO_2 with 1.5% methanol. Chromatogram B shows the separation of 0.2 ng per component in tomato extract. Mobile phase: CO_2 with 3.5% 2-propanol. (From Ref. 49.)

extracts. The slopes of the calibration curves were basically unchanged, indicating an absence of matrix effects. Limits of detection for the five compounds of interest ranged from 15 to 62 pg with 1.5% methanol in the mobile phase. Figure 8 shows the separation of the five compounds of interest in tomato and onion extracts.

Richter et al. [50] reported the analysis of carbamate pesticides in parsley extracts. Aldicarb, methomyl, Mesurol, oxamyl, carbofuran, and carbaryl were analyzed by capillary SFC (10 m × 50 μm ID SB-methyl-100 column) with nitrogen–phosphorus detection. The pesticides were present in the parsley extract at levels of approximately 2 ppb.

E. Photometric Detection

Lee et al. [51] investigated the use of a dual-flame photometric detector (FPD) with capillary SFC for the detection of sulfur- and phosphorus-containing compounds. Modifications were made to the detector to make it compatible with supercritical mobile phases. Oxamyl, parathion, chlorpyrifos, and Larvin were analyzed by capillary SFC (15 m × 75 μm ID). In the phosphorus mode, little baseline rise with pressure programming was noted over the range 50 to 200 atm (700 to 2900 psi). A detection limit of 0.5 ng ($S/N = 2$) was reported for parathion. In the sulfur mode more significant baseline rises with pressure programming were observed, requiring the use of a baseline correction program. Detection limits were poorer than those observed in the phosphorus mode, with a detection limit of 25 ng ($S/N = 2$) for benzo[b]thiophene. Figure 9 shows the separation of parathion, chlorpyrifos, and Larvin with detection in both the sulfur and phosphorus modes.

Sievers et al. [52] reported the use of sulfur chemiluminescence detection with capillary SFC for the analysis of malathion, carbophenothion, dioxathion, fenitrothion, and methyl and ethyl parathion. Chromatography was carried out on a 3.5 m × 100 μm DB-5 fused silica column. Positioning of the restrictor tip in relation to the chemiluminescence chamber was studied and was found to have a large effect on the quality of the chromatographic separation. The analysis of malathion from a commercial formulation was demonstrated. Detection limits for malathion were reported to be 4.5 ng or 77 pg/s. Detection limits were observed to be compound specific, with a detection limit of 39 ng (65 pg/s) for methyl parathion.

F. Atomic Emission Detection

Lee et al. [53] reported the use of a radio-frequency plasma (RFP) detector for the detection of sulfur and chlorine after sample introduction by capillary supercritical fluid chromatography. Emission spectra of carbon dioxide and nitrous oxide were studied to determine that these fluids should not create significant background or interference in the regions of interest for sulfur and chlorine. Short capillary columns (2 to 3 m × 50 μm) and low flow rates were used to prevent the introduction of too much mobile phase, and subsequent quenching of the plasma. The separation of two organophosphorus insecticides (i.e., methidathion and chlorpyrifos) and a carbamate insecticide (i.e., carbofuran) was demonstrated. When an FID was used for detection, all three compounds were detected. When the RFP detector was used, either the chlorpyrifos or both the chlorpyrifos

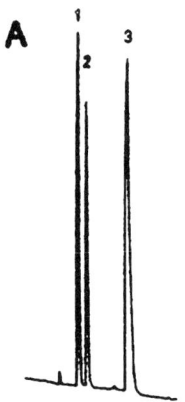

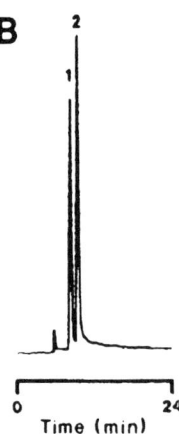

Fig. 9 Chromatogram of sulfur- and phosphorus-containing pesticides: (1) parathion; (2) chlorpyrifos; (3) Larvin. Detection: chromatogram A, flame photometric detection in the sulfur mode; chromatogram B, flame photometric detection in the phosphorus mode. CO_2 mobile phase at 120°C and 0.5 g/mL. (From Ref. 51.)

and the methidathion were detected, depending on whether the wavelength of chlorine emission or sulfur emission was being monitored. Chlordane was also analyzed by SFC with RFP detection at the emission wavelength for chlorine. The analysis of DDT in milk was also demonstrated. The FID response for the milk extract is a complex chromatogram due to the triglycerides present. However, the triglycerides are not detected by RPD, and a single peak for DDT is easily detected. A mixture of the insecticides α- and β-BHC, chlordane, and methoxychlor were also separated and detected by RPD in the chlorine mode. Nitrous oxide was used as the mobile phase in this separation. Detection limits for sulfur and chlorine were reported as pg/s as a function of the density of the mobile phase. Sulfur detection limits of 60 pg/s were obtained at 100 atm, while this limit rose to 178 pg/s at 400 atm.

Luffer and Novotny [54] investigated the use of a microwave-induced plasma for the detection of sulfur-containing pesticides. Capillary SFC (10 m × 50 μm, SB-cyanopropyl-50) was used to introduce the pesticide samples to the plasma. The pesticides investigated were phorate, Di-Syston, malathion, and ethion. This mixture of compounds was separated using nitrous oxide as the mobile phase, and detection at the sulfur line, as shown in Fig. 10. Baseline disturbances (either positive or negative slopes) were observed at differing helium flow rates. It was therefore determined that there is an optimum helium flow rate that depends on the sulfur line being monitored. The use of nitrous oxide as a mobile phase was explored, and interference from CN-band emission was found with most of the lines that were investigated. Sensitivities were compound and mobile-phase dependent. The sensitivity reported for sulfur with carbon dioxide as the mobile phase was 73 pg/s.

G. Fourier Transform Infrared Detection

Wieboldt and Smith [55] reported the separation of a four-component pesticide sample using capillary (10 m × 100 μm) SFC with FTIR (Fourier transform infrared) detection. The compounds of interest were aldicarb, methomyl, captan,

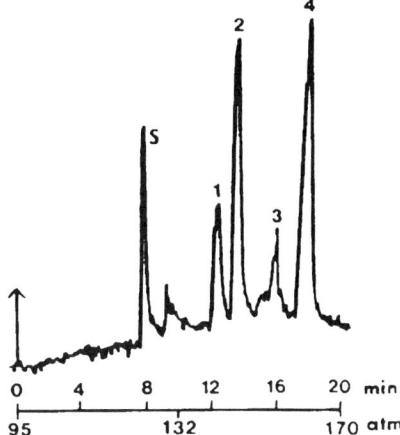

Fig. 10 Chromatogram of (1) phorate, (2) Di-Syston, (3) malathion, and (4) ethion, using element-specific detection (surfatron plasma) and monitoring the sulfur line. (From Ref. 54.)

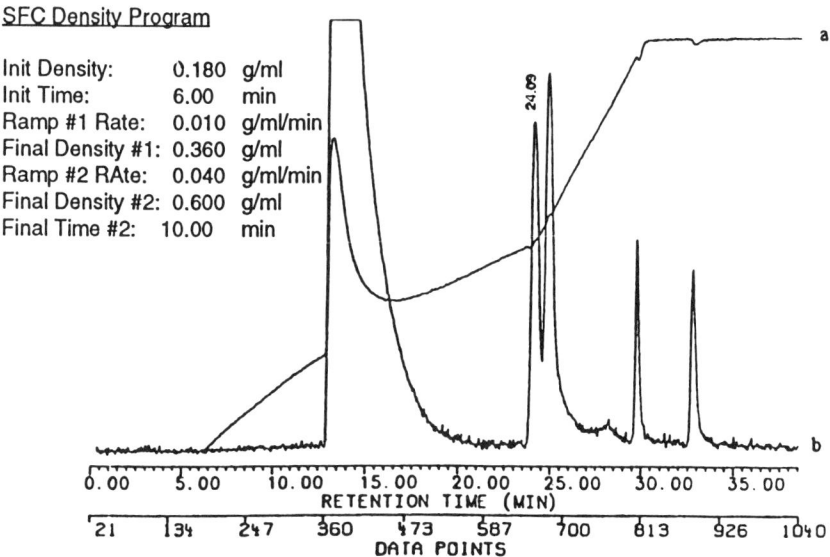

Fig. 11 Gram–Schmidt (a) and Gram–Schmidt Plus (b) reconstructed chromatograms of aldicarb, methomyl, captan, and phenmedipham. Column: SB-methyl-100, 10 m × 100 μm. Mobile phase: CO_2 at 100°C. (From Ref. 55.)

and phenmedipham. Detection was accomplished on-line through the use of a flow-cell interface. The flow cell is connected to the chromatographic column and to the restrictor by lengths of fused silica capillary tubing (100 μm). The concentrations of the pesticide samples were approximately 5 mg/mL per component. The injection volume was 200 nL with a 22:1 split ratio. "Gram–Schmidt Plus" reconstruction techniques were used to remove interference caused by the changing density of the carbon dioxide during pressure programming, and spectral subtraction techniques were used to remove CO_2 features from the obtained IR spectra. The quality of the spectra obtained was sufficient to provide structural information for the compounds of interest. Figure 11 shows the Gram–Schmidt Plus reconstructed chromatogram of the pesticide mixture.

Shah et al. [42] also demonstrated the use of the on-line FTIR detector for SFC; however, they reported the use of packed columns, as well as capillary columns, for the analysis of herbicide precursors. The column used in the urea separation was a 10 cm × 1 mm ID cyano column, while capillary columns were employed for the benzamide–anilide mixture and the benzamide and sulfonamide mixture. The concentrations of the compounds of interest were approximately 3 mg/mL and a splitless injection of 0.1 μL was used. Carbon dioxide was used as the mobile phase for all separations. The ureas studied were dimethylcarbanilide, (dimethylphenyl)urea, (diphenylmethyl)urea, monuron, diuron, and carbanilide. Baseline resolution of all compounds except for dimethylcarbanilide and (dimethylphenyl)urea, which were partially resolved, was obtained on the packed column. Peak assignments were made based on comparison with pure standards and by spectral interpretation. Capillary columns were used to separate the benzamide–anilide mixture after attempts to develop the separation on the packed CN column failed. The baseline resolution of six compounds of interest was achieved in approximately 25 min. On-line spectra with a high S/N ratio were obtained. A prototype chiral capillary column was employed in this work to separate positional isomers of some benzamides.

Wieboldt et al. [56] reported the separation of six pyrethrins, which are naturally occurring insecticides isolated from chrysanthemums. The compounds of interest were cinerin I and II, jasmolin I and II, and pyrethrins I and II. Capillary SFC was used to introduce the compounds of interest to the spectrometer. Detection was performed on-line through the use of a flow cell. Baseline resolution of the six compounds of interest was obtained on a biphenyl column, and enough spectral information was obtained through the FTIR spectra to distinguish between the structurally similar compounds of interest. The separation of the six compounds by GC resulted in the degradation of pyrethrins I and II; therefore, SFC/FTIR provided a way to analyze these compounds without degradation, while maintaining the structural information provided by GC/IR analysis.

H. Mass Spectrometric Detection

Wright et al. [57] reported the analysis of the labile pesticide aldicarb by SFC/MS employing ammonia chemical ionization. Capillary SFC was used for sample introduction into the MS instrument. Ammonia chemical ionization resulted in an $(M + 18)^+$ molecular ion for aldicarb. The separation of the thermally labile acid and carbamate pesticides propoxur, BPMC (fenobucarb), propachlor, carbofuran, alachlor, carbaryl, linuron, and diuron was achieved. Resolution of all compounds except BPMC and propachlor was obtained on a 10 m $\times$ 50 μm 5% phenyl methylpolysiloxane stationary phase. The spectra obtained for carbaryl with ammonia and methane as reagent gases were compared. Ammonia provided for much softer ionization with an $(M + 18)^+$ molecular ion and little fragmentation, while methane chemical ionization resulted in an $(M + 1)^+$ molecular ion and increased fragmentation.

Wright et al. [58] conducted further studies comparing ammonia and methane chemical ionization of carbamate and acid pesticides using capillary supercritical fluid chromatography as a means of sample introduction. The carbamate pesticides explored were aldicarb, aldicarb sulfoxide, aldicarb sulfone, carbaryl, BPMC, propoxur, chlorpropham, carbofuran, asulam, desmedipham, and phenmedipham. The acid pesticides explored were 2,4-D, 2,3-D methyl ester, dicamba, picloram, Silvex, and Silvex methyl ester. A short capillary column (2 m × 50 μm ID) and rapid pressure ramps (50 atm/min) were used to elute the compounds of interest. They noted that spectra obtained with ammonia as the reagent gas resembled those produced in thermospray HPLC/MS. For the carbamates studied, the general rule was that the ammonium adduct ion was the base peak. The presence of the molecular species indicated that thermal degradation was not occurring to any significant extent in these analyses. Methane CI spectra for the herbicides linuron, diuron, and alachlor following SFC have also been reported [59].

Wilkins et al. [60] reported the use of a Fourier transform mass spectrometer for the detection of a seven-component pesticide mixture separated by SFC. A dual-cell interface was used to try and minimize problems that had previously been observed due to the extreme difference in pressure requirements between the mass spectrometer and the supercritical fluid chromatograph. Capillary columns (20 m × 100 μm) were used for this work. The reconstructed FTMS chromatogram of lindane, aldrin, DDE, dieldrin, DDD, DDT, and methoxychlor is shown in Fig. 12. Typical chromatographic detection limits for this system were in the low-nanogram range, although the sensitivity for aldrin and dieldrin was not as great due to the extent of fragmentation.

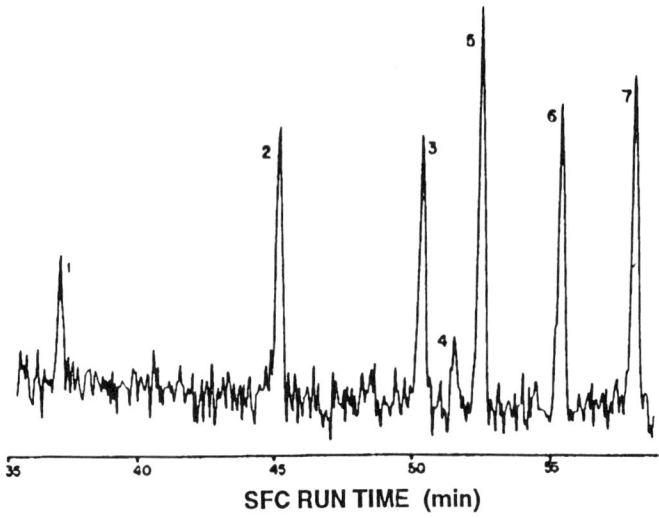

Fig. 12 Reconstructed SFC/FTMS chromatogram of (1) lindane, (2) aldrin, (3) DDE, (4) dieldrin, (5) DDD, (6), DDT, and (7) methoxychlor. Oven temperature 150°C, pressure programmed from 130 to 270 atm at 2 atm/min after a 20-min initial isobaric hold. (From Ref. 60.)

Kalinoski and Smith [61] demonstrated the use of microbore packed column supercritical fluid chromatography with mass spectrometric detection for the analysis of organophosphate pesticides. The interface between the SFC system and the mass spectrometer was the HFR (high flow rate) interface, which allows for the use of packed columns. Spectra were collected in the CI mode using ammonia or the 2-propanol modifier as the reagent gas. The compounds of interest were chlorpyrifos, chlorpyrifos methyl, iodofenphos, leptophod, methidathion, tetrachlorvinphos, phosmet, and famphur. Resolution of the compounds of interest was achieved on an amino column with 2% 2-propanol in the carbon dioxide mobile phase. With the exception of chlorpyrifos and chlorpyrifos methyl, baseline resolution was achieved in approximately 7 min. Spectra were obtained for the compounds of interest with both ammonia and 2-propanol as the reagent gas. Detection of chlorpyrifos in the selected ion monitoring mode at 94 pg on column with an S/N ratio of 21 was reported. An extract of cherries spiked with the mixture of organophosphorus insecticides was also analyzed by this method.

Lee et al. [62] reported the use of a double-focusing mass spectrometer for detection of a mixture of pesticides and herbicides separated by capillary (6 m × 50 μm) supercritical fluid chromatography. The interface developed for this work was a direct heated probe. The compounds of interest were carbofuran, α-BHC, ∂-BHC, β-BHC, chlordane, and DDT. Spectra were collected in the negative-ion chemical ionization mode with methane as the reagent gas. The mass on column per compound was in the subnanogram range.

Hawthorne and Miller [63] reported the analysis of a triazine herbicide metabolite by capillary SFC with a capillary-direct interface. The interface used required no modification of the mass spectrometer, therefore allowing very rapid conversion between an SFC/MS system and a GC/MS system. Spectra were collected in the CI mode with methane as the reagent gas.

Niessen et al. [64] examined the effect of repeller potential on spectra obtained by SFC/MS using a thermospray interface. The influence of the repeller potential on the degree of fragmentation of diuron was studied when the sample was introduced to the spectrometer by analytical scale (4.6 mm ID) packed column SFC. Spectra of diuron with 2% methanol in the carbon dioxide mobile phase were obtained at low, intermediate, and high repeller potentials. At low repeller potentials CI spectra were obtained, while at high repeller potentials the spectrum resemble EI spectra of diuron. The effect of vaporizer temperature on the spectra of diuron was also explored. At high vaporizer temperatures, thermal decomposition of diuron was observed.

Taylor et al. [65] reported the separation of ureas by packed column SFC followed by detection with a benchtop thermospray mass spectrometer. The separation of dimethylcarbanilide, (dimethylphenyl)urea, (diphenylmethyl)urea, monuron, diuron, and carbanbalide on a 1-mm ID cyanopropyl column was reported. Significant baseline rise with pressure programming was noted and attributed to hydrocarbon contamination in the carbon dioxide mobile phase.

I. Ion Mobility Detection

Morrissey and Hill [66] reported the use of an ion mobility detector for the analysis of 2,4-D and 2,4,5-T by capillary SFC. In the Fourier-transformed drift spectra for both 2,4-D and 2,4,5-T, two major ion peaks were observed instead of the one expected peak in each spectrum. It was hypothesized that dimer formation was occurring. As the temperature of the detector was increased, the peak attributed to the dimer decreased in intensity, and finally was not detected at a detector temperature of 250°C. However, it was found to be advantageous to monitor the dimer of 2,4-D at a detector temperature of 150°C in soil extracts due to the lack of interference.

J. SFE/SFC

McNally and Wheeler [67] reported the coupling of supercritical fluid extraction with supercritical fluid chromatography for the analysis of sulfonylurea herbicides and their metabolites. The matrices examined by SFE were soil, ground soybean, whole wheat kernels, wheat flour, wheat straw, and a cell culture medium. The compounds of interest were either radiolabeled or were detected by UV detection. Extraction efficiencies were found to be a function of extraction time, flow rate of the supercritical fluid, and the amount of modifier in the fluid. Extraction and the chromatographic separation were achieved with 2% methanol in the mobile phase. Chromatographic separations were done on packed columns.

McNally and Wheeler [8] also described the analysis of linuron and diuron in soil by SFE coupled with packed column SFC for analysis. The chromatographic behavior of linuron and diuron with varying temperature, pressure, and modifier was studied. The effects of the addition of methanol, ethanol, and acetonitrile and retention were determined. The effect of these variables on extraction efficiency was also studied. Methanol modified carbon dioxide allowed for more efficient extraction of diuron from soil than did acetonitrile-modified carbon dioxide. Recoveries of 81% were achieved with 10% methanol in carbon dioxide. The extraction of linuron from soil was more efficient when ethanol was used as a modifier rather than methanol. They concluded that the polarity of the mobile phase must be matched to the polarity of the analytes of interest to achieve efficient extraction. Extraction efficiencies of both compounds increased with increasing temperature. An optimum flow rate of 5 mL/min was determined for the efficient extraction of linuron and diuron from soil. Further work on SFE/SFC of diuron and linuron determined that using a microbore packed column for the analysis of linuron and diuron resulted in detection limits in the range 10 ppb [68].

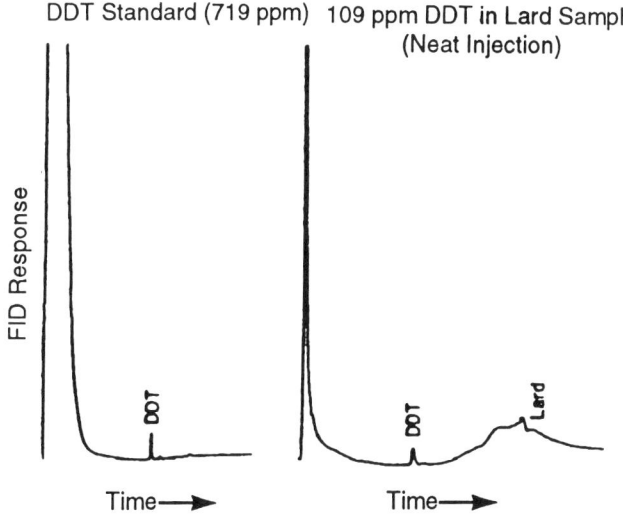

Fig. 13 Chromatograms of DDT standard and DDT in a lard sample. Instrument conditions: 150°C, pressure program: 100 atm (5 min), 100 to 485 atm in 30 min, 485 atm (10 min). (From Ref. 69.)

King [69] reported the SFC analysis of DDT in lard. In this work, a lard sample was spiked with 100 ppm of DDT. Neat injection of the undiluted lard sample onto a 100 × 1 mm ID packed alumina column was done. Pressure programming was done in order to selectively solubilize the pesticide sample and separate it from the fat. Flame ionization detection was used in the work. The resulting chromatogram is shown in Fig. 13.

Nishikawa [70] reported the SFE/SFC analysis of fenitrothion, esfenvalerate, and diniconazole from soil. Analysis of the extracted pesticides was carried out on an analytical-scale packed column (25 cm × 4.6 mm ID, C_{18}) with UV detection. Extraction efficiency was found to be extremely compound dependent, with fenitrothion being the most easily extracted compound (78.7% recovery). The extraction of esfenvalerate and diniconazole required the addition of water to the soil sample. These compounds were also extracted from straw and green tea.

IV. PHENOLS

The Environmental Protection Agency's (EPA) Method 604 monitors 11 phenols by gas chromatography because of their hazardous potential when released into the environment [71]. The method has been updated to include the use of capillary

GC columns, but as reported by Berger and Deye, remains difficult and cumbersome [72]. Phenols do offer a moderate polarity range in between nonpolar molecules most readily analyzed by GC and polar molecules typically characterized by reversed-phase liquid chromatography separations. Therefore, phenols represent transition solutes in supercritical fluid chromatographic separations, and several workers have included one or two phenols in supercritical chromatographic evaluations of standard test mixes. Still other workers have concentrated solely on the retention characteristics of phenols themselves. These studies, as outlined in the following section, provide representative methodologies for phenol analysis.

A. UV Detection

The majority of the chromatographic studies that have been conducted examining phenols utilized UV detection; since the phenols are typically UV active this is a legitimate selection. Packed column supercritical fluid chromatography was used with binary and ternary mobile phases while optimizing temperature, pressure, and mobile-phase composition [72]. The phenols examined were 2-nitrophenol, 4,6-dinitro-2-methylphenol, 2,4-dinitrophenol, 2-chlorophenol, 2,4-dimethylphenol, 2,4,6-trichlorophenol, 2,4-dichlorophenol, phenol, 4-chloro-3-methylphenol, pentachlorophenol, and 4-nitrophenol.

Separation of these 11 phenols was accomplished in less than 4 min with a Lichrosorb Diol column, 4.6 × 200 mm, 5 µm. Other examples used cyanopropyl and amino columns. The flow of a 4.8% methanol solution containing 0.025% trifluoroacetic acid (TFA) in carbon dioxide was 4.2 mL/min. The mobile phase was dynamically mixed at high pressure and delivered with either a Hewlett-Packard 1050 liquid pump or the Hewlett-Packard 1082 supercritical fluid chromatograph at 4°C. The column temperature was 40°C. A typical resultant chromatogram is shown in Fig. 14. Efficiencies reported for this separation were approximately 8000 plates, equivalent to those obtained with normal-phase liquid chromatography. The use of the TFA in this separation, labeled as an additive, not a modifier, because it is not directly miscible in the carbon dioxide of the eluant, causes more rapid elution and better peak shape. Modifications of the concentration from 0.025% to 0.1% of TFA did not change retention or peak shape significantly with the phenols examined, nor did substitution of the TFA moiety with trichloroacetic acid (TCA). Other more polar molecules did experience differences in retention when the TFA additive concentration was adjusted. The experimenters felt that the additive was adsorbed on the stationary phase as well as being present in the mobile phase. Additional analyses where the stationary phase had been treated with additive adjusted mobile phase (the phenols were eluted with methanol-modified mobile phase alone, no additive) showed some enhanced peak shape and decreased retention. However, changes in modifier

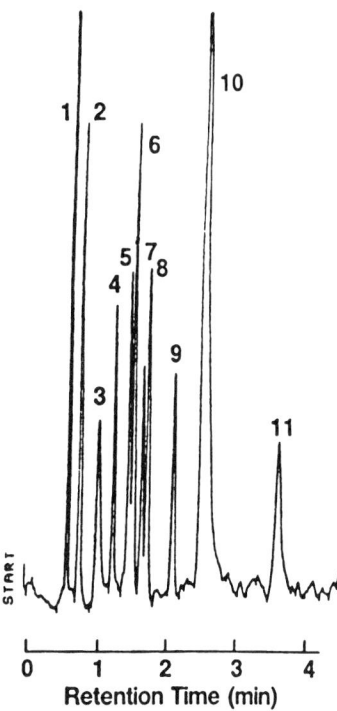

Fig. 14 Separation of the EPA Method 604 phenols on a Diol column with an additive in the mobile phase. Elution order: (1) 2-nitrophenol; (2) 4,6-dinitro-2-methylphenol; (3) 2,4-dinitrophenol; (4) 2-chlorophenol; (5) 2,4-dimethylphenol; (6) 2,4,6-trichlorophenol; (7) 2,4-dichlorophenol; (8) phenol; (9) 4-chloro-3-methylphenol; (10) pentachlorophenol; (11) 4-nitrophenol. Conditions: column 4.6 × 200 mm 5-μm Lichrosorb Diol; flow rate, 4.2 mL/min; mobile phase, 4.8% methanol (containing 0.025% TFA) in CO_2; temperature, 40°C; outlet pressure, 135 atm. (From Ref. 72.)

concentration influenced retention changes more readily and plate numbers were approximately 5000.

Pressure and temperature studies of these 11 phenol solutes indicated more drastic effects from pressure adjustments than from thermal changes. For this methanol-modified mobile-phase mixture, a single-phase system was visibly observed above 80 atm (1200 psi); below 80 atm, two phases were seen. An inlet pressure of 350 atm (5150 psi) was an instrumentation maximum; this inlet pressure corresponded to an outlet pressure of 240 atm (3500 psi). Therefore, the outlet pressure range for a one-phase system in these studies was from 240 down to 80 atm (3500 down to 1200 psi). At this pressure range, pressure was found to be a secondary control function compared to the control obtained with mobile phase and additive composition adjustments. No discontinuous retention changes were observed when temperatures were adjusted from 36.8°C to 50°C, which was below, through, and above the critical temperature region.

Citric acid was used as an additive to methanol-modified carbon dioxide by Giorgetti and co-workers for the packed column separation of phenols and other acids [73]. A Nucleosil 3-μm silica column was used at 100°C with pressure programming from 130 to 390 atm (1950 to 5750 psi) and peaks were detected at 230 nm with a UV detector. 2-Nitrophenol, phenylbenzoate, 2,4-dinitrophenol, phenol, sodium pentachlorophenolate, salicyclic acid, 3-bromobenzoic acid, 2-napthol, napthoic acid, 3-nitrophenol, and 4-nitrophenol were well resolved in less than 3 min as illustrated in Fig. 15. These authors studied the effects of five carboxylic acids as additives to carbon dioxide/methanol mobile-phase systems in packed column SFC. Of these five additives, citric acid reportedly showed the best modifying properties: better peak shape, decreased retention, enhanced selectivity, good mobile-phase solubility, and no corrosive effects on the delivery system. The less successful additives were formic, acetic, oxalic, and malonic

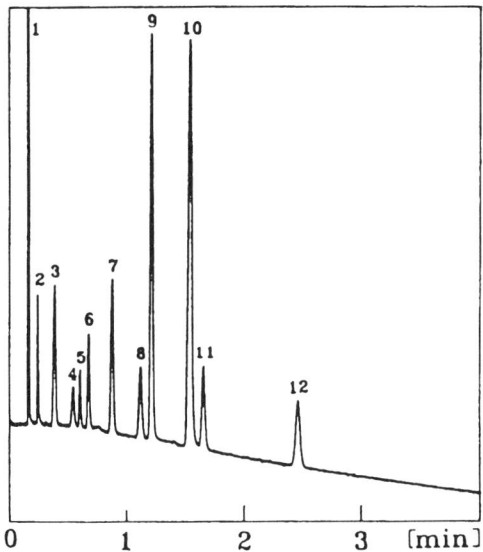

Fig. 15 Chromatogram of a synthetic mixture of acidic compounds. 1, Solvent (methanol/chloroform); 2, 2-nitrophenol; 3, phenyl benzoate; 4, 2,4-dinitrophenol; 5, phenol; 6, sodium pentachlorophenolate; 7, salicyclic acid; 8, 3-bromobenzoic acid; 9, 2-naphthol; 10, napthoic acid; 11, 3-nitrophenol; 12, 4-nitrophenol. Chromatographic conditions: column, 100 mm $\times$ 2 mm (Nucleosil Si, 3 μm); mobile phase, CO_2 (2 mL/min); modifier, 0.15% citric acid in methanol (0.04 mL/min); temperature, 100°C; pressure program, 130 to 350 bar in 4 min: detector, UV at 230 nm. (From Ref. 73.)

acids. In packed column SFC, the modifiers can interact with the potential stationary-phase adsorption sites. Citric acid interacted with the free silanol groups of the stationary phase and induced a controlled or desired activation. For the example in Fig. 15, when no citric acid was added to the methanol mobile phase, poor peak shape and strong tailing of the salicylic, 3-bromobenzoic, and naphthoic acids occurred.

In a fundamental study of the effects of modifiers in SFC, Janssen et al. [74] found the retention of phenol to decrease drastically initially but subsequently to decrease gradually with increasing concentration of THF added to the carbon dioxide mobile phase. The initial concentration of THF was 0.4%, with gradual increases to 4% by volume. The retention studies were conducted at 45°C with inlet and outlet pressures of 179 atm (2630 psi) and 162 atm (2380 psi), respectively.

Chlorodifluoromethane, Freon 22, has been reported as an effective mobile-phase eluant in the capillary SFC separation of a series of phenols using a 12.5-m RSL-300 fused silica column [75,76]. In these reports, Freon 22 was proposed as an alternative mobile phase or mobile-phase modifier for carbon dioxide. The phenols that were able to be separated are listed in Fig. 16. Example chromatograms showed the separation of up to 11 of these phenolic compounds in 20 min using a mobile phase of 5% Freon 22 and 95% carbon dioxide. Retention time, instead of capacity factor, comparisons were made under a range of pressure conditions from 100 to 300 atm (1450 to 4350 psi) for carbon dioxide and 50 to 100 atm (725 to 1450 psi) for Freon 22. The Freon 22 showed greatly reduced retention times of the phenols prior to any pressure adjustments (i.e., approximately an order of magnitude). With pressure increases, the retention using pure carbon dioxide decreased by a factor of 5; the retention using Freon 22 decreased by a factor of approximately 2. Clearly, the Freon illustrated better solvating properties for these phenolic compounds. Detection during the course of these studies was conducted with UV at 280 nm.

Aqueous and urine samples spiked with phenol and 4-chlorophenol were used to illustrate an on-line supercritical fluid extraction/supercritical fluid chromatography system [77]. The system employed a phase separator to remove the aqueous phase from the supercritical fluid extractant phase. The SFE conditions were temperature 40°C, CO_2 extraction fluid, and pump pressure 118 atm (1750 psi), at a flow rate of 0.15 mL/min. The SFC conditions were temperature 40°C, CO_2 mobile phase, column inlet pressure 150 atm (2200 psi), at a flow rate of 1.1 mL/min using a 5 μm 150 × 3.1 mm LiChrosorb RP-18 column with UV detection at 254 nm in a 320-μm-path length capillary cell. Recoveries averaged 85% for both compounds with a relative standard deviation of ±8%.

As can be seen in the report above, supercritical fluid extraction conditions are very similar to those used for SFC. Because of these similarities, and as has been suggested previously, references have been included illustrating extraction conditions for phenols and are listed in Table 2.

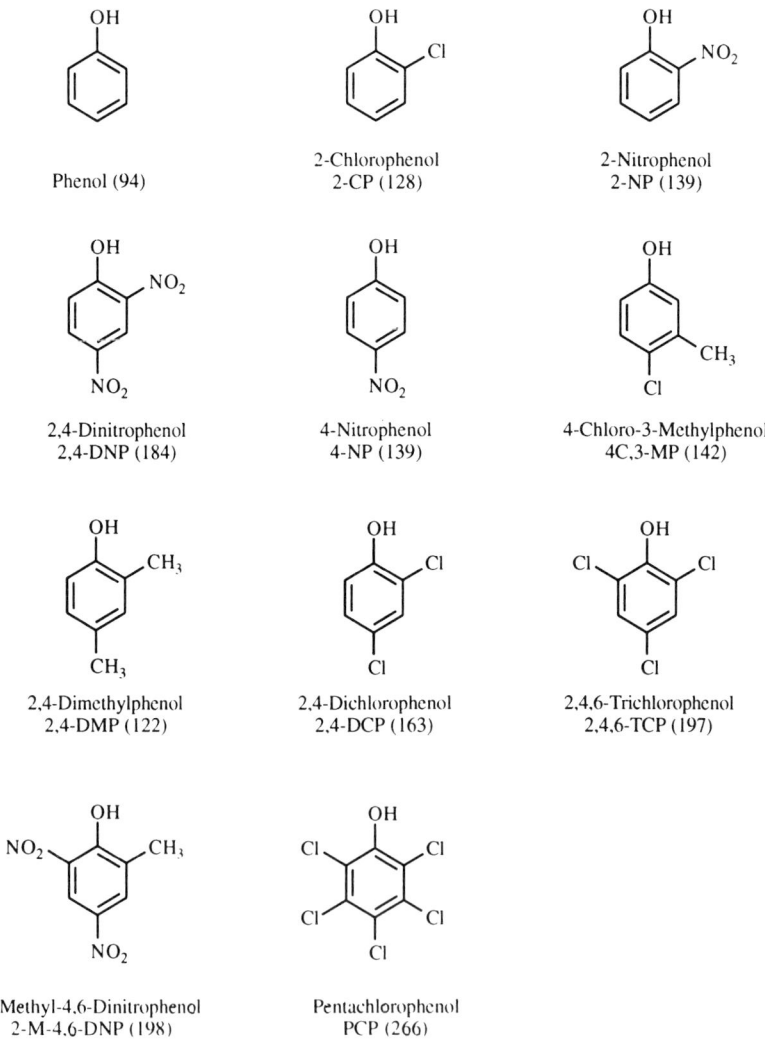

Fig. 16 Phenols separated using capillary SFC with Freon 22 mobile phase and a 12.5-m RSL-300 fused silica column.

B. Flame Ionization Detection

As reported by Ong and co-workers [75], who utilized chlorodifluoromethane, Freon 22, as the mobile phase in the supercritical fluid chromatographic analysis of phenols, attempts to utilize flame ionization detection were not satisfactory. Freon 22 decomposes at the detector even though it is reported to be nonflamma-

Table 2 Additional Supercritical Fluid Chromatography/Extraction Conditions for Phenols

Compound	Refs.
Phenol	79,80
Phenol, 2-nitrophenol, 4-nitrophenol	83
Phenol, 2-chlorophenol, 2,4-dichlorophenol, 2,4,6-trichlorophenol, pentachlorophenol, 2-fluorophenol, d6-phenol, 2,4,6-tribromophenol	81
Phenol, 2-chlorophenol, 2-nitrophenol, 2,4-dimethylphenol, 2,4-dichlorophenol, 4-chloro-3-methylphenol, 2,4,6-trichlorophenol, 2,4-dinitrophenol, 4-nitrophenol, 2-methyl-4,6-dinitrophenol, pentachlorophenol, 2-methylphenol, 4-methylphenol, 2,4,5-trichlorophenol	82

ble. At high temperatures and favorable oxidation conditions, chlorodifluoromethane forms weakly combustible mixtures with air, catalyzed by the metal components of the FID detector [78]. Reports illustrating other mobile phases coupled with FID detection for phenols were not discovered.

C. Miscellaneous Detection

Olesik has utilized a flame photometric detector at 393 nm for the determination of thiophenol and other organosulfur compounds from a synthetic mixture subsequent to supercritical fluid separation with a capillary column [84]. Thiophenol was eluted in less than 20 min of a total 90-min chromatographic elution profile when the density was linearly ramped from 88.5 to 190 atm (1300 to 2800 psi) at a rate of 1.7 atm/min, the temperature was also programmed from 70 to 150°C at 2°C/min. Two hundred twenty nanoliters of the mixture was injected directly under high pressure onto a BP-10 column made in the author's laboratory.

French and Novotny successfully illustrated the use of SFC/FTIR with supercritical fluid xenon mobile phase for the separation and detection of phenol and 2,6-di-*t*-butylphenol in a test mixture that also contained benzaldehyde, 2-napthol, 2-naphthaldehyde, and 9-anthraldehyde [85]. Xenon provided an ideal eluant for the infrared detector because of its transparency in the infrared region, but it has the disadvantage of being exorbitantly priced. The final resultant chromatogram was a Gram–Schmidt reconstruction. This same methodology, SFC/FTIR, was used for the separation of phenol and 2-nitrophenol from phenylbenzoate, salicyclic acid, and 3-bromobenzoic acid except that carbon dioxide modified with methanol and 0.15% citric acid was the mobile phase [86]. This example also used pressure programming from 150 to 320 atm (2200 to 4700 psi) at 24.3 atm/min to elute these compounds. Possible interference of methanol and citric acid with the IR detection regions was not illustrated.

SFC/MS was used to separate and detect p-chlorophenol as one of the components of a polarity test mixture for a comparison study [87]. This study illustrated the separations obtained using CO_2, and CO_2 modified with 1% methanol both using a C_{18} microbore column and CO_2 with a capillary column at high flow rates. With the microbore column and pure CO_2, significant peak tailing and excessive retention were exhibited; modifying the mobile phase reduced these chromatographic difficulties substantially. The capillary column offered improved chromatographic appearance; however, with a 30 m × 100 µm ID column the mass spectrometer was needed to separate p-chlorophenol from 1-decanol. The other components of the test mix were acetophenone, n-ethylanaline, and napthalene.

V. POLYNUCLEAR AROMATIC HYDROCARBONS

Polynuclear aromatic hydrocarbons (PAHs) are of particular economic and environmental concern, due primarily to their toxicity and wide distribution. As a result, several of these compounds have been classified as priority pollutants by the Environmental Protection Agency (EPA) [116]. The analysis of PAHs has routinely been accomplished by gas or liquid chromatography. Both techniques, however, have limitations in terms of volatility, retention time, and/or resolution. Supercritical fluid chromatography bridges the gap between GC and LC and offers a feasible separation alternative. In fact, PAHs have been analyzed successfully by SFC since studies in SFC began. A current review of PAH analyses by SFC is presented below.

A. UV Detection

Simpson et al. [88] demonstrated that carbon dioxide SFC could be carried out on a modified HPLC to which no permanent alterations were made. Separations of a PAH mixture were illustrated for each of the system's four elution modes: (1) isocratic–isobaric, (2) solvent gradient, (3) flow gradient, and (4) combined solvent–flow gradient. The PAH mixture contained toluene, naphthalene, fluorene, anthracene, terphenyl, and chrysene and was separated using an analytical-scale C_{18} packed column. Detection of these analytes was accomplished using a Perkin-Elmer LC-85B variable-wavelength detector monitoring 254 nm. Studies showed significant reductions in retention and improvement in detector response for the gradient elution techniques (modes 2 to 4) as compared to isocratic–isobaric separations. Preliminary investigations into pressure gradient programming were also discussed.

Levy and Ritchey [89] used CO_2 modified with methanol, 2-methoxyethanol, 1-propanol, THF (1,2,3,4-tetrahydro-9-fluorenone), dimethyl sulfoxide, acetonitrile, sulfur hexafluoride, and Freon 11 to study the effects of different modifiers and modifier concentration on retention and selectivity in SFC. A mixture of PAHs was prepared as the test solute and included 20 to 100 ppm each of naphthalene, fluorene, phenanthrene, fluoranthene, pyrene, chrysene, benzo[e]pyrene, benzo[ghi]perylene, and coronene. Experiments were run under constant-temperature, constant-pressure conditions on packed liquid chromatographic columns with diol, octyl, and octyl-endcapped stationary phases. The analytes were detected at 294 nm. Results indicated that selectivity and retention could be changed dramatically by varying the modifier identity and concentration. Possible mechanisms for the modifier effect were also suggested. In a later publication [90], Levy and Ritchey again investigated the effects of various modifiers on the separation of a PAH mixture. Experiments were run under constant temperature and pressure conditions on commercially available reversed-phase chromatographic columns. The columns were of equivalent length, internal diameter, and particle size and contained silica, diol-modified silica, and cyano-modified silica stationary phases. UV response was monitored at 294 nm. These studies also demonstrated that small amounts of modifier could affect peak shape, retention, and selectivity. However, the specific mechanisms of the modifier effect were shown to depend on the type of stationary phaese, mobile phase, and solute.

Blilie and Greibrokk [91] also reported that the addition of organic modifiers to supercritical CO_2 decreased retention times and improved peak shapes of PAHs. In this work, mixtures of both PAHs and nitrated PAHs were separated on commercially available microbore (250 mm × 1.3 mm ID) C_{18} columns. Detection was accomplished using a Perkin-Elmer LC-55 UV detector. Specifically, these studies showed that "retention decreased with increasing chain length of straight-chain alcohols up to hexanol and that straight-chained alcohols reduced retention more than branched alcohols." The effects of residual silanol groups on solute retention and peak shape were also examined by Blilie and Greibrokk by running a comparison study between two C_{18} columns, one of which had been specifically treated by the manufacturer to remove residual silanols.

Yonker and Smith [92] studied the effects of solvent modifiers on retention and selectivity in capillary SFC. The modifiers investigated included methanol, acetonitrile, and 2-propanol. The column used in this study was a 100-μm ID fused silica capillary coated with 50% phenyl methylpolysiloxane stationary phase. A small section of the column's polyimide coating was removed and was used as the UV absorbance cell. Analytes were detected by an ISCO V^4 variable-wavelength detector. Test solutes included myristophenone, phenanthrene, decylbenzene, phenanthridine, and perinapthenone. Data obtained by Yonker and Smith indicated that modifiers could be used to modify retention and selectivity in capillary SFC. Retention reversals and selectivity changes were noted with both the type and mole fraction of solvent modifier used.

Fields et al. [93] also studied the effects of polar modifiers on retention in capillary SFC. In this work, 2-propanol, acetonitrile, and dichloromethane were investigated as modifiers. The test solutes for the chromatographic retention studies included phenanthrene, chrysene, picene, pyrene, coronene, 1-aminopyrene, 4-hydroxypyrene, naphthalene acetamide, 1-chloromethyl naphthalene, and 1-naphthalene methanol. These PACs (polycyclic aromatic compounds) were separated on a 10 m × 50 μm ID deactivated fused silica capillary coated with a 0.25-μm film of 100% methylpolysiloxane. Analyte response was monitored by a UV detector. Although data from this study were not compared directly to data in the literature, results were consistent with those reported by Yonkers.

Blilie and Greibrokk [94] used flow-pressure gradients, modifier gradients, and combined pressure-modifier gradients in the separation of a group of PAHs and a group of nitro-PAHs. The compounds of interest are listed in Table 3. The samples contained 50 to 400 ng of each component and were separated on a microbore C_{18} reversed-phase column. UV absorbance was measured at 254 nm. The combined pressure-modifier gradient separation of 11 nitro-PAHs is shown in Fig. 17. This combined gradient provided the best combination of high resolution and detectability for the group of PAHs and resolved the 11 components in less than 13 min. Short-term retention time and peak height reproducibility were determined for the various gradients; all coefficients of variation, based on 5 to 10 measurements, were below 2 to 3% RSD.

Klesper et al. [95] used consecutive and/or simultaneous pressure, temperature, and composition gradients to "tune" chromatographic conditions to a specific separation. In this work, the sample of interest consisted of a polystyrene standard to which naphthalene, anthracene, pyrene, and chrysene were added. The

Table 3 PAHs and Nitro-PAHs Superated by a Combined Pressure-Modifier Gradient[a]

Peak	Compound	Peak	Compound
1	Naphthalene	12	1-Nitronaphthalene
2	Fluorene	13	2-Nitrofluorene
3	Anthracene	14	3-Nitrofluoranthene
4	Fluoranthene	15	1-Nitrotriphenylene
5	Pyrene	16	2-Nitrotriphenylene
6	Bena[a]anthracene	17	1-Nitrobenzo[e]pyrene
7	Chrysene	18	6-Nitrobenzo[a]pyrene
8	Benzo[a]pyrene	19	9-Nitrodibenz[a,c]anthracene
9	Benzo[a]pyrene	20	3-Nitrobenzo[e]pyrene
10	Dienz[a,c]anthracene	21	7-Nitrobenzo[ghi]perylene
11	Benzo[ghi]perylene	22	5-Nitrobenzo[ghi]perylene

[a]SFC conditions outlined in Fig. 17 caption.

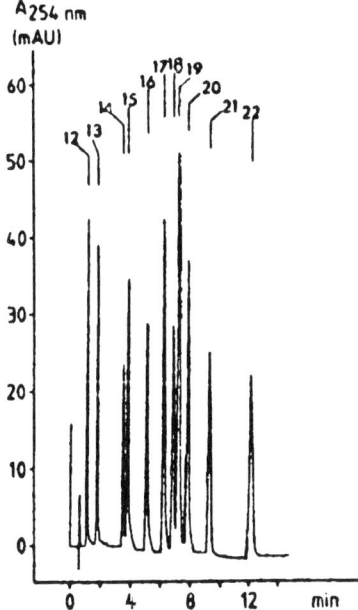

Fig. 17 Combined pressure-modifier gradient separation of 11 nitro-PAHs (see Table 3). The flow rate of carbon dioxide was increased from 0.5 mL/min to 1.0 mL/min over a 5-min period, simultaneously with an increase in flow rate of methanol from 15 μL/min to 50 μL/min. Pressure, 155 to 264 atm. (From Ref. 94.)

mixture was separated on a 25 cm × 4.5 mm ID stainless-steel column slurry packed with LiChrosorb Si 100 (10 μm). UV detection was performed at 254 nm. Separation of this polystyrene mixture was accomplished in approximately 120 min by simultaneously programming a pressure and temperature gradient [p = 100 to 500 atm (1500 to 7350 psi), T = 220 to 270°C] followed by a 1,4-dioxane composition gradient (5 to 60%) and a positive temperature gradient (T = 270 to 300°C).

Anton et al. [96] described a mixed mobile-phase delivery system for capillary SFC. This instrument was capable of pressure, composition, or combined pressure–composition gradients. A 12-component mixture, including naphthalene, anthracene, p-terphenyl, and pyrene, was prepared to test the selectivity of the system. The mixture contained 0.1% of each compound in $CHCl_3$ and was separated on a 10 m × 50 μm ID SB-biphenyl-30 capillary column. CO_2 was used as the primary mobile-phase component and 2-propanol as the modifier. The UV detector monitored 262 nm. Separation of the compounds of interest was complete in less than 18 min.

Saito et al. [97] developed a constant-mass-flow SFC system which employed a pressure-regulating valve based on high-speed flow switching. A mixture of PAHs including naphthalene, biphenyl, fluorene, anthracene, pyrene, chrysene, and benz[a]pyrene was prepared and used to evaluate the instrument. The mixture was separated on an analytical-scale column packed with silica gel. UV detection was monitored at 230 nm. Baseline separation of the components of interest was achieved in approximately 5 min using a linear pressure program from 135 to 345 atm (1950 to 5000 psi). Retention time, peak area, and peak height reproducibilities were calculated from 10 consecutive runs and were better than 1.7% RSD for most peaks.

Schoenmakers and Uunk [98] studied the effecrts of pressure drop on retention and efficiency for packed columns of various lengths and internal diameters. A dilute solution, which included naphthalene, was used as the test sample. Separation of this test mixture was illustrated using a 250 mm $\times$ 4.6 mm ID C_{18} packed column at 50°C and 150 atm (2200 psi) of pressure. UV detection was monitored at 210 nm. The separation was completed in approximately 3 min and produced more than 20,000 theoretical plates, suggesting that fast, efficient separations were possible with packed column SFC. Schoenmakers' and Uunk's studies also showed, however, that the performance of packed column SFC is limited by the maximum allowable pressure drop over the column.

Janssen et al. [99] examined the effects of column pressure drop on retention and efficiency in both packed and capillary supercritical fluid chromatography. However, the test mixture, which included naphthalene was evaluated only on an analytical-scale reversed-phase HPLC column. Separation of this mixture was achieved in less than 2 min at a high flow rate and low density [T, 50°C; P_{in}, 110 atm (1600 psi); P_{out}, 92 atm (1350 psi)] but tailing peaks resulted. An exact cause for this phenomenon was not established.

Jinno and Kuwajima [100] developed a Retention Prediction System (REPRES) for PAHs in SFC which was based on the effects of column temperature and column pressure. PAHs such as naphthalene, fluorene, pyrene, chrysene, and benzo[a]pyrene were used as the standard material for constructing REPRES and in evaluating the system's predictive ability. Studies showed excellent agreement between UV-identified and predicted retention times for a mixture of nine PAHs at a variety of pressure/temperature conditions.

Malik and Jinno [101] investigated the use of cyclodextrin (CD) stationary phases in SFC. Two PAH mixtures, 2-methylnaphthalene/biphenyl and ethylbenzene/naphthalene, were prepared and used as test samples. The mixtures were chosen because of the difficulty associated with their separation by HPLC on a C_{18} column. Separation by SFC for these mixtures was accomplished successfully using a 250 mm $\times$ 1 mm ID CD column with CO_2 as the mobile phase. Pressure and temperature were held constant at 116 atm (1710 psi) and 44°C, respectively. Solutes were monitored with a Jasco model 875 UV–visible detector. For both solute pairs, SFC provided high selectivity and baseline separations in 10 to 30 min.

Jinno et al. [102] evaluated an inorganic synthetic clay material (SC) as a stationary phase in packed column SFC. Resolution of a five-component mixture of PAHs was achieved in approximately 180 min on a 0.53 mm ID × 50 mm long column packed with synthetic clay. CO_2 was used as the mobile phase, and pressure and temperature were held constant at 194 atm (2850 psi) and 40°C, respectively. Jinno et al. also showed this synthetic clay stationary phase to be promising for molecular planarity recognition. In an earlier publication, Jinno and Mae [103] conducted a more thorough evaluation into the molecular planarity recognition of PAHs in SFC. A variety of polymeric and monomeric C_{18} columns were investigated using methanol-modified CO_2 as the mobile phase. A mixture of triphenylene and o-terphenyl was prepared to represent planar and nonplanar molecules, respectively. Separations were accomplished over a range of pressures, 126 to 165 atm (1850 to 2400 psi), and temperatures, 33 to 80°C, in as little as 50 to 100 min.

B. Mass Spectrometric Detection

Lee et al. [104] interfaced a HRMS (high-resolution double-sector mass spectrometer) to a capillary supercritical fluid chromatograph. A heated direct insertion probe (DIP) was used as the interface. No modifications of the pumping system or the ion source of the HRMS were necessary. Operation of the SFC/HRMS was evaluated using mixtures of PAHs as test solutes. The compounds of interest were eluted from a 5 m × 50 µm ID SE-54 fused silica column. Mass spectra were collected for pyrene by CI (chemical ionization) SFC/HRMS and for phenanthrene, pyrene, and chrysene using CE (charge exchange) SFC/HRMS. Experiments using electron impact ionization (EI) were not possible, due to poor ionization efficiency. Solute transfer and the stability of the ion source during pressure programming were also investigated.

Hawthorne and Miller [63] used a direct capillary interface to couple a supercritical fluid chromatograph to a commercially available quadrupole mass spectrometer. No modification of the MS was required. Spectra for two PAHs, chrysene and rubrene, were collected in the CI mode using methane as the reagent gas. Separations were performed using a 10 m × 50 µm ID SB-phenyl-5 capillary column and CO_2 as the mobile phase. Hawthorne and Miller [105] later used this same SFC/MS to obtain CI mass spectra of 29 standard PAHs and heteroatom-containing PAHs. Each of the spectra was generated using approximately 50 ng of the test solute and showed significant ions at M + 1 and at the M + 29 adduct. Most of the PAHs, however, showed little fragmentation. The capillary direct interface was also used to obtain SFC/MS TIC (total ion current) chromatograms of standard PAH mixtures. Chromatographic peak shapes were acceptable, and full-scan spectra were obtained at the low-nanogram level. Detection limits were lowered to approximately 25 pg using selected-ion monitoring.

Wilkins et al. [60] used a dual-cell FTMS (Fourier transform mass spectrometer) as a detector for PAHs in SFC. The dual-cell interface was designed to minimize problems associated with the significant pressure differential between the SFC and the MS. In this work, a six-component mixture of PAHs, including o-dichlorobenzene, naphthalene, 1-bromonaphthalene, fluorene, anthracene, and pyrene, was eluted from a 20 m × 100 µm ID DB-5 column pressure programmed from 85 atm (1250 psi) (10-min hold) to 200 atm (3000 psi) at 2 atm/min. Temperature was maintained at 100°C. Ionization was accomplished by charge exchange with CO_2. A segment of the reconstructed SFC/FTMS chromatogram is shown in Fig. 18. Detection limits for this system were in the low-nanogram range, although sensitivity was shown to decrease at the expense of resolution. Wilkins et al. [106] later developed a probe interface for SFC/FTMS. This interface was designed to be convenient, to have good temperature control, and to fit entirely within the standard FTMS probe. An evaluation of this combined system demonstrated that a simple mixture of PAHs, fluorene, phenanthrene, pyrene, and perylene, could easily be separated and analyzed. Detection

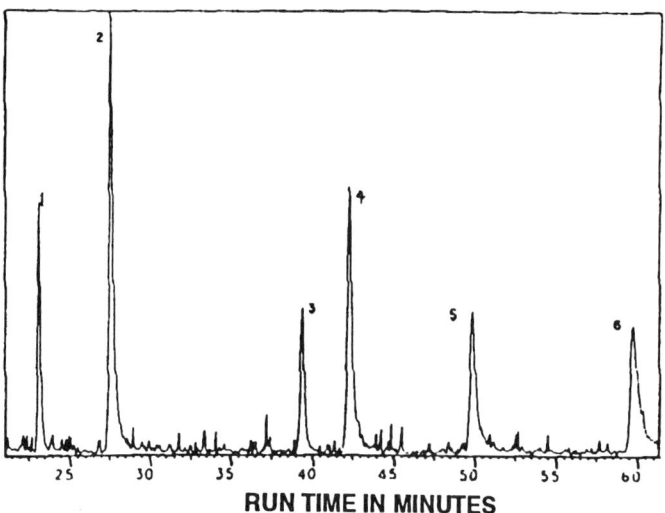

Fig. 18 Segment of reconstructed SFC/FTMS chromatogram. 1, o-Dichlorobenzene; 2, naphthalene; 3, 1-bromonaphthalene; 4, fluorene; 5, anthracene; 6, pyrene. SFC conditions were as follows: oven temperature 100°C and CO_2 pressure programmed from 85 to 204 atm at 2 atm/min after an initial 10-min isobaric period. (From Ref. 60.)

limits were estimated at 100 pg. The authors also noted that the probe interface should be suitable for use in other types of mass spectrometers.

Niessen et al. [64] used a SFC/MS with a thermospray interface to study the effects of repeller potential on the appearance of mass spectra. A mixture of anthracene, phenanthrene, and pyrene was used in this study and was separated using an analytical-scale packed column and methanol-modified CO_2. Spectra were obtained at low (20 V) and high (120 V) repeller potentials. At the low repeller potentials, proton transfer CI effects were observed, while at the higher potentials, charge-exchange CI effects predominated. Studies also showed that the degree of fragmentation for a particular potential was related to the methanol content in the mobile phase and that spectral intensity was significantly reduced at high repeller potentials.

C. Miscellaneous Detection

Markides et al. [51] coupled a supercritical fluid chromatograph to a FPD for the detection of sulfur- and phosphorus-containing compounds. The detector was modified for use in the SFC system. A three-component polynuclear aromatic sulfur-containing hydrocarbon mixture was analyzed by capillary SFC/FPD in approximately 26 min using CO_2 as the mobile phase at 120°C. Density was programmed from 0.55 to 0.65 g/mL at 0.005 g/mL per minute after an 8-min isoconfertic period. Baseline correction was required, however, to compensate for an increase in baseline noise and a rising baseline associated with the increase in density. A detection limit of 25 ng was reported for benzo[*b*]thiophene ($S/N = 2$) for the FPD in the sulfur mode.

West and Lee [107] evaluated a TID for capillary SFC of nitrated polycyclic aromatic compounds. The TID required no modifications for SFC. This detector was studied in three different modes of operation. Sensitivity, linearity, and stability were determined for each of these modes using peak area measurements of 1-hydroxy-2-nitronaphthalene with respect to naphthalene. The nitro-selective mode, TID-1-N_2, of the detector performed the best and chromatograms of two nitro-PAC standards using this detector were provided. Separations were accomplished using a 15 m × 50 μm ID SE-54 coated cpillary column and CO_2 (pressure programmed) at 101°C. Detector sensitivity, estimated at three times baseline noise, was 20 pg injected for *p*-nitrophenol. In addition, the nitro-selective detector was found to be stable during density programming. The linear dynamic range, however, was limited. This setup was later used to analyze a subfraction of a diesel particulate sample.

Sim et al. [108] evaluated the applicability of a PID (photoionization detector) as a detector in packed column supercritical fluid chromatography. A standard mixture of 13 PAHs was prepared and used to evaluate the detector. Sensitivities were calculated to be on the order of a few hundred picograms (S/N

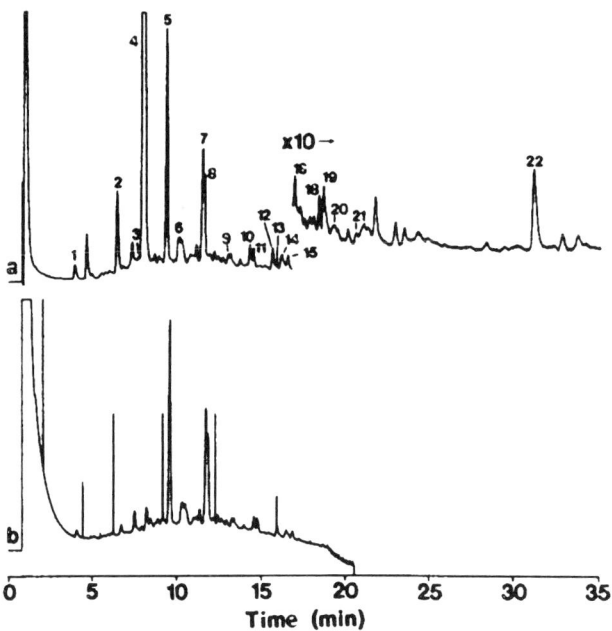

Fig. 19 SFC of a sediment extract. 1, Naphthalene; 2, acenaphthalene; 3, fluorene; 4, sulfur; 5, phenanthrene; 6, 2-methylanthracene; 7, pyrene; 8, fluoranthene; 9, benzo[*a*]fluorene; 10, benz[*a*]anthracene; 11, chrysene; 12, benzo[*b*]fluoranthene; 13, benzo[*j*]fluoranthene; 14, benzo[*k*]fluoranthene; 15, benzo[*a*]pyrene; 16, perylene; 17, 9,10-diphenylanthracene; 18, indeno[1,2,3-*cd*]pyrene; 19, benzo[*ghi*]perylene; 20, dibenz[*a,h*]anthracene; 21, coronene; 22, anthraquinone. SFC conditions were as follows: oven temperature of 85°C and CO_2 pressure programmed from 140 atm (5-min hold) to 340 atm (5-min hold) in 25 min: (a) PID trace; (b) FID trace. (From Ref. 108.)

= 3) and dynamic ranges spanned 3.5 orders of magnitude. A baseline separation of this mixture was achieved using CO_2, pressure programmed from 140 atm (2000 psi) (5-min hold) to 340 atm (5000 psi) (5-min hold) in 25 min at a column temperature of 85°C. A 25 cm × 1 mm ID Spheri-5 silica column was used in this analysis. The separation of PACs in sediment matrices was also investigated, and a sample chromatogram is shown in Fig. 19. All conditions were as described previously. Separation, in this case, of 22 PAHs was accomplished in less than 35 min. A comparison between the PID trace (Fig. 19a) and a FID (Fig. 19b) trace show the PID to have less detector spiking, a smoother and flatter baseline, and a molar response that is less dependent on the analytes.

Raynor et al. [109] investigated xenon as a mobile phase for on-line capillary SFC-FTIR (Fourier transform infrared spectrometry). Xenon was of particular interest as a mobile phase because of its transparency in the IR region of the spectrum. Initial solubility predictions using the Peng–Robinson equation of state indicated that xenon would be able to separate high-molecular-weight PAHs. This is illustrated in Fig. 20 by the baseline separation of a six-component mixture of PAHs at 100°C on a 10 m × 50 μm ID SB-biphenyl-30 capillary column. For this separation, the mobile phase was pressure programmed from 120 atm (1760 psi) (15-min hold) to 350 atm (5100 psi) at 3 atm/min and a FID used for detection. FTIR detection of the PAHs was not presented in this paper; however, a

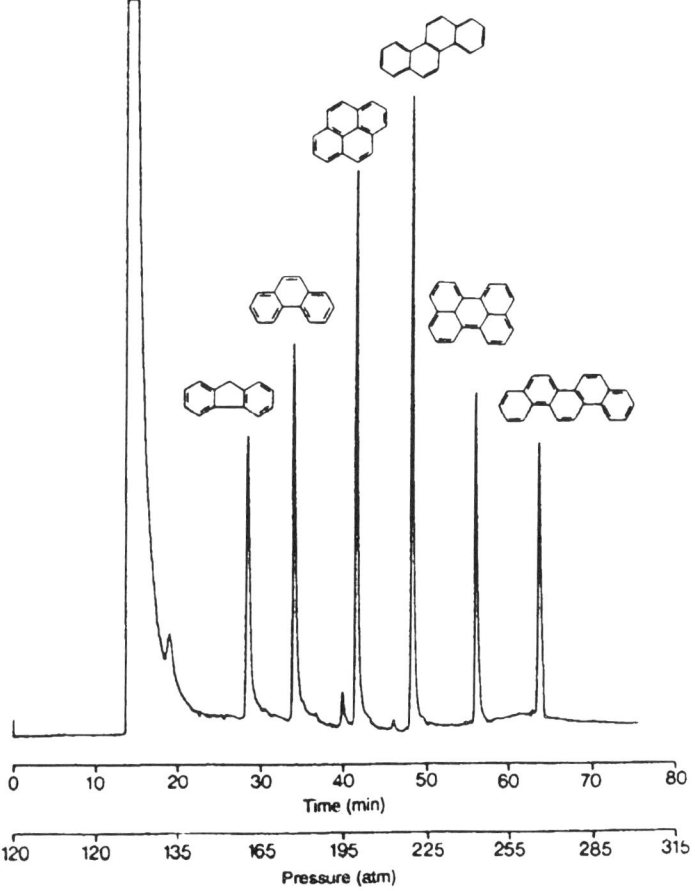

Fig. 20 Separation of PAHs with xenon as a mobile phase and flame ionization detection. SFC conditions were as follows: oven temperature 100°C and xenon pressure programmed from 120 atm (5-min hold) to 350 atm at 3 atm/min. (From Ref. 109.)

Gram–Schmidt reconstructed chromatogram was shown for a mixture of aromatic isocyanate oligomers. To accommodate any weakly absorbing compounds, a stop-flow system was developed which effectively stopped the mobile-phase flow in the column and flow cell and allowed "spectral accumulation of the components of interest." No significant increase in peak width was reported.

Sievers et al. [52] interfaced a SCD (sulfur chemiluminescence detector) to capillary column SFC. The interface was accomplished by inserting the capillary through a heated transfer line and into the chemiluminescence reaction chamber. Correct positioning of the capillary within this chamber was found to be critical to achieving good peak shape. The feasibility of the SFC-SCD was demonstrated by the separation of a six-component standard mix of PAHs. The mixture included naphthalene, fluorene, phenanthrene, chrysene, benzo[a]pyrene, and 1,2,5,6-dibenzanthracene. Separation was achieved in less than 40 min using a 3.5 m × 0.10 mm ID DB-5 fused silica capillary and a CO_2 pressure program, 100 to 300 atm (1500 to 4500 psi) in 40 min. Temperature was maintained at 100°C.

Leren et al. [110] studied the properties of sulfur dioxide as a mobile phase in SFC. Separations were attempted on a variety of columns, both packed and capillary, using pure SO_2 (99.87%) and 20% w/w SO_2/CO_2 mixed mobile phases. Progress was limited, as no column was found to be compatible with the SO_2 and SO_2/CO_2 mixture. In all cases, the stationary phases were stripped from the columns and the restrictors plugged. Restrictor plugging was delayed, however, by switching to the diluted SO_2/CO_2 mixture and a few separations were achieved. An FID was used to monitor analyte response.

Yonker and Smith [111] studied the effects of a fluid modifier gradient on solute retention and selectivity for a mixture of PAHs in capillary SFC. The mixture, which contained phenanthrol, chrysene, and 6-aminochrysene, was separated in a SE-54 capillary column at constant temperature (126.5°C) and density (0.26 g/mL). The mole fraction of the modifier, IPA (isopropyl alcohol), was varied from 0 to 0.081. Detection was accomplished with a fluorescence detector. Experiments showed both retention and selectivity of 6-aminochrysene to chrysene to decrease with an increase in modifier concentration. The decrease in retention was attributed to density changes and to modification of the solute–solvent interaction as solvent modifier was introduced.

Sandman and Grayeski [112] developed an interface that permits the use of conventional low-pressure flow cells in SFC. The interface was designed to decrease pressure gradually from the column to the detector flow cell and used a postcolumn solvent to dissolve the CO_2 and analytes of interest. 9,10-Diphenylanthracene was used to evaluate the interface and chromatograms of the three detection techniques: (a) conventional SFC/UV (250 nm), (b) interface SFC/UV (260 nm), and (c) interface fluorescence (260 nm) of Fig. 21. Separations were achieved using a Deltabond CN small-bore (1.0 mm ID) column and pure CO_2 pressure programmed from 98 atm (1450 psi) (2-min hold) to 300 atm (4350 psi)

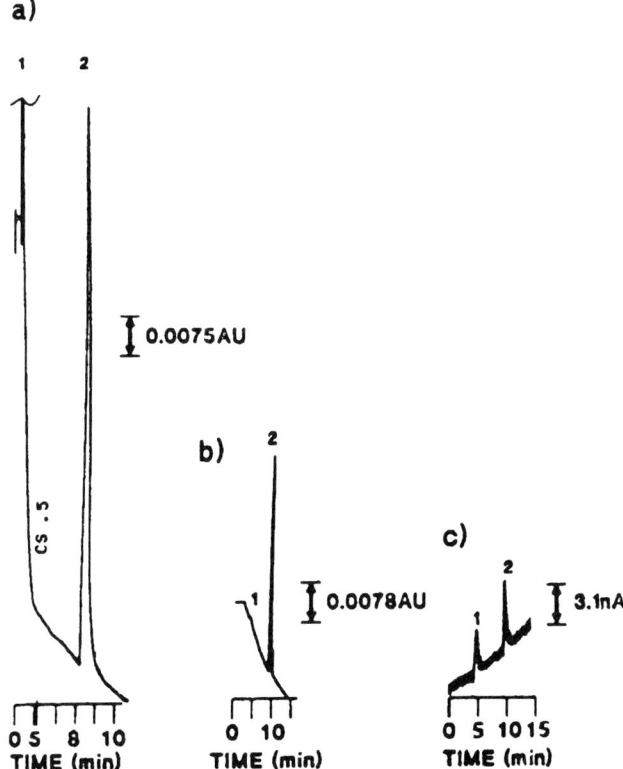

Fig. 21 Chromatograms of three detection techniques for 9,10-diphenylanthracene: (a) conventional SFC/UV detection of 56 ng injected; (b) interface SFC/UV detection of 56 ng injected; (c) interface fluorescence detection of 43 pg injected. 1, Ethyl acetate; 2, 9,10-diphenylanthracene. SFC conditions were as follows: oven temperature 75°C and CO_2 pressure programmed from 98 atm (2-min hold) to 300 atm at 20 atm/min. (From Ref. 112.)

at 20 atm/min. Temperature was held constant at 75°C. Ethyl acetate was used as the postcolumn solvent. The difference in the UV signals for equal amounts of solute injected was attributed to the dilution effects of the interface. Detection limits ($S/N = 3$) for these techniques were calculated to be (a) 250 pg, (b) 110 pg, and (c) 15 pg, respectively. The effect of the interface on band broadening was also investigated, but no statistically significant difference between the conventional and interface detectors was observed.

Lee et al. [113] evaluated an IMD (ion mobility detector) with a modifid SFC/IMD interface for open-tubular column supercritical fluid chromatography.

The new interface was designed to minimize the formation of ion clusters and to enhance ionization efficiency. A mixture of anthracene, pyrene, 1-phenylphenanthrene, benzo[a]pyrene, and benzo[b]chrysene was prepared and used as the test solute. Separation was achieved using a 4 m × 50 μm ID SB-phenyl-50 fused silica capillary and acetonitrile-modified CO_2 pressure programmed from 80 atm (1200 psi) (5-min hold) to 200 atm (2900 psi) at 4 atm/min. Temperature was held constant at 80°C. Detection was investigated in both the positive product-ion-monitoring mode and positive reactant-ion-monitoring mode. The minimum detectable quantity (MDQ) for pyrene using 5% acetonitrile-modified CO_2 was determined to be 2.1 ng at a signal-to-noise ratio of 3.

D. Multidimensional Techniques

Levy et al. [114] developed an on-line SFC/GC. This technique allowed not only for direct (100%) sample transfer from the SFC to the GC but also for selective or multistep heartcutting of various sample peaks as they eluted from the supercritical fluid chromatograph. A test mixture which included 1-methyl naphthalene and anthracene was prepared and used to evaluate this system. Resolution of the four-component mixture was achieved by SFC/FID in less than 10 min on an analytical-scale packed column with CO_2 as the mobile phase. Temperature and pressure were held constant at 50°C and 200 atm (3000 psi), respectively. Direct transfer of this mixture from the SFC to the GC was also illustrated. SFC/GC direct transfer reproducibilities (peak area) were calculated and reported at values less than or equal to 2.0% RSD. This same mixture was used to determine SFC/GC heartcut repeatability. The RSDs based on area counts were less than 4.0%.

Lee et al. [37] constructed a two-dimensional supercritical fluid chromatographic (SFC/SFC) system that used a flow switching interface between two open-tubular columns: a biphenyl primary column and a liquid crystal secondary column. A standard mixture of PAHs was used to demonstrate the potential of the instrument for separating complex mixtures. The solution contained benzo[a]fluoranthene, benzo[b]fluoranthene, benzo[j]fluoranthene, benzo[k]fluoranthene, benzo[a]pyrene, benzo[e]pyrene, and perylene. Resolution of this mixture was achieved in just under 5 h using a CO_2 density program, 0.20 g/mL (hold 5 min) to 0.76 g/mL at a ramp rate of 0.005 g/mL per minute. The oven temperatures for the biphenyl and liquid crystal columns were maintained at 100°C and 120°C, respectively. A FID was used to monitor analyte response. In this particular separation, the biphenyl column was used to separate the two groups of compounds, the benzofluoranthenes and the benzopyrenes/perylene, while the sample was transferred to the liquid crystal column for resolution of the individual components.

Taguchi et al. [115] used a SFE/SFC system to examine the extraction efficiency of PAHs from absorbent materials using CO_2 as a supercritical fluid. The PAHs of interest, naphthalene, anthracene, pyrene, and 1-nitropyrene, were extracted from glass wool, filter paper, and activated charcoal absorbents at pressures and temperatures ranging from 100 to 300 atm (1500 to 4500 psi) and 40 to 100°C, respectively. These extracts were separated by SFC in approximately 25 min on a 10 m × 50 μm ID SB-biphenyl-30 column. For these separations, pressure was programmed from 100 atm (1500 psi) (hold 5 min) to 400 atm (5900 psi) at 12 atm/min and temperature maintained at 100°C. Analytes were monitored by FID.

Ong et al. [116] used on-line SFE/SFC for the determination of selected PAHs in aqueous environmental samples. The setup used an extraction chamber, constructed from a 35 mm × 2 mm ID brass tube and filled with Partisil-5-C_{18}-3 packing material, which was connected to the SFC system between the sampling valve and the analytical column. Samples to be extracted and analyzed were injected directly into the system using the injection valve. Extractions were carried out isobarically at a pressure of 110 atm (1600 psi) and an oven temperature of 60°C. A chromatogram of the PAH extract from a spiked aqueous sample is shown in Fig. 22a. Separation of this extract, which included fluoranthene, benzo[*b*]fluoranthene, benzo[*a*]pyrene, and benzo[*ghi*]perylene, was achieved in less than 60 min on a 9 m × 100 μm ID SE-52 fused silica capillary column. Chromatographic conditions were identical to those used for extraction. A UV detector monitored 254 nm. In Ong's investigation, environmental samples were also monitored for the four PAHs. However, as Fig. 22b illustrates, only fluoranthene was detected in the wastewater samples.

VI. CONCLUSIONS

Supercritical fluid chromatography for the analysis of environmentally pertinent compounds is under rapid development, as has been illustrated with the various examples in this chapter. SFC provides the means to couple a fast, versatile separation process to a wide variety of detectors. With the rapid advancement of commercial instrumentation, automation of this supercritical fluid chromatographic analyses is equivalent to what can be found in other, better-established chromatographic methods. Currently, systematic approaches to method development are still being developed. As yet, they are not at the point where conditions can be plugged into a computer program and the next analysis step suggested. However, there is no doubt that separations which use supercritical fluids as the mobile phase will substitute for HPLC and GC in specific applications. Ultimately, the niche technique will replace these more conventional analytical technologies in the more routine quality control environment. This is especially

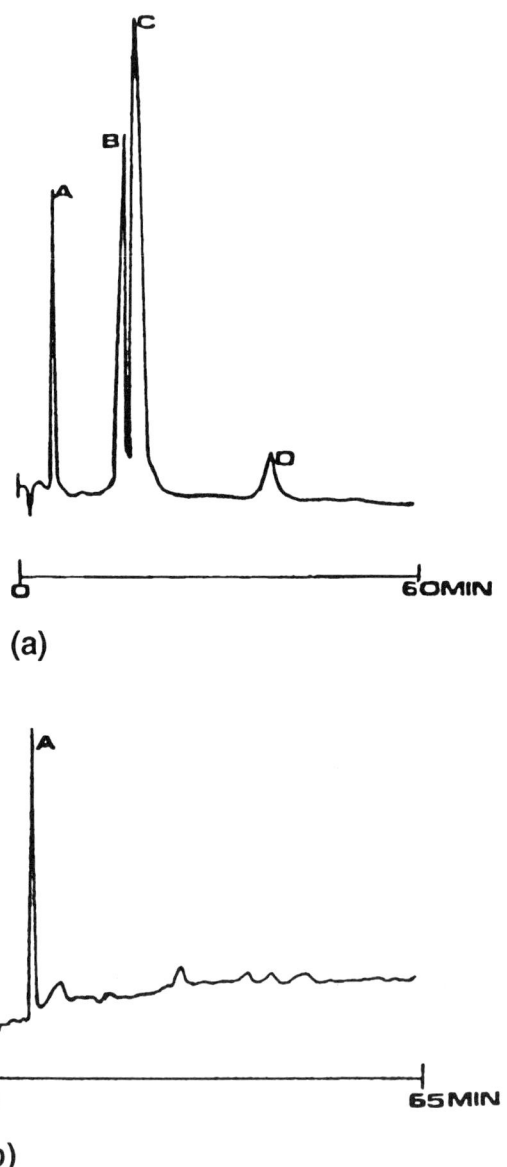

Fig. 22 SFE/SFC determination of (a) a spiked aqueous sample and (b) an environmental aqueous sample. A, Fluoranthene; B, benzo[*b*]fluoranthene; C, benzo[*a*]pyrene; D, benzo-[*ghi*]perylene. SFC conditions were as follows: oven temperature 60°C and CO_2 at 110 atm. (From Ref. 116.)

true as the issue of waste disposal becomes more prevalent in terms of environmental safety and cost. SFC has proven to be extremely useful for separations of compounds routinely analyzed in the environment. We believe that this chapter illustrates SFC as a continually growing field.

REFERENCES

1. M. Yoshioka, S. Parvez, T. Miyazaki, and H. Parvey, Eds., *Supercritical Fluid Chromatography and Micro-HPLC,* in the series *Progress in HPLC,* Vol. 4, VBP, The Netherlands, 1989.
2. K. P. Johnston and J. M. L. Penninger, Eds., *Supercritical Fluid Science and Technology,* ACS Symposium Series, 406, American Chemical Society, Washington, D.C., 1989.
3. B. A. Charpentier and M. R. Sevenants, Eds., *Supercritical Fluid Extraction and Chromatography Techniques and Applications,* ACS Symposium Series, 336, American Chemical Society, Washington, D.C., 1988.
4. R. M. Smith, Ed., *Supercritical Fluid Chromatography,* RSC Chromatography Monographs, The Royal Society of Chemistry, London, 1988.
5. F. V. Bright and M. E. McNally, Eds., *Supercritical Fluid Technology: Theoretical and Applied Approaches in Analytical Chemistry,* ACS Symposium Series, 488, American Chemical Society, Washington, D.C., 1992.
6. S. A. Westwood, Ed., *Supercritical Fluid Extraction and Its Use in Chromatographic Sample Preparation,* Blackie Academic & Professional, CRC Press, Boca Raton, Fla., 1993.
7. T. J. Bruno and J. F. Ely, Eds., *Supercritical Fluid Technology: Reviews in Modern Theory and Applications,* CRC Press, Boca Raton, Fla., 1993.
8. M. E. McNally and J. R. Wheeler, *J. Chromatogr., 447:* 53 (1988).
9. R. L. Grob and M. A. Kaiser, *Environmental Problem Solving Using Gas and Liquid Chromatography,* Journal of Chromatography Library, Vol. 21, Elsevier, Amsterdam, 1982.
10. O. Hutzinger, S. Safe, and V. Zitko, *The Chemistry of PCB's,* CRC Press, Cleveland, Ohio, 1974.
11. P. W. Albro, J. T. Corbett, and J. L. Schroeder, *J. Chromatogr., 205:* 103 (1981).
12. J. R. Gordon, J. Szita, and E. J. Faedler, *Anal. Chem., 54:* 478 (1982).
13. J. Brosky and K. Ballschmiter, *Fresenius Z. Anal. Chem., 335:* 817 (1989).
14. Z. Juvancz, K. M. Payne, K. E. Markides, and M. L. Lee, *Anal. Chem., 62:* 1384 (1990).
15. F. I. Onuska, K. A. Terry, S. Rokushika, and H. Hatano, *J. High Resolut. Chromatogr., 13*(5): 317 (1990).
16. K. Cammann and W. Kleibohmer, *J. High Resolut. Chromatogr., 14*(5): 327 (1991).

17. K. Cammann and W. Kleibohmer, *J. Chromatogr., 522:* 267 (1990).
18. R. D. Smith and H. R. Usdeth, *Anal. Chem., 55:* 2266 (1983).
19. S. B. Hawthorne, D. J. Miller, and J. J. Langenfeld, in *International Symposium on Supercritical Fluid Chromatography and Extraction Abstracts,* Park City, Utah, Jan. 1991, p. 91.
20. B. E. Richter, E. R. Campbell, A. F. Rynaski, B. J. Murphy, R. B. Nielsen, and N. L. Porter, in *International Symposium on Supercritical Fluid Chromatography and Extraction Abstracts,* Park City, Utah, Jan. 1991, p. 121.
21. C. A. Craig, T. Kruger, S. Prashar, A. Munoz, C. Oliveros, S. Jones, B. E. Richter, and B. J. Murphy, in *International Symposium on Supercritical Fluid Chromatography and Extraction Abstracts,* Park City, Utah, Jan. 1991, p. 141.
22. F. I. Onuska and K. A. Terry, *J. High Resolut. Chromatogr., 12*(8): 527 (1989).
23. S. B. Hawthorne and D. J. Miller, *J. Chromatogr., 408:* 63 (1987).
24. B. O. Brady, C.-P. C. Kao, K. M. Dooley, F. C. Knopf, and R. P. Gambrell, *Ind. Eng. Chem. Res., 26:* 261 (1987).
25. K. M. Dooley, C.-P. C. Kao, R. P. Gambrell, and F. C. Knopf, *Ind. Eng. Chem. Res., 26:* 2058 (1987).
26. M. R. Anderson, J. T. Swanson, N. L. Porter, and B. E. Richter, *J. Chromatogr. Sci., 27*(7): 371 (1989).
27. R. C. Burk, P. Kruus, I. Ahmad, and G. Crawford, *J. Environ. Sci. Health, B25*(5): 553 (1990).
28. K. S. Nam, S. Kapila, D. S. Viswanath, and A. F. Yanders, *Chemosphere, 19*(1–6): 33 (1989).
29. K. S. Nam, S. Kapila, A. F. Yanders, and R. K. Puri, *Chemosphere, 20*(7–9): 873 (1990).
30. K. S. Nam, S. Kapila, A. F. Yanders, and R. K. Puri, *Chemosphere, 23*(8–10): 1109 (1991).
31. B. W. Wright and R. D. Smith, *J. High Resolut. Chromatogr., 8:* 8 (1985).
32. B. W. Wright and R. D. Smith, *J. High Resolut. Chromatogr., 9:* 73 (1986).
33. H. T. Kalinoski, H. R. Udseth, B. W. Wright, and R. D. Smith, *J. Chromatogr., 400:* 307 (1987).
34. B. E. Richter, *Chromatogr. Forum,* Nov.–Dec., p. 52 (1986).
35. J. R. Wheeler and M. E. McNally, *J. Chromatogr., 410:* 343 (1987).
36. C. Borra, F. Andreolini, and M. Novotny, *Anal. Chem., 61:* 1208 (1989).
37. I. L. Davies, B. Xu, K. E. Markides, K. D. Bartle, and M. L. Lee, *J. Microcolumn Sep., 1:* 71 (1989).
38. S. Ashraf, K. D. Bartle, A. A. Clifford, I. L. Davies, and R. Moulder, *Chromatographia, 30:* 618 (1990).

39. A. J. Berry, D. E. Games, and J. R. Perkins, *J. Chromatogr., 363:* 147 (1986).
40. J. E. France and K. J. Voorhees, *J. High Resolut. Chromatogr., 11:* 692 (1988).
41. M. E. McNally, J. R. Wheeler, and W. R. Melander, *Liq. Chromatogr. Gas Chromatogr., 6:* 816 (1988).
42. S. Shah, M. Ashraf-Khorassani, and L. T. Taylor, *J. Chromatogr., 505:* 293 (1990).
43. S. Shah and L. T. Taylor, *J. High Resolut. Chromatogr., 12:* 599 (1989).
44. S. Kennedy and R. J. Wall, *Liq. Chromatogr. Gas Chromatogr., 6:* 930 (1988).
45. H.-C. K. Chang and L. T. Taylor, *J. Chromatogr. Sci., 28:* 29 (1990).
46. L. Mathiasson, J. A. Jonsson, and L. Karlsson, *J. Chromatogr., 467:* 61 (1989).
47. S. Ashraf, K. D. Bartle, A. A. Clifford, and R. Moulder, *J. High Resolut. Chromatogr., 14:* 29 (1991).
48. D. E. Knowles and B. E. Richter, *J. High Resolut. Chromatogr., 14:* 689 (1991).
49. J. G. J. Mol, B. N. Zegers, H. Lingeman, and U. A. Th. Brinkman, *Chromatographia, 32:* 203 (1991).
50. B. E. Richter, M. R. Anderson, D. E. Knowles, E. R. Campbell, N. L. Porter, L. Nixon, and D. W. Later, in *Supercritical Fluid Extraction and Chromatography: Techniques and Applications,* ACS Symposium Series 366, B. A. Charpentier and M. R. Sevenants, Eds., American Chemical Society, Washington, D.C., 1988, p. 189.
51. K. E. Markides, E. D. Lee, R. Bolick, and M. L. Lee, *Anal. Chem., 58:* 740 (1986).
52. W. T. Foreman, C. L. Shellum, J. W. Birks, and R. E. Sievers, *J. Chromatogr., 465:* 23 (1989).
53. R. J. Skelton, P. B. Farnsworth, K. E. Markides, and M. L. Lee, *Anal. Chem., 61:* 1815 (1989).
54. D. R. Luffer and M. Novotny, *J. Chromatogr., 517:* 477 (1990).
55. R. C. Wieboldt and J. A. Smith, in *Supercritical Fluid Extraction and Chromatography: Techniques and Applications,* ACS Symposium Series 366, B. A. Charpentier and M. R. Sevenants, Eds., American Chemical Society, Washington, D.C., 1988, p. 229.
56. R. C. Wieboldt, K. D. Kempfert, D. W. Later, and E. R. Campbell, *J. High Resolut. Chromatogr., 12:* 106 (1989).
57. B. W. Wright, H. T. Kalinoski, H. R. Udseth, and R. D. Smith, *J. High Resolut. Chromatogr., 9:* 145 (1986).
58. H. T. Kalinoski, B. W. Wright, and R. D. Smith, *Biomed. Environ. Mass Spectrom., 13:* 33 (1986).

59. H. T. Kalinoski, H. R. Udseth, B. W. Wright, and R. D. Smith, *J. Chromatogr., 400:* 307 (1987).
60. D. A. Laude, S. L. Pentoney, P. R. Griffiths, and C. L. Wilkins, *Anal. Chem., 59:* 2283 (1987).
61. H. T. Kalinoski and R. D. Smith, *Anal. Chem., 60:* 529 (1988).
62. E. C. Huang, B. J. Jackson, K. E. Markides, and M. L. Lee, *Anal. Chem., 60:* 2715 (1988).
63. S. B. Hawthorne and D. J. Miller, *Fresenius Z. Anal. Chem., 330:* 235 (1988).
64. W. M. A. Niessen, R. A. M. Van Der Hoeven, M. A. G. de Kraa, C. E. M. Heeremans, U. R. Tjaden, and J. Van Der Greef, *J. Chromatogr., 478:* 325 (1989).
65. C. W. Saunders, L. T. Taylor, J. Wilkes, and M. Vestal, *Am. Lab.,* Sept., p. 46 (1990).
66. M. A. Morrissey and H. H. Hill, *J. Chromatogr. Sci., 27:* 529 (1989).
67. M. E. McNally and J. R. Wheeler, *J. Chromatogr., 435:* 63 (1988).
68. J. R. Wheeler and M. E. McNally, *J. Chromatogr. Sci., 27:* 534 (1989).
69. J. W. King, *J. Chromatogr. Sci., 27:* 355 (1989).
70. Y. Nishikawa, *Anal. Sci., 7:* 567 (1991).
71. EPA Method 604, *Fed. Regist., 49*(209): 59 (1984).
72. T. A. Berger and J. F. Deye, *J. Chromatogr. Sci., 29*(2): 54 (1991).
73. A. Giorgetti, N. Pericles, H. M. Widmer, K. Anton, and P. Datwyler, *J. Chromatogr. Sci., 27*(6): 318 (1989).
74. J. G. M. Janssen, P. J. Schoenmakers, and C. A. Cramers, *J. High Resolut. Chromatogr., 12*(10): 645 (1989).
75. C. P. Ong, H. K. Lee, and S. F. Y. Li, *Anal. Chem., 62*(14): 1389 (1991).
76. S. K. Yeo, C. P. Ong, H. K. Lee, and S. F. Y. Li, *Environ. Monit. Assess., 19:* 47 (1991).
77. D. Thiebaut, J. P. Chervet, R. W. Vannoort, G. J. De Jong, U. A. Th. Brinkman, and R. W. Frei, *J. Chromatogr., 477:* 151 (1989).
78. J. R. Sand, D. L. Andrjeski, and J. Ashraf, *J. Chromatogr. Sci., 24*(5): 38 (1982).
79. R. E. Roop, R. K. Hess, and A. Akgerman, in *Supercritical Fluid Science and Technology,* ACS Symposium Series 406, K. P. Johnston and J. M. L. Penninger, Eds., American Chemical Society, Washington, D.C., 1989, p. 468.
80. J. L. Hedrick and L. T. Taylor, *J. High Resolut. Chromatogr., 13:* 312 (1990).
81. M. Richards and R. M. Campbell, in *International Symposium on Supercritical Fluid Chromatography and Extraction Abstracts,* Park City, Utah, Jan. 1991, p. 125.
82. W. D. Spall, A. M. Martinez, and B. F. Smith, in *International Sympo-*

sium on *Supercritical Fluid Chromatography and Extraction Abstracts,* Park City, Utah, Jan. 1991, p. 111.
83. K. Anton, in *SFC Applications* (compiled by K. E. Markides and M. L. Lee), *1989 Symposium/Workshop on Supercritical Fluid Chromatography,* Snow Bird, Utah, June 13–15, 1989, p. 346.
84. S. V. Olesik, in *SFC Applications* (compiled by K. E. Markides and M. L. Lee), *1989 Symposium/Workshop on Supercritical Fluid Chromatography,* Snow Bird, Utah, June 13–15, 1989, p. 345.
85. S. B. French and M. Novotny, *Anal. Chem., 58:* 164 (1986).
86. A. Giorgetti and N. Pericles, in *SFC Applications* (compiled by K. E. Markides and M. L. Lee), *1989 Symposium/Workshop on Supercritical Fluid Chromatography,* Snow Bird, Utah, June 13–15, 1989, p. 348.
87. R. D. Smith, B. W. Wright, and H. T. Kalinoski, in *Progress in HPLC,* M. Yoshioka, S. Parvez, T. Miyazaki, and H. Parvey, Eds., Vol. 4, VBP, St. Augustine Neiderberg, DE. The Netherlands, 1989, p. 111.
88. R. C. Simpson, J. R. Gant, and P. R. Brown, *J. Chromatogr., 371:* 109 (1986).
89. J. M. Levy and W. M. Ritchey, *J. High Resolut. Chromatogr., 8:* 503 (1985).
90. J. M. Levy and W. M. Ritchey, *J. Chromatogr. Sci., 24:* 242 (1986).
91. A. L. Blilie and T. Greibrokk, *Anal. Chem., 57:* 2239 (1985).
92. C. R. Yonker and R. D. Smith, *J. Chromatogr., 361:* 25 (1986).
93. S. M. Fields, K. E. Markides, and M. L. Lee, *J. Chromatogr., 406:* 223 (1987).
94. A. L. Blilie and T. Greibrokk, *J. Chromatogr., 349:* 317 (1985).
95. B. Gemmel, F. P. Schmitz, and E. Klesper, *J. Chromatogr., 455:* 17 (1988).
96. K. Anton, N. Pericles, S. M. Fields, and H. M. Widmer, *Chromatographia, 26:* 224 (1988).
97. M. Saito, Y. Yamauchi, H. Kashiwazaki, and M. Sugawara, *Chromatographia, 25:* 801 (1988).
98. P. J. Schoenmakers and L. G. M. Uunk, *Chromatographia, 24:* 51 (1987).
99. H. Jenssen, H. M. J. Snijders, J. Rijks, C. A. Cramers, and P. J. Schoenmakers, *J. High Resolut. Chromatogr., 14:* 438 (1991).
100. K. Jinno and M. Kuwajima, *Chromatographia, 23:* 631 (1987).
101. A. Malik and K. Jinno, *Chromatographia, 31:* 561 (1991).
102. K. Jinno, H. Mae, M. Yamaguchi, and Y. Ohtsu, *Chromatographia, 31:* 239 (1991).
103. K. Jinno and H. Mae, *J. High Resolut. Chromatogr., 13:* 512 (1990).
104. E. C. Huang, B. J. Jackson, K. E. Markides, and M. L. Lee, *Chromatographia, 25:* 51 (1988).
105. S. B. Hawthorne and D. J. Miller, *J. Chromatogr., 468:* 115 (1989).

106. E. R. Baumeister, C. D. West, C. F. Ijames, and C. L. Wilkins, *Anal. Chem., 63:* 251 (1991).
107. W. R. West and M. L. Lee, *J. High Resolut. Chromatogr., 9:* 161 (1986).
108. P. G. Sim, C. M. Elson, and M. A. Quilliam, *J. Chromatogr., 445:* 239 (1988).
109. M. W. Raynor, G. F. Shilstone, K. D. Bartle, A. A. Clifford, M. Cleary, and B. W. Cook, *J. High Resolut. Chromatogr., 12:* 300 (1989).
110. E. Leren, K. E. Landmark, and T. Greibrokk, *Chromatographia, 31:* 535 (1991).
111. C. R. Yonker and R. D. Smith, *Anal. Chem., 59:* 727 (1987).
112. B. W. Sandmann and M. L. Grayeski, *J. Chromatogr. Sci., 31:* 49 (1993).
113. M. X. Huang, K. E. Markides, and M. L. Lee, *Chromatographia, 31:* 163 (1991).
114. J. M. Levy, J. P. Guzowski, and W. E. Huhak, *J. High Resolut. Chromatogr., 10:* 337 (1987).
115. M. Taguchi, T. Hobo, and T. Maeda, *J. High Resolut. Chromatogr., 14:* 140 (1991).
116. C. P. Ong, H. K. Lee, and S. F. Y. Li, *Environ. Monit. Assess., 19:* 63 (1991).

6
HPLC of Homologous Series of Simple Organic Anions and Cations

Norman E. Hoffman *Marquette University, Milwaukee, Wisconsin*

I.	INTRODUCTION	310
	A. Homologous Series	310
	B. Methylene Selectivity	312
	C. Mobile Phase and Selectivity	312
	D. Stationary Phase and Selectivity	313
	E. Effect of Temperature	314
	F. Micellar Chromatography	315
II.	REVERSED-PHASE CHROMATOGRAPHY OF ORGANIC IONS	316
	A. Polymeric Stationary Phases	316
	B. Bonded Silica Stationary Phases	318
	C. Ion Exchange on Reversed-Phase Columns	320
III.	ION-EXCHANGE CHROMATOGRAPHY OF ORGANIC IONS	324
	A. Cation Exchange	325
	B. Anion Exchange	329
IV.	ION-EXCLUSION CHROMATOGRAPHY OF ORGANIC IONS	340

I. INTRODUCTION

In this chapter we present the observations and interpretation of these observations that have been made in studies of the retention of organic ions in liquid chromatography. Many investigations of the chromatography of organic ions have been published, especially therapeutic drug ions. This chapter is restricted to investigations of homologous series of simple organic ions.

A. Homologous Series

Early studies in reversed-phase chromatography showed that retention increased from increasing the size of the nonpolar part of the eluite. Sleight [1] took this concept one step further and showed that retention increased regularly with increasing carbon number of the alkyl group of alkylbenzenes when the eluant was methanol–water and the stationary phase was octadecyl bonded silica (ODS). He plotted the log k' versus alkyl group carbon number and obtained a straight line.

In a classic paper, Horvath et al. [2] presented their solvophobic theory of retention in reversed-phase liquid chromatography. They introduced the idea of the need for a free-energy change of cavity formation in the mobile phase. The size of this cavity was related to the hydrocarbonaceous surface area of the eluite. Increasing this surface area led to greater solvophobicity or increasing retention. Plots of log k', which is proportional to the standard free-energy change, versus hydrocarbonaceous surface area were linear for amides, amino acids, and aromatic carboxylic acids under reversed-phase conditions.

A homologous series of organic eluites, being a series in which the members differ by a methylene group, CH_2, form a class that differs in the members' hydrocarbonaceous surface area in a regular manner. Thus solvophobic theory predicts a linear relationship between log k' and the number of methylene carbons in the eluite; and this has been observed, generally, experimentally. The linearity has sometimes been said to obey Martin's rule or law [3].

In early work Colin et al. [4] found linear relationships between log k' and the carbon number of homologous series of alkyl halides and alcohols. If benzene and toluene were excluded, alkylbenzenes also gave a straight line. The mobile phase was methanol, and the stationary phase was modified carbon black. Nakae et al. [5–7] found straight lines in log k' versus carbon number plots for a variety of eluites with methanol–water mobile phases and a stationary phase of microporous styrene–divinylbenzene copolymer. Uchida and Tanimura [8] showed that fatty acids with methylene groups of 1 to 17 gave straight-line plots of log k' versus carbon number when a microporous styrene–divinylbenzene copolymer was the stationary phase and mobile phases of 40 to 100% methanol in water were used.

Hoffman and Liao [9] found a linear relationship between log $(V_r - V_0)$ and carbon number for n-alkanes with an ODS column and a variety of aqueous mobile phases having organic modifiers of acetonitrile, methanol, ethanol, 2-propanol, dioxane, and tetrahydrofuran. But when acetonitrile–water mobile phases were used, polar eluites (i.e., alcohols, amides, and ketones) did not give a straight line unless members of the homologous series with seven to eight carbons or less were excluded.

Dufek and Smolkova [10] found deviations from linearity in retention–carbon number plots for N-n-alkylphthalimides when the alkyl group carbon number was less than about 4. An ODS stationary phase and methanol–water eluants of 50 to 80% methanol were used. Figure 1 shows the linearity plots. This kind of deviation has been observed by others. Dufek and Smolkova argued that the deviation resulted from small alkyl groups providing "ineffective shielding" of the —CO—N—CO group of phthalimides.

The deviations from linearity for low-molecular-weight members of homologous series are violations of Martin's rule. It has been argued that silanol groups furnishing a second mode of retention show their effect more strongly for homologous series members with few methylene groups [9]. A qualitative

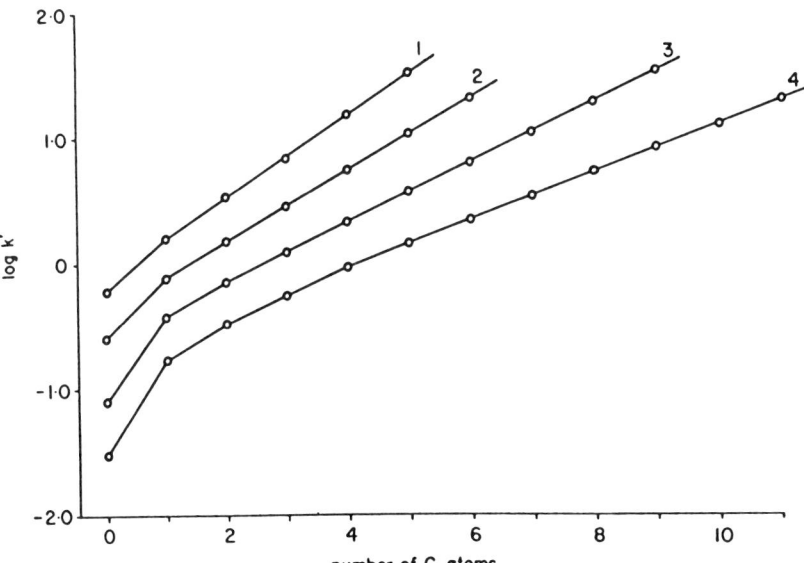

Fig. 1 Dependence of log k' on the number of carbon atoms in the alkyl groups of N-n-alkylphthalimides. Column, ODS; mobile phase, methanol–water in the ratios 50:50 (1), 60:40 (2), 70:30 (3), 80:20 (4) by volume. (From Ref. 10.)

interpretation is that as the molecular weight decreases, solvophobicity decreases to a point where it disappears. For example, one can argue that although 1-decanol is solvophobic in methanol–water solutions, ethanol in these solutions can hardly be considered solvophobic. Somewhere between 1-decanol and ethanol the homologs change from solvophobic to solvophilic, and this change would appear in a log k' versus carbon number plot as a deviation from linearity in the direction of a drop in retention compared to the linear extrapolated retention.

B. Methylene Selectivity

The linearity of log k' versus carbon number plots provides a method for measuring retention selectivity. The selectivity, α, for two members of a homologous series differing by a methylene group has been called the α-methylene selectivity, α_{CH_2}.

$$\alpha_{CH_2} = \frac{k'_n}{k'_{n-1}} \tag{1}$$

where k'_n is the capacity factor of the homolog with n carbons and k'_{n-1} is the capacity factor of the next-lower homolog. From the linear retention-carbon number curve,

$$\log k' = mn + b \tag{2}$$

where m is the slope, n the carbon number of the homolog, and b a constant. For a change in carbon number from $n - 1$ to n the log change is log k'_{n-1} to log k'_n. The slope then is

$$m = \log \alpha_{CH_2} \tag{3}$$

Thus the α-methylene selectivity can be obtained by taking the antilog of the slope of a plot of Eq. (2).

The log α_{CH_2} or, more commonly, α_{CH_2} has been viewed as the nonspecific selectivity for a homologous series. In Eq. (2), b is sometimes expressed as log β, where β is considered the specific selectivity or the capacity factor of the homologous series functional group. Colin et al. have shown that α_{CH_2} and log β are linearly related when they are a function of the composition of water–methanol and water–acetonitrile mobile phases [11].

C. Mobile Phase and Selectivity

Many workers have found that selectivity, α, decreases as the organic modifier fraction increases in modified aqueous mobile phases. Colin et al. [12] found that log α_{CH_2} was independent of the homologous series and the type of ODS stationary phase. The dependence of log α_{CH_2}, called solvophobicity by these workers, on the organic modifier fraction in the mobile phase was used to define eluotropic strength of mobile phases. Binary organic modifier–water solutions showed no linearity of solvophobicity with volume percent water except for solutions of about 30 to 100% methanol.

Johnson et al. [13] also found linearity between log α_{CH_2} and percent methanol in water–methanol mobile phases but nonlinearity for acetonitrile–water mobile phases. However, both methanol–water and acetonitrile–water mobile phases gave linearity in plots of log α_{CH_2} versus E_T [30] polarity. These four plots were obtained with an ODS stationary phase. Of the four plots of data obtained with these mobile phases and a styrene–divinylbenzene copolymer column, only the volume percent methanol–log α_{CH_2} plot was linear. Also significant in these workers' results was that for a given organic modifier concentration, the log α_{CH_2} was markedly higher for the styrene–divinylbenzene copolymer column than for the ODS column. As did Colin et al. [12], these workers found little change in log α_{CH_2} when the functional group of a homologous series was changed. Similarly, Varughese et al. [14] found very little change in log α_{CH_2} when the mobile-phase composition was constant and the functional group was changed. This group studied nonionic surfactant homologous eluites and ethyl alkanoic ester eluites.

D. Stationary Phase and Selectivity

In addition to using homologous series to study the role of the mobile phase in reversed-phase chromatography, homologs have been used in the investigation of the stationary phase. Krstulovic et al. [15] found that for a given mobile-phase composition, the log α_{CH_2} varied with the alkyl chain length of the bonded stationary phase. This was true for a variety of functional groups in the homologs. In continuing this investigation, Tchapla et al. [16] found that when homologous series were extended to very long chain lengths, log k' versus carbon number plots lost their linearity. The plots appeared to be two linear curves with an intersection point.

The intersection point varied with the chain length of the stationary-phase bonded alkyl groups. Monomeric stationary phases were used. These workers viewed these results as indicating that the eluite penetrated into the stationary phase vertically. When the bonded phase was considered a brush phase, the homolog sorbed on a bristle through contact of its methylenes. When the homolog became too long to permit contact of all its methylenes, sorption was described by a new linear equation with a different slope and intercept. The linear curve for longer-chained members of the homologous series gave an α_{CH_2} value that was smaller than that of the shorter-chained members.

E. Effect of Temperature

Understanding the retention process can be helped by studying the effect of temperature on retention. For example, the driving force of reversed-phase retention can be enthalpic, entropic, or both, and the degree of each may possibly be measured in temperature studies. In using homologs as eluites the magnitude of $\Delta H°$ in adding to the eluite chain length can be evaluated. These enthalpy changes might result from mobile-phase composition changes or changes in the length of the bonded group of the stationary phase.

The equilibrium constant, K, for mobile phase/stationary phase equilibrium is related to the standard free-energy change by the familiar expression

$$\Delta G° = -RT \ln K \qquad (4)$$

Since

$$\Delta G° = \Delta H° - T \Delta S° \qquad (5)$$

$$\ln K = -\frac{\Delta H°}{RT} + \frac{\Delta S°}{R} \qquad (6)$$

If

$$k' = K\Phi \quad \text{where } \Phi \text{ is the phase ratio} \qquad (7)$$

then

$$\ln k' = -\frac{\Delta H°}{RT} + \frac{\Delta S°}{R} + \ln \Phi \qquad (8)$$

According to Eq. (8) a plot of $\ln k'$ or $\log k'$ versus $1/T$, called a van't Hoff plot, should give a straight line from whose slope $\Delta H°$ is readily calculated. This enthalpy is the standard enthalpy for transfer of a mole of eluite from the mobile to the stationary phase. If the van't Hoff curve is not linear, it is likely that the mechanism of retention changed with temperature or a phase transition has occurred in the stationary phase [17].

Vigh and Varga-Puchony [18] showed that for straight-chain 1-alkanols, alkanal dinitrophenylhydrazones, and 2-alkanone dinitrophenylhydrazones, van't Hoff plots were linear in reversed-phase chromatography with water–methanol mobile phases. They also observed a decrease in $\log \alpha_{CH_2}$ with increasing temperature for all eluites. Their enthalpy changes were negative, showing exothermicity for the transfer of eluite from the mobile to the stationary phase. Furthermore, $-\Delta H°$ increased linearly with increasing carbon number in all three homologous series.

Grushka et al. [19] in studying the reversed-phase chromatography of n-alkylbenzene homologs also found that $\log \alpha_{CH_2}$ decreased as the temperature increased. They also found that log α_{CH_2} increased with increasing water content of the mobile phase at all temperatures studied.

Morel et al. [20], with a limited number of n-alkane and n-alkylbenzene homologs as eluites, studied temperature effects. They observed, for ODS columns and methanol as the mobile phase, a stationary-phase transition at 17.5°C. Their van't Hoff plots were not linear but appeared to be two linear plots intersecting at the transition temperature. They also studied n-alkanes with a large number of methylenes. Their results were similar to those previously discussed [16] in that the log k' versus carbon number plot appeared to be two straight lines intersecting. They showed that the intersection point moved to lower carbon numbers as the temperature was lowered.

On ODS columns Cole and Dorsey [21] found that stationary-phase transitions occurred when the C_{18} bonding densities were greater than 2.84 μmol/m² with a mobile phase of 60:40 acetonitrile–water. At lower bonding densities linear van't Hoff plots were obtained. The slopes and intercepts of these straight lines showed that for the transfer of the eluite from the mobile phase to the stationary phase, $\Delta H°$ and $\Delta S°$ were both negative. The absolute value of $\Delta H°$ was always greater than that of $T \Delta S°$, and thus the enthalpy played a greater role in retention than did entropy. They also found that $\Delta S°$ increased with respect to $\Delta H°$ as the stationary phase bonding density increased, or, in other words, enthalpy's dominance in retention was present but was decreasing as the bonding density of the stationary phase was increased.

F. Micellar Chromatography

Khaledi et al. [22] found that micellar chromatographic systems do not produce linear log k' versus carbon number plots for homologous series of n-alkylbenzenes and n-alkyl phenyl ketones. ODS and octyl bonded columns with mobile phases containing the cationic surfactant cetyltrimethylammonium bromide or the anionic surfactant sodium dodecyl sulfate were used. The mobile phases of neat water or water containing up to 20 vol % 2-propanol gave log k' versus carbon number plots that were concave down and were best described by a quadratic equation. If k', not log k', was plotted against carbon number, a linear curve resulted.

Borgerding et al. [23] observed a linear relationship between k' and carbon number in micellar liquid chromatography of n-alkylbenzene homologs. Their mobile phases were aqueous solutions of either of two nonionic surfactants, poly[oxyethylene(10 or 23)]dodecanol, of varying concentration. An ODS stationary phase was used. Their interpretation of the k'–carbon number linearity was that there was a direct transfer of the eluite located in the micelle to the stationary phase which itself was modified by the surfactant (called by them a "hemimicellar" stationary phase). The interpretation was based on partition coefficients, water solubilities, and eluite selectivities.

Another view of nonlinearity in micellar chromatography is based on the thermodynamic arguments of Hinze and Weber [24]. For low-molecular-weight water-soluble homolog eluites occupying primarily the nonmicellar part of the mobile phase, transfer is from the water of the mobile phase to the surfactant-modified stationary phase. As methylenes are added to the eluites, solubility in water decreases, and the eluites are found mostly in the micelles of the mobile phase. Here the transfer is from the micelles directly to the surfactant-modified stationary phase. Martin's rule leads to equations that produce plots of log k' versus carbon number that are not linear.

II. REVERSED-PHASE CHROMATOGRAPHY OF ORGANIC IONS

Short retention times of organic ions in reversed-phase chromatography have long been recognized. Being electrically charged leaves these ions solvophilic or poorly solvophobic. There are many examples of carboxylic acids that have extremely short retention times at high-mobile-phase pH, where they are in their conjugate base form. Subsequent lowering of the pH increases retention because of the increasing mole fraction of the acid form in equilibrium. Similarly, amines have small capacity factors at low pH and large ones at high pH. Studies of homologous series of organic ions have been conducted with styrene–divinylbenzene copolymeric stationary phases and alkyl-bonded silica phases.

A. Polymeric Stationary Phases

n-Alkylbenzyldimethylammonium chlorides and N-n-alkylpyridinium chlorides or bromides with alkyl groups of 10, 12, 14, 16, and 18 carbons gave zero capacity factors with mobile phases of 100% methanol or 90% methanol–10% water [6]. A stationary phase of styrene–divinylbenzene copolymer was used. When hydrochloric or perchloric acids or sodium perchlorate was added to the mobile phase of neat methanol, retention increased. Perchloric acid at the same molarity had a markedly greater effect on increasing retention than did hydrochloric acid. Sodium perchlorate was nearly as effective as perchloric acid in increasing retention. Increasing the perchloric or hydrochloric acid concentration increased retention. Alkylpyridinium bromides and chlorides gave the same capacity factors with a mobile phase containing perchloric acid. This fact, along with studies using a methyl orange/N-n-hexadecylpyridinium salt led Nakae et al. to conclude that the ion counter to the eluite ion was exchangeable to a great extent.

Further interpretation was not offered by these workers. It seems likely that a free ion-paired ion equilibrium existed in the mobile phase. With no additive in the methanol mobile phase, the free ion dominated the equilibrium and retention was neglibile. With a pairing ion present, the paired ion, having its charge neutralized, was more solvophobic than the free ion. Increasing the perchloric or hydrochloric acid concentration favored formation of the paired ion. Perchlorate from the perchloric acid was more effective than chloride from the hydrochloric acid because it pairs better than chloride. Plots of log k' versus alkyl chain length of these quaternary ammonium ions gave linear curves with increasing slope (log α_{CH_2}) as the concentration of perchloric acid increased. As has already been stated, log α_{CH_2} has been considered a measure of solvophobicity [12]. An increase in pairing ion concentration would increase solvophobicity, and thus the perchloric acid concentration effect is consistent with the ion-pair hypothesis.

In the temperature range of 30 to 50°C the quaternary ammonium ions gave linear van't Hoff plots. The slopes for all members of the homologous series were positive. Therefore, transfer of the eluite ion from the mobile to the stationary phase was exothermic.

In continuing their investigation of high-molecular-weight organic ions, Nakae and Kunihiro [7] examined the retention of several homologous series of alkylbenzenesulfonates. Their ionic eluites were derivatives of decane, undecane, dodecane, tridecane, tetradecane, and pentadecane. A homologous series formed by substitution of phenylsulfonate groups in the 2-position, $CH_3(CH_2)_n$CH-$(C_6H_4SO_3^-)CH_3$, was labeled 2Φ; in the 3-position, $CH_3(CH_2)_n$CH-$(C_6H_4SO_3^-)CH_2CH_3$, was labeled 3Φ, and so on.

They found that a mobile phase of 100% methanol and a stationary phase of styrene–divinylbenzene copolymer gave no retention. Perchloric acid (0.5 M), sodium perchlorate (0.5 M), or lithium chloride (0.5 M) in methanol provided measurable retention. Alkylbenzenesulfonates had larger capacity factors in the perchloric acid mobile phase than in the sodium perchlorate or lithium chloride mobile phase. The sodium and lithium salts produced more comparable retentions with the lithium salt consistently giving less retention. One can again argue that ion pairs formed and that hydrogen ion was more effective in ion-pair formation than lithium or sodium.

Among isomers the 2Φ isomer always had the highest k'. In many other cases the capacity factors for isomers were identical, for example, $k' = 2.00$ for both 5Φ and 7Φ $C_{14}H_{29}$ with a mobile phase of 0.5 M perchloric acid. For a given homologous series, for example, the 3Φ series, a plot of log k' versus carbon number of the alkyl group gave a straight line. Figure 2 presents the plots of the 2Φ, 3Φ, and 4Φ series along with higher Φ value series. The log k' values for this last group fell on the same line. It can be seen that as the symmetry of the ion increases, going to higher Φ values, log α_{CH_2} decreases.

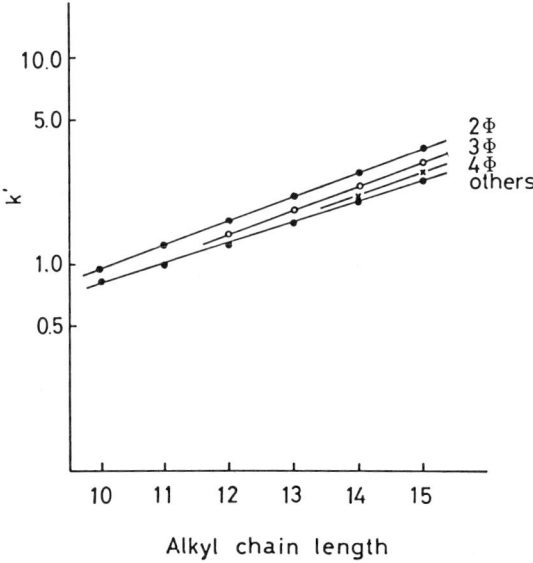

Fig. 2 Relationship between log k' and alkyl chain length for alkylbenzenesulfonate anions differing in the position of the sulfophenyl group. (From Ref. 7.)

B. Bonded Silica Stationary Phases

In an extensive study of quaternary ammonium compounds, Abidi [25] used mobile phases of acetonitrile–water and tetrahydrofuran–water and silica-based stationary phases of bonded octadecyl, cyanopropyl, and phenylpropyl groups. Eluites were n-alkyldimethylbenzylammonium ions with 12, 14, 16, and 18 carbons in the alkyl group. Regardless of which stationary or mobile phase used plots of capacity factor versus volume percent water in the mobile phase had the shape normally found in reversed-phase chromatography. For all columns for a given percent water and a given eluite, the order of k' was methanol > acetonitrile > tetrahydrofuran.

For all members of the homologous series sodium perchlorate in the mobile phase had an effect on retention with all three stationary phases. With cyanopropyl and phenylpropyl phases, k' decreased as the concentration of the sodium salt in 50% acetonitrile increased, and the change was very great at concentrations below 20 mM. ODS with 90% acetonitrile showed a small rise in the capacity factor as the sodium perchlorate concentration was increased, although for the C_{12} ammonium ion derivative the rise was negligible. Abidi was unable to rationalize these results without further work.

When organic anions such as hexanesulfonate were added to the mobile phase, retention of the ammonium ions increased, as would be expected from ion interaction. Left unexplained was the comparatively high retention observed when sodium dihydrogenphosphate was added to the mobile phase. It caused greater retention with all eluites than did n-octanesulfonate. As the pH of the mobile phase was raised from 2 to 6, retention increased, although not linearly.

For the mobile and stationary phases studied, Abidi found linear relationships between log k' and carbon number of the alkyl group. The linearity is shown in Fig. 3. The log α_{CH_2} was greatest when the ODS stationary phase was used, and greater when phenylpropyl was used than when cyanopropyl was used.

In developing their ion-interaction mechanism for ion-pair chromatography, Bidlingmeyer et al. [26] studied the retention of n-alkanesulfonates on an ODS

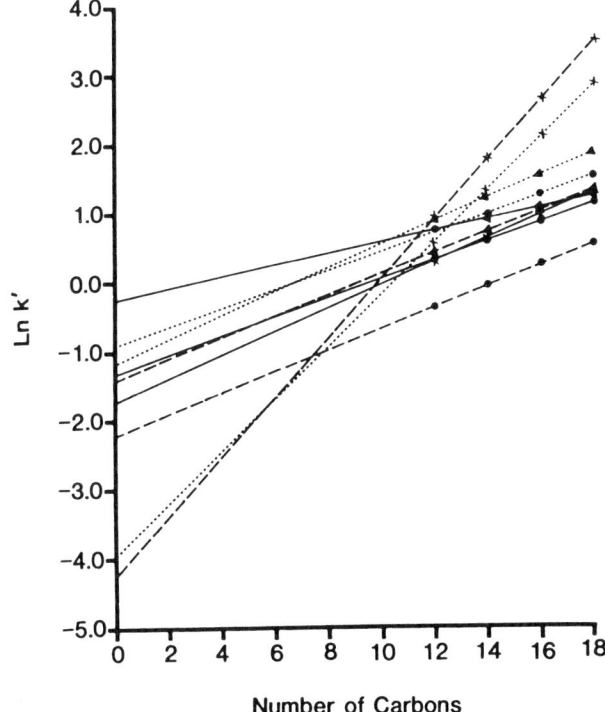

Fig. 3 Relationship between log k' and alkyl chain length in n-alkybenzyldimethylammonium chlorides. Mobile phase: ·····, acetonitrile–water; ----, methanol–water; ——, tetrahydrofuran–water. Stationary phases: X, ODS; ▼, cyanopropylsilica; ●, phenylpropylsilica. (From Ref. 25.)

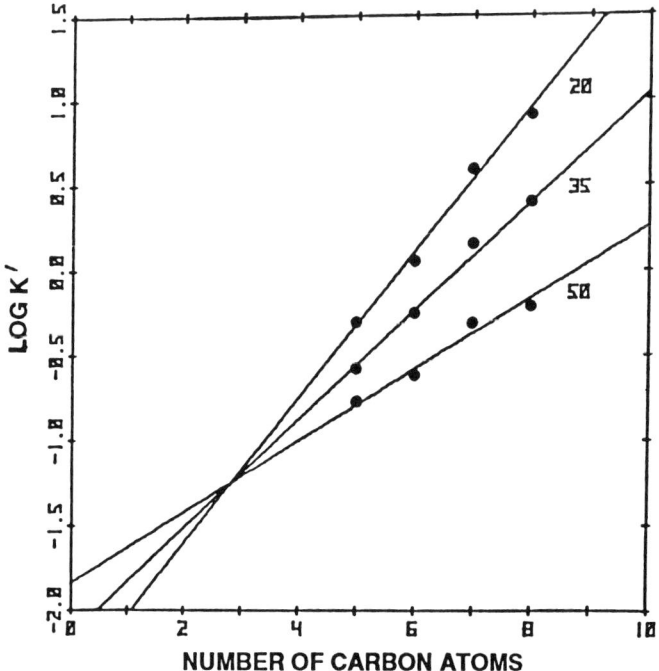

Fig. 4 Relationship between log k' and the number of carbon atoms in the alkyl chain of n-alkanesulfonate anions. Numbers besides the straight lines indicate the percent methanol in the mobile phase. (From Ref. 26.)

column. Their mobile phases were methanol–water. Figure 4 shows that sulfonate homologs behave on silica bonded stationary phases as they do on styrene–divinylbenzene resins. The effect of methanol content of the mobile phase is shown in Fig. 5. Noteworthy are the small k' values even for n-octanesulfonate.

As expected, n-octylammonium ion in the mobile phase significantly increased k' for all the sulfonate homologs. Interestingly, these workers found that n-octanesulfonate in the mobile phase reduced k' for the sulfonate homologs. They attributed this reduction to coulombic repulsion between the sorbed primary layer of negatively charged octanesulfonate ions from the eluant and the eluite negatively charged homolog ions. In experiments with both positive and negative eluant ions, the order of n-alkanesulfonate retention was that followed in Fig. 4.

C. Ion Exchange on Reversed-Phase Columns

Silica-based bonded stationary phases can behave as ion exchangers because of the acidic silanol groups present. Hoffman and Liao [27] studied ion-exchange

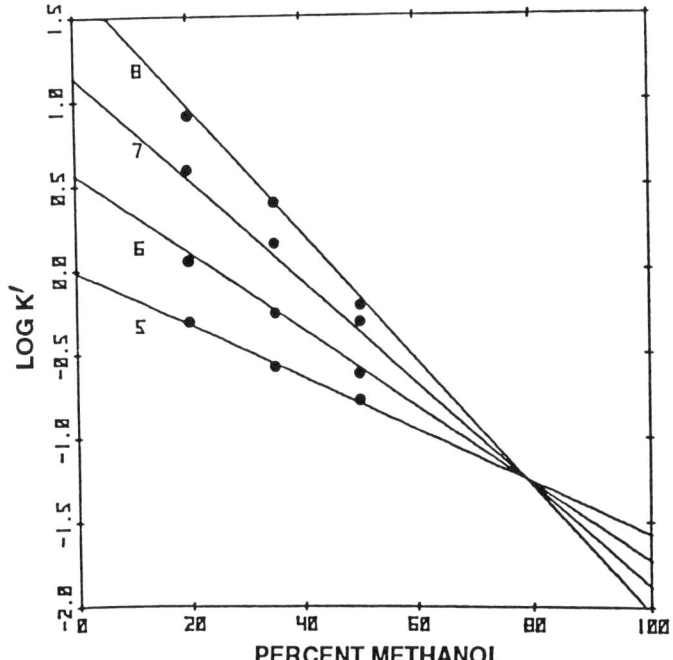

Fig. 5 Relationship between log k' of n-alkanesulfonate anions and percent methanol in the mobile phase. Numbers beside the straight lines indicate the number of carbons in the alkyl chain of the n-alkanesulfonate anions. (From Ref. 26.)

participation in the reversed-phase chromatography of the phenylalkylammonium ion series, $C_6H_5(CH_2)_nN^+H_3$. In addition to this study with homologous series, many other papers have described evidence of ion exchange in reversed-phase chromatography of cations [28–33].

In cation exchange a competitive equilibrium between eluant ion (Z^+) and the eluite ion (N^+) for coulombic interaction with the bound anionic site of the stationary phase is the basis for the transfer of N^+ from the mobile phase (m) to the stationary phases (s).

$$\mathrm{SiO^-Z_s^+} + N_m^+ \rightleftharpoons -\mathrm{SiO^-}N_s^+ + Z_m^+ \tag{9}$$

The ion-exchange equilibrium constant is

$$K_{ie} = \frac{[\mathrm{SiO^-}N_s^+][Z_m^+]}{[\mathrm{SiO^-}Z_s^+][N_m^+]} \tag{10}$$

The equilibrium constant for partitioning of the eluite is

$$K = \frac{[\text{SiO}^- N_s^+]}{[N_m^+]} = K_{ie} \frac{[\text{SiO}^- Z_s^+]}{[Z_m^+]} \qquad (11)$$

The injection of a small amount of eluite leaves $[-\text{SiO}^- Z_s^+]$ constant. Therefore,

$$K = \frac{C}{[Z_m^+]} \qquad (12)$$

where C is a constant. Since

$$k' = K\Phi \qquad (7)$$

$$k' = \frac{\Phi C}{[Z_m^+]} = \frac{C'}{[Z_m^+]} \qquad (13)$$

Thus plotting k' versus $[Z_m^+]^{-1}$ gives a straight line with zero k' intercept for ideal ion-exchange chromatography. If reversed-phase interaction is the primary mode of retention and ion exchange is secondary, k_0' at infinite $[Z_m^+]$ would be the capacity factor for pure reversed-phase chromatography. At any eluant ion concentration,

$$k' = \frac{C'}{[Z_m^+]} + k_0' \qquad (14)$$

For the phenylalkylammonium ions plots of k' versus $[Z_m^+]^{-1}$ for 30 to 60% methanol in water on an ODS column gave straight lines. Table 1 shows the results. Ammonium ion was the eluant ion (Z^+). The k_0' values followed the normal change with volume percent change of the organic modifier. The slope, C', can be considered the tendency toward ion exchange. For a given eluite it followed a pattern having a maximum value when the mobile phase had about 40% methanol. This observation is not well understood and requires further study.

There was a consistent rise in C' as the carbon number of the eluite ion increased. This was interpreted to mean that solvophobicity aided ion exchange. Linear plots of log k_0' versus carbon number from which α_{CH_2} and β could be calculated were obtained. These values are shown in Table 2.

For comparison to a reversed-phase column, a silica stationary phase and water mobile phase were used for the chromatography of the phenylalkylammonium ion homologs. The values for C' were much larger than for a reversed-phase column, but exact comparison cannot be made because of the unusual dependence of C' on methanol content in the reversed-phase case. The results did show an increase in both C' and k_0' with increasing carbon number, although significantly less than in reversed phase. The methylene effect could be the result of solvophobic interaction, even with a normal-phase column [34].

Table 1 Phenylalkylammonium Ion Data from Plots of Eq. (14)

Carbon number	Vol % methanol[a]	k'_0	C' (mol/L)
1	30	1.5	0.030
	40	1.1	0.047
	50	0.91	0.046
	60	0.84	0.032
2	30	2.3	0.044
	40	1.6	0.067
	50	1.4	0.046
	60	1.0	0.039
3	30	4.0	0.072
	40	2.7	0.088
	50	1.7	0.074
	60	1.2	0.056
4	40	3.8	0.35
	50	3.0	0.13
	60	1.6	0.095

Source: Data from Ref. 27.
[a]Eluant ion, NH_4^+; pH, 7.00; column, ODS.

The retention of anilines on an ODS column was also investigated. Table 3 presents the data for aniline, *N*-methylaniline, and *N,N*-dimethylaniline at pH 3.00, where these eluites are cations. In all cases, the tendency toward ion exchange increased as methyl groups were added to the protonated aniline. This increase again showed solvophobicity enhancement of ion exchange.

Within experimental error k'_0 for an aniline was the same whether potassium or ammonium ion was the eluant ion. This result is consistent with the interpreta-

Table 2 Selectivities from log k'_0 Versus Carbon Number Plots[a]

Vol % methanol	α_{CH_2}	β
30	1.63	0.90
40	1.54	0.71
50	1.46	0.63
60	1.23	0.66

Source: Data from Ref. 27.
[a]k'_0 values from Table 1.

Table 3 Protonated Aniline Data from Plots of Eq. (14)[a]

	Aniline	N-Methylaniline	N,N-Dimethylaniline
20% Methanol			
C'	0.012	0.025	0.14
k'_0	1.0	1.6	2.5
30% Methanol			
C'	0.012	0.033	0.085
k'_0	0.79	1.1	1.5
40% Methanol			
C'	0.016	0.037	0.075
k'_0	0.60	0.82	1.2
50% Methanol			
C'	0.015	0.035	0.070
k'_0	0.52	0.68	1.1
30% Methanol			
C'	0.0074	0.016	0.038
k'_0	0.75	1.2	1.7

Source: Data from Ref. 27.
[a] Eluant ion was NH_4^+ except for the last 30% entries, in which K^+ was used. pH 3.00.

tion that k'_0 was measuring pure reversed-phase interaction. On the other hand, C' was always smaller for a given aniline when potassium ion was used in place of ammonium ion. Normally, potassium ion is a stronger eluant than ammonium ion [35]. The result then strengthens the argument for the occurrence of ion exchange.

Using Eq. (14) and the results in Tables 1 and 2, one can readily calculate the contribution of reversed phase (k'_0) and ion exchange (C'/molarity) to k'. When the mobile phase was 60% methanol, for example, ion exchange made up 70% of the retention of 4-phenylbutylamine if the eluant was 25 mM ammonium ion. The corollary is that comparatively high concentrations of cations with good eluant strengths are needed in the mobile phase when only reversed-phase interaction is desired in the chromatography of organic cations.

III. ION-EXCHANGE CHROMATOGRAPHY OF ORGANIC IONS

Clearly, the behavior of homologous cations gives strong evidence that ion exchange can occur during their reversed-phase chromatography. Is there, on the other hand, strong evidence for reversed-phase interaction during ion-exchange chromatography of organic ions? To answer this question, homologous series of cations and anions have been investigated.

A. Cation Exchange

Rahman and Hoffman [36] studied the ion-exchange chromatography of phenylalkylammonium ions in which the alkyl group had one to four methylene groups. The stationary phase was sulfonated styrene–divinylbenzene polymeric resin and mobile phases were acetonitrile–water solutions containing alkali metal and organic cations.

With lithium, sodium, and potassium eluant ions plots of the reciprocal of the eluant ion concentration against the capacity factor gave straight lines. This linearity is illustrated in Fig. 6, which shows plots for the eluant sodium ion and eluite ions benzylammonium ion (BA^+), 2-phenylethylammonium ion (PEA^+), 3-phenylpropylammonium ion (PPA^+), and 4-phenylbutylammonium ion (PBA^+). Similar curves were obtained by using the other alkali metal ions. The results follow Eq. (14).

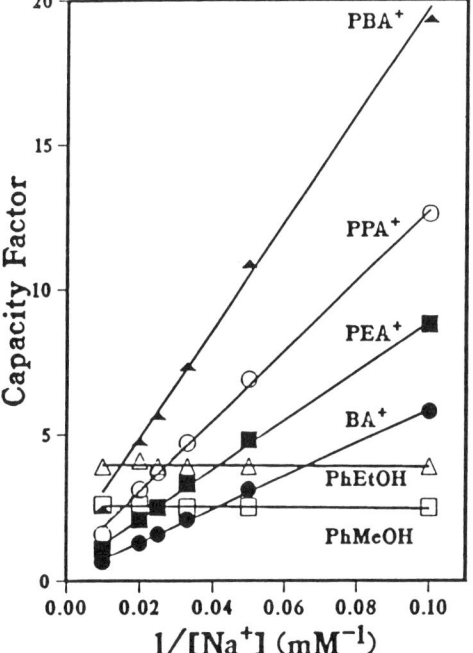

Fig. 6 Relationship between k' of phenylalkylammonium ions or phenylalkanols and the reciprocal of the sodium ion concentration in the aqueous fraction of the mobile phase. Mobile phase, acetonitrile–water, 30%–70%. (From the *Journal of Chromatographic Science,* Vol. 28, April 1990, by permission of Preston Publications, a Division of Preston Industries, Inc.)

It can be seen that the slope of the linear curve varies with the homolog, increasing with the number of methylene groups. The y-intercept also increases with the increasing number of methylene groups. Also plotted in Fig. 6 are the results from using two electrically neutral eluites, benzyl alcohol (PhMeOH) and 2-phenylethyl alcohol (PhEtOH). These eluites were well retained despite the stationary phase being an ion exchanger, and PhEtOH was retained more strongly than PhMeOH. Salt concentration had a negligible effect on the retention of the alcohols.

Figure 7 shows how the retention of PBA^+ changes with the kind of eluant alkali ion. The order of eluant strength was $K^+ > Na^+ > Li^+$. This is the order normally found in ion exchange of inorganic ions [35].

The curves in Fig. 8 show the relationship between capacity factor and acetonitrile content of the mobile phase. The curve shapes for the cations are the

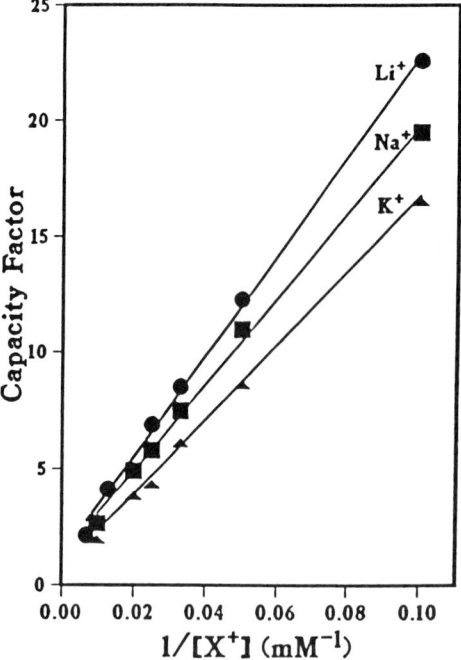

Fig. 7 Relationship between k' of phenylbutylammonium ion and the reciprocal of alkali metal ion concentration. Mobile phase as in Fig. 6. (From the *Journal of Chromatographic Science,* Vol. 28, April 1990, by permission of Preston Publications, a Division of Preston Industries, Inc.)

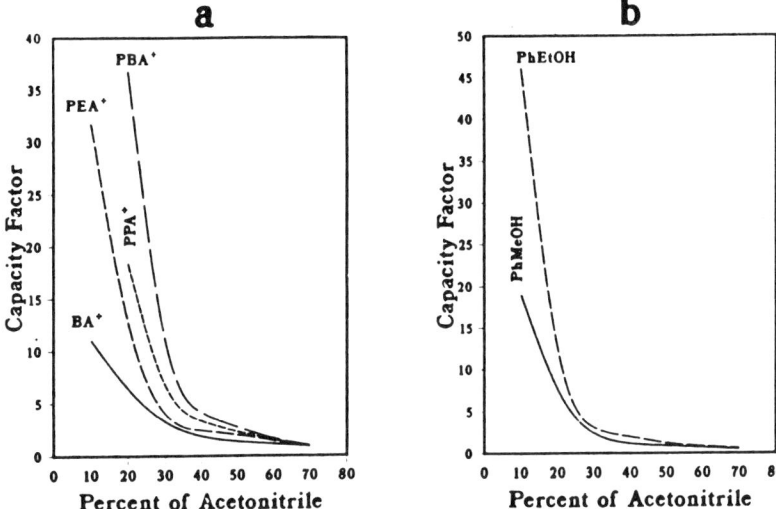

Fig. 8 Relationship between k' of (a) phenylalkylammonium ions and (b) phenylalkanols and the percent acetonitrile in the mobile phase; 20 mM potassium ion in the mobile phase. (From the *Journal of Chromatographic Science,* Vol. 28, April 1990, by permission of Preston Publications, a Division of Preston Industries, Inc.)

same as those for the electrically neutral eluites, the alcohols. If one accepts the contention that the alcohols were retained by reversed-phase interaction, the curve shapes offer additional evidence that reversed-phase interaction occurred with the ions also.

To show the solvophobic effect indicated by the slopes in the reciprocal ion concentration plots in Fig. 6 more clearly, the logarithm of the slopes was plotted against carbon number. Figure 9 shows straight lines were obtained. The slope of a curve in Fig. 6 is C' from Eq. (14). Because C' is related to K_{ie} from Eqs. (11) to (13), it can be shown that the equation for the curves in Fig. 9 is

$$\ln K_{ie} = mn + d \tag{15}$$

where d is a collection of constants, and

$$m = \ln \alpha_{CH_2} \tag{16}$$

The parallelism of the curves in Fig. 9, then, indicated that the $\ln \alpha_{CH_2}$ was independent of the alkali metal ion used as eluant. This independence is not difficult to accept because α_{CH_2} should not be related to coulombic forces between

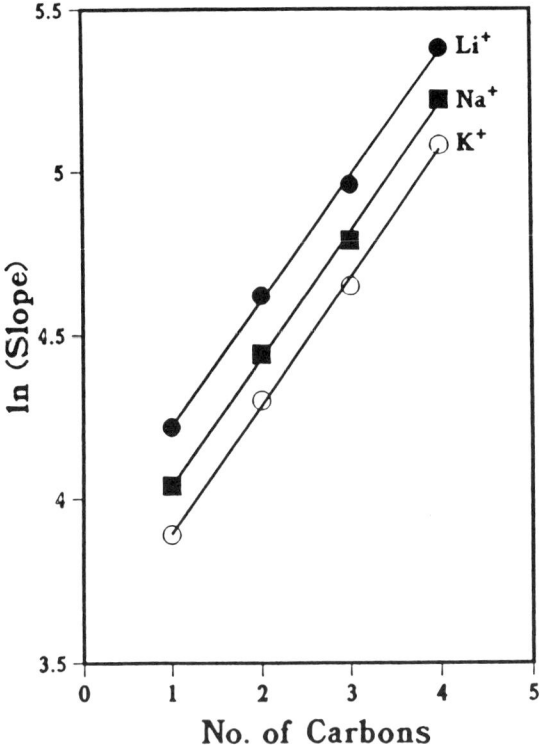

Fig. 9 Relationship between ln of the slopes of the curves of the type shown in Fig. 6 and the carbon number of the alkyl group in the phenylalkylammonium ions. (From the *Journal of Chromatographic Science*, Vol. 28, April 1990, by permission of Preston Publications, a Division of Preston Industries, Inc.)

ions. Methylene selectivity should, however, be dependent on the acetonitrile/water ratio in the mobile phase.

Rahman and Hoffman argued that solvophobic interaction operated in two ways in the retention of organic cation eluites. The ability to retain the electrically neutral phenylalkanol eluites indicated that the stationary phase had electrically neutral sites to accept eluites. These sites also accepted the organic ions. The measure of this retention was the k'_0 for the ion. In addition, retention by the main process, ion exchange, was aided by solvophobicity, as indicated by a slope greater than zero in Fig. 9.

Organic eluant cations should also be solvophobic and interact with the stationary phase at both types of sites. Trialkylammonium ions were chosen as

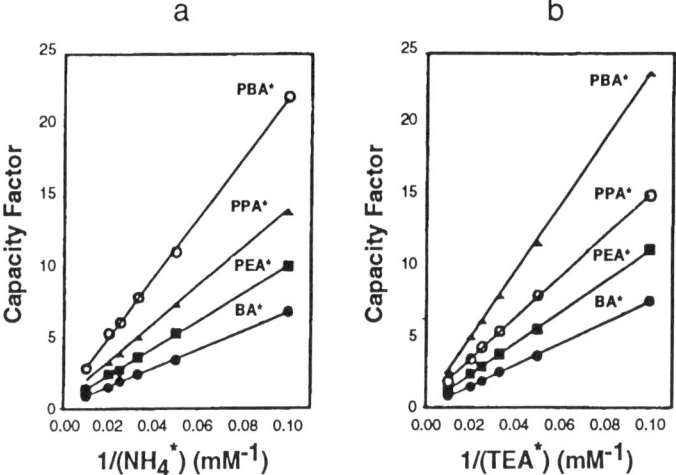

Fig. 10 Relationship between k' of phenylalkylammonium ions and the reciprocal of the concentration of (a) ammonium and (b) triethylammonium ion in the aqueous fraction of the mobile phase. Mobile phase as in Fig. 6. (From the *Journal of Chromatographic Science*, Vol. 28, April 1990, by permission of Preston Publications, a Division of Preston Industries, Inc.)

eluant ions, and their eluting strength was compared to that of the ammonium ion. The results of chromatography with these ions are shown in Figs. 10 and 11. Triethylammonium ion (TEA^+) was a weaker eluant than tripropylammonium ion (TPA^+), which in turn was weaker than tributylammonium ion (TBA^+). Ammonium ion was weakest of the four eluant ions. The results correlate with the solvophobicity of the ions. Again a $\ln C'$ versus carbon number plot gave a straight line. Furthermore, the lines for different eluant ions were parallel. Thus α_{CH_2} was independent of the solvophobicity of the eluant ion.

B. Anion Exchange

Solvophobicity and Retention

The anion-exchange chromatography of a homologous series of phenylalkanoate anions [$C_6H_5(CH_2)_nCO_2^-$] was studied on a styrene–divinylbenzene copolymeric resin containing trimethylbenzylammonium ion groups as nonexchangeable ions by Lee and Hoffman [37]. The mobile phases were solutions of acetonitrile and water containing eluant inorganic and organic anions.

Figure 12 shows the variation of k' with the reciprocal of nitrate ion concentration. The nitrate was in 30% acetonitrile–70% water. The relationship

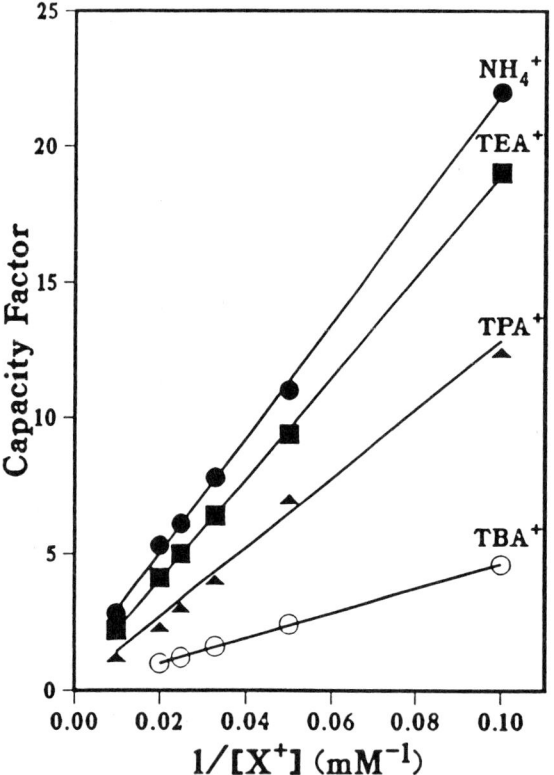

Fig. 11 Relationship between k' of phenylbutylammonium ion and the reciprocal of the concentration of the ammonium and trialkylammonium ions in the aqueous fraction of the mobile phase. Mobile phase as in Fig. 6. (From the *Journal of Chromatographic Science*, Vol. 28, April 1990, by permission of Preston Publications, a Division of Preston Industries, Inc.)

was linear, and linear curves were also obtained with chloride and bromide as eluant ions. The curves are described by the anionic analog of Eq. (14). The k'_0 values, ranging from 0 to 1.3, were small, as was the case with the values obtained from cations. The small values again show that reversed-phase interaction is a minor retention process compared to ion exchange.

When k'_0 is small compared to $C'[X_m^-]^{-1}$ (X_m^- = eluant ion), the following equation can be shown to apply [35]:

$$\log k' = \frac{a}{b} \log C - \frac{a}{b} \log [X_m^-] + \text{constant} \tag{17}$$

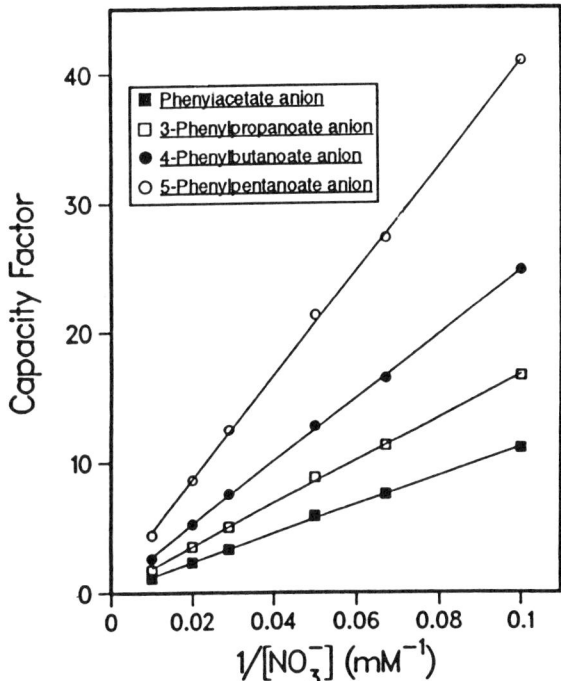

Fig. 12 Relationship between k' of phenylalkanecarboxylate anions and the reciprocal of the nitrate ion concentration in the aqueous fraction of the mobile phase. Mobile phase as in Fig. 6. (From the *Journal of Chromatographic Science*, Vol. 30, March 1992, by permission of Preston Publications, a Division of Preston Industries, Inc.)

where C is the resin capacity, a the charge on the eluite ion, and b is the charge on the eluant ion. For phenylalkanoate ions and the monovalent nitrate, chloride, and bromide eluant ions, the a/b ratio should be 1. From plotting log k' versus log $[X_m^-]$, the straight lines gave the experimental values of 0.96 to 1.04, in excellent agreement with the theoretical value.

Phenylalkanols $[C_6H_5(CH_2)_nOH]$, with the same number of methylene groups as the phenylalkanoates, gave k' versus $[X_m^-]^{-1}$ plots with negligible slope but with substantial capacity factors; for example, 4-phenyl-1-butanol had a k' value of 15. Therefore, although electrically charged, the stationary phase had considerable retentive capacity in the reversed-phase mode.

The slope change with the number of methylene carbons shown in Fig. 12 is depicted in Fig. 13 in a log k' versus carbon number plot. The lines are parallel at the 95% confidence level. This parallelism shows that as was the case with cation exchange, α_{CH_2}, was independent of the kind of inorganic eluant ion used.

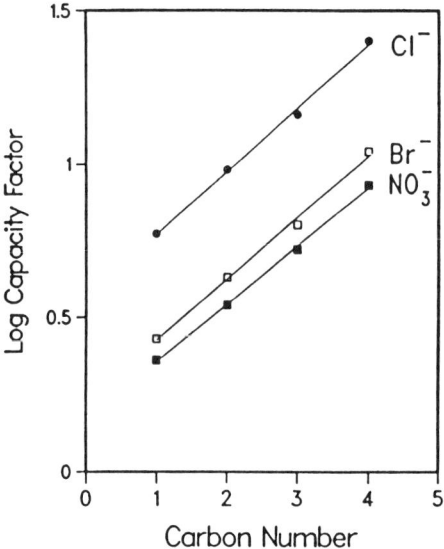

Fig. 13 Relationship between log k' and the carbon number of the alkyl group in phenylalkanecarboxylate anions when different inorganic eluant ions were used. Eluant ions are 50 mM in the aqueous fraction of the mobile phase; mobile phase as in Fig. 6. (From the *Journal of Chromatographic Science*, Vol. 30, March 1992, by permission of Preston Publications, a Division of Preston Industries, Inc.)

Curves showing the relationship between k' and % acetonitrile in the mobile phase had shapes like those in Fig. 8. This shape identity was the case for both phenylalkanoate ions and phenylalkanols, and the shapes were independent of the kind and concentration of the inorganic eluant ion.

To extend the study of the solvophobic interaction indicated in Fig. 13, a homologous series of alkanesulfonate ions, $CH_3(CH_2)_nSO_3^-$, was used as eluant ions. Of particular interest was the effect of the number of methylenes in the sulfonate on the retentions of phenylalkanoate ions.

Plots of k' versus $[X_m^-]^{-1}$ were linear for all the phenylalkanoate ions when the number of methylene carbons in the alkansulfonate was varied from 0 to 6. The order of eluotropic strength of the sulfonate homologs was $C_7H_{15}SO_3^- > C_5H_{11}SO_3^- > C_2H_5SO_3^- > CH_3SO_3^-$. The order is that expected on the basis of solvophobicity of the eluant ions. The a/b ratios of Eq. (17) were 0.87 to 0.99 in reasonably good agreement with the value 1.

Figure 14 shows the alkanesulfonates have the same kind of relationship between k' and acetonitrile content of the aqueous mobile phase as did the

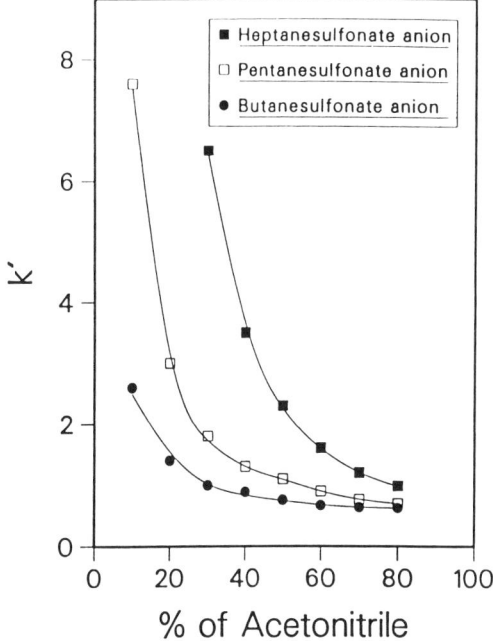

Fig. 14 Relationship between k' of alkanesulfonate anions and the percent acetonitrile in the aqueous mobile phase; 70 mM nitrate ion in the mobile phase. (From Ref. 42.)

solvophobic phenylalkanols, phenylalkanoate ions, and the phenylalkylammonium ions in the presence of a constant concentration of inorganic eluant ion. However, when a constant concentration of organic eluant ion was used in the chromatography of phenylalkanoate homologs, the shapes of k' versus acetonitrile mobile-phase content can change [38]. When heptanesulfonate was the eluant ion, the curves shown in Fig. 15 were obtained. When pentanesulfonate was the eluant ion the curves shown in Fig. 16 were obtained. Butanesulfonate as the eluant ion gave the shapes shown in Fig. 8. Noteworthy in Figs. 15 and 16 are the curves that show a rise in retention with an increase in acetonitrile content.

The drop in eluotropic strength with increasing acetonitrile content of the mobile phase was interpreted by Lee and Hoffman to result from a change in solvophobicity of both the eluant and eluite ion. An increase in acetonitrile content decreased the solvophobicity of the heptanesulfonate ion more than it decreased the solvophobicity of the phenylacetate, 3-phenylpropanoate, and 4-phenylbutanoate ions. Only 5-phenylpentanoate's solvophobicity change was greater than heptanesulfonate's, and its curve was the "normal" reversed-phase

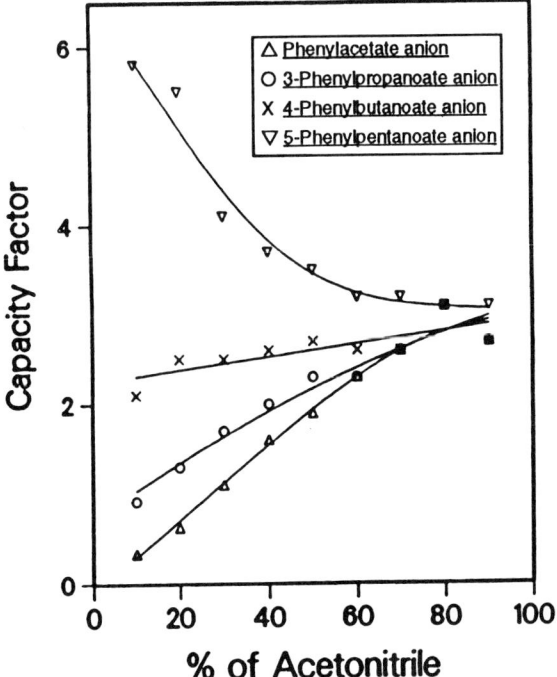

Fig. 15 Relationship between k' of phenylalkanoate anions and the percent acetonitrile in the aqueous mobile phase; 14 mM heptanesulfonate anion in the mobile phase. (From the *Journal of Chromatographic Science*, Vol. 30, October 1992, by permission of Preston Publications, a Division of Preston Industries, Inc.)

type of curve. As the number of methylenes of the alkanesulfonate ion decreased, the drop in solvophobicity of the eluant ion with increasing acetonitrile content was less than that of the phenylalkanoate ion. Only in the case of phenylacetate was the decrease in solvophobicity greater for pentanesulfonate ion than for the eluite ion as the acetonitrile content increased.

The acetonitrile content of the mobile phase, if high enough, can have an effect on the linearity of log k' versus carbon number plots. Figure 17 shows that when the acetonitrile content reached 80%, the alkanesulfonate log k' versus carbon number curve lost its linearity at the low-carbon-number end of the curve. Figure 18 shows this effect more forcefully for the phenylalkanoate ion homologs. Here linearity was lost at the low-carbon end of the curve for 60% acetonitrile and even more so at 70% acetonitrile.

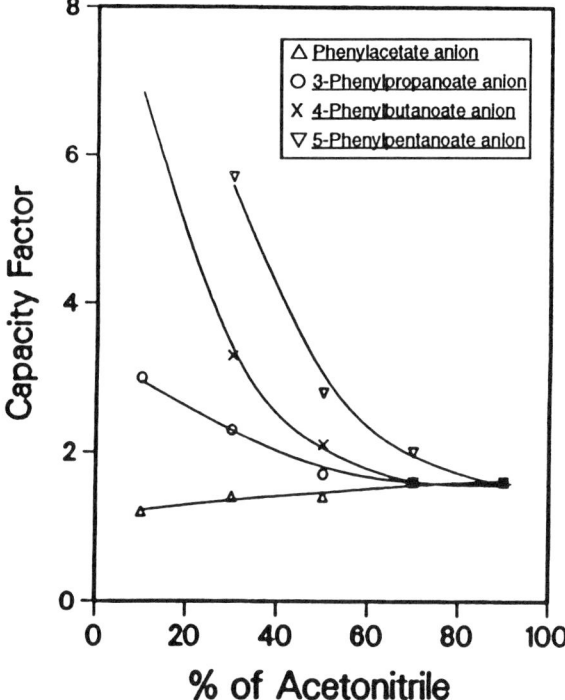

Fig. 16 Relationship between k' of phenylalkanoate anions and the percent acetonitrile in the aqueous mobile phase; 35 mM pentanesulfonate anion in the mobile phase. (From the *Journal of Chromatographic Science,* Vol. 30, October 1992, by permission of Preston Publications, a Division of Preston Industries, Inc.)

The interpretation was that solvophobicity was being lost in going to high acetonitrile levels in the mobile phase. This loss appeared in the least solvophobic ions first, that is, the ions with the least number of methylenes.

Effect of Temperature

When the retention of the homologous series of phenylalkanoate anions was studied as a function of temperature, all k' versus $[NO_{3m}^-]^{-1}$ curves were linear over the range 0 to 55°C [39]. Phenylalkanols using the anion-exchange column gave linear van't Hoff plots. The magnitude of $\Delta H°$ for the alcohols was 2-phenylethanol < benzyl alcohol < 3-phenyl-1-propanol < 4-phenyl-1-butanol. In all cases transfer of the alcohol from the mobile phase to the stationary phase was exothermic. The $\Delta H°$ values were about 1 kcal/mol but varied slightly with the type and concentration of the eluant ion. The variation was 15% at most.

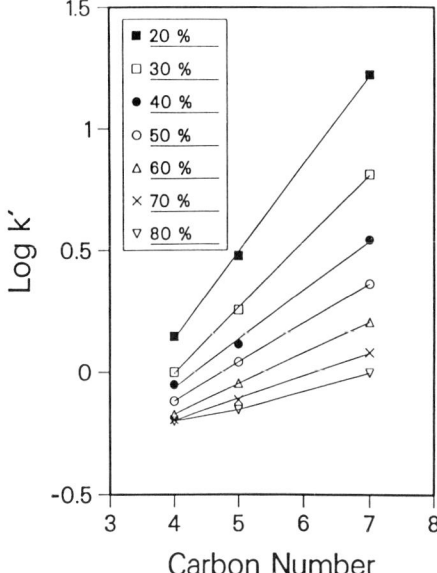

Fig. 17 Relationship between log k' and the carbon number of the alkyl group in alkanesulfonate anions when the acetonitrile content of the aqueous mobile phase was varied; 70 mM nitrate ion in the mobile phase. (From Ref. 42.)

The phenylalkanoate ions, on the other hand, did not give linear van't Hoff curves. Figures 19 to 21 show the types of curves obtained. There appears to be two maxima when the eluotropically strong nitrate ion was used. When the nitrate ion concentration was varied, the curves had almost identical shape to those in Fig. 19, but they were offset on the ln k' axis. With a less strong ion, bromide, the high-temperature maximum shifted to lower temperature. With the least strong ion, chloride, the maximum shifted to even lower temperature.

The lack of linearity cannot be attributed to a stationary-phase transition because the phenylalkanols with the same stationary phase and over the same temperature range gave linear van't Hoff plots. The phenylalkanoate nonlinearity is probably related to a difference in temperature effects on the transfer eluant versus eluite ions from the mobile phase to the stationary phase. Nevertheless, the irregular shapes of these curves must have other yet unexplained factors contributing to them.

Injection Solvent Effects

Using an injection solvent that is eluotropically weaker than the mobile phase in reversed-phase chromatography has been shown by Hoffman and Chang [40] to

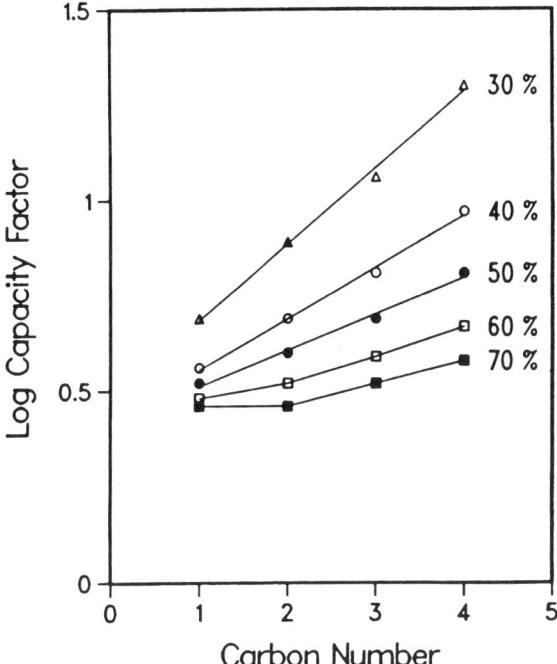

Fig. 18 Relationship between log k' and the carbon number of the alkyl group in phenylalkanoate anions when the acetonitrile content of the aqueous mobile phase was varied; 20 mM bromide ion in the mobile phase. (From the *Journal of Chromatographic Science,* Vol. 30, March 1992, by permission of Preston Publications, a Division of Preston Industries, Inc.)

enhance the peak height of eluites that have capacity factors below about 3. The enhancement is not extremely large, but in cases where a sensitivity increase of 20 to 30% is needed, use of a weak injection solvent may meet this requirement. Another practical aspect of using a weak injection solvent is in developing calibration curves. If the injection solvent strength differs between analyte samples and calibration standards, some error is introduced in the analysis.

A study of the effect of injection solvent strength on peak height in ion exchange was made by Lee and Hoffman [38]. A styrene–divinylbenzene resin anion exchanger was the stationary phase. The mobile phase was a 30% acetonitrile–70% water solution containing different eluant ion types and concentrations. Phenylacetate anion was the eluite.

Typical results are shown in Fig. 22 for the case of injection into a relatively strong mobile phase of 100 mM nitrate ion with an injection solvent of 10 mM

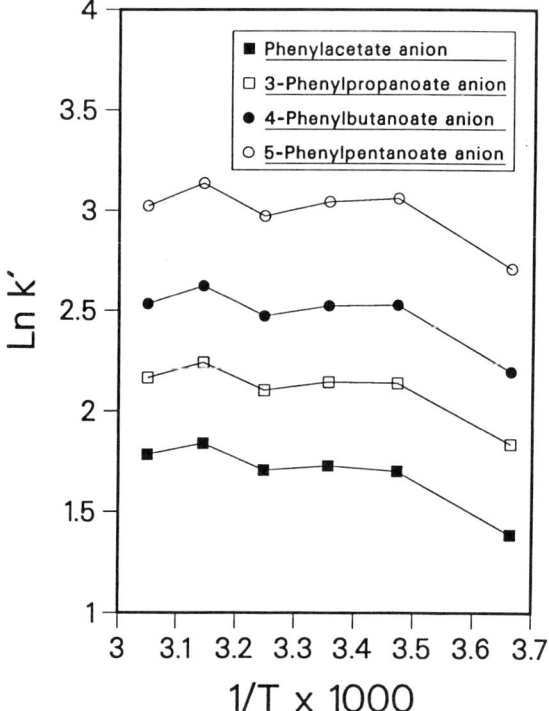

Fig. 19 Van't Hoff plots for phenylalkanoate ions. Mobile phase 30% acetonitrile–70% water; 20 mM nitrate ion in the aqueous fraction of the mobile phase. (From Ref. 39.)

chloride. When injecting the same amount in increasing injection volume, both the mobile phase as injection solvent and the chloride solution as injection solvent gave a decreasing peak height, as expected. But the peak height was not reduced as much with the weak chloride injection solvent. The peak height improvement was most notable when the injection volume was large.

The results above were obtained when the strength of the injection solvent was changed by changing the ion type and concentration. The injection solvent strength can also be changed by varying the acetonitrile content. A typical case found 26% peak height improvement when the acetonitrile content of the injection solvent was reduced to 10% and the injection volume was 200 µL. The mobile phase and injection solvent were 70 mM in nitrate ion.

Hoffman et al. [41] showed in a systematic study that the opposite result, decreased peak height, occurred in reversed-phase chromatography when an injection solvent stronger than the mobile phase was used and k' was less than

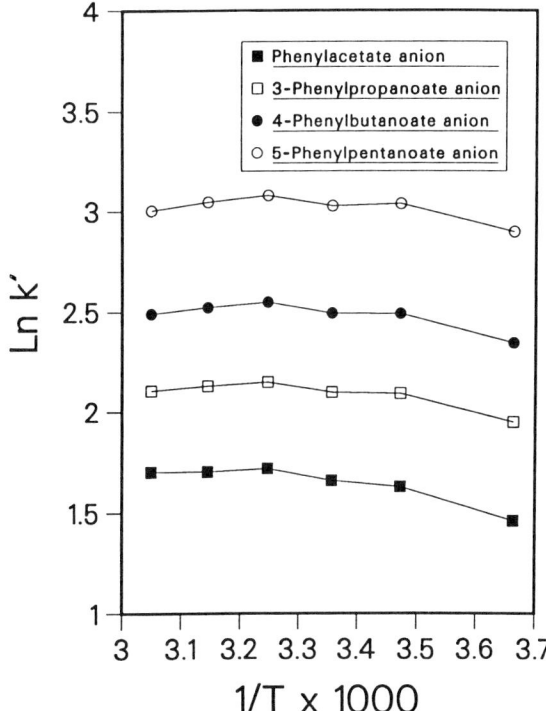

Fig. 20 Van't Hoff plots for phenylalkanoate ions. Mobile phase, 30% acetonitrile–70% water; 30 mM bromide ion in the aqueous fraction of the mobile phase. (From Ref. 39.)

about 3. Lee and Hoffman showed similar results in ion exchange by increasing the acetonitrile content of the injection solvent over the acetonitrile content of the mobile phase. By using 70% acetonitrile in the injection solvent and 30% acetonitrile in the mobile phase, a significant relative reduction in peak height occurred for the injection of a constant amount of phenylacetate ion. The percent reduction was greatest for the largest injection volumes.

Hoffman et al. rationalized the injection solvent effects in terms of eluite focusing. When retention is not great ($k' < 3$), injecting in a weak injection solvent focuses the eluite in a thin band at the column entrance because the stationary phase is so effective in sorbing the eluite from a weak solvent. With a strong injection solvent a broad band is produced because the eluite capacity factor for the injection solvent is very small and the stationary phase sorbs less

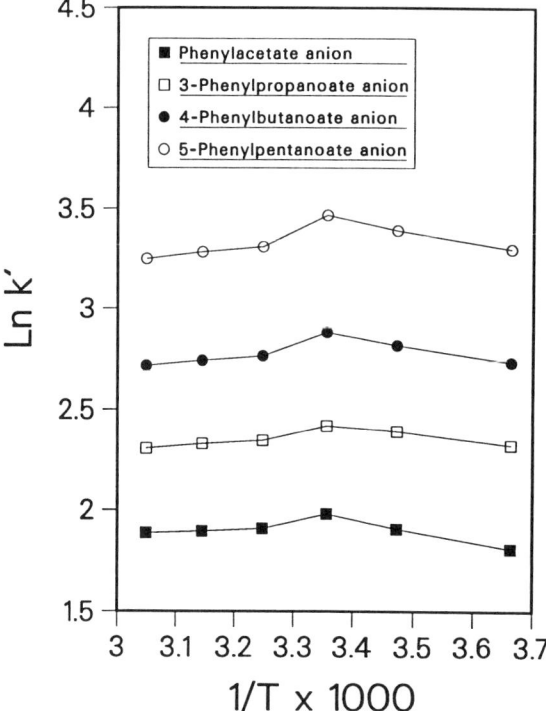

Fig. 21 Van't Hoff plots for phenylalkanoate ions. Mobile phase, 30% acetonitrile–70% water; 50 mM chloride ion in the aqueous fraction of the mobile phase. (From Ref. 39.)

than it would have had it been contacted with a mobile-phase solution of the eluite.

IV. ION-EXCLUSION CHROMATOGRAPHY OF ORGANIC IONS

In the chromatography of a cation eluite with an anion-exchange column, a column in which the nonexchangeable ions are cationic, the eluite should experience a repulsion by the stationary phase. Therefore, retention should be low. The ionic strength of the mobile phase should affect this repulsion and should therefore affect retention.

Lee studied ion exclusion of homologous organic cations to determine the effect of salts in the mobile phase and the relationship between retention and

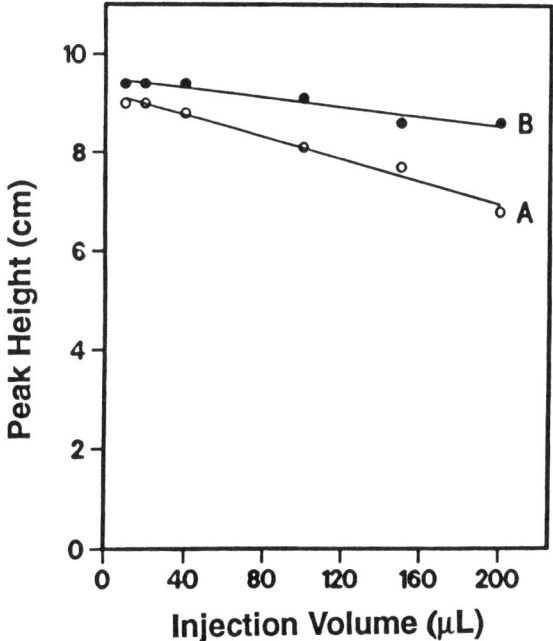

Fig. 22 Relationship between peak height of phenylacetate ion and injection volume. Mobile phase 30% acetonitrile–70% 100 mM nitrate in water. A, Injection solvent was mobile phase; B, injection solvent was 30% acetonitrile–70% 10 mM chloride ion in water. Injection amount 2.0 µg. (From the *Journal of Chromatographic Science*, Vol. 30, October 1992, by permission of Preston Publications, a Division of Preston Industries, Inc.)

carbon number [42]. The eluites were phenylalkylammonium ions: $C_6H_5(CH_2)_n$-N^+H_3, and the stationary phase was a styrene–divinylbenzene resin functionalized with benzyltrimethylammonium ion groups.

When a 10% acetonitrile–90% water mobile phase was used with no ions added except enough hydrochloric acid to give a pH of 3.0, the retention volumes of all the homologs were less than the void volume measured with deionized water. A negative retention volume was attributed to the repulsion between the eluite and the stationary-phase functional group preventing the eluite from occupying all the volume used by deionized water. Increasing the ionic strength of the mobile phase increased the retention volumes of the eluites. Slais [43] similarly observed an increase in k' of aromatic acid anions with increasing ammonium sulfate concentration in the mobile phase.

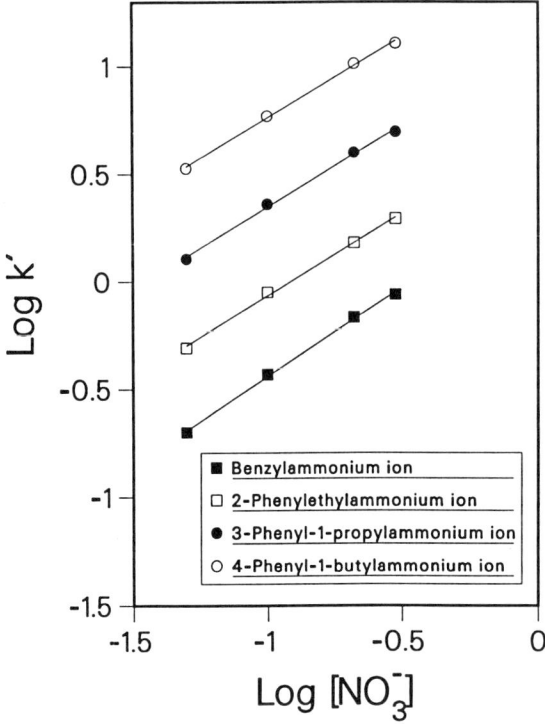

Fig. 23 Relationship between log k' for phenylakylammonium ions and the log of the nitrate ion concentration of the aqueous fraction of the mobile phase. Mobile phase, 10% acetonitrile–90% water; stationary phase, styrene–divinylbenzene polymeric anion exchanger. (From Ref. 42.)

The relationship between retention and the anion concentration in the mobile phase is shown in Fig. 23. The linear curves are described by the equation

$$\log k' = B \log [X_m^-] + D \tag{18}$$

where B and D are constants. Equation (18) can be written as

$$\log \left[\frac{k'}{[X_m^-]^B} \right] = D \tag{19}$$

Then

$$k' = A[X_m^-]^B \tag{20}$$

where A is a constant.

Jandera et al. [44] found an equation similar to Eq. (18), $\ln k' = A + B \ln c$, where A and B are constants and c was the mobile-phase salt concentration. However, they were studying the retention of sulfonic and carboxylic acids on a reversed-phase column with a mobile phase of varying salt concentration.

Figure 24 shows the curves obtained with the eluite 4-phenylbutylammonium ion when using different mobile-phase anions. The capacity factor increased with the eluant ion order $Cl^- < Br^- < NO_3^-$, the same order found in eluant ion eluotropic strength in ion exchange.

The parallel curves in Fig. 23 indicate that B of Eq. (20) was independent of the number of methylene groups in the eluite, or, in other words, the solvophobicity of the eluite. The parallel curves in Fig. 24 show that B was not influenced by the type of anion in the mobile phase. The constant A in Eq. (20), on the other hand, was dependent both on the number of homolog methylenes and the type of anion in the mobile phase.

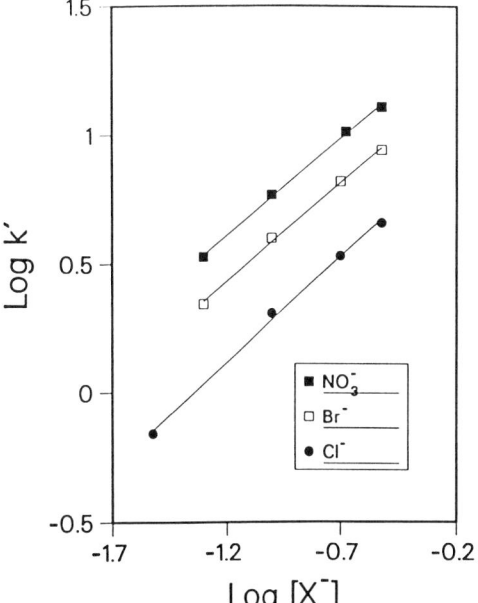

Fig. 24 Relationship between log k' for phenylakylammonium ions and the log of the inorganic anion concentration of the aqueous fraction of the mobile phase. Mobile and stationary phases as in Fig. 23. (From Ref. 42.)

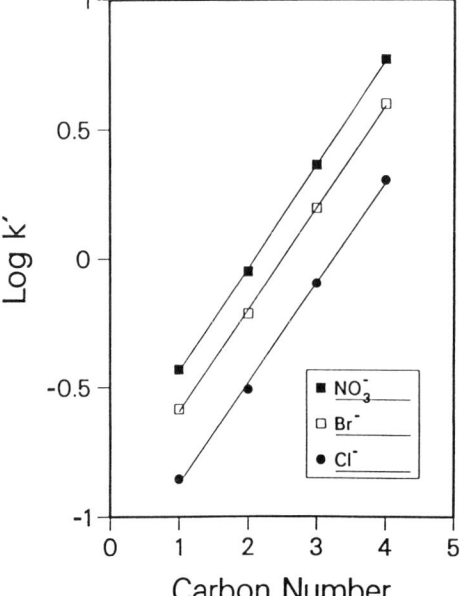

Fig. 25 Relationship between log k' for phenylakylammonium ions and the carbon number of the alkyl groups when various inorganic anions at 0.100 M concentration were in the aqueous fraction of the mobile phase. Mobile phase and stationary phase as in Fig. 23. (From Ref. 42.)

When log k' was plotted against carbon number for different mobile phase anions, the curves shown in Fig. 25 were obtained. The parallel nature of these curves shows that α_{CH_2} was independent of the kind of mobile-phase anion used in ion exclusion. Figure 26 shows that changing the concentration of the ion in ion exclusion produced no change in α_{CH_2}.

The interpretation of these results is that increasing the concentration of the anions in the mobile phase increased their concentration in the layer surrounding the fixed resin charges. The increased concentration reduced the repulsive action of the fixed resin charges toward the eluite ion. But these coulombic forces did not affect the solvophobicity of the eluite ion.

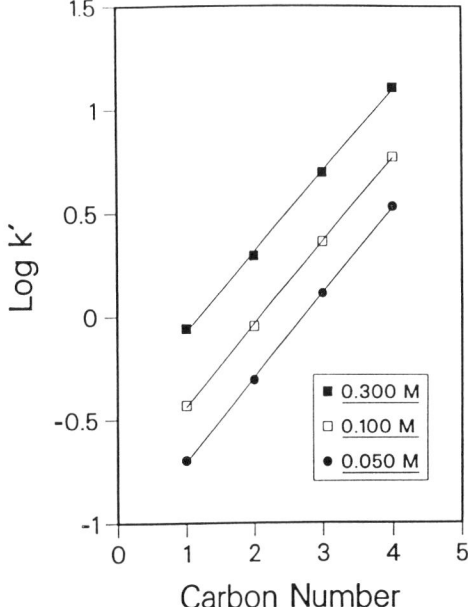

Fig. 26 Relationship between log k' of phenylakylammonium ions and the carbon number of the alkyl group for various nitrate ion concentrations in the mobile phase. Mobile phase and stationary phase as in Fig. 23. (From Ref. 42.)

REFERENCES

1. R. B. Sleight, *J. Chromatogr., 83:* 31 (1973).
2. C. Horvath, W. Melander, and I. Molnar, *J. Chromatogr., 125:* 129 (1976).
3. A. J. P. Martin, *Biochem. Soc. Symp., 3:* 4 (1949).
4. H. Colin, C. Eon, and G. Guiochon, *J. Chromatogr., 119:* 41 (1976).
5. A. Nakae and G. Muto, *J. Chromatogr., 120:* 47 (1976).
6. A. Nakae, K. Kunihiro, and G. Muto, *J. Chromatogr., 134:* 459 (1977).
7. A. Nakae and K. Kunihiro, *J. Chromatogr., 152:* 137 (1978).
8. M. Uchida and T. Tanimura, *J. Chromatogr., 138:* 17 (1977).
9. N. E. Hoffman and J. C. Liao, *Anal. Lett., A11:* 287 (1978).
10. P. Dufek and E. Smolkova, *J. Chromatogr., 257:* 247 (1983).
11. H. Colin, A. M. Krstulovic, M. F. Gonnord, G. Guiochon, Z. Yun, and P. Jandera, *Chromatographia, 17:* 9 (1983).
12. H. Colin, G. Guiochon, Z. Yun, J. C. Diez-Masa, and J. Jandera, *J. Chromatogr. Sci., 21:* 179 (1983).
13. B. P. Johnson, M. G. Khaledi, and J. G. Dorsey, *J. Chromatogr., 384:* 221 (1987).

14. P. Varughese, M. E. Gangoda, and R. K. Gilpin, *J. Chromatogr., 499:* 469 (1990).
15. A. M. Krstulovic, H. Colin, A. Tchapla, and G. Guiochon, *Chromatographia, 17:* 228 (1983).
16. A. Tchapla, H. Colin, and G. Guiochon, *Anal. Chem., 56:* 621 (1984).
17. D. Morel and J. Serpinet, *J. Chromatogr., 248:* 231 (1982).
18. G. Vigh and Z. Varga-Puchony, *J. Chromatogr., 196:* 1 (1980).
19. E. Grushka, H. Colin, and G. Guiochon, *J. Chromatogr., 248:* 325 (1982).
20. D. Morel, J. Serpinet, J. M. Letoffe, and P. Claudy, *Chromatographia, 22:* 103 (1986).
21. L. A. Cole and J. G. Dorsey, *Anal. Chem., 64:* 1317 (1992).
22. M. G. Khaledi, E. Peuler, and J. Ngeh-Ngwainbi, *Anal. Chem., 59:* 2738 (1987).
23. M. F. Borgerding, F. H. Quina, W. L. Hinze, J. Bowermaster, and H. M. McNair, *Anal. Chem., 60:* 2520 (1988).
24. W. L. Hinze and S. G. Weber, *Anal. Chem., 63:* 1808 (1991).
25. S. L. Abidi, *J. Chromatogr., 324:* 209 (1985).
26. B. A. Bidlingmeyer, S. N. Deming, W. P. Price, Jr., B. Sachok, and M. Petrusek, *J. Chromatogr., 186:* 419 (1979).
27. N. E. Hoffman and J. C. Liao, *J. Chromatogr., 28:* 428 (1990).
28. O. A. G. J. Van Der Houwen, R. H. A. Sorel, A. Halshoff, J. Teeuwsen, and A. W. M. Indemans, *J. Chromatogr., 209:* 393 (1981).
29. S. G. Weber and W. G. Tramposch, *Anal. Chem., 55:* 1771 (1983).
30. E. M. Thurman, *J. Chromatogr., 185:* 625 (1979).
31. D. J. Mackey, *J. Chromatogr., 242:* 275 (1982).
32. K. Sugden, G. B. Cox, and C. R. Loscombe, *J. Chromatogr., 149:* 377 (1978).
33. H. J. C. F. Nelis and A. P. DeLeenheer, *J. Chromatogr., 195:* 35 (1980).
34. G. B. Cox and R. W. Stout, *J. Chromatogr., 384:* 315 (1987).
35. D. T. Gjerde and J. S. Fritz, *Ion Chromatography,* Hüthig, Heidelberg, 1987, Chapter 4.
36. A. Rahman and N. E. Hoffman, *J. Chromatogr. Sci., 28:* 157 (1990).
37. H. K. Lee and N. E. Hoffman, *J. Chromatogr. Sci., 30:* 98 (1992).
38. H. K. Lee and N. E. Hoffman, *J. Chromatogr. Sci., 30:* 415 (1992).
39. H. K. Lee and N. E. Hoffman, *J. Chromatogr. Sci.,* in press.
40. N. E. Hoffman and J. H. Y. Chang, *J. Liq. Chromatogr., 14:* 651 (1991).
41. N. E. Hoffman, S. Pan, and A. B. Rustum, *J. Chromatogr., 465:* 189 (1989).
42. H. K. Lee, Doctoral dissertation, Marquette University, 1993.
43. K. Slais, *J. Chromatogr., 469:* 223 (1989).
44. P. Jandera, J. Churacek, and J. Bartosova, *Chromatographia, 13:* 485 (1990).

7
Uncertainty Structure, Information Theory, and Optimization of Quantitative Analysis in Separation Science

Yuzuru Hayashi and Rieko Matsuda *National Institute of Hygienic Sciences, Setagaya, Tokyo, Japan*

I.	INTRODUCTION	348
II.	STOCHASTIC PROPERTIES OF SIGNALS	351
	A. Precision of Independent Peaks	352
	B. Precision of Interfered Peaks	355
	C. Accuracy and Precision	357
III.	UNCERTAINTY STRUCTURE OF QUANTITATIVE ANALYSIS	361
	A. Error Prediction in HPLC and CE	361
	B. Uncertainty Structure in Analytical Apparatus	366
	C. Uncertainty Principle in Separation Science	367
	D. Literature on Precision and Reproducibility	368
IV.	INFORMATION THEORY AND QUANTITATIVE ANALYSIS	371
	A. Precision and Information	371
	B. Throughput of Analysis	373
	C. Resolution and Information	374
V.	FUNDAMENTALS OF OPTIMIZATION	375
	A. Practical Optimization Strategies	375
	B. Nonlinear Programming Problems	376
VI.	APPLICATIONS OF FUMI AND MEI	378
	A. Model Optimization	378

	B.	Optimal Mobile-Phase Composition in Pesticide Analysis	383
	C.	Optimization of Wavelength and Amount of Internal Standard	387
	D.	Merit of Short Columns	390
	E.	Evaluation of Columns for Optical Resolution	391
	F.	Simplex Optimization for HPLC	392
	G.	Optimization of MEKC Analysis	394
VII.		TOTAL CHROMATOGRAPHIC OPTIMIZATION	396
	A.	TOCO with FUMI and MEI for Antipyretics Mixture	396
	B.	Throughput for Dissolution Test Using Robotics	399
	C.	Validity of Internal Standard	402
	D.	Simplified TOCO with FUMI	406
VIII.		FACTORS AFFECTING PRECISION AND THROUGHPUT	407
	A.	Sample Size	407
	B.	Peak-Resolving Power of Data Processing	407
	C.	Area Ratio of Adjacent Peaks	410
	D.	Unresolvable Peaks	414
	E.	Interference of Unnecessary Peaks	414
	F.	System Miniaturization	414
IX.		OUTLOOK	415

I. INTRODUCTION

Uncertainty is omnipresent in data derived from experimental measurements [1]. The origins of indeterminate error can be found in every step of the measurement process. Figure 1 illustrates the measurement steps in high-performance liquid chromatography (HPLC) and capillary electrophoresis (CE), comprising injection, separation, and detection. The output of an analytical apparatus, a chromatogram or electropherogram, cannot provide the concentration or amount of analyte, called the analytical data, until the data are treated mathematically by signal processing. A commercial integrator then provides a peak area or height that is related to the analytical data in a straightforward way. Thus the errors from the overall measurement steps and signal processing sum up to influence the final analytical data.

A mathematical expression of the total error or overall imprecision of the result is the standard deviation (SD) or relative standard deviation (RSD) of the analytical data [1]. The total variance (SD_{tot}^2) of the error propagated during the progressive steps is equal to the sum of all the error variances, assuming that all the steps are probabilistically independent of each other:

$$SD_{tot}^2 = SD_{injection}^2 + SD_{separation}^2 + SD_{detection}^2 + SD_{signal\ processing}^2 \qquad (1)$$

The contribution of each step to the total error (SD_{tot}) will be different from step to step and is, on the whole, referred to as the uncertainty structure of quantitative analysis.

Much effort has been expended in improving the statistical reliability of data at every step. As is well known in HPLC, loop injection replaced microsyringe injection. Stable temperature control and smooth pumping are also necessary to reduce the possible error produced through a column. Electronic filter has come a long way toward improving analytical sensitivity by reducing the random noise in the electronic circuits of the detection step. A digital filter can elicit a certain amount of potential information contained in the output of the analytical apparatus. This discipline of analytical chemistry is called chemometrics [2–4].

The aims of this chapter are to elucidate the uncertainty structure of quantitative analysis and to describe application of this concept to system optimization in separation science. First, a method for predicting the total error of measurement and signal processing, expressed in Eq. (1), is reviewed. To the author's knowl-

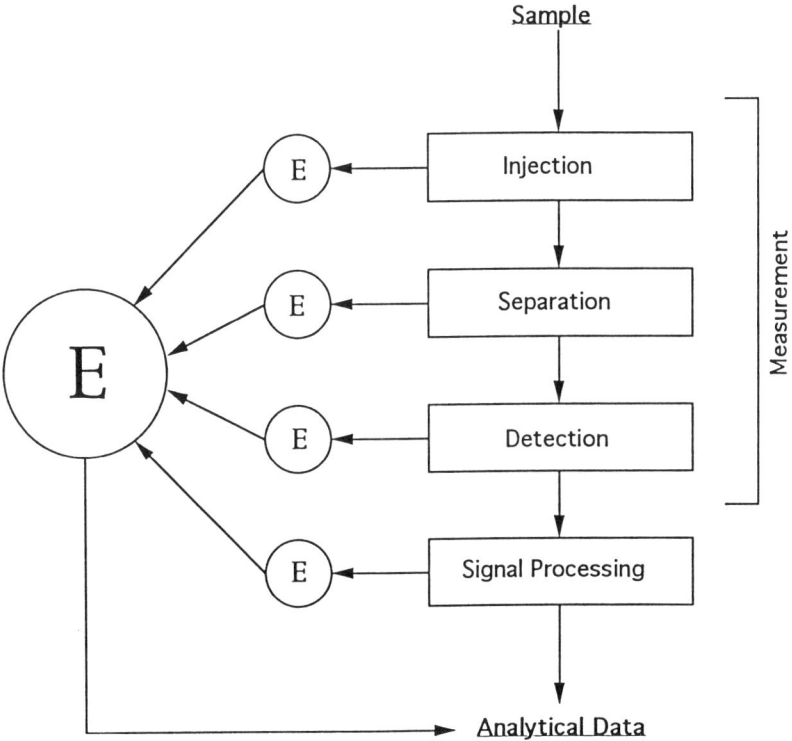

Fig. 1 Errors (E) from measurement and signal processing steps in HPLC and CE.

edge, the experimental RSD of analytical data was believed to be difficult or impossible to predict theoretically, because different analytes usually have different peak areas (or amounts), widths, and retention times, and the RSD varies depending on these peak parameters. In Section III we demonstrate that this RSD prediction is fairly simple and exact over a wide range of analyte amounts. The stochastic properties of the simplest model in which white noise is the only error source are fundamental to the error prediction (Section II).

The most fruitful application of error prediction will be in optimization of separation science. In HPLC, the optimal operating conditions are searched for by selecting a stationary phase and by manipulating chromatographic variables such as mobile-phase composition and detection wavelength over a wide range of the variables [5–16]. The reasonable and natural definition of "optimum" is the set of the operating conditions that provides the highest precision (or the lowest RSD values) among all the examined conditions [5–8]. Error prediction is quite useful for finding this statistical optimum with the minimal expenditure of time and effort. Without the theoretical prediction of RSD, a large number of experiments must be repeated at each set of candidate conditions to obtain statistically valid results.

Information theory underlies the above-mentioned statistical optimization (Section IV). In this light, the precision is represented as mutual information (Φ_f or Φ_m) and analytical throughput as the transmission speed of the information (ϑ_f or ϑ_m). The highest throughput means the most rapid analysis with sufficient precision. The two optima, the most precise analysis and the highest throughput analysis, which are characterized by the maxima of the mutual information and its transmission speed, respectively, can be separately located in the optimization. Examples of the optimization in HPLC and CE are given in Sections VI and VII. Some important factors affecting the precision and throughput are considered in Section VIII. The terminology used in this chapter is listed in Table 1.

A great many papers concerning chromatographic optimization have been published [5–31]. Most optimization techniques put forward so far deal with what is known as optimization of "separation." The aim of this methodology is to acquire good separation of every peak pair within a tolerable analysis time [14,16,19–21]. A new shape-up veers toward desirability functions applied by Deming [29,30] and by Bourguignon and Massart [31]. In this chapter we introduce readers to the literature that refers not only to separation and time but also the statistical aspects of optimization in separation science. In addition, it is shown that the mutual information (or precision) can be related mathematically to the well-known resolution function Rs in special cases (Section IV.C).

Information theory and its application to analytical chemistry, as proposed by Eckschlager and Štěpánek, has been studied extensively [32–45]. Their information for quantitative analysis [e.g., Eq. (7) of Ref. 32] bears some similarity to the mutual information treated in this chapter: both involve the precision of a method.

Table 1 Terminology

FUMI	Function of mutual information [Eqs. (5) and (6)]
FUMI theory	Theory concerning information theory of optimization and error prediction on the basis of FUMI and MEI
MEI	Measurement-elicited information [Eqs. (7) and (8)]
TOCO	Total chromatographic optimization
Precision	1/RSD or information FUMI or MEI [= log(1/RSD)]
Throughput	Information/time
$\phi_f(j)$	FUMI for peak j [Eq. (5)]
Φ_f	FUMI for multipeaks [Eq. (6)]
$\phi_m(j)$	MEI for peak j [Eq. (7)]
Φ_m	MEI for multipeaks [Eq. (8)]
ϑ_f	Throughput based on FUMI [Eq. (9a)]
ϑ_m	Throughput based on MEI [Eq. (9b)]

The most prominent differences are: (1) the mutual information treated here is based on the theoretical prediction of RSD from the peak shape (area and width), overlap, and noise level in the output of a measurement system; (2) it is the information provided by the Kalman filter that lays the foundation of the information theory of optimization reviewed in this chapter [5].

In the mid-1970s, the mutual information of the Kalman filter was derived and its applicability to optimization problems was formulated in the engineering area [46,47]. Thijssen et al. introduced the information of the filter into analytical chemistry [48–52].

II. STOCHASTIC PROPERTIES OF SIGNALS

The influences of peak overlap, shape, and noise on the precision of analytical data are considered with recourse to computer simulations. The well-known properties of signals that affect the precision of data are [8]:

1. As the peak area increases, the precision increases (i.e., RSD decreases).
2. As the peak broadens, the precision decreases.
3. As the noise increases, the precision decreases.
4. As the peak overlaps with another, the precision decreases.

Properties 1 to 3 are related to the signal/noise ratio (S/N), which is indigenous to an individual peak, but property 4 refers to the relationship between neighboring peaks. Therefore, these conditions should be examined differently. Accuracy is also discussed.

A. Precision of Independent Peaks

Figure 2 illustrates the procedures used in the computer simulations cited below. Each data set of experiments 1, 2, . . . , n contains the same Gaussian peak signals over which white noise is superimposed. The white noise is a random noise with a zero mean and normal distribution, and can easily be constructed from uniform random numbers according to the central limit theorem [53]. The data sets differ from each other only in noise appearance, while the areas, widths (standard deviation), and retention times (mean) of the Gaussian peaks and standard deviation (SD) of the white noise are kept constant. Signal processing provides an estimate, $\hat{X}(i)$, of the analytical quantity, such as concentration in each data set, i, and the estimates, $\hat{X}(1), \ldots, \hat{X}(n)$, will therefore vary from each other due to the random noise. Thus the mean and variance should result from a sufficiently large number of data sets, n. The statistics of Fig. 2 are called ensemble mean and variance (not time average or variance). This simple measurement model describes the ideal situation in which the chromatograms or electropherograms obtained from the same samples under the same operating conditions are the same except for the random pattern of the white noise.

Figure 3 shows the effect of the area of the Gaussian peak on the results of signal processing, while the other peak parameters and SD of the white noise remain invariant. Linear least-squares curve fitting is used as a data processing method to obtain the area estimates, $\hat{X}(1), \ldots, \hat{X}(n)$, as mentioned above ($n = 100$). The peak model (width and position of a Gaussian peak) is assumed to be available and the peak area is the quantity to be estimated. The S/N ratio is defined as the height of the peak maximum divided by the SD of the white noise.

The RSD of the estimates decreases with increasing peak area. In Fig. 3, S/N is directly proportional to the peak area. Peaks with an S/N of 10 are illustrated in Fig. 2 and the RSD obtained from the measurements of these peaks is about 5%. Improving the S/N is effective, especially with respect to noisy data ($S/N < 10^2$), but does not provide much improvement for smooth peaks ($S/N > 10^3$) (see Fig. 3).

The effect of the width of the Gaussian peak on the area estimates of the data processing is shown in Fig. 4. The RSD increases with increasing peak width. Since the S/N increases with decreasing peak width in this example, we can conclude again that the precision of the estimates increases with increasing S/N. We can easily imagine that the RSD will be infinitesimal for signals such as the Dirac delta function and infinitely large for an infinitely broad peak. The delta function assumes an infinite signal intensity exclusively at a data point.

The adverse effect of peak broadening on the precision is very easy to grasp [8]. Assume two rectangular peaks with different widths but the same areas. One is sharp and spreads over five data points. The other is broad and lies over 20 data points. The sum of the raw data over the peak areas contains 5 and 20 white

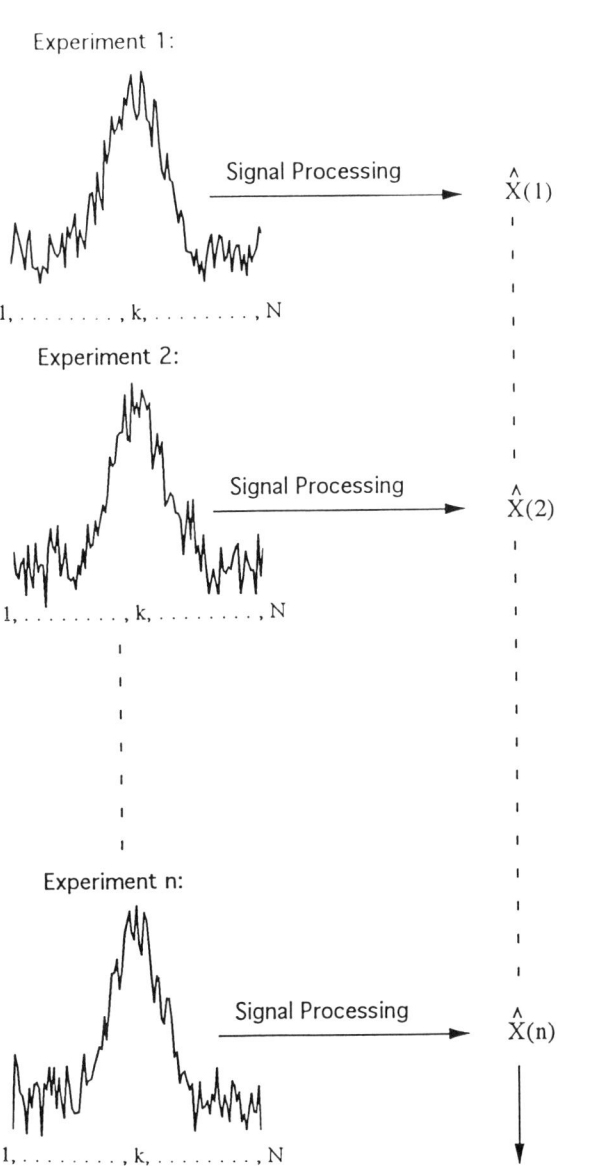

Fig. 2 Ensemble mean and variance. Area of the Gaussian peak is 1000; width is 10; the position is 50; the SD of the white noise is 4; S/N is 10; the number of data points N is 100.

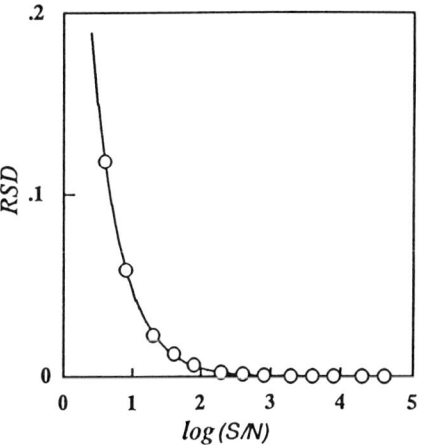

Fig. 3 Effect of peak area on signal processing. S/N is defined as the signal intensity at the peak maximum divided by the SD of the white noise. $n = 100$ for each peak area (○); the solid line denotes the theoretical prediction based on Eq. (2). The SD ($= \tilde{W}$) of the white noise is fixed at 2. The simulation was conducted as in Fig. 2.

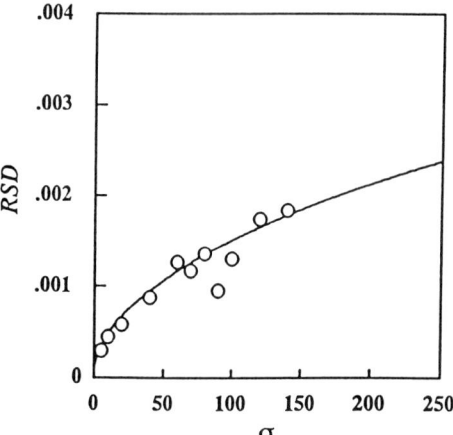

Fig. 4 Effect of peak width on signal processing. σ denotes the width (SD) of the Gaussian peak. $n = 100$ for each width (○); the solid line denotes the theoretical prediction based on Eq. (2). The SD ($= \tilde{W}$) of the white noise is fixed at 2. The simulation was conducted as in Fig. 2.

noises, respectively, as well as the same signal intensities. Recall that the repeatability is verified in the same way as for Fig. 2. The averages of the data summation for both peaks are equal to the true value because the average of the white noise is zero. However, since the variances of the noises are positive and finite, the variance of the area estimates for the sharp peak contains 5 variances of the white noise and that for the broad peak, 20 variances. Thus the RSD is lower for the sharp peak with the high S/N. Note that S/N exerts an influence on the precision, not on the accuracy.

B. Precision of Interfered Peaks

The RSD of the area estimates is plotted against the resolution Rs for two Gaussian peaks (see Fig. 5). The first eluted peak is fixed at a position and the second eluted peak is moved with its width changed according to the fundamental equation of chromatography: $N = (t_j/\sigma_j)^2$, where N denotes the number of theoretical plates (constant), σ_j the width of Gaussian peak j, and t_j its position.

As the resolution decreases from Rs = 1.0, the RSD of the area estimates for the moving peak increases abruptly because of the mathematical difficulty of the data processing in resolving the strongly overlapped peaks. The reduction in S/N arising from peak broadening as Rs increases from 1.0 is the direct cause of the gradual increase in the RSD. Excessive separation is not recommended for

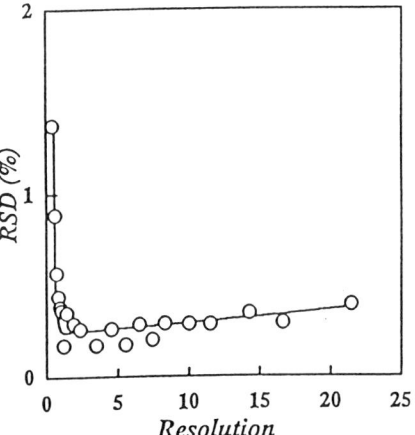

Fig. 5 Signal processing error as a function of resolution (Rs). The RSD is for the area estimates obtained by the least-squares curve fitting of the moving peak. ○, 100 simulations are repeated each separation; —, theoretical curve of Eq. (2). $\tilde{W}$ is kept constant in all the simulations and S/N is ca. 400 at the peak maximum. The plate number N is fixed at 10000. (From Ref. 82.)

chromatographic analysis. We should note that high precision (RSD ≤ 0.26% in Fig. 5) always leads to good separation (ca. 5 ≥ Rs ≥ ca. 1.0), but good separation does not necessarily mean high precision. For example, the RSD at Rs = 23 is even higher (≈ 0.4%) than the lowest RSD at Rs = 1.0 (≈ 0.26%). This difference is enhanced under actual observation (see Section III).

Without noise, the estimates would be emancipated from the error and the same results could always be obtained from repeated experiments such as Fig. 2 (RSD = 0). In the examples above, however, the white noise induces an appreciable uncertainty in the data. For convenience, the RSD of the measurements obtained from the simple model shown in Fig. 2 is called signal processing error and is symbolized as $SD_{\text{signal processing}}/A_j$.

The RSD values observed from the simulations above (Figs. 3 to 5) can be predicted theoretically. The square of the RSD ($= SD_{\text{signal processing}}/A_j$) of the area estimates obtained from Gaussian peak j with width σ_j and area A_j is described as [5,54–57]

$$\left(\frac{SD_{\text{signal processing}}}{A_j}\right)^2 = \frac{2\pi^{1/2}\sigma_j \Delta T \tilde{W}^2}{A_j^2} 10^{2\delta\phi_f(j)} \qquad (2)$$

where $\tilde{W}$ denotes the standard deviation of the white noise in the output, ΔT the sampling interval of an analog-to-digital (A/D) converter, and $\delta\phi_f(j)$ the information loss caused by the overlap. The loss $\delta\phi_f(j)$ is a nonnegative function of peak overlap (positions of peak j itself and adjacent peaks $j - 1$ and $j + 1$, their peak areas, etc.). A detailed explanation of the information loss is given in Section VIII.B and a computer program for the loss and FUMI is available directly from the authors.

Without peak overlap, the information loss, $\delta\phi_f(j)$, takes the lowest value (zero) and the error variance of the estimates [Eq. (2)] relies only on the peak shape (area and width) and noise variance, $\tilde{W}^2$. The solid lines of Figs. 3 to 5 (Rs > 1.0 for Fig. 5) are calculated by Eq. (2) [$\delta\phi_f(j) = 0$]. The error prediction is excellent in the simple models. If two peaks are interfering with each other in a data set, the loss, $\delta\phi_f(j)$, begins to increase from zero and spoils the precision with increasing overlap (here, Rs < 1.0). The Eq. (2)-based prediction of RSD is also exact for the overlapped peaks (see the solid line in Fig. 5).

Equation (2) is derived from the theory of another least squares, the Kalman filter [54–57], and is known to predict the RSD of the filter estimates in the same situations exactly as Figs. 3 to 5 [58–60]. The common least squares used for Figs. 3 to 5 and the Kalman filter have almost an equal capacity for mathematical signal resolution. Analytical applications of the Kalman filter have been described in detail [61–64].

At this point it is necessary to distinguish between estimation and prediction. According to a book of signal processing [65], the estimation (also called filtering) means to assess the present state from the previous and current data, whereas prediction is a procedure for assessing the future state from the previous and current data. The values $\hat{X}(1), \ldots, \hat{X}(n)$ for the analyte concentration shown in Fig. 2 result from estimation, because data processing provides the results just after the updated (and previous) information has been input.

On the other hand, Eq. (2) can be used for prediction. Usually, statistical values such as the mean and variance cannot be calculated until a sufficient number of experiments are carried out. However, no experiments are needed for determining the variance for the area estimates by Eq. (2) if all the parameters for Eq. (2) are known. In practice, if the peak shape and noise level can be estimated from experiment 1 of Fig. 2, we can predict the statistic from the single experiment.

The information loss, $\delta\phi_f(j)$, in Eq. (2) involves a constant, $\hat{R}s$. It specifies the lowest acceptable resolution below which the data processing employed cannot provide exact data. In Fig. 5, the lowest resolution, $\hat{R}s$, is set at 1.03. In general, the lower limit of Rs often coincides with the observed RSD minimum in some situations and is called the optimal resolution. $\hat{R}s$ depends on the peak-resolving power of signal processing (see Section VIII.B). The determination of this constant requires tiresome repetitions of experiments on various separations, and $\hat{R}s$ is usually set at 1.03 for the least-squares or Kalman filter and at 1.53 for the perpendicular dropping. As stated by Kowalski, all the chemometrical tools (including the optimal resolution, $\hat{R}s$) should be used to achieve a proper balance between chromatographic resolution and mathematical resolution [66].

Posener proposed a theory to predict the precision of the amplitude (height at the peak maximum) and position (center) of Gaussian and Lorentzian peaks in the same situations as in Fig. 2 [67]. The least-squares estimation was used to verify the theory with computer simulation. This theory has been applied to some practical problems [68–70]. Although Posener's equation for the error prediction apparently differs from Eq. (2) and lacks the effects of signal overlap, the reliability of the prediction seems to be commensurate with that of Eq. (2). It has yet to be investigated whether Posener's equation is equivalent to Eq. (2) under some assumptions.

C. Accuracy and Precision

There have been several publications concerning the evaluation of accuracy in chromatographic situations where the peaks overlap [71–75]. This subsection demonstrates that the precision and accuracy (bias) are themselves not entirely independent [58–60]. Overlapped chromatograms result from successive injections of a series of samples at short regular intervals in HPLC [76–80]. This successive-injection method can spare much analysis time. The degree of peak overlap is easily adjustable by changing the injection intervals. Figure 6 illustrates an example of overlapped chromatograms derived from five injections [77]. Each of the constituent chromatograms contains four peaks (a–d). Peak c from the first

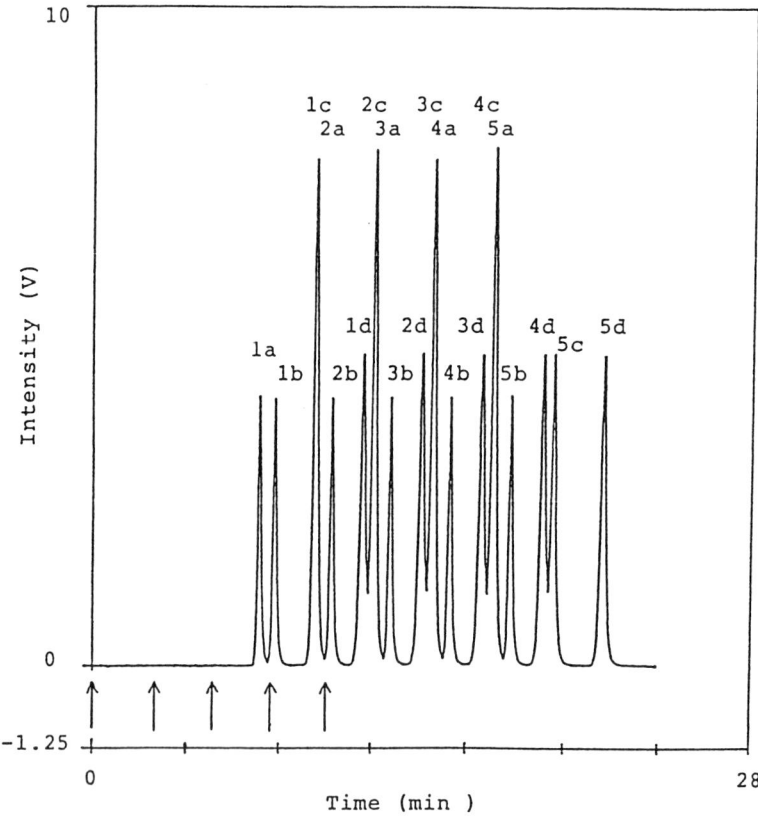

Fig. 6 Overlapped chromatograms derived from the successive-injection method. Peaks: a, Phenetol; b, biphenyl; c, pyrene; d, perylene. Injection intervals are 150 s and the injection points (1–5) are indicated by the arrows. (From Ref. 77.)

injection strongly overlaps peak a from the second injection (1c–2a), but peaks b and d remain separated.

The statistical results for the peaks are listed in Table 2. The Kalman filter was used for signal processing, and the strongly overlapped peaks (1c–2a, etc.) could be quantified more exactly than by the well-known perpendicular dropping. For the separated peaks b and d, the bias and RSD are both less than 0.2%. These statistics are much larger for the overlapped peaks a and c: bias is ca. 10% and RSD is ca. 6%. Precision and accuracy display similar trends toward peak overlap.

Figure 7 shows the influences of peak overlap and noise level on the accuracy (A) and precision (B) of the estimates obtained by the adaptive Kalman filter [58].

Table 2 Observed Precision and Accuracy of Overlapped Chromatograms of Fig. 6

Peak[a]	Mean[b]	RSD (%)
a	112.01	6.23
b	100.05	0.17
c	91.54	6.32
d	100.15	0.12

Source: Ref. 77.
[a]For peaks a to d, see the legend of Fig. 6.
[b]For bias, the true value is assumed to be 100.

The adaptive Kalman filter [64] is a sophisticated version of the regular Kalman filter used in Table 2, but in this context, it may be considered equivalent to the common least squares. The situations of these computer simulations are similar to those in Fig. 5. The only difference is that the fixed peak is unknown (i.e., no model is available) and the moving peak is known (i.e., a model is available such as in Figs. 3 to 5).

If the moving target peak is completely separated from the other disturbing peak fixed at 600 s, the average area estimate for the target peak is very close to the true value ($=1$). As the peak position approaches that of the fixed peak from 625 s (Rs = ca. 1), the average of the estimates begins to exceed the true value because of the intrusion of the other signals. If the two peaks coelute (Rs = 0), the average estimate of the target peak is equal to the sum of the estimates of the two peaks ($=2$). The precision also decreases with decreasing Rs, as in Fig. 5. The observed RSD has a maximum at 602 s (RSD = 1.23%; Rs = 0.083). The lower RSD values near Rs = 0 at 600 s are due to the two fused peaks, which appear as one peak with the same shape but twice the area. No distinction is made between the target and unknown peaks in this situation. The effect of the peak broadening is difficult to recognize from Fig. 7B.

The increase in the white noise affects the precision over the entire range of the possible peak positions and the RSD curve shifts upward (see Fig. 7B). On the other hand, the noise effect on the accuracy is restricted within the region of strong overlap. The precision possesses all the signal properties 1 to 4, but the peak overlap is the only influential factor with respect to accuracy. The precision of both the moving known peak (shown here) and fixed unknown peak (not shown here) is also predictable based on Eq. (2) [60].

Here, we discuss whether precision alone can be an appropriate criterion for system optimization rather than both accuracy and precision. There are four reasons for this proposition. First, error is classified into two groups: determinate (systematic) and indeterminate (random) [1]. Systematic error can be ascribed to

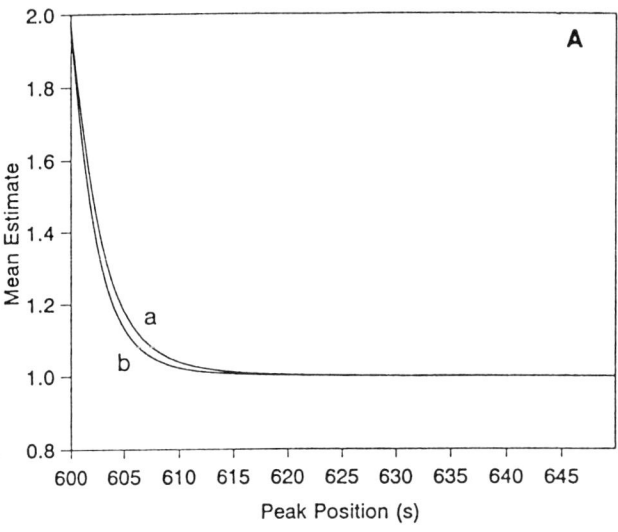

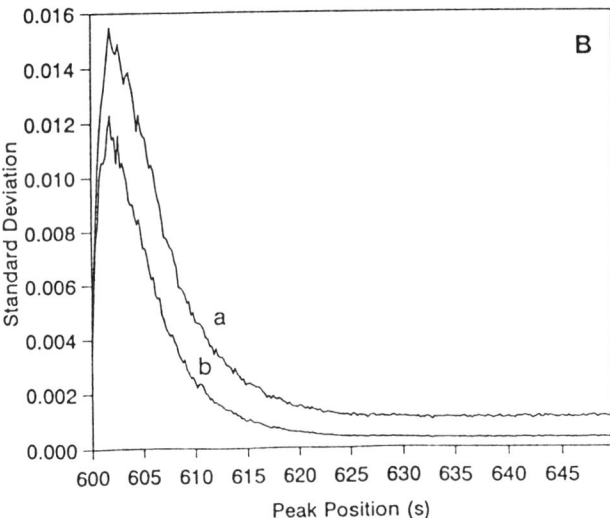

Fig. 7 Influences of peak overlap and noise level on (A) accuracy and (B) precision of the adaptive Kalman filter (simulation). $\tilde{W} = 10$ (a) or 1 (b); $n = 500$ for each separation (sampling intervals, $\Delta T = 0.5$ s). The unknown peak is fixed at 600 s; the abscissa indicates the position of the target peak to which the statistics are calculated. The true values of these peaks are both 1. The plate number is 10000. (From Ref. 58.)

definite causes, such as error due to the ineptitude of an experimenter in weighing a hygroscopic substance [1]. This error can be traced down and eliminated, or at least substantially reduced, by suitable quantitative techniques. On the other hand, indeterminate error cannot be eliminated or corrected, which places an ultimate limitation on the measurement [1]. Thus our interest is in indeterminate error. It is not the precision but the accuracy that is mainly affected by avoidable systematic error.

The second reason concerns the correlation between accuracy and precision, as mentioned above. For example, results with high precision are always very accurate. However, improved accuracy does not always result in improved precision (e.g., the reduced reliability of the analysis coming from peak broadening at late elution times is not represented by the accuracy but by the precision). It seems unlikely that results with even acceptable precision can be very unreliable, owing to poor accuracy in the measurement steps.

Third, the theoretical prediction of the RSD for analysis [Eq. (2)] is based on Kalman filter theory. Unfortunately, the accuracy is not accounted for explicitly by the theory, and in practice, accuracy is difficult to determine experimentally [81]. Finally, the precision (= 1/RSD) can, theoretically, be related to the mutual information function and thus is an important criterion for optimization in separation science.

III. UNCERTAINTY STRUCTURE OF QUANTITATIVE ANALYSIS

The data processing error [Eq. (2)] underestimates the actual error in HPLC and CE in most real cases, although it does provide exact prediction for the simple situations described in Figs. 2 to 5. For example, under some circumstances, the observed error in HPLC (RSD = 0.12 to 0.25%) was 10 times as high as the signal processing error [RSD = $SD_{\text{signal processing}}/A_j$ = ca. 0.02%; Eq. (2)] [5,6,8]. Therefore, it is necessary to take into account the other error sources in HPLC and CE (e.g., injection, temperature control, etc.) to reach an exact prediction of the total measurement error.

A. Error Prediction in HPLC and CE

Figure 8 shows the chromatograms of naphthalene (peak 1), acenaphthene (2), pyrene (3), and perylene (4) at 500-fold different amounts [82] and Fig. 9 the observed RSD for peak j (RSD_j) as a function of the amount of analyte injected [82]. Six to nine replicate samples of the same mixture were injected at each amount. From Fig. 9 we can see that the RSD of measurements depends greatly on the sample size.

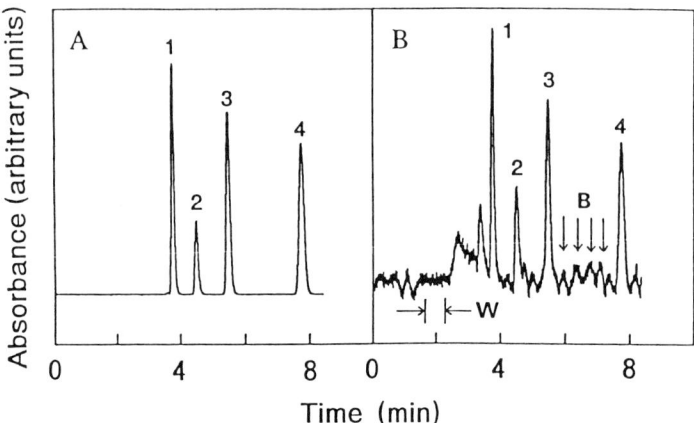

Fig. 8 Chromatograms at different sample amounts. (A) Peaks: 1, naphthalene (0.05 mg/mL); 2, acenaphthene (0.05 mg/mL); 3, pyrene (0.02 mg/mL); 4, perylene (0.03 mg/mL). (B) 500-fold diluted sample of (A). All the reagents were of analytical grade or equivalent and used without further purification. The samples of various concentrations were determined on a Shimadzu liquid chromatograph equipped with an LC-6A pump, SPD-6A variable-wavelength detector (tuned at 254 nm), and type 7125 loop injector (20 μL; Rheodyne). The temperature of an Inertsil ODS column (4.6 × 250 mm; Gasukuro Kogyo) was maintained at 35°C by Shimadzu CTO-6A column oven. The mobile phase was HPLC-grade methanol and the flow rate was 1 mL/min. The data acquisition and computer system have been described elsewhere [169].

The experimental RSD_j, although including many types of indeterminate errors, can be approximated by the following simple equation [82]:

$$RSD_j^2 = \left(\frac{s_{\text{signal processing}}}{A_j}\right)^2 + \left(\frac{s_{\text{measurement}}}{A_j}\right)^2 \quad (3a)$$

$$= \frac{2\pi^{1/2}\sigma_j \Delta T \tilde{W}^2}{A_j^2} 10^{2\delta\phi_f(j)} + I^2 + \frac{\sigma_j^3 B^2}{A_j^2} \quad (3b)$$

where the first term is equal to Eq. (2) and I, B, and $\tilde{W}$ are observable constants. The white noise is attributable to the dark current of the detection system. Constant I denotes the RSD, which is independent of the signals (e.g., injection volume error).

The third term represents the relative variance for the random noise observed between the peaks in Fig. 8B, which has a wavelength as wide as the widths of the target peaks (indicated by arrow B). The numerator ($\sigma_j^3 B^2$) of the third term is an approximation of the variance of the area under the path of Brownian motion with B as the standard deviation [82]. Thus the third term is referred to here as

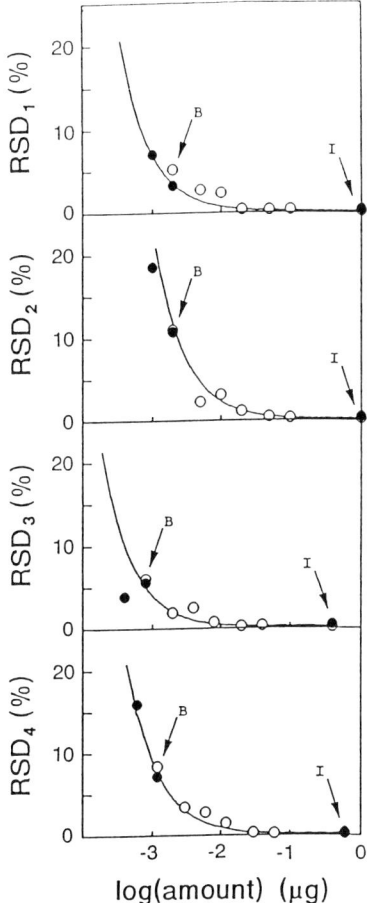

Fig. 9 Dependence of RSD_j on among of injected analyte. j denotes the peak number of Fig. 8. ○, Observed RSD_j (n = 6 to 9); —, theoretical curve by Eq. (3b) ($\delta\phi_j$ = 0). Seven analyte amounts were used (original sample and 10-, 20-, 50-, 100-, 200-, and 500-fold dilutions). The measurements were obtained by the least squares of the chromatograms. ●, RSD_j (n = 8) observed 80 days after the first experiments (○) and plotted with the same constants I, B, and $\tilde{W}$. Original sample and 500- and 1000-fold dilutions were used. Some of the right and middle closed circles overlap with the open circles of the same analyte amount. (From Ref. 82.)

Brownian noise. If peak j lies over the ridge of the Brownian noise, the area estimate of the peak will be increased by the area under the path of the Brownian noise. If the peak is located over the valley of the Brownian noise, the area estimate will be decreased by the area over the path. The Brownian noise and peak signals are mutually independent. Thus the variance of the area estimates will be the sum of the variance of the area estimates without the Brownian noise and the variance of the area created by the Brownian path.

The power spectrum of an HPLC detector noise shows the type of $1/f$ noise (f is frequency) [2,84]. The Brownian noise may originate from the fluctuation of a detector or light source, because the HPLC and CE systems examined displayed similar noise characteristics, even if the flow of the measurement systems is stopped. The driving force of the flow for HPLC is pumping and that for CE is mainly electroosmotic flow [85]. Thus the flow itself is excluded from error sources.

The solid lines in Fig. 9 demonstrate that the prediction of RSD_j based on Eq. (3b) is amazingly good for analytes of different peak areas and widths. There is no peak overlap in this situation and the simplified form of Eq. (3b) [$\delta\phi_f(j) = 0$] can be used for prediction. The slight underestimation of RSD_1 is attributable to the interference of the unknown peaks adjacent to the target peak (see Fig. 8B). The errors from all the measurement steps and signal processing shown in Fig. 1 are included in Eq. (3): injection (I), separation [$\delta\phi_f(j)$], detection ($\tilde{W}$ and B), and signal processing [the first term of Eq. (3b)].

Constants I, B, and $\tilde{W}$ were determined experimentally and are not arbitrary. Constant I was the average of the four RSD_j values at the highest sample concentration (indicated by arrow I in Fig. 9). The standard deviation $\tilde{W}$ of the white noise is calculated from the noisy but flat baseline of a chromatogram (indicated by W in Fig. 8B).

Figure 10 shows the plot of

$$RSD_j^2 - \frac{2\pi^{1/2}\sigma_j \Delta T \tilde{W}^2}{A_j^2} - I^2 \tag{4a}$$

against

$$\frac{\sigma_j^3}{A_j^2} \tag{4b}$$

for the RSD_j values of the analytes at the lowest sample concentration in the series of experiments (indicated by arrow B in Fig. 9) [82]. The square of constant B is equal to the slope of the straight line of this plot; the already determined values of I and $\tilde{W}$ are used. Constants I, B, and $\tilde{W}$ were observed to be 0.0024, 0.11, and 0.055, respectively [82].

The final decision of the reliability of the prediction model [Eq. (3b)] should refer to the goodness of fit of Fig. 9. However, it must be iterated that the theoretical curves of Fig. 9 are not derived from the least-squares fit of this figure, and neither are constants $\tilde{W}$, I, and B. In statistics, RSD_j is known as a quantity obtained from repeated experiments, but in probability theory, it is predictable from a probability space (or a model) [53].

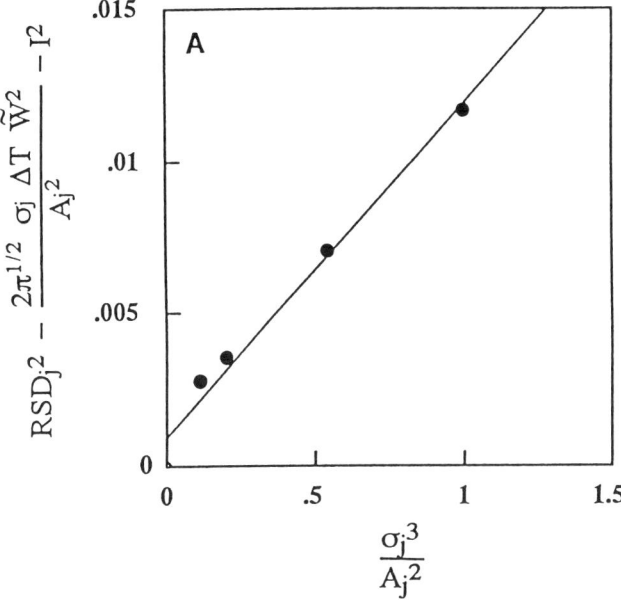

Fig. 10 Determination of constant B of Eq. (3b). ●, Sample of the lowest amount (500-fold dilution), which is indicated by arrow B in Fig. 9 (○ in Fig. 9). (From Ref. 82.)

The similar experiments were carried out 80 days after the first experiments [82]. The chromatograph had not been used during this period. The RSD_j of the materials in the new experiments is still predictable accurately from the original curve (the solid lines in Fig. 9). Month-to-month variation of the constants is negligible. Most of the RSD_j values for the 1000-fold dilution (the leftmost solid dot) lie on the theoretical lines that were calculated from the 500-fold diluted and original samples. The extrapolation of the theoretical curves is also quite effective.

The constants will be inherent to a specific apparatus, and system validation is feasible on the basis of the constants. The smaller the constants, the higher the reliability of the analytical system. Once the constants are determined experimentally for an analytical system, RSD of each component in a given sample can be predicted by Eq. (3b) from a single measurement to determine the peak area and width of every target peak. If a quantification limit is set, for example, at 20% RSD, the lowest acceptable amount of a target material (or sample) can also be calculated from Eq. (3b). The robustness of the constants against various environments of operation should be verified by further experiments.

To summarize, the uncertainty (RSD) can be predicted by Eq. (2) for the signal-processing step and by Eq. (3b) for the measurement (HPLC or CE) and signal-processing steps without resorting to repeated experiments. In practice, knowledge concerning the peak area and width (and position) is the only requirement for the error prediction in the HPLC or CE measurement if the constants inherent to the analytical apparatus, I, B, and $\tilde{W}$, are determined by prior experiments. The prediction will work well unless the peak parameters estimated from the single experiment are far from the averages (or true values). An example of CE analysis is given in Fig. 30.

B. Uncertainty Structure in Analytical Apparatus

The uncertainty structure of HPLC and CE will be in good agreement with the common experience of analytical chemists with regard to errors and error sources (see Fig. 11A). At a high analyte amount (large A_j), the first (a) and third (c) terms of Eq. (3b) become relatively small and the experimental RSD_j is subject to the injection error (b). The RSD_j is almost the same for any analyte in macro analysis ($\approx 0.25\%$; see Fig. 9).

As the analyte amount decreases, the mathematical error (a) and Brownian noise (c) become more conspicuous than the injection error because of small A_j (see Fig. 11A). In trace analysis, unlike macro analysis, the RSD_j varies from peak to peak depending on the area and width [= 2 to 20%; see Fig. 9 and Eq. (3b)]. Figure 11B demonstrates that if the peaks overlap strongly, the mathematical

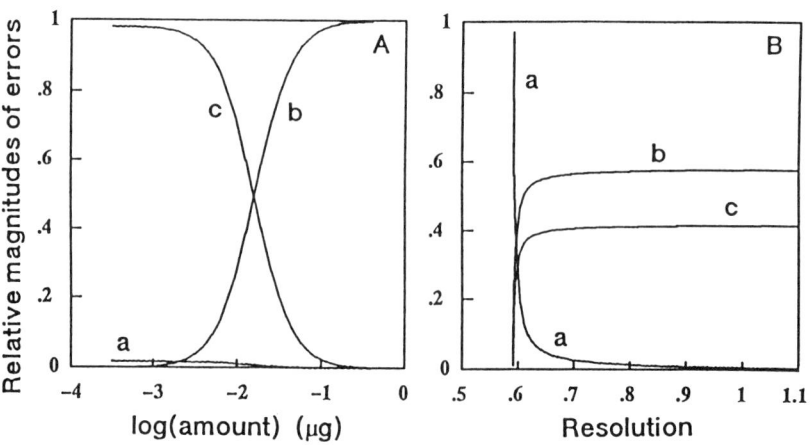

Fig. 11 Uncertainty structure of HPLC: contributions of the three errors of Eq. (3b) to the observed total error (RSD_j). Curves: a, signal processing error [first term of Eq. (3b)]/RSD_j^2; b, injection error [second term of Eq. (3b)]/RSD_j^2; c, Brownian error [third term of Eq. (3b)]/RSD_j^2. The same constants I, B, and $\tilde{W}$ as in Fig. 9 were used. (A) Peak 3 in Fig. 8; (B) peak 3 in Fig. 8 (10-fold dilution), which overlaps with a fictitious peak.

error is overwhelming and will predominate in the observed RSD_j. Equation (2) is a good approximation in the case of strong peak overlap.

A CE analysis of sulfa drugs yielded similar results about the goodness of fit in Figs. 9 and 10 and the uncertainty structure in Fig. 11. Constants I, B, and $\tilde{W}$ were 0.025, 0.027, and 0.0094, respectively, for the CE analysis [82]. The cause of the larger I in the CE system used will result from the lack of a forced cooling unit and a sample injection system that is considered inferior to the loop injector used in HPLC.

C. Uncertainty Principle in Separation Science

Technological advances over the last two decades probably eliminated many of the mechanical errors in separation science. On the other hand, although the new method, the Kalman filter, was developed only 30 years ago and has some advantages, such as its recursive property, the filter could not exceed the traditional limitations of least-squares curve fitting in the particular situations above [see Figs. 3 to 5 and Eq. (2)]. This result suggests that the mathematical error originating from the data processing step has remained invariant for more than a century since the discovery of the least squares by Carl F. Gauss [86]. The mathematical error must have a different profile from that of the technological error and can be considered to yield the lower limit of error which any actual analytical method cannot exceed, even if modern technology were able to remove all the mechanical and electronic error in the highest degree (uncertainty principle in separation science).

Virtually, any type of error involved in the analysis can set the limit of uncertainty for the entire analytical procedure, since the observed (total) variance of measurements is the sum of the error variance of every step that is positive and finite. The following conditions are indispensable for making the definition of the uncertainty principle clear and meaningful: It should be universal irrespective of the state of the art of technology; and it should take a single mathematical expression for various types of analytical apparatus, such as HPLC and CE. As far as the classical and new least-squares methods are concerned, it is apparent that the mathematical error as expressed in Eq. (2) is the unique solution of the principle.

The fundamental principle also provides information on many possible situations in quantitative analysis. In macro analysis, for example, reducing mechanical error (I) is the best way to improve the reliability of the system to near the limit of uncertainty [Eq. (2)], because the limit is far lower than the observed RSD_j (see Fig. 11). In the case of peak overlap, chromatographic peak separation (e.g., acquired by changing the chemistry of the mobile phase) is the only efficacious means for collecting more information, because the limit of uncertainty is comparable to the RSD_j observed.

The error prediction and uncertainty principle can be applied successfully to the various branches of separation science, including HPLC, CE, gas chromatography, micellar electrokinetic chromatography, and so on. In the addition to Gaussian peaks, signals of any shape can be treated in a similar manner [5].

D. Literature on Precision and Reproducibility

It is obvious that the precision with which quantitative measurements can be made is exceedingly important in analytical chemistry. However, relatively few papers have been published on the factors controlling the statistics of one of the most prevalent analytical systems, HPLC. Notable exceptions include papers by Grushka and Zamir [87] and by Scott and Reese [88] and a book by Katz [89]. Scott and Reese first investigated the dependence of the precision of retention time, peak width, area, and so on, on the sampling intervals of an analog-to-digital (A/D) converter, mobile-phase composition, pumping, and column temperature [88].

Grushka and Zamir studied some errors and their propagation during HPLC analysis with the aid of a mathematical technique [87]. They expressed the precision of retention times as a function of flow rate, mobile-phase composition, temperature, and chemical nature of solutes. Theoretical treatment of the influential factors on the precision of peak area estimates differs, in some respects, from that for the retention times. The conclusion obtained from the uncertainty structure of HPLC is in good agreement with their proposition that the major error source in the area estimation is found in the process of data processing (in trace analysis). Both of the groups above emphasized the important role of data processing in quantitative analysis [87,88].

The equation of Grushka and Zamir [Eq. (38) of Ref. 87] expressing the precision of area estimates is noteworthy. The performance of a commercial integrator is taken into account for the derivation. It can also describe the stochastic signal properties mentioned in Section II except for peak overlap. Their equation and Eq. (2) for predicting the signal-processing error share the things important in analytical chemistry, although they have totally different theoretical backgrounds. Grushka and Zamir also paid much attention to the significant effects of the sample size on the precision, as shown in Fig. 9.

The precision of the retention time, peak area and height, and 80-day reproducibility in an HPLC analysis are listed in Tables 3 and 4. The experimental conditions are the same as in Figs. 8 to 11. By "reproducibility" we mean the statistical result from identical experiments run in two different occasions or laboratories [87]. A cursory examination of Tables 3 and 4 yields the following remarks: (1) the retention times can be measured more precisely than the peak area; and (2) the reproducibility is satisfactory. Point 1 coincides with the conclusion of the previous work [87,88]. Scott and Reese observed that the peak

Table 3 Precision of Retention Time, Peak Area, and Height in HPLC ($n = 8$)[a]

	Peak			
	1	2	3	4
	Concentrated sample			
Retention time				
Mean (s)	225.5	269.9	328.9	465.6
RSD (%)	0.094	0.079	0.098	0.086
Peak area				
Mean	5814.5	2186.1	6255.5	6992.9
RSD (%)	0.17	0.23	0.38	0.26
Peak height				
Mean	956.2	304.7	759.7	627.8
RSD (%)	0.29	0.53	0.25	0.29
	Diluted sample (500-fold dilution of the sample above)			
Retention time				
Mean (s)	225.7	270.1	328.8	465.4
RSD (%)	0.11	0.17	0.17	0.15
Peak area				
Mean	11.53	4.904	13.54	13.34
RSD (%)	8.8	20	7.1	14
Peak height				
Mean	1.932	0.6786	1.551	1.218
RSD (%)	4.3	8.1	4.6	6.9

[a]For the peak number, see the legend of Fig. 8. The experimental conditions for the concentrated sample are the same as shown in Fig. 8A; those for the diluted sample Fig. 8B. About 60 data points over the $\pm 3\sigma$ period around the peak maximum were used for the integration (summation) with a desktop computer.

height data were more precise than the peak area [88]. However, the results of Tables 3 and 4 are negative in the concentrated sample and seem to be supportive in the diluted sample. This affirmative result is surprising as stated by Scott and Reese [88], and the reason is unclear.

The major sources of the imprecision of the retention time are fluctuation of flow rate, mobile-phase composition, and temperature [87,88]. The experiments of Tables 3 and 4 were designed to cut down on the foregoing error sources. The temperature was controlled by a column oven and further stringently by wrapping a material of low thermal conductivity around every outside part of tubing. Pure methanol was used as a mobile phase. As far as the results in the concentrated data are concerned, the RSD values of the retention time and peak area in the literature

Table 4 Eighty-Day Reproducibility of Retention Time, Peak Area, and Height in HPLC ($n = 8$)

	Peak			
	1	2	3	4
Concentrated sample				
Retention time				
Mean (s)	223.1	266.9	325.5	461.3
RSD (%)	0.063	0.053	0.043	0.023
Peak area				
Mean	5769.6	2166.0	6101.2	6587.7
RSD (%)	0.42	0.56	0.42	0.50
Peak height				
Mean	954.9	303.7	752.0	596.0
RSD (%)	0.54	0.48	0.60	0.49
Diluted sample (500-fold dilution of the sample above)				
Retention time				
Mean (s)	223.1	266.8	325.6	462.3
RSD (%)	0.091	0.20	0.084	0.12
Peak area				
Mean	12.77	4.617	12.60	10.45
RSD (%)	4.6	17	8.0	17
Peak height				
Mean	2.093	0.6656	1.544	0.9908
RSD (%)	3.2	9.2	4.1	6.2

[87] are twice to 10 times those of Tables 3 and 4. This difference will possibly be ascribed to the lower fluctuation mentioned above or the sample size effect. A more critical comparison of the data in Tables 3 and 4 with those of the previous work will make no sense for the same reasons.

The imprecision of the retention time will in turn influence the statistics of the area estimates. In macro analysis (see the concentrated sample of Tables 3 and 4), the RSD value of the retention times is relatively small (<0.1%), only one-tenth that of the area estimates on the average ($\approx 0.25\%$). The former will be overshadowed by the injection error. In trace analysis, the small RSD value of the retention time ($\leq 0.2\%$) will also be negligible compared with the predominant signal-processing error and Brownian error ($\approx 10\%$). These facts corroborate omission of the retention-time error from Eq. (3b). In CE analysis, however, the effect of the migration time will play a critical role in the precision of area estimates, since a change in the migration time causes a change in peak area. The prediction of the RSD in the CE analysis (see Fig. 30), nevertheless, seems to be of quality comparable to that of the HPLC analysis shown in Figs. 9 and 29. It is likely that the deleterious effects of the migration time change are allotted to the three terms of Eq. (3b).

IV. INFORMATION THEORY AND QUANTITATIVE ANALYSIS

Until now, this chapter has focused on the precision for a single peak. However, we still are without a solid basis for answering the following question concerning the statistic of a multicomponent chromatogram: Which is better, the sum or the product of RSD over the peaks? The answer to this question lies in information theory.

A. Precision and Information

The data processing error [Eq. (2)] can be related directly to the Shannon mutual information, $\phi_f(j)$, for peak j [5]:

$$\phi_f(j) = -\log\left(\frac{s_{\text{signal processing}}}{A_j}\right) \tag{5a}$$

$$= \frac{1}{2}\log\left(\frac{A_j^2}{2\pi^{1/2}\sigma_j \Delta T \tilde{W}^2}\right) - \delta\phi_f(j) \tag{5b}$$

This relationship is referred to as FUMI (function of mutual information). The first term of Eq. (5b), called the intact information, denotes the information indigenous to peak shape. Without peak overlap [$\delta\phi_f(j) = 0$], the information depends only on the peak shape and noise level (i.e., the intact information). Peak overlap [$\delta\phi_f(j) > 0$] adversely affects the information and precision.

According to information theory in the Kalman filter expression [46], if the target peaks are independent of each other, the total information, Φ_f, provided by all the target peaks in a chromatogram can be given as the sum of the individual peak information, $\phi_f(j)$:

$$\Phi_f = \sum_{j=1}^{q} \phi_f(j) \tag{6}$$

This is one of the objective functions for optimization used in this chapter. The total information is also called FUMI. A high total information Φ_f means a high precision for the target peaks on the average. If the precision is defined as the reciprocal of RSD ($= s_{\text{signal processing}}/A_j$), the information is the logarithm of the precision [see Eq. (5a)]. Thus mutual information, precision, and FUMI are essentially equivalent concepts.

A natural generalization of FUMI into the experimental RSD_j [Eq. (3)] takes the form [82]: For individual peak information,

$$\phi_m(j) = \log\left(\frac{1}{RSD_j}\right) \tag{7a}$$

$$= -\frac{1}{2}\log\left(\frac{2\pi^{1/2}\sigma_j \Delta T \tilde{W}^2}{A_j^2} 10^{2\delta\phi_f(j)} + I^2 + \frac{\sigma_j^3 B^2}{A_j^2}\right) \tag{7b}$$

and for the total information,

$$\Phi_m = \sum_{j=1}^{q} \phi_m(j) \tag{8}$$

Hereafter, Eqs. (7) and (8) are referred to as MEI (measurement-elicited information). Equation (8) is also an optimization criterion.

FUMI [$\phi_f(j)$ and Φ_f] and MEI [$\phi_m(j)$ and Φ_m] are both nonnegative. The poorest precision within reach of FUMI and MEI is 100% RSD [$\phi_f(j) = \phi_m(j) = 0$]. The following description might help judge the statistical reliability of a system from the viewpoint of information amount, $\phi_m(j)$:

If $\phi_m(j) = 3$, then $RSD_j = 0.1\%$ and the precision of peak j is excellent.
If $\phi_m(j) = 2$, then $RSD_j = 1\%$ and the precision of peak j is not bad.
If $\phi_m(j) = 1$, then $RSD_j = 10\%$ and the precision of peak j is poor.

In our previous papers, the natural logarithm has been used for theoretical purposes [5–8]. From now on, it is replaced by the common logarithm for practical considerations.

The concept of the binary logarithm and bit is rather effective in information theory for qualitative or structural analysis [32,36,39], where only two possibilities are concerned: a material either is or is not present above the detection limit of a test. On the other hand, in FUMI theory, the RSD of measurements takes positive real numbers less than 100%.

FUMI [Eq. (5b)] and the total information yield used by Thijssen et al. [48–52] are based on the same principal concepts of the Kalman filter. The most prominent advantage of FUMI is that it is a scalar function of the signal parameters and noise level, as shown in Eqs. (5b) and (6). We can predict the precision of signals from the values above without repeated experiments or simulations. The total information yield, however, requires stepwise calculation of the multidimensional determinants for multicomponent chromatograms for the same purpose, as the filter proceeds with raw data.

B. Throughput of Analysis

The ultimate purpose of quantitative analysis is not only to achieve the highest precision but also to minimize the run time, t_q. The throughput, which is defined as the average transmission speed of the mutual information, is introduced to express the efficiency of analysis numerically [6,7]. Two types of throughput are defined as

$$\vartheta_f = \frac{\Phi_f}{t_q} \tag{9a}$$

$$\vartheta_m = \frac{\Phi_m}{t_q} \tag{9b}$$

A high throughput indicates a markedly rapid analysis with satisfactory precision as long as the information takes a large value. A low throughput is ascribed to either of two failures: a precise analysis with too long a run time or a very rapid analysis suffering from insufficient separation. We should note that the highest throughput (the maximum of ϑ_f or ϑ_m) is not necessarily equivalent to the most rapid analysis but to the most rapid analysis with satisfactory precision. The definition of throughput above coincides in concept with the time-information performance, which was defined primarily by Danzer [45].

For convenience and clear presentation of figures, the individual peak throughput is often used:

$$\vartheta_f(j) = \frac{\phi_f(j)}{t_q} \tag{10a}$$

$$\vartheta_m(j) = \frac{\phi_m(j)}{t_q} \tag{10b}$$

The throughputs ϑ_f and ϑ_m are the sum of the individual throughputs, $\vartheta_f(j)$ and $\vartheta_m(j)$, respectively, over all the target peaks [see Eqs. (6), (8), and (9)].

In liquid chromatography, the operating conditions for the minimum-time analysis with sufficient precision often coincide with those for the analysis of the highest throughput. Even if the run time for the former is shorter than for the latter, more information can be picked up from the highest throughput analysis in a fixed time period.

Unfortunately, time itself cannot be a good optimization criterion. Actually, a rapid assay can be regarded as valuable if it meets the conditions of both rapidity and precision. The throughput involves not only the run time but also precision and thus is concluded to be a suitable objective function for optimization.

The consumption of organic solvent in a mobile phase poses problems associated with environmental pollution. If the throughput is defined as (total information)/(total volume of used solvent), the condition of the lowest solvent use will be identified by the throughput-based optimization. The cost–performance characteristics of micro-column chromatography can be evaluated in a similar way. We may safely say that the throughput is a general description of the rapidity, conservation, and economy of analysis.

Besides FUMI and MEI, a bewildering number of optimization criteria have been put forward so far in the literature. The detailed explanation can be found in Refs. 9, 12, 13, 16, 17, and 29. A critical comparison of FUMI and MEI with the other criteria is outside the purview of this chapter, but a brief examination is given in Ref. 90. The philosophy of FUMI and MEI is disparate from that of most of the others, which explicitly refer to the separation and time alone. To the authors' knowledge, an exception was the criterion propounded by Wegscheider et al., which incorporated noise level [13,91].

C. Resolution and Information

As long as the peaks are overlapped strongly, resolution always increases with decreasing RSD or increasing information (see Section VI.A). Thus there must exist a close correlation between Rs and the mutual information. It is proved mathematically that more peak separation (increase in Rs) is equivalent to an increase in the mutual information $\phi_f(j)$ of peak j only if peak j strongly overlaps another peak. The mathematical expression for this statement is [92–94]

$$\frac{\partial \phi_f(j)}{\partial Z} = C \frac{\partial Rs}{\partial Z} \tag{11}$$

where Z denotes a chromatographic variable to be changed (e.g., mobile-phase composition) and coefficient C (> 0) is a function of Rs. ∂Rs denotes the increase in Rs and $\partial \phi_f(j)$ the increase in FUMI. If the increase ∂Rs is positive, the increase $\partial \phi_f(j)$ should also be positive because of the positive coefficient C [$\partial \phi_f(j) = C\, \partial$Rs] irrespective of the sign of ∂Z. The more separated the peaks (∂Rs > 0), the more precise [$\partial \phi_f(j) > 0$] the analysis in the situation of peak overlap. However, this relationship does not hold true in the case of sufficient peak separation, where further separation (∂Rs > 0) causes peak broadening and spoils the precision [$\partial \phi_f(j) < 0$]. That is, the signs of the changes, $\partial \phi_f(j)$ and ∂Rs, are opposite in this situation. Equation (11) is the information-theoretical interpretation of the most commonly used separation function, Rs.

The discussion about the uncertainty structure of an analytical apparatus (Section III.B) and the differential formalism of optimization [93,94] lead us to the conclusion that peak separation is the most efficacious means in chromatography to collect information [5]. The philosophy of separation science comprising chromatography, electrophoresis, and so on, will be condensed in the conclusion above and in Eq. (11).

In general, the differential formalism is important in comprehending the mathematical structure of optimization [5,93,94]. Let X be the mobile-phase composition, with attention confined to a microscopic scale around the optimum. If $\partial \phi_f(j)/\partial X > 0$, a slight increase in X leads to a corresponding increase in information, indicating that this particular procedure of X brings the optimization process closer to the optimum. If $\partial \phi_f(j)/\partial X < 0$, an adequate procedure to gain upon the optimum is not the increase in X, but the decrease in X. Let L be the column length. If $|\partial \phi_f(j)/\partial X| > |\partial \phi_f(j)/\partial L|$, the manipulation of the mobile-phase composition is more effective than the selection of a new column of another length in the current operating conditions.

The signs and magnitudes of the derivatives of FUMI are useful indicators for optimization. Grushka et al. implied the importance of the differential formalism of optimization by presenting the derivatives of the logarithm of the selectivity with respect to mobile-phase composition and temperature in an HPLC analysis of nucleotides [95].

V. FUNDAMENTALS OF OPTIMIZATION

A. Practical Optimization Strategies

We have understood the mathematical technique to calculate the mutual information of the output of a measurement system considered here. The prime demand of practical optimization is a description of the criteria [Eqs. (5)–(9)] as a function of variables to be optimized over a wide range. Many empirical equations are very helpful for assessing the behaviors of retention times and peak shape under a variety of conditions [9,96–99]. Coupled with these methodologies, the optimum can be found with a few experiments that can establish the relationship between the variables and responses of the criteria over a wide range of the variables.

The logarithm of capacity factor k_j of peak j is well approximated by the polynomial of mobile-phase composition, X, in HPLC [9,97].

$$\log (k_j) = a_j X^2 + b_j X + c_j \tag{12}$$

where a_j, b_j, and c_j are the constants to be determined by the least-squares regression of experimental data. An instance of X is the methanol volume fraction (%) of aqueous mobile phase in reversed-phase HPLC. Three experiments conducted under different X values are the least requirement for determining constants a_j, b_j, and c_j. The retention time and width of peak j can be estimated with Eq. (12) [$k_j = (t_j - t_0)/t_0$] and the relationship that $N = (\sigma_j/t_j)^2$, respectively. If area and theoretical plate number N are assumed to be invariant (or known for any material at any composition), the information, FUMI or MEI, can be calculated for every composition in question, and the maximum of the criterion can be spotted over the range examined.

The optimization of other variables, such as detection wavelength, flow rate, column length, and amount of internal standard, can be done similarly [8]. For example, the information can be calculated from the ultraviolet-visible spectra of target materials which are directly proportional to the analyte concentration (Lambert–Beer law). The optimal column length can also be chosen by assuming that the retention time and plate number are both directly proportional to the column length. If a practical situation is far different from the idealistic ones above, empirical equations that can more exactly describe the retention behaviors will be recommended.

The above initial-stage strategy to locate the optimum is called regression design [16]. The other designs include the lattice design and sequential simplex. A detailed explanation of these optimization procedures is beyond the scope of this chapter. A good introduction can be found in Refs. 9, 16, and 100.

In practice, when two or more variables are to be manipulated concurrently, the combined effects of these multiple variables may make it difficult to predict the retention profiles [14,101]. These interactions of the variables are complicated for complex analytes that have multiple functional groups such as nucleotides [95,101]. An experimental design that is useful for discovering the interactions is the factorial design [9,14]. Grushka et al. have investigated the effects of pH, temperature, and content of an organic modifier (methanol) on the retention behaviors of deoxyribonucleotides in reversed-phase HPLC [95,101]. They found a simple rule for the concurrent influences of the variables on the retention behaviors and suggested a successful optimization strategy.

The quantitative structure–retention relationship (QSRR) is an algorithm for predicting the retention times of chemicals [98,99,102–105]. Applications of QSRR to the statistical optimization have been described [106,107]. An optimization technique that dispenses with the foregoing consideration over the entire variable space is simplex optimization (see Section VI.F).

B. Nonlinear Programming Problems

The optimization theory explained in this chapter is formulated as a nonlinear programming problem well known in the engineering area [108]. The goal of nonlinear programming is to find a solution that yields the maximum of an objective function (criterion), while some constraints are met. A set of operating conditions that satisfies all the constraints is called a feasible solution. The optimum is among the feasible solutions.

In the optimization with FUMI, the objective functions are the precision Φ_f and throughput ϑ_f. Constraints might seem to be trivial for the FUMI optimization, since FUMI includes the four signal properties listed in Section II. However, the following four constraints are necessary to establish the versatility [109]:

$$\Phi_f \geq c_1 \quad \text{(lower limit of precision)} \tag{13}$$

$$\vartheta_f \geq c_2 \quad \text{(lower limit of throughput)} \tag{14}$$

$$\delta\Phi_f \leq c_3 \quad \left[\text{total information loss} = \sum_{j=1}^{q} \delta\phi_f(j)\right] \tag{15}$$

$$\begin{aligned}\Delta\Phi_f \leq c_4 \quad &\text{(information range} \\ &= \text{Max}[\phi_f(1), \phi_f(2), \ldots, \phi_f(q)] \\ &\quad - \text{Min}[\phi_f(1), \phi_f(2), \ldots, \phi_f(q)])\end{aligned} \tag{16}$$

First, high precision is essential for quantification [Eq. (13)]. With the threshold precision, Φ_f ($= c_1$), a maximum permissible analysis time t assigns the least sample throughput ($c_2 = c_1/t$). Peaks should interfere as little as need be with each other. If the resolution of peaks with equal areas falls below the acceptable lower limit, $\hat{R}s$, the information loss takes a positive value [$\delta\phi_f(j) > 0$]. Then the constraint that $\delta\Phi_f$ is zero or a very small value is adopted. Furthermore, the dispersion of the precision, $\phi_f(j)$, of target peaks has a natural limit to afford them the equal opportunity for precise analysis. The constraint for the information range [Eq. (16)] is critical in optimizing wavelength and the amount of internal standard. If $\Delta\Phi_f = 0$, every peak can be quantified with equal precision. Similar constraints can also be defined for MEI.

In Section VII.A, the constraints ($\Phi_f \geq 12.0$, $\vartheta_f \geq 0.0065$, $\delta\Phi_f \leq 0.0043$, $\Delta\Phi_f \leq 0.43$) are adopted in a TOCO of four peaks. These values imply that every peak can be quantified with an RSD less than 0.1% on the average (calculated from $\Phi_f \geq 12.0$); the analysis with the limit RSD ($= 0.1\%$) is complete within 30 min (from $\vartheta_f \geq 0.0065$); there is no peak overlap (Rs ≥ 1) in the chromatograms (from $\delta\Phi_f \leq 0.0043$ and $\hat{R}s = 1.03$); the ratio of the maximum to the minimum RSD of the peaks is less than e ($= 2.72$) (from $\Delta\Phi_f \leq 0.43$).

In nonlinear programming problems, the variables are divided into two categories: controllable variables and environmental parameters. The former involves all the variables within reach of the operator (e.g., mobile-phase composition, wavelength, etc.). The latter concerns sample injection and temperature control. The influence of these environmental parameters on the precision is considered to be invariant for a given analytical system in usual laboratory circumstances. On the other hand, the controllable variables affect the precision of measurements through peak separation and shape.

Analytical interest is in the controllable variables, the optimum of which varies from sample to sample depending on the chemistry of the samples. The environmental parameters are independent of the chemistry. That is, a good injector and good temperature controller always work well for any sample. The entire optimization considering the controllable variables and environmental parameters would usually yield quite similar results to the optimization of the controllable variables alone (FUMI). Evidence for this is given in Sections VI.A and VII.A. Environmental parameters such as detection apparatus and temperature controller can also be evaluated by MEI.

VI. APPLICATIONS OF FUMI AND MEI

FUMI refers to the data processing error and underestimates the observed RSD in HPLC and CE, but MEI exactly predicts the real RSD. The similarities and differences between FUMI and MEI are described in the model optimization. Some examples of practical optimization with FUMI as a criterion are given.

A. Model Optimization

In the model optimization of mobile-phase composition (Figs. 12 and 13) and column length (Figs. 14 and 15), peak overlap is the only unfavorable factor on the information. In Figs. 12 and 14, the conditions that meet sufficient separation (Rs $\geq$ 1.5) are represented by solid lines (feasible solutions); insufficient conditions are represented by dashed lines. It must be stressed that the solution to the optimization problems varies depending on the sample size. Examples of 40-fold different sample concentrations are presented in Figs. 12 and 14.

Figure 12A shows a typical elution model of peaks j and $j + 1$ (A) in reversed-phase liquid chromatography where the retention time decreases with increasing volume fraction, X, of an organic solvent in the mobile phase. A simplified equation is used in which that log $(k_j) = b_j X + c_j$ [96]. Resolution (B), information at a high sample concentration (C), information at a low concentration (D), throughput at the high concentration (E), and throughput at the low concentration (F) are also plotted against the volume fraction X. The peak area and plate number are assumed to be constant. Chromatograms corresponding to points A to F of Fig. 12 are illustrated in Fig. 13 with the same letters.

At the low sample concentration (Fig. 12D and F), FUMI, $\phi_f(j)$, for peak j is almost equal to MEI, $\phi_m(j)$, because the signal processing error that is calculated by FUMI is predominant in the total error of measurement [Eq. (3)] in trace analysis. As X increases from 0 to 15.25% (point B), the information increases gently because of the gained peak sharpness, while Rs decreases. Peak overlaps are absent in this region (Rs $\geq$ 1.5). Above point B, the overlap (Rs $<$ 1.5) causes the abrupt decrease in information, but Rs continues to decrease smoothly. At X = 18.25% (point C), the peaks fuse into one and the information and Rs take the minimum ($= 0$). A further increase in X releases the overlap, adding to the information and Rs. After complete separation at $X = 21.25\%$ (point D), peak sharpening again leads to a gradual increase in information (points D to F), but Rs takes the maximum value (point E). The small difference between FUMI and MEI at the small values of X in Fig. 12D is due to the peak width σ_j [compare σ_j of the first term of Eq. (3b) [$=$ Eq. (2)] and σ_j^3 of the third term of Eq. (3b)].

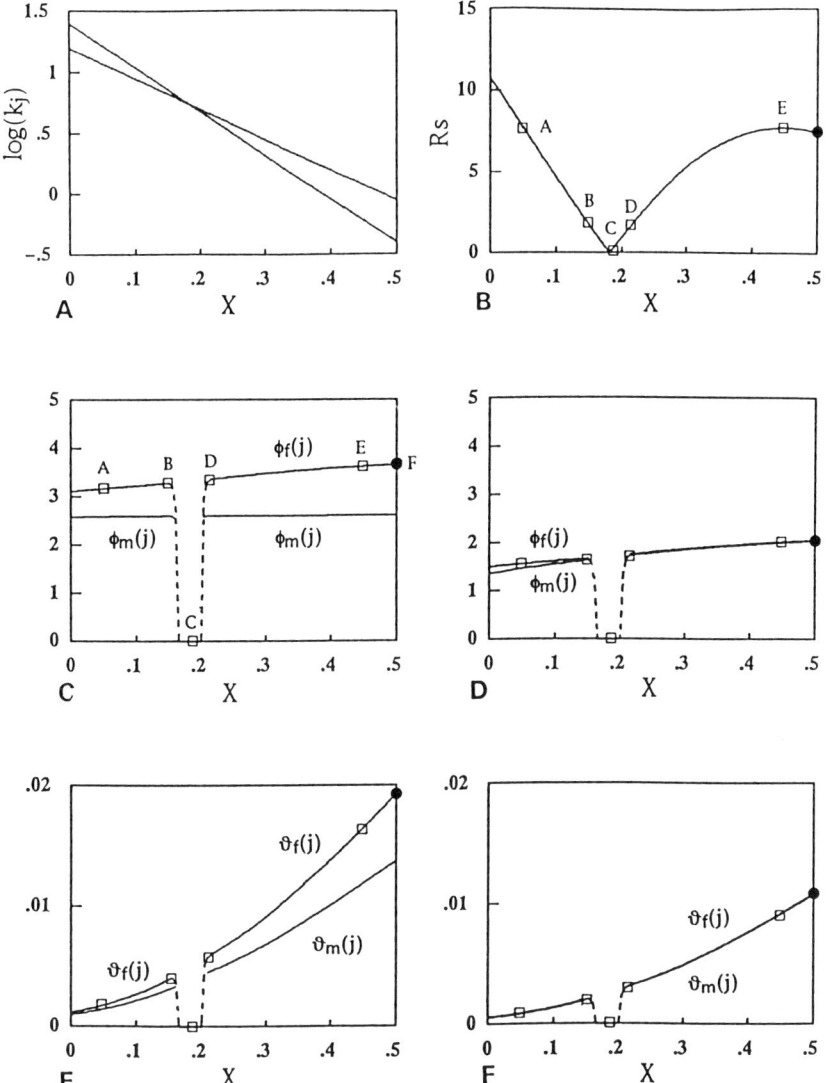

Fig. 12 Dependence of the precision $\phi_f(j)$ and $\phi_m(j)$, throughput $\vartheta_f(j)$ and $\vartheta_m(j)$, and resolution Rs on volume fraction X with crossing log (k_j) lines. $\hat{R}s = 1.53$. ----, Peaks are overlapped (Rs < $\hat{R}$s); —, peaks are separated (Rs ≥ $\hat{R}$s). Point F (●) is the precision optimum and throughput optimum. It is assumed that $t_{j+1} > t_j$. $X = 0$ to 50%; $N = 10,000$; $t_0 = 100$ s. The sample concentration in (C) and (E) is 40 times that in (D) and (F).

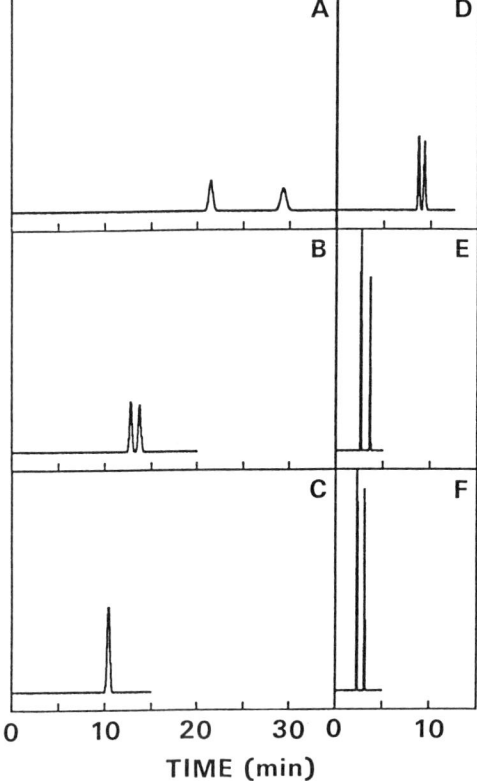

Fig. 13 Simulated chromatograms A to F with the same letters as points A to F in Fig. 12. X: (A) 0.05; (B) 0.15; (C) 0.19; (D) 0.215; (E) 0.45; (F) 0.5.

The other peak, $j + 1$, exhibits virtually the same behavior with respect to mobile-phase composition. We might expect that point F, where the precision for peak j takes the highest value, to be the precision optimum over the X range examined. Without overlap, the throughput is more sensitive than the precision to changes in observation time t_q (see Fig. 12F). In this model optimization, the maxima of the precision and throughput are both located at point F. That is, the precision optimum is identical to the throughput optimum.

At high sample concentration (Fig. 12C and E), precision $\phi_f(j)$ and throughput $\vartheta_f(j)$ based on FUMI exhibit the same trends toward the variable X as those at low sample concentration (Fig. 12D and F). However, the injection error is the major contributor to the total measurement error [Eq. (3)] in macro analysis and MEI yields smaller values than FUMI. The high, horizontal region of the MEI plot demonstrates that in macro analysis, every operating condition yields almost the same precision, as long as the peaks are sharp enough and separated sufficiently.

Only if the peaks overlap do the precision $\phi_f(j)$ and resolution Rs assume similar behaviors (see Fig. 12). An increase in FUMI is always accompanied by an increase in Rs. This is why the mathematical relationship between FUMI and Rs [Eq. (11)] exists exclusively in the case of peak overlap.

Figure 14 shows the influences of column length on resolution (A), information at a high sample concentration (B), information at a low concentration (C), throughput at a high concentration (D), and throughput at a low concentration (E). Figure 15 illustrates the chromatograms corresponding to points A to E of Fig. 14 with the same letters. At a low concentration, as the column lengthens from $L = 0$, the precision, $\phi_f(j)$ and $\phi_m(j)$, initially increases abruptly due to the chromatographic resolution of overlap (see points A to C), but then the peak broadening without overlap exerts the untoward influence on the precision (see points C to E).

The information curve reaches a maximum at the precision optimum (point C), but Rs is a monotonously increasing function of L. For the throughput maximum, the strong overlap violates one of the constraints $\delta\Phi_f = 0$, and prevents this condition of L from being the throughput optimum. Point C is also the throughput optimum. At the low concentration, there is no significant difference in the predicted precision between FUMI and MEI, but the difference in macro analysis (see Fig. 14B and D) is due to the same reason as the example given for the mobile-phase composition (Fig. 12).

In macro analysis, the major cause of the measurement error is the injection error and any condition will yield nearly the same precision as long as all the peaks are sharp enough and sufficiently separated. Any separation is acceptable as the optimum (see the horizontal lines of MEI in Figs. 12C and 14B), if the rapidity of analysis is of no great concern. Then the throughput is a preferable criterion for the optimization of macro analysis.

In micro analysis, the optimum must be selected more carefully. The RSD value varies substantially from condition to condition. For example, if the sample concentration is high (see Fig. 12C), the precision and resolution at the optimum mobile-phase composition (Fig. 13F; Rs = 7.5 at $X = 0.5$; RSD = 0.25%) are comparable to those at the maximal resolution (Rs = 10.8 at $X = 0$; RSD = 0.27%). These RSD values are calculated by MEI [Eq. (3b)] with the same constants determined in Section III. If the concentration is reduced by 40 times (see Fig. 12D), an appreciable difference appears between the conditions above: RSD = 4.3% at $X = 0$ and RSD = 1.1% for Fig. 13F (optimum). This is a typical example indicating that the resolution should not be relied on too much in the optimization.

In trace and macro analyses, FUMI and MEI identify the same conditions as the throughput optimum in the optimization of mobile phase and column length. This is also true for the precision optimum in micro analysis. In macro analysis, the precision optimum selected by FUMI is involved in the wide range of the

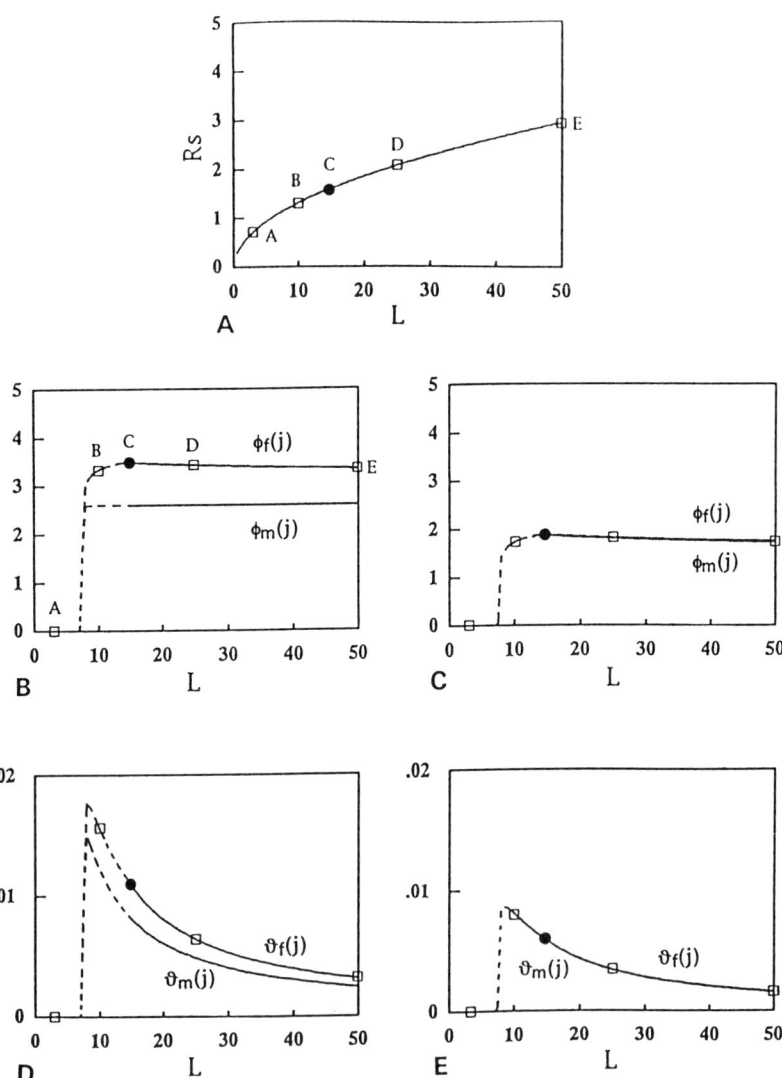

Fig. 14 Dependence of the precision $\phi_f(j)$ and $\phi_m(j)$, throughput $\vartheta_f(j)$ and $\vartheta_m(j)$, and resolution Rs on column length L. $\hat{R}s = 1.53$. ----, Peaks are overlapped (Rs < $\hat{R}$s); —, peaks are separated (Rs ≥ $\hat{R}$s). Point C (●) is the precision optimum and throughput optimum, and the resolution Rs at point C corresponds to the preset optimal resolution $\hat{R}$s. $N = 10,000$ and $t_0 = 100$ sat. L of 15 cm; $k_j = 2$; $k_{j+1} = 2.2$. The sample concentration in (B) and (D) is 40 times that in (C) and (E).

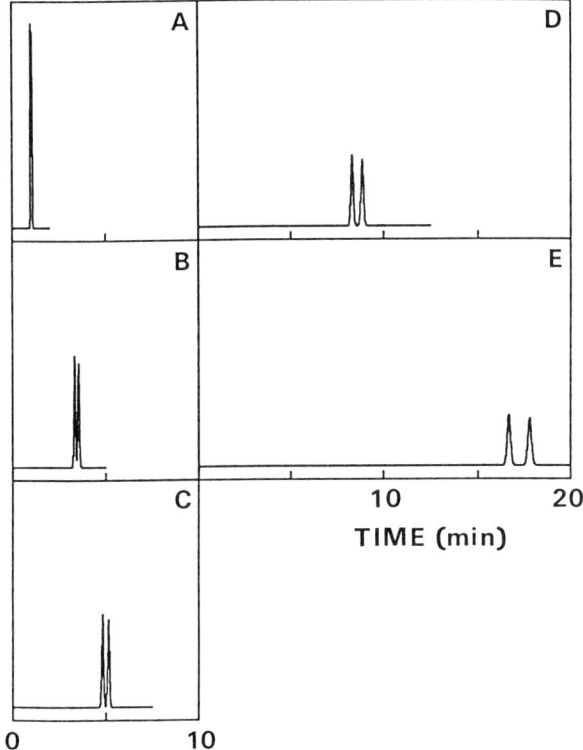

Fig. 15 Simulated chromatograms A to E with the same letters as points A to E in Fig. 14. L (cm): (A) 3; (B) 10; (C) 14.5; (D) 25; (E) 50.

precision-optimum conditions chosen by MEI. Therefore, FUMI is a simple but reliable optimization criterion irrespective of the sample size.

B. Optimal Mobile-Phase Composition in Pesticide Analysis

Residues of pesticides and their toxic metabolites in foods of plant origin are of serious concern to society. Liquid chromatographic methods for determining residue levels of the pesticides in crops such as lettuce, tomatoes, and so on, have been reported from the viewpoints of chemical analysis [110] and governmental regulation [111].

N-Methylcarbamates are widely used pesticides in agriculture. The HPLC analysis for these compounds in this subsection is not as practical as that of Blaß [110] or Krause [111], who used postcolumn derivatization and fluorometric detection for determining more than 10 pesticides at one time. However, the simplified experiments presented here [112] will suffice to demonstrate the applicability of the information measure, FUMI, to the agricultural problem.

The following pesticides were analyzed: peak 1, metolcarb; 2, propoxur; 3, carbofuran; 4, carbaryl; 5, xylylcarb; 6, ethiofencarb; 7, 3,5-xylyl methylcarbamate; 8, pirimicarb; 9, 4-ethyl thiophenyl methylcarbamate; 10, isoprocarb. The model used for the retention prediction is based on the relationship between the capacity factor, k_j, for peak j and the volume fractions of acetonitrile (X_1) and methanol (X_2) in water (mobile phase) [112]:

$$\log (k_j) = b_0 + b_1 X_1 + b_2 X_2 + b_3 X_1^2 + b_4 X_2^2 + b_5 X_1 X_2 \qquad (17)$$

The coefficients b_0 to b_5 were determined by the linear least-squares regression on the measured capacity factors, k_j, at more than six different conditions. Thus FUMI, Φ_f, is now a function of X_1 and X_2.

Figure 16 illustrates the response surface of the total information, Φ_f, for the pesticides against the mobile-phase compositions, X_1 and X_2. The complicated response surface comes from the fact that the behaviors of the solutes toward the two organic solvents in the mobile phase differ greatly. Therefore, the solutes

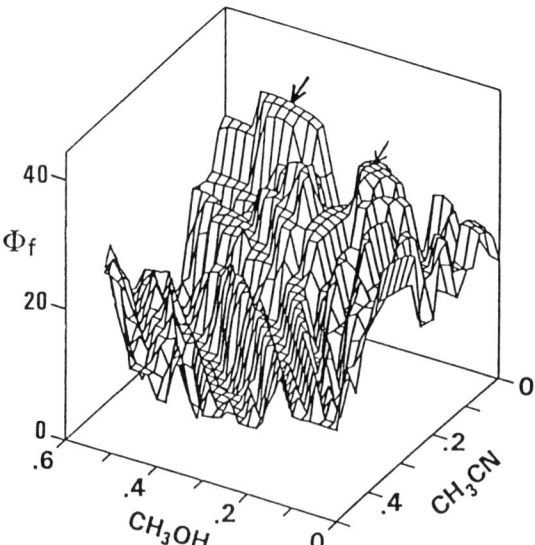

Fig. 16 Influence of the mobile-phase compositions on the information FUMI. An LC-6A liquid chromatograph (Shimadzu) was operated at the following conditions: column, Inertsil ODS-2 (250 × 4 mm) GL Sciences Co.; mobile phase, acetonitrile–methanol–water; flow rate, 1 mL/min; detection, UV 210 nm (SD6A, Shimadzu); column temperature, 40°C; injection volume, 20 μL (5 to 20 ng). The lowest resolution limit, $\hat{R}_s$, was set at 1.03. (From Ref. 112.)

show entangled behaviors with the reversal of the peak elution order throughout the examined ranges of the variables X_1 and X_2. This complex problem can be solved relatively simply by use of the precision, Φ_f (and throughput, ϑ_f) as the criteria.

Without peak overlap, the chromatograms or operating conditions make a high ridge on the response surface for the total information. The height of the ridge depends on the peak shape, especially peak height (or width σ_j). However, strong peak overlap caused by the peak crossings adds to the total information loss, $\delta\Phi_f$, deepening the corresponding valley of the response surface. The response surface of the total loss is also complicatedly undulated [112].

The optimum (maximal Φ_f) of the present experiments is indicated by the bold arrow in Fig. 16 and corresponds to the condition of an acetonitrile–methanol–water mixture of 0:45:55. Figure 17 illustrates (A) observed and (B)

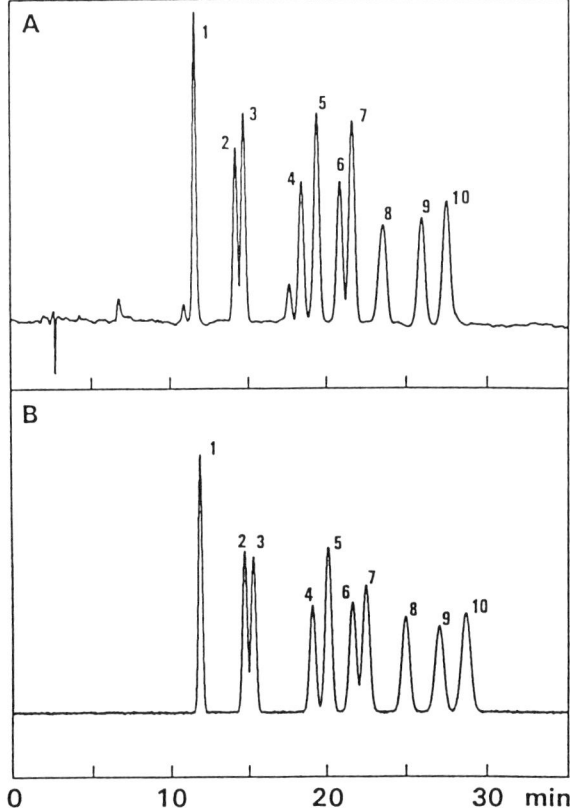

Fig. 17 Optimum chromatograms: (A) observed; (B) predicted. Mobile phase: acetonitrile–methanol–water (0:45:55). The peak number is given in the text. (From Ref. 112.)

predicted chromatograms at the precision optimum. The prediction of the solute retention behaviors by Eq. (17) is choice. Every peak pair is separated by an Rs value above 1.0. The analysis time of the optimum condition (= 28 min) is shortest among all the overlap-free chromatograms (Rs ≥ 1.0 and $\delta\Phi_f = 0$). The optimum selected above provides a typical example in which the most precise analysis coincides with the most rapid analysis and the highest throughput analysis.

Figure 18 shows a chromatogram under nonoptimum conditions (acetonitrile–methanol–water = 8:22:70). The position of this chromatogram is located on another lower ridge of the response surface (indicated by the plain arrow) and can be considered a local optimum. Although every peak pair is separated sufficiently in this chromatogram, the total information under this condition is much less ($\Phi_f = 29.9$) than the global optimum shown in Fig. 17 ($\Phi_f = 32.1$)

Comparing the chromatograms in Figs. 17 and 18, we can see how the peak height affects the individual peak information, $\phi_f(j)$, or precision. Note that every peak in the nonoptimum chromatogram is much lower and wider than that in the precision optimum and the S/N at the peak maximum is poorer for the former. From the discussion in Section VI.A, we can easily understand how this difference in the precision could be fatal in a trace analysis, which is almost always the level of analysis with regard to pesticide residues.

To find the highest point of the response surface is the only requirement needed to find the precision optimum in this example. The constraints [Eqs. (13) to (16)] can be dispensed with. This is because the information loss caused by the peak overlap is strongly correlated with the change in the total information during

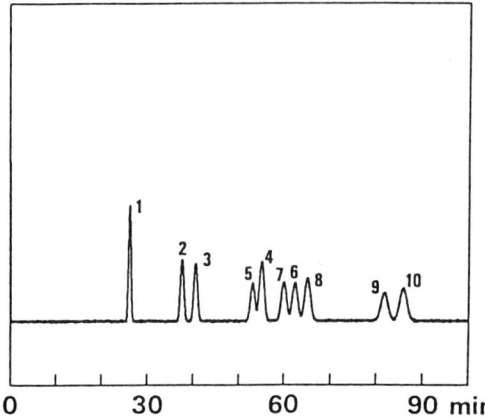

Fig. 18 Predicted chromatogram at a nonoptimal condition. Mobile phase: acetonitrile–methanol–water (8:22:70). (From Ref. 112.)

the optimization of mobile-phase composition [113]. That is, always in this case, the constraints [Eqs. (13) to (16)] are violated by the strong overlap alone.

C. Optimization of Wavelength and Amount of Internal Standard

The optimization of the amount of internal standard contained in sample solutions has attracted little attention in chromatography. Various reasons have been given for this nonchalance: an appropriate amount of standard material may be added to sample solutions after the other optimization procedures have been completed, because a standard material can be selected with wide freedom; visual inspection often seems to be effective for the best selection of a standard; widely used quality criteria based on the resolution Rs or separation $S_{i,i-1}$ [114] from the definition cannot cope with this quantity optimization. The last reason also holds true for the optimization of wavelength.

Simultaneous optimization of the amount of internal standard and detection wavelength for an antipyretics mixture in reversed-phase liquid chromatography is introduced here [115]. The interference of an inessential solvent peak is also taken into account (see Section VIII.E). Unlike the optimization of mobile-phase composition shown in Section VI.B, some constraints are claimed for the foregoing purpose.

A high FUMI value is a prerequisite for precise analysis. However, even if the total information, Φ_f, assumes a large value, the following problems can still occur [115,116]: (1) a high level of information can appear for a large peak despite considerable information loss; (2) some peaks can contain large amounts of information, $\phi_f(j)$, while others contain relatively little information. Point 1 mistakenly identifies a chromatogram with strongly overlapped peaks of high absorbance as the optimum, but can be circumvented by introducing a constraint on the information loss such that $\delta\Phi_f \leq c_3$ [Eq. (15)]. Point 2 results from the wide range of precision from large to small peaks, but a limit on the information range such that $\Delta\Phi_f \leq c_4$ [Eq. (16)] prevents such a condition from being chosen as the optimum.

Figure 19 shows the optimal chromatogram (simulation) for the analysis of (from left) acetaminophen, caffeine, salicylamide, guaifenesin (internal standard), and ethenzamide. A solvent peak appears at the 2-min retention point near the first target peak (not shown in Fig. 19). Ultraviolet spectra of the four antipyretics and the internal standard as well as the related experimental conditions are given elsewhere [115,117].

The total information Φ_f, the information loss $\delta\Phi_f$, and the information range $\Delta\Phi_f$ are plotted as response surfaces in three-dimensional spaces (see Fig. 20). For simplicity, the amount m_s of internal standard chemical is denoted by

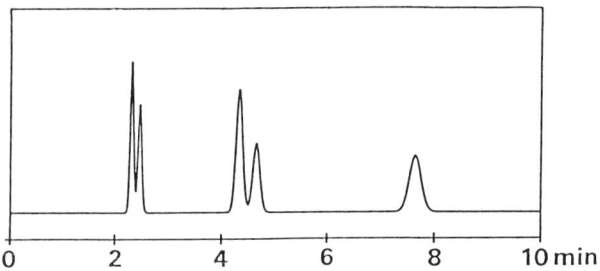

Fig. 19 Simulated chromatogram at the optimal wavelength and amount of internal standard. Peaks (from left): acetaminophen, caffeine, salicylamide, gauifenesin (internal standard), ethenzamide. The chromatographic profiles are $\lambda = 285$ nm, $m_s = 0.2$; $N = 5000$. (From Ref. 115.)

"unit" (1 unit = 0.4 mg/mL). FUMI is calculated at intervals of 0.2 unit of m_s and 5 nm of wavelength λ. The constant $\hat{R}s$ involved in FUMI is set at 1.0.

As the wavelength becomes shorter, FUMI increases, waving slightly, according to the increase in the absorbance of the analytes (see Fig. 20A). The decrease in information, Φ_f, in the short-wavelength region (≤ 245 nm) is attributable to the strong overlap between the inessential solvent peak and the essential peak adjacent to it. The information increases gradually as the amount of internal standard increases.

The strong overlap of the solvent peak and large adjacent peak is the direct cause of the abrupt increase in information loss $\delta\Phi_f$ ($\lambda \leq 245$ nm; see Fig. 20B). Without such critical information loss, the slight increase in the information loss at the wavelength from 280 to 250 nm results from the overlap of neighboring target peaks with different peak areas. The response surface of the information range displays strong undulation along the axis of wavelength because of the diverse spectral patterns of the analytes. The more different the absorbances, the larger the information range, $\Delta\Phi_f$.

The optimum should be chosen from among the conditions that meet the constraints. The constraints ($\Phi_f \geq 17.4$, $\delta\Phi_f \leq 0.0217$, $\Delta\Phi_f \leq 0.434$) adopted here imply that the averaged RSD of measurements should be less than 0.03%, the peak separation should be more than Rs = 1.1, and the ratio of maximal to minimal RSD for measurements does not exceed 2.7 ($= e$). The optimal conditions selected by FUMI are a wavelength of 285 nm and the amount (m_s) of internal standard is 0.2 unit (see Fig. 19).

Rutan et al. studied the thin-layer chromatographic analysis of polyaromatic hydrocarbons by use of the Kalman filter to resolve overlapped chromatographic and fluorescent responses [118]. They used information theory principles to select subsets of an excitation-emission matrix that yielded the highest information.

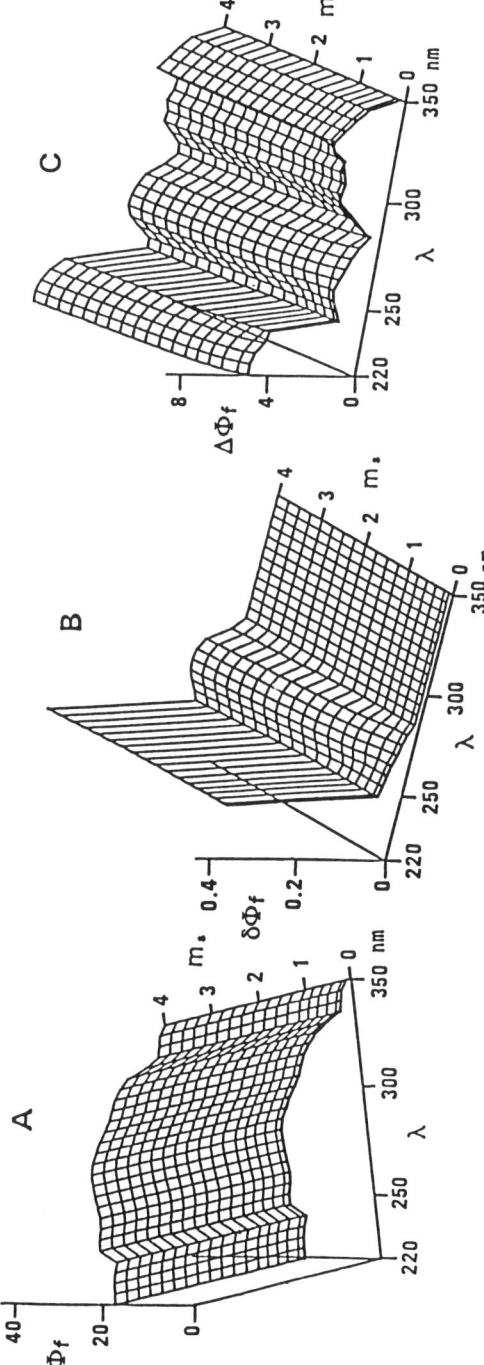

Fig. 20 Response surfaces as a function of the standard quantity m_s and wavelength λ: (A) total information Φ_f; (B) information loss $\delta\Phi_f$; (C) information range $\Delta\Phi_f$. For the amount m_s, see the text. (From Ref. 115.)

D. Merit of Short Columns

The plate number N and resolution R_s are often used as criteria for evaluating chromatographic performance. A high column efficiency N and appropriate separation always lead to high-precision measurements. However, column shortening can result in rapid analysis without significantly degrading the precision in a special case, although the plate number and resolution decrease [119–121]. In this subsection we offer a theoretical and experimental proof for short-column chromatographic assays, recently studied widely [122–127].

Figure 21 shows chromatograms for an antipyretics mixture [119]. Column shortening from 25 (A) to 15 cm (B) accounts for the sharper peaks but reduces

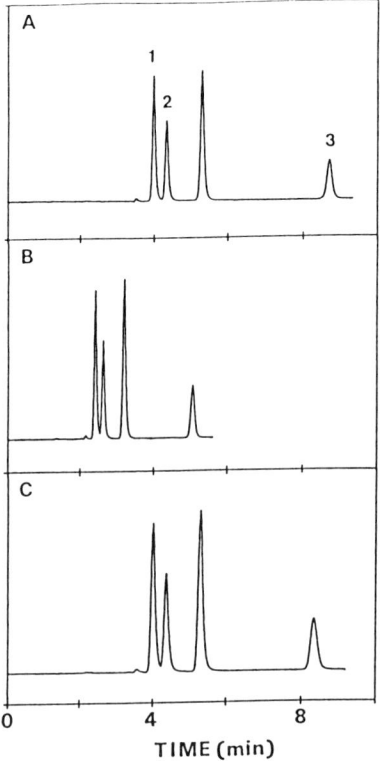

Fig. 21 Merits of short column. Peaks (from left): solvent peak; acetaminophen (30 μg/mL); caffeine (6.25 μg/mL); 8-chlortheophylline (internal standard) (0.6 μg/mL); salicylamide (25 μg/mL). (A) Column length, 25 cm; flow rate, 1 mL/min; (B) column length, 15 cm; flow rate = 1 mL/min; (C) column length = 15 cm; flow rate = 0.6 mL/min. Measured plate number was ca. 12,000 for the 25-cm column (A) and ca. 7000 for the 15-cm column (B and C). (From Ref. 119.)

the resolution and plate number. Peak areas are kept constant in this procedure. Precision increases slightly [$\Phi_f(A) = 13.5$; $\Phi_f(B) = 13.8$] and the throughput increases markedly [$\vartheta_f(A) = 0.0279$; $\vartheta_f(B) = 0.0473$].

When the flow rate is decreased from 1 (B) to 0.6 mL/min (C), the peak areas, A_j, in chromatogram B increase to $5/3A_j$ for chromatogram C, but the resolution and plate number were observed to remain unchanged. The precision increases greatly from chromatogram B ($\Phi_f = 13.8$) to C ($\Phi_f = 14.2$), but the throughput decreases from B ($\vartheta_f = 0.0473$) to C ($\vartheta_f = 0.0293$).

The error arising from sample injection will be the most predominant one in the foregoing macro analysis and every peak will exhibit almost the same precision. In fact, the RSD values for all peaks of Fig. 21 were observed to be 0.2 to 0.4% and the bias less than 2%. According to the uncertainty structure discussed in Section III, if the experiments shown in Fig. 21 were carried out at a far lower concentration of analytes, the statistical merits of the short column would be obtained.

Of course, the results above do not mean simply that a column with a low plate number is preferable. The increase in plate number is essential for resolving overlapped peaks. We should note that the improvement in plate number by column elongation has a disparate perspective from that with the column length kept constant, such as the development of a superior stationary phase [5,94]. The latter always improves the information $\phi_f(j)$, but column elongation spoils the information in cases where there is no peak overlap. In other words, column length attenuates the information.

E. Evaluation of Columns for Optical Resolution

In HPLC, optical resolution techniques based on chiral stationary phases have attracted attention recently, and many chiral stationary phases (CSPs) have come into prominence [128–130]. Next, naphthylethylurea multiple-bonded chiral stationary phases prepared for the optical resolution of *p*-bromophenylcarbamyl (Br-PC) derivatives of enatiomeric amino acids [131–135] are examined from the viewpoint of information theory.

Four naphthylethylurea multiple-bonded chiral stationary phases to be examined have different numbers (n) of polyethyleneamine spacers: diethylenetriamine ($n = 1$), triethylenetetramine ($n = 2$), tetraethylenepentamine ($n = 3$), and pentaethylenehexamine ($n = 4$) [125]. The method for preparing the naphthylethylurea multiple-bonded chiral stationary phases has been described in detail [135].

Figure 22 shows the influence of the number of polyethyleneamine spacers ($n = 1$ to 4) on (A) FUMI and (B) Rs for the pairs of Br–PC amino acid enantiomers [136]. FUMI in Fig. 22 is the total information Φ_f of each peak pair in the Br–PC amino acid enantiomers. The retention times increase with increasing spacer

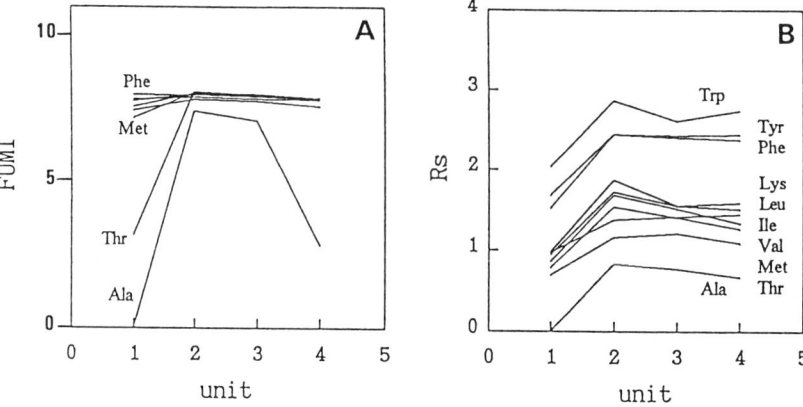

Fig. 22 Dependence of (A) FUMI and (B) Rs on the number of spacers. FUMI is the total information of each pair of Br–PC amino acid enatiomers. Unit denotes the number of polyethyleneamine spacers. The mobile phase was the mixture of 0.15 M acetic acid buffer (pH 5.0) and acetonitrile and the flow rate was 1.0 mL/min. Detection wavelength was 250 nm. The column temperature was maintained at room temperature. Amino acids are Ala, Ser, Thr, Met, Val, Leu, Ile, Phe, Tyr, Lys, and Trp. (From Ref. 136.)

number. The enantiomers of the Br–PC amino acids, except Phe, produce the highest precision, Φ_f, at $n = 2$. Some enantiomers show more resolution Rs for the larger spacers ($n = 3$ or 4). That is, a larger Rs value does not necessarily give more precision, as is apparent from direct experience and Figs. 5 and 12 to 15. For a column in which $n = 1$, FUMI for Ala and Thr has a very small value, due to the peak overlap (Rs $<$ 1.03).

The Rs values for the enantiomers are so widely scattered that we cannot choose the best spacer number by simple inspection. However, the FUMI values for all peak pairs are very similar, except Ala and Thr. This is because FUMI depends not only on the peak overlap but also the peak shape in this situation, where all the peak pairs are sufficiently separated except Ala and Thr. It can thus be concluded that the stationary phase of $n = 2$ is the best solution, because it provides the highest information overall.

F. Simplex Optimization for HPLC

Simplex optimization has been used successfully in HPLC [9–11,19,25, 137–140], since its introduction into analytical chemistry by Long in 1967 [141]. Details of the simplex algorithm and concepts have been written elsewhere [142,143]. Simplex optimization is a systematic trial-and-error method and every step is accompanied by an experiment. In the initial stage, the simplex optimization needs experimental results from some sets of different conditions, and the values of an objective function for these experiments are recorded. Subsequent experiments are selected so that the new results are expected to yield an increasingly better response of the criterion to reach the optimum in the minimal possible number of experiments. Two of the most important advantages of the sequential simplex methods are [16,25,139] that (1) no preconceived model of the retention behavior of solutes is required, and (2) the identification or recognition of solutes in individual separations is unnecessary. That is, the simplex optimization is an alternative option to the development of theoretical and simulation models of retention behavior [144].

As pointed out by Berridge [25], a stumbling block of simplex optimization is the lack of a universally acceptable objective function. Particularly at the early stage of the simplex optimization, fused peaks (only one apparent peak) can appear and the number of solutes in a sample is unknown. In this situation, some separation-based criteria will mistakenly select chromatograms comprising fused or strongly overlapped peaks as the optimum [12]. Thus an ideal objective function will have to satisfy the following prerequisite: If a subsequent experiment provides better separation and the proof that a previously observed peak is made up of two components, the criterion should provide much better response for this updated experiment than for the previous experiment, where the coexistence of the two components was not yet recognizable. FUMI meets the abovementioned prerequisite for an ideal criterion [145].

If a peak-search routine that can recognize peaks with an Rs of more than 0.25 is used for detecting individual peaks, the following situations occur [145]:

1. *Fused* or *pure,* if Rs < 0.25
2. *Strongly overlap,* if $0.25 \leq$ Rs < 1.5
3. *Separate,* if Rs ≥ 1.5

As long as the peaks can be recognized individually by the routine, FUMI can be calculated exactly. However, the following equivocation remains in situation 1: (a) FUMI $= 0$ if the peak is known to be composed of two components; (b) FUMI > 0 if the peak is known to be pure or the coexistence of two peaks is unknown.

Figure 23 demonstrates the model dependence of the total information Φ_f for two peaks on the volume fraction X when the identification of peaks is unknown during the entire optimization process [145]. Chromatograms B to D are illustrated in Fig. 13 with the same letters and retention behaviors as in Fig. 12A. The corresponding response surface of FUMI when the retention behaviors of all the peaks are known is shown in Fig. 12C. The difference between these response surfaces is the existence of the hillock around $X = 18\%$ (Fig. 23), where the elution order of the peaks is reversed and strong peak overlap occurs. The hillock represents the information of the "single" large peak, but this information would be zero if the coexistence of the two peaks were recognized, as shown in Fig. 12C.

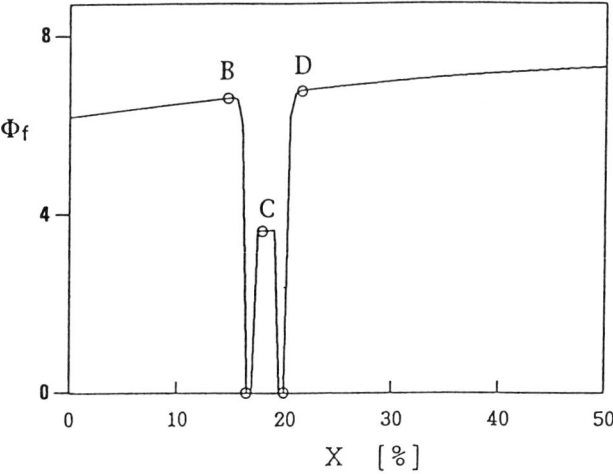

Fig. 23 Dependence of the apparent total information on the mobile-phase composition, X, for two peaks. The conditions are the same as in Fig. 12. (From Ref. 145.)

In any case, the information on the large peak (Fig. 13C) is zero or almost half that of the divided peaks (Fig. 13B or D). That is, the division of a large peak into two peaks produces much more information than does the single large peak, even if the area of the large peak is equal to the sum of the areas of the individual peaks. The simplex optimization is a hill-climbing method and can easily achieve the better conditions (the high ridge where the optimum should be located) than the hillock if the initial conditions of mobile-phase composition (here, two points) range widely enough in the simplex algorithm. Successful application of FUMI to simplex optimization in reversed-phase HPLC analysis of an antipyretics mixture is given elsewhere [145].

G. Optimization of MEKC Analysis

Micellar electrokinetic chromatography (MEKC) has achieved separation of electrically neutral chemicals in the operation mode of capillary electrophoresis (CE) since Terabe and his associates introduced micelle as a pseudostationary phase [146–148]. MEKC retains the merits of CE, such as high column efficiency and micro-scale separation. Details of MEKC concepts and applications can be found elsewhere [146–153]. Some optimization techniques have been published with special regard to MEKC [154–158].

Figure 24 shows the influence of the capacity factor k_j on the information Φ_f, throughput ϑ_f, and resolution Rs for two Gaussian peaks with the separation factor α (ratio of capacity factors) kept constant [158]. Peaks without information

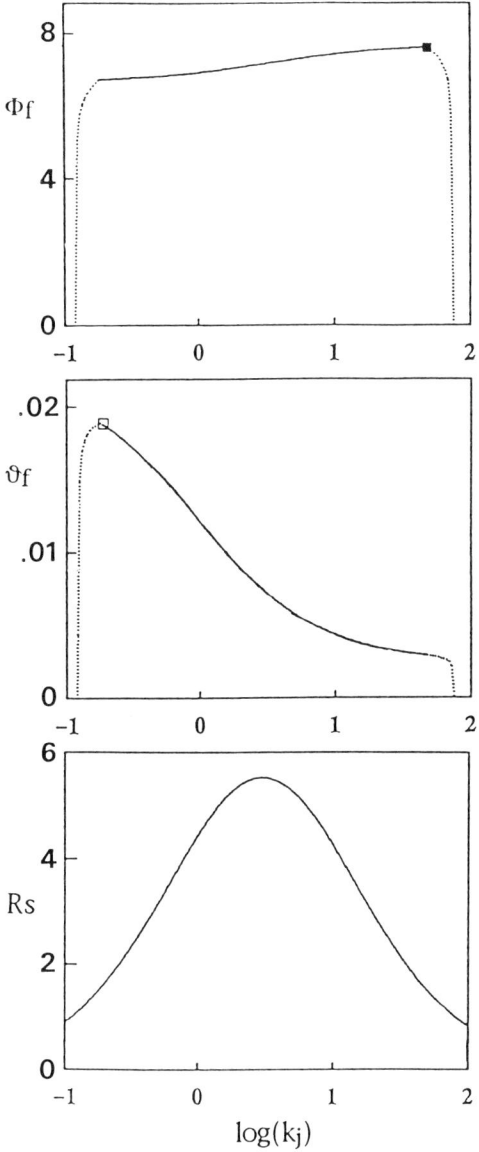

Fig. 24 Precision Φ_f, throughput ϑ_f, and resolution Rs as a function of capacity factor k_j, with separation factor α, kept constant in MEKC ($t_{mc}/t_0 = 10$). $\alpha = 1.05$. The dashed line denotes the peak overlap ($\delta\Phi_f > 0$; Rs < 1.5); the solid line, no overlap ($\delta\Phi_f = 0$; Rs $\geq$ 1.5). The closed square marks the precision optimum (the maximum of Φ_f), and the open square, the throughput optimum (the maximum of ϑ_f). (From Ref. 158.)

loss ($\delta\Phi_f = 0$) are denoted by a solid line, and peaks suffering from overlap ($\delta\Phi_f > 0$), by a dashed line. As the capacity factor increases, the information first takes a zero value and further separation causes an abrupt increase in FUMI. After sufficient separation (Rs = 1.5), the information increase gradually because of the gradual increase in peak area, then reaches a maximum. In the on-column detection of MEKC, the peak area of an analyte increases with increasing migration time, while the height at the peak maximum remains invariant, like the change in flow rate of HPLC. When the capacity factor increases further, the migration times for the two peaks approach the upper limit of the elution range ($= t_{mc}/t_0$) and peak overlap reduces the information. The maximum in FUMI indicates the most precise analysis (precision optimum).

The run time t_q is such a strong influence on the throughput ϑ_f that it acts opposite to the precision Φ_f in the same situation; despite a sufficient amount of information, too long a run time is the immediate cause of a decrease in throughput with increasing capacity factor. The maximum of the throughput with a sufficient amount of precision corresponds to the analysis of the highest throughput. The low throughput at the left side of the throughput optimum results from poor peak separation, although the run time is very short. The separation of the maximal resolution is not tipped as either of the highest precision or throughput.

VII. TOTAL CHROMATOGRAPHIC OPTIMIZATION

Total chromatographic optimization (TOCO) is an attempt to optimize as many important variables as possible in a consistent way. The flowchart of the computer program for TOCO is annexed in Ref. 159.

A. TOCO with FUMI and MEI for Antipyretics Mixture

Here we describe the TOCO with FUMI and MEI as the criteria for many controllable variables in reversed-phase HPLC analysis of an antipyretics mixture [6]. The following six variables are totally optimized: acetonitrile volume fraction in the mobile phase, column length, mobile-phase velocity, detection wavelength, amount of internal standard, and choice of the best internal standard from six candidates. The retention behaviors and ultraviolet spectra of the materials are given in detail elsewhere [6].

All the required criteria and constraints [Eqs. (13) to (16)] are the functions of these six variables. MEI uses the same values of constants I, B, and $\tilde{W}$ as in Fig. 9. The lowest acceptable resolution, $\hat{R}s$, is set at 1.0. This optimization problem is classified into macro analysis.

Figure 25 illustrates the chromatograms observed at the optima of the most precise analysis (A; maximum of FUMI, Φ_f) and of the highest throughput

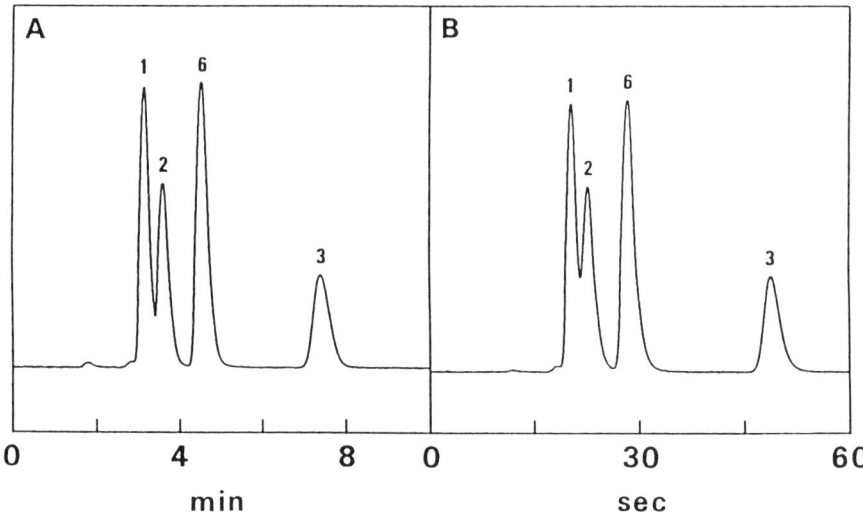

Fig. 25 Chromatograms of (A) precision optimum and (B) throughput optimum in TOCO. Peaks: 1, acetaminophen; 2, caffeine; 3, salicylamide; 6, 8-chlorotheophylline. The best internal standard which was selected from among the six cadidates: 4, guaifenesin; 5, phenol; 6, 8-chlorotheophyline; 7, methyl p-hydroxybenzoate; 8, benzoic acid; 9, p-hydroxybenzoic acid. $\hat{R}s$ = 1.03. A solvent peak appears at the left of peak 1 and is also taken into account by the TOCO. The simulated chromatograms correspond well to the observed chromatograms except for the solvent peaks [8,18]. (From Ref. 5.)

analysis (B; maximum of throughput, ϑ_f). The most significant difference is the observation time: 8 min and 50 s, respectively. Table 5 lists the optimal conditions under which the optimal chromatograms were obtained. The experimental RSD values for the analytes are excellent (= 0.12 to 0.24%; n = 8; see Table 5) at the precision optimum where the theoretical RSD based on FUMI is lowest among all the conditions examined ($\approx$ 0.02%). This optimum is confirmed to be one of the optima selected by the similar TOCO with MEI; the MEI-based theoretical RSD (= 0.25 to 0.26%) is very close to the RSD observed. For the throughput optimum, FUMI and MEI select the same optimal conditions and the RSD values predicted by MEI (= 0.25 to 0.26%) coincide with the observed values (= 0.26 to 0.35%; see Table 5). We conclude that the optima chosen by FUMI yields the best analysis in theory and practice.

The slight overlap (Rs $\approx$ 1.0) appearing in the precision optimum (Fig. 25B) holds for the type of data processing that exhibits high peak-resolving power, such as the Kalman filter. The rather short column (L = 5 cm) for both the precision and throughput optima demonstrates that a simple determination like

Table 5 Conditions for Precision Optimum and Throughput Optimum

	Φ_f optimum	ϑ_f optimum
Column length (cm)	5	5
Mobile phase velocity (mm/s)	0.5	5
Acetonitrile volume fraction (%)	18	19
Wavelength (nm)	280	280
Internal standard	6	6
Amount of internal standard (μg/mL)	9.6	9.6
Analyte	Observed RSD (%) ($n = 8$)[a]	
1	0.12	0.26
2	0.24	0.33
3	0.23	0.35

Source: Ref. 6.
[a] n denotes the number of experiments. For the number of the internal standard, see the legend of Fig. 25.

that in the example above can be carried out more efficiently with a short column than with a more commonly used long column (e.g., 15 cm) [6].

One of the most outstanding advantages of TOCO leads to the general consideration that the optimum reached by the total optimization is so high that the one-variable-at-one-time optimization methods cannot emulate it [14,16]. For example, if a fixed column length or a fixed internal standard is a wrong choice, the TOCO for the remaining five variables proposes optimal conditions that differ from the six-variable TOCO, and the precision of the former is lower than that of the latter [6].

Mobile-phase velocity is the most significant difference between the two optima, as shown in Table 5. Readers might suspect that an operator only had to manipulate this velocity to obtain the throughput optimum from the precision optimum without attention to the other variables. In the example above, this is true, but in general, it is not the case. The TOCO shown here is based on the observation that the plate number was not significantly affected by variations in the velocity. A functional relationship between the plate number and velocity would yield more complicated results from TOCO.

Every chromatogram can be characterized by the numerical values of the precision Φ_f and throughput ϑ_f and can be identified as a point on the space spanned by the precision and throughput. Then continuous change in a controllable variable such as mobile-phase composition draws a curved line on the space.

The pattern of this line will be different from variable to variable, and we can recognize how the variable affects the ultimate analytical aims, the precision and throughput, from the slope of the $\Phi-\vartheta$ line. This plot is called the $\Phi-\vartheta$ plot.

Figure 26 illustrates the $\Phi-\vartheta$ plots for mobile-phase composition (A), column length (B), and mobile-phase velocity (C) at a concentrated sample (the same as Fig. 25) and for mobile-phase composition (D), column length (E), and mobile-phase velocity (F) at a diluted sample. These $\Phi-\vartheta$ lines are drawn based on the experimental results of Fig. 25. The values of the variables and the two types of information, FUMI and MEI, are indicated in the figure. Detailed interpretation of these lines can be found elsewhere [5,7].

With the diluted sample, FUMI and MEI show the same pattern of the $\Phi-\vartheta$ lines, and the precision and throughput optima chosen by FUMI and MEI are the same, except that the flow rate for the precision optimum displays a slight difference between FUMI and MEI (see Fig. 26F). This difference is due to Brownian noise: A broad peak is more susceptible to Brownian noise [σ_j^3 of the third term of Eq. (3b)] than to the signal-processing error of FUMI [σ_j of the first term of Eq. (3b)].

With the concentrated sample, the precision indicated by MEI is almost constant in every condition examined, unless the peaks are overlapped. The most precise analysis identified by FUMI is included in the wide range of the precision optimum shown by the solid line for MEI. On the other hand, the throughput changes drastically depending on the operating conditions. The throughput optima identified by FUMI and MEI are consistent in this situation. Thus FUMI can again be concluded to be a simple but statistically reliable criterion. This fact demonstrates that if peak separation is easily achieved, as in some drug analyses, we have a much greater chance to improve the throughput rather than the precision. Industrial process control engineers who treat copious samples every day will prefer maximal throughput rather than maximal precision.

B. Throughput for Dissolution Test Using Robotics

The kinetic profile of the dissolution of active ingredients from a solid dosage formulation into surrounding aqueous solution is a valid means of assessing the effectiveness of a drug. In general, the dissolution speed varies not only from formulation to formulation, but also from ingredient to ingredient [160]. The analytical applications of robotics to dissolution test and other chemical analyses can be found elsewhere [161–165].

A dissolution test introduced here is made up of four steps [165]:

1. Each formulation is delivered into a glass vessel in the dissolution apparatus.
2. The apparatus is started by stirring the dissolution medium with a paddle.

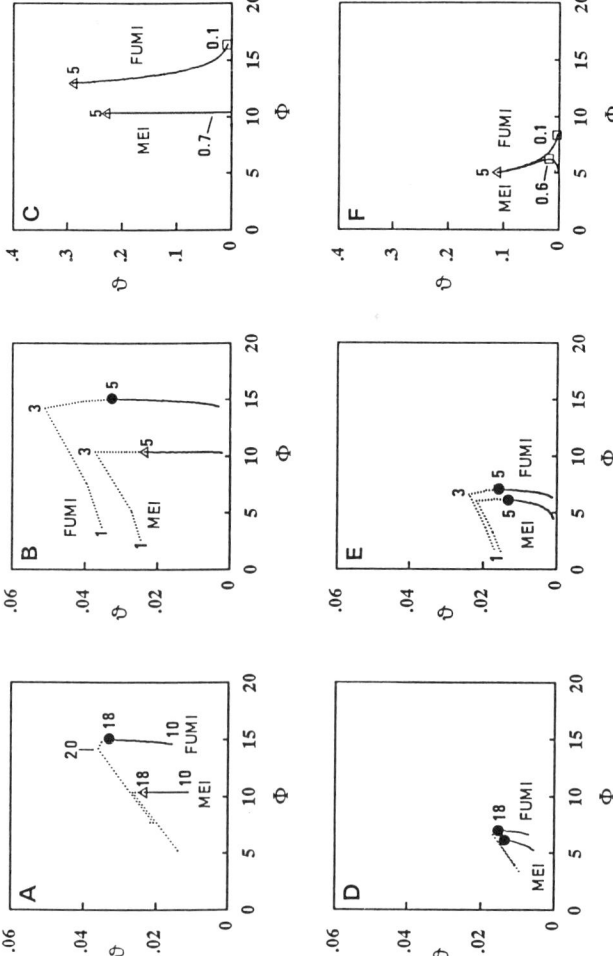

Fig. 26 Precision–throughput plots (Φ–ϑ plots) for mobile-phase composition (A), column length (B), and mobile-phase velocity (C) at a high sample concentration and for mobile-phase composition (D), column length (E), and mobile-phase velocity (F) at a low sample concentration. The analyte concentration of (A)–(C) is 100 times that of (D)–(F). ●, Precision and throughput optimum; □, precision optimum; △, throughput optimum.

3. The samples are taken and prepared for measurement by filtering the sample solution and adding an internal standard (8-chlorotheophyline).
4. The system performs HPLC measurement.

A robotics system (Zymate II) is used for achieving a fully automated dissolution test. The formulations used are slow-releasing capsules consisting of acetaminophen, salicylamide, and caffeine.

The dissolution test is a macro analysis in many cases. As mentioned in Section VII.A, the throughput is worth considering in the test. The appropriate sampling intervals are 15 min to follow the kinetics, and three formulations are observed simultaneously. Then the run time of the HPLC analysis should fall below 5 min each formulation for the sake of efficiency.

For this problem, the TOCO with FUMI selects the analysis of the highest throughput, which is similar to the results of Section VII.A [165]. Figure 27 illustrates the chromatogram observed at the throughput optimum. The limit resolution $\hat{R}s$ is fixed at 1.5 for a commercial integrator. The analysis time is quite short ($= 1.1$ min). The RSD observed ($\approx 0.2\%$; $n = 6$) is satisfactory for dissolution testing. The dissolution kinetics for capsules prepared from slow-releasing granules is shown in Fig. 28. The different kinetics of the ingredients can be distinguished clearly.

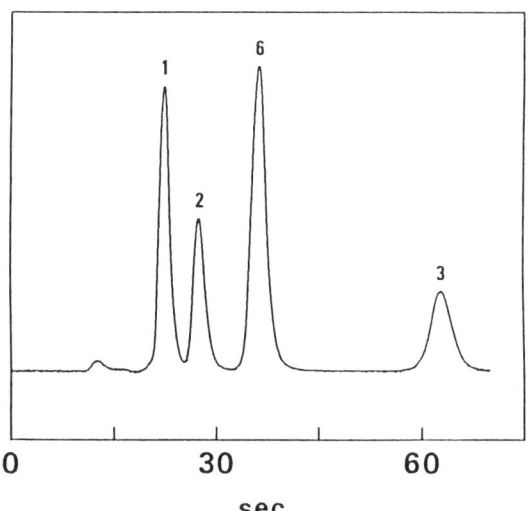

Fig. 27 Chromatogram of throughput optimum with another $\hat{R}s$ ($= 1.53$). This is the result from the same TOCO as Fig. 25 except for $\hat{R}s$. $\Phi_f = 29.62$; $\vartheta_f = 0.603$. (From Ref. 5.)

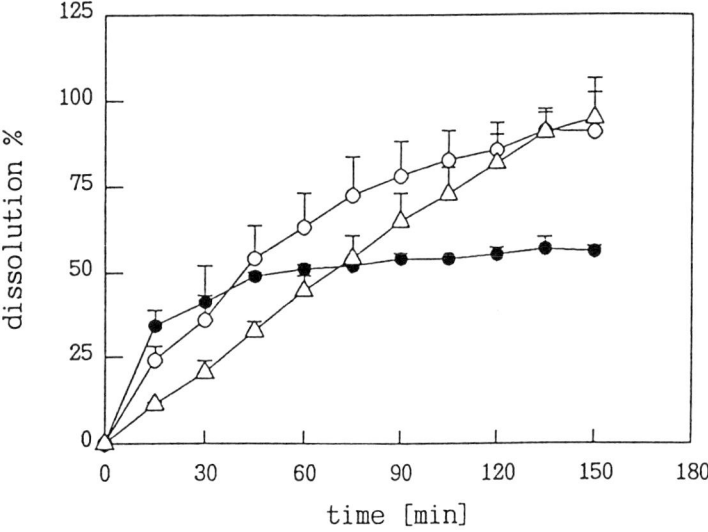

Fig. 28 Dissolution kinetics of acetaminophen (○), caffeine (●), and salicylamide (△) contained in slow-releasing capsules. (From Ref. 165.)

The HPLC measurement could be the rate-determining step for the dissolution test. However, the sufficiently short run time (below 4 min for the three measurements) is due to the short column (5 cm long) and slightly high mobile-phase velocity (5 mm/s) chosen as optimum by the TOCO. If the chromatography was conducted with a conventional 15-cm column at the usual mobile-phase velocity (2 mm/s), the analysis time would be prolonged and the sampling intervals would have to be increased.

C. Validity of Internal Standard

The internal standard technique has been used widely in chromatography [166,167]. Haefelfinger studied this technique statistically with various kinds of experiments [166]. The conclusion of this work is that in practice, some form of external calibration is often advantageous compared to the use of the internal standard technique [166]. Here we demonstrate that the usefulness of an internal standard added just before the measurement step depends greatly on the amount of analyte. The validity of internal standard can be deduced from the uncertainty structure in analytical systems.

In the internal standard technique, the quotient (= the area ratio of peak j to internal standard peak I) is used rather than the absolute values for these peaks. The law of error propagation tells us that if the numerator and denominator are probabilistically independent, the squared RSD of the quotient is equal to the sum of the squared RSD values of the target peak (RSD_j) and internal standard (RSD_I). However, if these terms are correlated, the squared RSD of the quotient is reduced from the squared sum above by $2rRSD_jRSD_I$, where r denotes the correlation coefficient [166].

The three types of errors existing in the measurement and signal-processing steps as shown in Eq. (3b) are taken into account. The signal-processing error [the first term of Eq. (3b)] and Brownian error [the third term of Eq. (3)] are probabilistically independent from peak to peak, but the injection error, I, will show a strong, if not complete, correlation between different peaks. Therefore, based on the law of error propagation, we can see that the RSD of measurements for peak j with peak I as an internal standard can be approximated by

$$(RSD_{j/I})^2 = \frac{2\pi^{1/2}\sigma_j \Delta T \tilde{W}^2}{A_j^2} + \frac{\sigma_j^3 B^2}{A_j^2} + \frac{2\pi^{1/2}\sigma_I \Delta T \tilde{W}^2}{A_I^2} + \frac{\sigma_I^3 B^2}{A_I^2} \qquad (18)$$

where no peak overlap is assumed and $RSD_{j/I}$ denotes the RSD for the quotient of the two measurements. The relative variance of the internal standard technique [$= (RSD_{j/I})^2$] is less by $2I^2$ than the sum of the relative variances for the individual measurements because of the correlation: $(RSD_{j/I})^2 < RSD_j^2 + RSD_I^2$. The correlation coefficient is $I^2/(RSD_jRSD_I)$.

Figure 29 shows the plots of the observed RSD_j values against the analyte amount injected into an HPLC system with peak 3 as an internal standard. The experimental conditions are the same as in Fig. 9. The theoretical curves are obtained from Eq. (18) and the dashed lines are the same as the solid theoretical lines in Fig. 9 without internal standard. Similar experiments in CE analysis are shown in Fig. 30. The detailed experimental conditions are given in the figure legend.

If the sample amount is large, the major source of experimental error is the injection and each term of Eq. (18) will be even smaller than the injection error I. Thus the internal standard technique will be quite effective compared with the external calibration, which involves the relatively larger imprecision from the injection [Eq. (3b)]: $RSD_{j/I} < RSD_j$. In trace analysis, however, the internal standard itself suffers a large imprecision and the external calibration is more profitable. Equation (18), which involves twice the signal-processing error and Brownian error predominant in this situation, will be larger than Eq. (3b). Thus $RSD_{j/I} > RSD_j$. Note that the theoretical RSD lines of internal standard and of no internal standard cross over the concentration range in the CE analysis (see Fig. 30). This is not so noticeable in Fig. 29.

Unfortunately, the validity of the internal standard technique is not experimentally evident for the CE analysis in Fig. 30, possibly because of the variation in the migration times of analytes, which directly affects the peak area. The loop

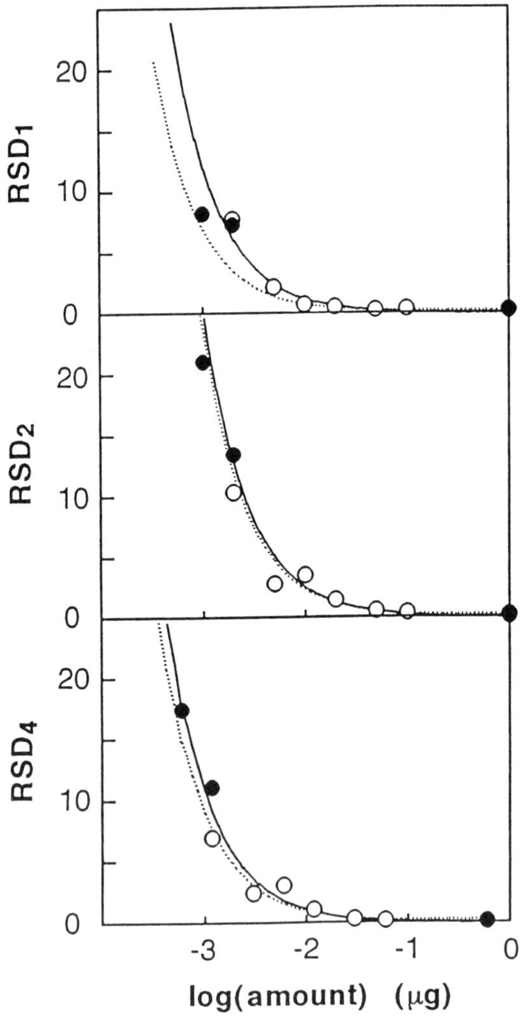

Fig. 29 RSD for the same HPLC analysis as Fig. 9 except for peak 3 used as the internal standard.

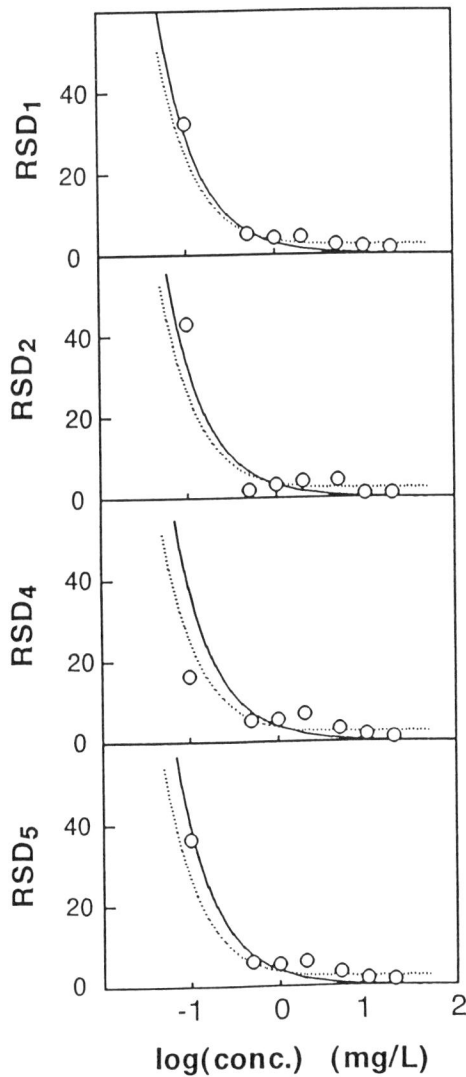

Fig. 30 RSD for CE analysis for sulfa drugs (peak 1, sulfaquinoxaline; 2, sulfamonomethoxine; 3, sulfadimidine; 4, oxolinic acid; 5, piromidic acid) with peak 3 as the internal standard ($n = 8$). The CE analysis was performed with Quanta 4000 system (Milipore Ltd.) equipped with 75 μm ID $\times$ 65 cm (effective length 60 cm; Milipore Ltd.) fused silica capillary. The separation solution was mixture of 25 mM sodium borate and 100 mM sodium monobasic phosphate adjusted at pH 8.0. Detection wavelength was 240 nm. Room temperature was maintained at 26 to 31°C. Applied voltage was 10 kV.

injector of HPLC is regarded as superior ($I = 0.24\%$) to the gravimetrical injector of CE ($I = 2.5\%$). Then the technique for HPLC will not be so effective as for CE (see Fig. 29). Of course, the foregoing consideration may change if the errors in the premeasurement step, such as sample preparation, are also taken into account.

D. Simplified TOCO with FUMI

A facile technique for TOCO with FUMI is explained here [168]. All the necessary quantities for calculating the information are peak area A_j, height h_j (or width σ_j), and retention time t_j. The calculation of the information loss and noise variance [see Eq. (5b)] is dispensed with [168]. Then the quantities provided by a commercial integrator will suffice for the optimization.

The foregoing technique is similar in a sense to the Rs-minimum method [9]. The latter is an optimization strategy with the run time as a criterion and the constraint is good separation (e.g., Rs $\geq$ 1.5 for every peak pair). The optimum is defined as the conditions that provide the shortest run time with sufficient separations. The constraint of the simplified TOCO is the same as that of the Rs-minimum method, but the objective function is the mutual information (FUMI).

The entire process with the simplified TOCO is:

1. Determine the lowest acceptable resolution, $\hat{R}s$ (e.g., $\hat{R}s = 1.5$).
2. Select the conditions (or chromatograms) under which every peak pair has Rs more than the limit, $\hat{R}s$.
3. For one of the conditions selected, calculate the mutual information for all the peaks according to one of the following equations:

$$\phi_f(j) = \frac{1}{2} \log \left(\frac{A_j^2}{2\pi^{1/2} \sigma_j} \right) \tag{19a}$$

$$= \frac{1}{2} \log \left(\frac{A_j h_j}{2^{1/2}} \right) \tag{19b}$$

Assuming that $\Delta T \tilde{W}^2 = 1$ and $\delta \phi_f(j) = 0$ and noticing that $h_j = A_j/[(2\pi)^{1/2}\sigma_j]$, we can obtain these equations from Eq. (5b).

4. For the condition above, calculate the precision Φ_f and throughput ϑ_f according to Eqs. (6) and (9a), respectively.
5. Repeat the calculation of steps 3 and 4 for all the conditions selected in step 2.
6. Choose the most precise analysis of the maximal Φ_f and the highest throughput analysis of the maximal ϑ_f.

The method described above is useful for the optimization, especially of the mobile-phase composition and column length. The optima identified by the method coincided with those identified by the orthodox FUMI optimization for an HPLC analysis of an antipyretics mixture [168].

VIII. FACTORS AFFECTING PRECISION AND THROUGHPUT

Various kinds of factors affect the solution to optimization problems in separation science. The instances given in this section cannot be ignored in establishing the versatility of an optimization system.

A. Sample Size

As noted in Sections III, VI, and VII, the precision of measurements depends greatly on sample size. Therefore, the experimental design and system optimization should be performed with careful reference to the sample size. Unfortunately, the traditional optimization methods based on the resolution, called optimization of separation, cannot draw the foregoing conclusion, because Rs does not include peak area and bears no overall correlation with the statistic.

To recapitulate:

1. In macro analysis, the throughput [Eq. (9)] is worth considering, but the precision [Eq. (6) or (8)] is not critical, if the peaks are separated sufficiently and are sharp enough. The throughput optimum is suitable for an assay of numerous samples such as for industrial process control, where the highest time efficiency is more important than the highest precision.
2. In trace analysis, the operating conditions should be selected carefully according to the precision rather than the throughput.

B. Peak-Resolving Power of Data Processing

Any type of data processing such as the least squares has its own inherent lowest acceptable resolution, $\hat{R}s$. Signal processing cannot provide exact data for a separation below this lowest resolution because such a strong overlap is beyond the peak-resolving power of the mathematical method. Far above the limit $\hat{R}s$, peak broadening spoils the precision of estimates in the chromatographic situations. In some situations the lower limit, $\hat{R}s$, coincides with the resolution observed at the optimal conditions and is called the optimal separation. As would be imagined, $\hat{R}s$ has relevance to the accuracy for quantitation and the purity and recovery for preparative chromatography.

Data handlings of different powers cannot rely on the same lowest acceptable separation, $\hat{R}s$ [159]. Table 6 demonstrates that the Kalman filter and perpendicular dropping provide different accuracy and precision from the same data of

Table 6 Accuracy and Precision of Various Data Processing Methods

	Kalman filter		Perpendicular dropping	
	Mean[a]	RSD	Mean[a]	RSD
Naphthalene	100.2	0.40	90.9	0.80
Diphenyl	100.1	0.64	110.6	0.70

Source: Ref. 169.
[a]One hundred is assumed to be the true value.

naphthalene and diphenyl (Rs ≈ 1) [169], although the difference in the latter statistic is not very clear. The bias of the Kalman filter is excellent (≈ 0.2%) but is not acceptable for perpendicular dropping (≈ 10%).

In order to describe the different powers of various signal processings in the FUMI theory, we have to know the effect of the peak overlap on FUMI. Figure

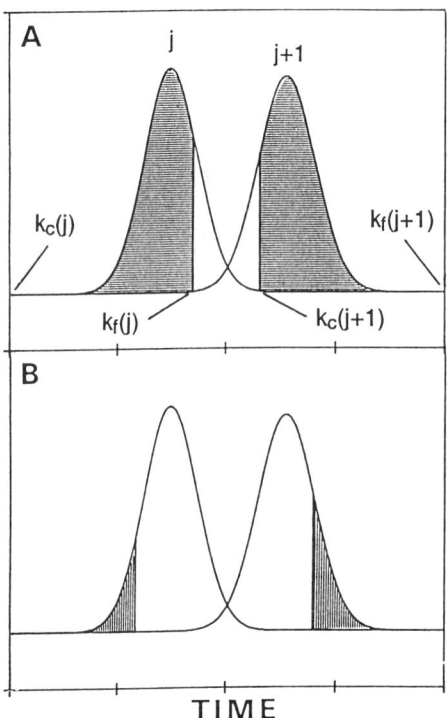

Fig. 31 Peak separation, information $\phi_f(j)$ and information loss $\delta\phi_f(j)$: (A) $\hat{R}s = 1.03$; (B) $\hat{R}s = 1.53$. (From Ref. 159.)

31A demonstrates that the shaded region, $(k_c(j), k_f(j))$, and the clear region, $(k_f(j), k_c(j + 1))$, correspond, though indirectly, to the amounts of FUMI, $\phi_f(j)$, and information loss, $\delta\phi_f(j)$, respectively, for peaks j and $j + 1$. The shape of the entire peak denotes the intact information [the first term of Eq. (5b)], which is independent of the overlap.

In theory, FUMI, $\phi_f(j)$, consists of the sum of the squared signal intensities of peak j over the shaded region where no interfering signals from other peaks exist. In Fig. 31A, signals from the adjacent peak $j + 1$ greater than 0.05% of the apex of peak j are excluded from the overlap-free region, $(k_c(j), k_f(j))$. Therefore, if the second peak $j + 1$ elutes later and overlaps less strongly with peak j, the overlap-free region, $(k_c(j), k_f(j))$, for peak j widens. This increases FUMI and accordingly decreases the loss, $\delta\phi_f(j)$. If the second peak elutes earlier, the shaded region, $(k_c(j), k_f(j))$, narrows and FUMI decreases. If peak j overlaps with another earlier peak $j - 1$, FUMI is delineated by a shaded Gaussian peak lacking both edges.

The upper limit of the interfering signal intensity ($=$ 0.05% of the peak apex in the example above) can be connected with the lowest acceptable resolution, $\hat{R}s$, for peaks having the same areas. For example, $\hat{R}s = 1.03$ if the upper limit of the interfering signals is 0.05%; $\hat{R}s = 1.53$ if it is 0.000002%.

Inferior data processing can be considered more susceptible to signal interferences, and a smaller value for the interfering signal limit and a larger optimal resolution, $\hat{R}s$, should be assigned to such signal processing. Figure 31B shows the information (shaded region) and information loss (clear region) for the optimal resolution of 1.53. Although resolution is the same for Fig. 31A and B, the large value of the optimal resolution produces a large information loss and relatively little information.

The different settings of the optimal resolution, $\hat{R}s$, lead to different solutions to optimization problems in chromatography. Figure 32 illustrates the optimal chromatograms for $\hat{R}s$ of 1.03 (A) and for $\hat{R}s$ of 1.53 (B) in the optimization of mobile-phase composition for an antipyretics mixture [159]. Optimal chromatogram A, comprising slightly overlapped peaks, is well suited for the Kalman filter with its excellent peak-resolving power. On the other hand, FUMI, with its large value of $\hat{R}s$ ($= 1.53$), selects chromatogram B, involving baseline-separated peaks as the optimum, which can be evaluated exactly by inferior data processing, such as a commercial integrator.

Different amounts of mutual information are collected by the various methods described above. Chromatogram A provides a FUMI value of 15.7 and chromatogram B, one of 15.2. This demonstrates that inferior data processing can elicit only less information, even from its own optimum (chromatogram B), than a superior form of data processing can from its own optimum (chromatogram A). Optimal chromatogram A may be considered an efficient condensation of the useful information for the superioir data handling, but may be too condensed for the inferior paradigm.

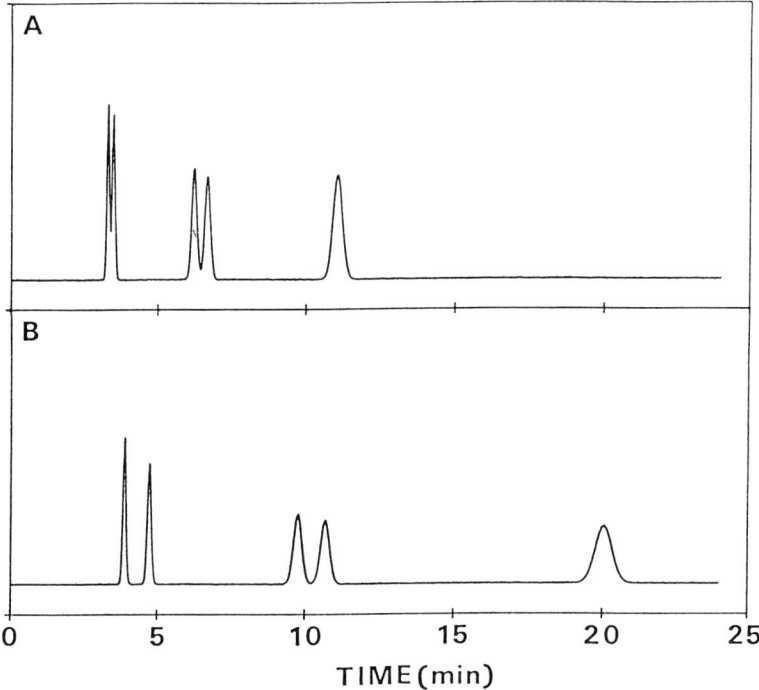

Fig. 32 Optimal chromatograms (simulation) of an antipyretics mixture in mobile-phase composition. A: the optimum for $\hat{R}s = 1.03$; B: the optimum for $\hat{R}s = 1.53$. The solutes are the same as those in Fig. 19. (From Ref. 159.)

Even if the peaks are separated sufficiently, different signal processings provide different precision and information. Figure 33A shows correlation of the RSD values of estimates between the least-squares method and integration (summation of raw data over a fixed time period) in the HPLC analysis in Figs. 8 and 9. When the sample concentration is high and the RSD values are less than 2%, the two types of signal processing are mutually commensurate. However, in trace analysis, data from the least squares (e.g., RSD = 10%) are much more reliable than those from integration (RSD ≈ 15 to 20%). That is, the information (= log [1/(observed RSD)]) obtained through least squares is higher than that obtained through the integration in trace analysis (see Fig. 33B). Chemometrics will play an important role in trace analysis.

C. Area Ratio of Adjacent Peaks

In chromatography, situations often arise where it is necessary to quantify a small peak adjacent to a large inessential (disturbing) peak (e.g., an assay of a small

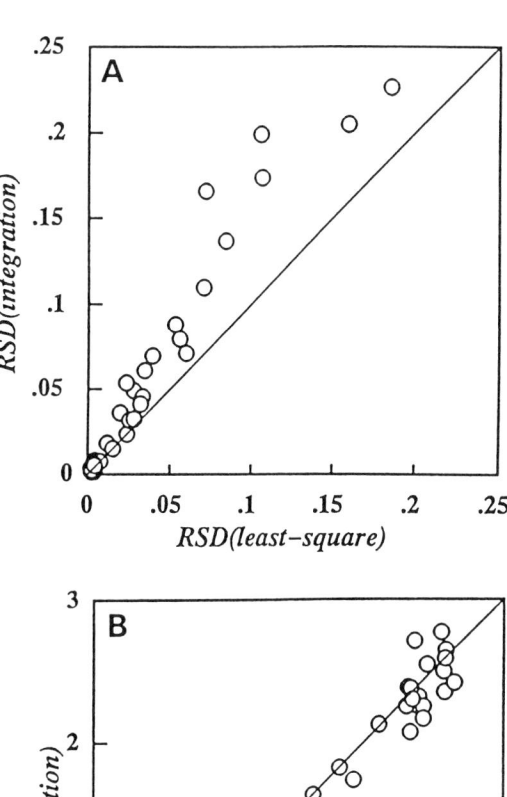

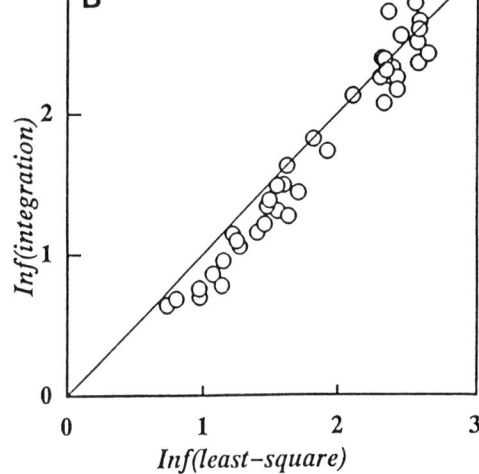

Fig. 33 Correlation of RSD (A) and information (B) between least squares and integration. The experimental conditions are the same as in Figs. 8 and 9. The information is defined as log (1/RSD). The integration is the summation of the raw data over a fixed time period ($\pm 3\sigma_j$), which can vary if an unknown peak is near the target peak but which is invariant for the same peaks irrespective of concentration. The model peak of the least squares comes from the smooth shape of the peaks at the highest sample concentration.

impurity of an optical isomer present only in a small amount). Also, analysis of the composition and decomposition products or impurities in drug or food is often required, especially in pharmacology and toxicology [170]. There are some approaches to the "relative size" effects on the analytical purposes of chromatography (i.e., the accuracy and precision for quantitation and the purity and recovery for preparative chromatography) [71–75,138,171–174].

It is well known that peaks of widely different areas must be separated to a greater extent for successful determination than peaks of similar areas. Figure 34A and C illustrate peaks having different areas but the same resolution Rs ($= 1.2$) [174]. The peaks in Fig. 34A constitute a satisfactory separation, but the separation in Fig. 34C can hardly be recommended for precise analysis. Therefore, it is essential that we fully understand the quantitative relationship between the optimal peak separation $\hat{R}s$ and the area ratio A_2/A_1 of adjacent peaks 1 and 2 for quantitative analysis.

First, the highest acceptable limit of the interfering signal intensities of the other peaks should be set at an appropriate value according to the peak-resolving

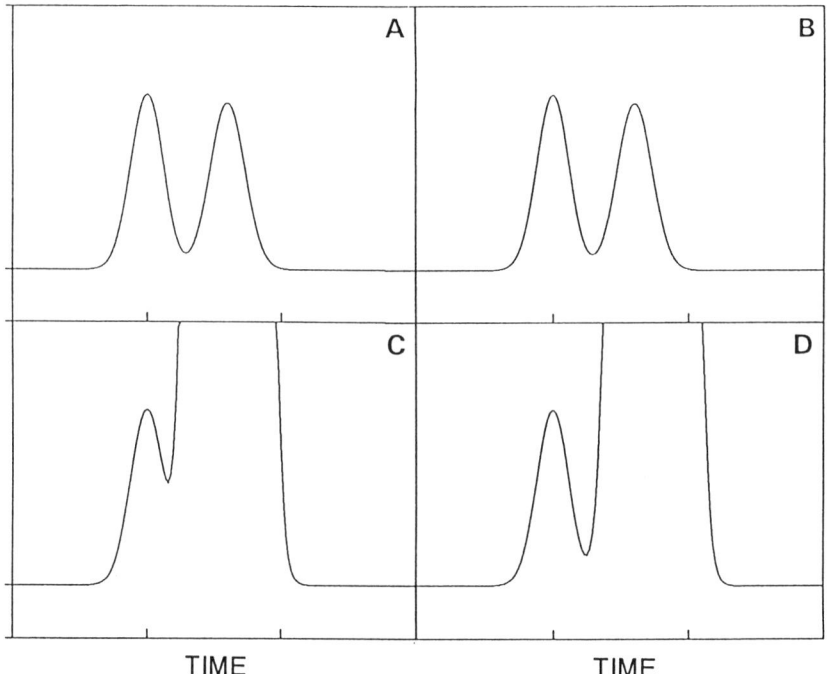

Fig. 34 Peak separations with the same Rs ($= 1.22$) but different areas [(A) and (C)] and their optimal separations [(B) and (D)]. The area ratio of peak 2 to peak 1 is 1 [(A) and (B)] or 100 [(C) and (D)]. (From Ref. 174.)

power of a data-handling method used. This value, in turn, sets a lower resolution limit $\hat{R}s$ for the peaks of the same areas (here, $\hat{R}s = 1.2$).

Let the left peak ($j = 1$) be the target peak and the right peak ($j = 2$) be the inessential (disturbing) peak. An earlier elution of the right inessential peak from the optimal resolution causes a little more overlap with the target peak, reducing the information of the target peak. The optimal separation is defined to be the least resolution without information loss. If the area of the disturbing right peak is larger and if the highest acceptable limit of the disturbing signals is kept constant, the optimal resolution should be larger because of the larger interference of the disturbing peak signals. The larger the area of the disturbing peak, the larger the optimal peak resolution. The quantitative relationship between the optimal resolution and peak area (Fig. 35) is obtained along this line [174].

Figure 34B and D show the optimal peak separations in Fig. 34A and C, respectively. The position of the left peak is fixed and that of the right peak at the optimal condition is determined according to the quantitative relationship in Fig. 35. The optimal separation $\hat{R}s$ is 1.2 (B); $A_2/A_1 = 1$) and 1.5 (D; $A_2/A_1 = 100$). The optimal resolution in Fig. 34B is equal to the initial setting of $\hat{R}s$ ($= 1.2$), but the optimal separation in Fig. 34D is clearly different from the separation in Fig. 34C.

The FUMI theory guarantees that the signal processing used will be able to collect the highest amount of information or precision from the optimal chromatogram (Fig. 34D). The quantitative relationship between the peak area ratio and optimal separation will be a useful indicator for many problems mentioned at the

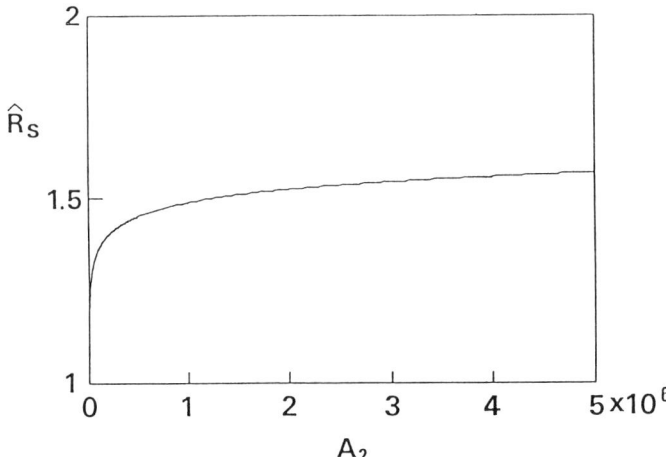

Fig. 35 Dependence of the optimal resolution $\hat{R}s$ on peak area. $A_1 = 10,000$; $N = 10,000$. $\hat{R}s = 1.22$ for peaks of the same areas. (From Ref. 174.)

beginning of this subsection. If the preset optimal resolution $\hat{R}s$ assumes a different value, the quantitative relationship will be shifted.

D. Unresolvable Peaks

The appearance of unresolved peaks even in the "best" chromatogram is not uncommon in practical situations. The constraint ($\delta\Phi_f = 0$) should be ignored because of a large amount of information loss, $\delta\Phi_f$, remaining even under the best conditions. The FUMI theory can also be applied successfully to this nonlinear programming problem, lacking the constraint ($\delta\Phi_f = 0$) that is essential for successful optimization of wavelength and amount of internal standard [106].

E. Interference of Unnecessary Peaks

The total information Φ_f is the sum of the individual information $\phi_f(j)$ for all the target peaks [see Eq. (6)]. If both peaks in Fig. 31A are to be quantified, $\Phi_f = \phi_f(j) + \phi_f(j + 1)$. If the early peak j is the target and the late peak $j + 1$ is not, $\Phi_f = \phi_f(j)$. The overlap effect due to the interfering peak $j + 1$ remains in the information loss of the essential peak j, although the information of peak $j + 1$ is eliminated from the total information, $\Phi_f [= \phi_f(j)]$. This is the reason why FUMI can easily be applied to optimization problems involving interfering peaks. A solvent peak can also be taken into account if the overlap-free region of a peak neighboring the solvent peak is determined. Of course, the information on the solvent peak does not contribute to the total information.

F. System Miniaturization

Interest in the miniaturization of column liquid chromatography has increased. It has many advantages, as discussed by Ishii and Takeuchi [175,176], including increased resolution, decreased solvent consumption, lower heat capacity, increased mass sensitivity, and higher compatibility with various detectors and secondary chromatography systems. Microbore columns, microcolumns, capillary columns, narrow-bore columns, and small-bore columns are the subjects of this kind of study [177–180].

A chromatograph should be scaled down without spoiling the analytical integrity of every module. Suppose that a micro-scale HPLC system has a very low solvent consumption but poor "precision" for target peaks because of band broadening or overlap caused by a failure in the instrumental requirements. In this situation, solvent consumption is no longer a useful indicator for characterizing the social merits of the system.

The influence on the precision and throughput of reducing the column diameter and length was studied in reversed-phase HPLC analysis for paraben food additives [181]. The consumption of organic solvent, which is often associated with environmental pollution, was expressed as the solvent throughput [Eq.

(9a)]. Limitations on the degree of system miniaturization have also been discussed [181].

IX. OUTLOOK

To date, system optimization and evaluation have been described primarily in terms of information theory and the uncertainty structure of quantitative analysis. Many controllable variables can be optimized totally (e.g., mobile-phase composition, detection wavelength, column length, flow rate, etc.). Furthermore, environmental parameters such as the detection unit and temperature control can also be evaluated by FUMI theory [MEI; Eq. (3b)].

It is true that analytical chemists strongly desire formal strategies for optimal decision making at all levels of analytical chemistry [144] and that identification of the error sources and assessment of the error magnitude in each source are essential in any chemical analysis from a statistical viewpoint [182]. However, our attention has been confined to the system optimization shown in Fig. 1. Further development of FUMI theory is likely to have two directions. Longitudinal development comprises the probabilistic error prediction of every step in complete quantitative analysis, including sample preparation. Our next goal will be the theoretical prediction of total error from sample preparation to the final step of analysis, when the use of pipettes, volumetric flask, and so on, is specified and a suitable measurement system is selected for an experimental design. Complete quantitative analysis could be optimized according to precision and throughput.

In many cases the uncertainty of a sample preparation step could not be estimated until the substances in a sample were determined. The errors from sample preparation and measurement steps have been treated so far as a lumped value [166,167]. The uncertainty structure of HPLC and CE [Eq. (3)] is well suited to the task of evaluating individually the elementary steps in chemical analysis.

Application of FUMI theory to other methods of measurement in analytical chemistry, such as titrimetry, flow injection analysis, and electronic balance, is referred to as latitudinal development. The error from a typical example of titration was expressed, qualitatively, by FUMI [183]. The bidirectional development described above is now under investigation by our group. The availability of FUMI is expected to extend to expert systems [21,184–197], validation [198], and so on.

ACKNOWLEDGMENT

The authors wish to express their gratitude to T. Maruyama of Waters, Chromatography Division, for supporting the CE equipment used to generate the data in Fig. 30.

REFERENCES

1. R. A. Day, Jr. and A. L. Underwood, *Quantitative Analysis,* Prentice Hall, Englewood Cliffs, N.J., 1991.
2. D. L. Massart, B. G. M. Vandeginste, S. N. Deming, Y. Michotte, and L. Kaufman, *Chemometrics: A Textbook,* Elsevier, Amsterdam, 1988.
3. M. A. Sharaf, D. L. Illman, and B. R. Kowalski, *Chemometrics,* Wiley, New York, 1986.
4. S. D. Brown, R. S. Bear, Jr., and T. B. Blank, *Anal. Chem., 64:* 22R (1992).
5. Y. Hayashi and R. Matsuda, *Chemometr. Intel. Lab. Syst., 18:* 1 (1993).
6. Y. Hayashi, R. Matsuda, and A. Nakamura, *Chromatographia, 30:* 85 (1990).
7. Y. Hayashi and R. Matsuda, *Chromatographia, 30:* 171 (1990).
8. Y. Hayashi and R. Matsuda, *Analytical Methods & Instrumentation,* submitted.
9. P. J. Schoenmakers, *Optimization of Chromatographic Selectivity,* Elsevier, Amsterdam, 1986.
10. J. C. Berridge, *Chemometr. Intel. Lab. Syst., 3:* 175 (1988).
11. J. C. Berridge, *Chemometr. Intel. Lab. Syst., 5:* 195 (1989).
12. H. J. G. Debets, B. L. Bajema, and D. A. Doornbos, *Anal. Chim. Acta, 151:* 131 (1983).
13. C. E. Goewie, *J. Liq. Chromatogr., 9:* 1431 (1986).
14. S. N. Deming, J. G. Bower, and K. D. Bower, *Adv. Chromatogr., 24:* 35 (1984).
15. H. J. Issaq, *Adv. Chromatogr., 24:* 55 (1984).
16. L. de Galan and H. A. H. Billiet, *Adv. Chromatogr., 25:* 63 (1986).
17. Y. Hayashi and R. Matsuda, *Bunseki, 1990:* 901 (1990).
18. Y. Hayashi and R. Matsuda, *Bunseki, 1991:* 57 (1991).
19. S. N. Deming, J. M. Palasota, J. Lee, and L. Sun, *J. Chromatogr., 485:* 15 (1989).
20. E. P. Lankmayr, W. Wegscheider, J. C. Gfeller, N. M. Djordjevic, and B. Schreiber, *J. Chromatogr., 485:* 183 (1989).
21. P. J. Schoenmakers and N. Dunand, *J. Chromatogr., 485:* 219 (1989).
22. K. Jinno, M. Yamagami, and M. Kuwajima, *J. Chromatogr., 485:* 461 (1989).
23. J. L. Glajch and J. J. Kirkland, *J. Chromatogr., 485:* 51 (1989).
24. H. A. H. Billiet and L. de Galan, *J. Chromatogr., 485:* 27 (1989).
25. J. C. Berridge, *J. Chromatogr., 485:* 3 (1989).
26. P. M. J. Coenegracht, A. K. Smilde, H. J. Metting, and D. A. Doornbos, *J. Chromatogr., 485:* 195 (1989).

27. D. R. Van Hare and L. B. Rogers, *Anal. Chem., 57:* 628 (1989).
28. L. R. Snyder, J. W. Dolan, and D. C. Lommen, *J. Chromatogr., 485:* 65 (1989).
29. S. N. Deming, *J. Chromatogr., 550:* 15 (1991).
30. J. A. Palasota, I. Leonidou, J. M. Palasota, H.-L. Chang, and S. N. Deming, *Anal. Chim. Acta, 270:* 101 (1992).
31. B. Bourguignon and D. L. Massart, *J. Chromatogr., 586:* 11 (1991).
32. K. Eckschlager and V. Štěpánek, *Anal. Chem., 54:* 1115A (1982).
33. K. Eckschlager, V. Štěpánek, and K. Danzer, *J. Chemometr., 4:* 195 (1990).
34. K. Eckschlager and V. Štěpánek, *Information Theory as Applied to Chemical Analysis,* Wiley, New York, 1979.
35. K. Eckschlager and V. Štěpánek, *Chemometr. Intel. Lab. Syst., 1:* 273 (1987).
36. D. E. Clegg and D. L. Massart, *J. Chem. Educ., 70:* 19 (1993).
37. C. H. Malissa, J. Rendl, and I. L. Marr, *Talanta, 22:* 597 (1975).
38. K. Eckschlager and V. Štěpánek, *Mikrochim. Acta, 1981:* 143 (1981).
39. K. Eckschlager, *Anal. Chem., 49:* 1265 (1977).
40. K. Danzer and K. Eckschlager, *Talanta, 25:* 725 (1978).
41. P. Cleij and A. Dijkstra, *Fresenius Z. Anal. Chem., 298:* 97 (1979).
42. F. H. Heite, P. F. Dupuis, H. A. Van't Klooster, and A. Dijkstra, *Anal. Chim. Acta, 130:* 313 (1978).
43. G. Mazerolles, D. Matthew, R. Phan Tan Luv, and A. M. Siouffi, *J. Chromatogr., 485:* 433 (1989).
44. R. Cela, C. G. Barroso, and J. A. Perez-Bustamante, *J. Chromatogr., 485:* 477 (1989).
45. K. Danzer, *Z. Chem., 15:* 326 (1975).
46. S. Arimoto, *Kalman Filter,* Sangyo Tosho, Tokyo, 1977.
47. S. Omatsu, Y. Tomita, and T. Soeda, *IEEE Trans. Inform. Theory, 22:* 593 (1975).
48. P. C. Thijssen, S. M. Wolfrum, and G. Kateman, *Anal. Chim. Acta, 156:* 87 (1984).
49. P. C. Thijssen, G. Kateman, and H. C. Smit, *Anal. Chim. Acta, 157:* 99 (1984).
50. P. C. Thijssen, *Anal. Chim. Acta, 162:* 253 (1984).
51. P. C. Thijssen and G. Kateman, *Anal. Chim. Acta, 173:* 265 (1985).
52. P. C. Thijssen, L. T. M. Prop, and G. Kateman, *Anal. Chim. Acta, 174:* 27 (1985).
53. M. Nishio, *Probability Theory,* Jikkyo Shuppan, Tokyo, 1985.
54. Y. Hayashi, S. Yoshioka, and Y. Takeda, *Anal. Sci., 5:* 329 (1989).
55. Y. Hayashi and R. Matsuda, *Anal. Chim. Acta, 222:* 313 (1989).

56. Y. Hayashi, *Anal. Sci., 6:* 15 (1990).
57. Y. Hayashi, S. Yoshioka, and Y. Takeda, *Anal. Chim. Acta, 212:* 81 (1988).
58. Y. Hayashi and S. C. Rutan, *Anal. Chim. Acta, 271:* 91 (1993).
59. Y. Hayashi, R. S. Helburn, and S. C. Rutan, the *Proceedings of the 4th Symposium on Computer-Enhanced Analytical Spectroscopy,* in press.
60. Y. Hayashi, S. C. Rutan, R. S. Helburn, and J. M. Pompano, *Chemometr. Intel. Lab. Syst., 20:* 163 (1993).
61. S. D. Brown, *Anal. Chim. Acta, 181:* 1 (1986).
62. S. C. Rutan, *Chemometr. Intel. Lab. Syst., 6:* 191 (1989).
63. S. C. Rutan, *J. Chemometr., 4:* 103 (1990).
64. S. C. Rutan, *Anal. Chem., 63:* 1103A (1991).
65. S. Minami, M. Kiri, and K. Sakurai, *Introduction to Computers for Instrumental Analysis,* Kodansha, Tokyo, 1982.
66. B. R. Kowalski, *Trends Anal. Chem., 1:* 210 (1982).
67. D. W. Posener, *J. Magn. Reson., 14:* 121 (1974).
68. L. Chen, C. Cottrell, and A. Marshall, *Chemometr. Intel. Lab. Syst., 1:* 51 (1986).
69. S. Ebel and A. Karger, *Chemometr. Intel. Lab. Syst., 6:* 301 (1989).
70. A. Karger, *Chemometr. Intel. Lab. Syst., 8:* 217 (1990).
71. L. R. Snyder, *J. Chromatogr. Sci., 10:* 200 (1972).
72. J. P. Foley, *J. Chromatogr., 384:* 301 (1987).
73. N. Dyson, *Chromatographic Integration Methods,* Royal Society of Chemistry, Cambridge, 1990, pp. 50–55.
74. A. Westerberg, *Anal. Chem., 41:* 1770 (1969).
75. W. Wei, N. S. Wu, and X. H. Jiang, *J. Chromatogr., 623:* 366 (1992).
76. Y. Hayashi, T. Shibazaki, and M. Uchiyama, *Anal. Chim. Acta, 201:* 185 (1987).
77. Y. Hayashi, T. Shibazaki, R. Matsuda, and M. Uchiyama, *Anal. Chim. Acta, 202:* 187 (1987).
78. Y. Hayashi, T. Shibazaki, and M. Uchiyama, *J. Chromatogr., 411:* 95 (1987).
79. R. Matsuda, Y. Hayashi, M. Ishibashi, and Y. Takeda, *J. Chromatogr., 462:* 13 (1989).
80. R. Matsuda, Y. Hayashi, M. Ishibashi, and Y. Takeda, *J. Chromatogr., 462:* 23 (1989).
81. W. J. Youden and E. H. Steiner, *Statistical Manual of the Association of Official Analytical Chemists,* Association of Official Analytical Chemists, Arlington, Va., 1990.
82. Y. Hayashi and R. Matsuda, *Anal. Chim. Acta.,* submitted.
83. W. Feller, *An Introduction to Probability Theory and Its Applications,* Vol. 2, Wiley, New York, 1971.

84. H. C. Smith and H. L. Walg, *Chromatographia, 8:* 311 (1975).
85. R. A. Wallingford and A. G. Ewing, *Adv. Chromatogr., 29:* 1 (1989).
86. Japanese Society of Mathematics, *Dictionary of Mathematics,* Iwanami, Tokyo, 1980.
87. E. Grushka and I. Zamir, *Chem. Anal., 98:* 529 (1989).
88. R. P. W. Scott and C. E. Reese, *J. Chromatogr., 138:* 283 (1977).
89. E. Katz, *Quantitative Analysis Using Chromatographic Techniques,* Wiley, Chichester, West Sussex, England, 1987.
90. Y. Hayashi and R. Matsuda, *Anal. Lett., 23:* 1765 (1990).
91. W. Wegscheider, E. P. Lankmayr, and K. W. Budna, *Chromatographia, 15:* 498 (1982).
92. Y. Hayashi, *Anal. Sci., 6:* 257 (1990).
93. Y. Hayashi and R. Matsuda, *Chromatographia, 31:* 367 (1991).
94. Y. Hayashi and R. Matsuda, *Chromatographia, 31:* 374 (1991).
95. E. Grushka, N.-I. Jang, and P. R. Brown, *J. Chromatogr., 458:* 617 (1989).
96. L. R. Snyder, J. W. Dolan, and J. R. Gant, *J. Chromatogr., 165:* 3 (1979).
97. P. J. Schoenmakers, H. A. H. Billiet, and L. De Galan, *J. Chromatogr., 185:* 179 (1979).
98. K. Jinno, *Computer-Assisted Chromatography System,* Hüthig, Heidelberg, 1990.
99. K. Jinno, *Adv. Chromatogr., 30:* 123 (1989).
100. D. L. Massart, L. Kaufman, and A. Dijkstra, *Evaluation and Optimization of Laboratory Methods and Analytical Procedures,* Elsevier, Amsterdam, 1978.
101. E. Grushka, P. R. Brown, and N.-I. Jang, *Anal. Chem., 60:* 2104 (1988).
102. K. Jinno, *Chromatographia, 20:* 743 (1985).
103. K. Jinno, M. Yamagami, and M. Kuwajima, *Chromatographia, 25:* 974 (1988).
104. K. Jinno and M. Yamagami, *Chromatographia, 27:* 417 (1989).
105. K. Jinno, M. Yamagami, and M. Kuwajima, *J. Chromatogr., 485:* 461 (1989).
106. Y. Hayashi, R. Matsuda, and K. Jinno, *Chromatographia, 31:* 554 (1991).
107. Y. Hayashi, R. Matsuda, and K. Jinno, *Anal. Lett., 24:* 2083 (1991).
108. J. Kondo, *Optimization Techniques,* Corona, Tokyo, 1984.
109. Y. Hayashi and R. Matsuda, *J. Chromatogr. Sci., 29:* 60 (1991).
110. W. Blaß, *Fresenius J. Anal. Chem., 339:* 340 (1991).
111. R. T. Krause, *J. Assoc. Off. Anal. Chem., 68:* 726 (1985).
112. R. Matsuda, Y. Hayashi, T. Suzuki, and Y. Saito, *Bunsekikagaku, 42:* 881 (1993).
113. Y. Hayashi and R. Matsuda, *Chromatographia, 29:* 446 (1990).

114. J. L. Glajch, J. J. Kirkland, and J. J. Minor, *J. Liq. Chromatogr., 10:* 1727 (1987).
115. Y. Hayashi, R. Matsuda, and A. Nakamura, *J. Chromatogr. Sci., 28:* 628 (1990).
116. Y. Hayashi and R. Matsuda, *Anal. Sci., 5:* 459 (1989).
117. R. Matsuda, Y. Hayashi, M. Ishibashi, and Y. Takeda, *Anal. Chim. Acta, 222:* 301 (1989).
118. S. C. Rutan, D. D. Gerow, and G. Hartmann, *Chemometr. Intel. Lab. Syst., 3:* 61 (1988).
119. Y. Hayashi and R. Matsuda, *Anal. Sci., 7:* 329 (1991).
120. Y. Hayashi and R. Matsuda, *Anal. Sci., 6:* 131 (1990).
121. Y. Hayashi and R. Matsuda, *Anal. Sci., 7:* 9 (1991).
122. M. Franklin, *J. Chromatogr., 526:* 590 (1990).
123. J. P. Hart and P. H. Jordan, *Analyst, 114:* 1633 (1989).
124. R. S. Markin, M. C. Wadman, P. L. Bottjen, M. C. Haven, and J. A. Huth, *J. Chromatogr., 525:* 464 (1990).
125. C. Humpel, C. Haring, and A. Saria, *J. Chromatogr., 491:* 235 (1989).
126. S. Murai, H. Saito, H. Nagahama, H. Miyate, Y. Masuda, and T. Itoh, *J. Chromatogr., 497:* 363 (1989).
127. S. W. Kennedy and A. L. Maslen, *J. Chromatogr., 493:* 53 (1989).
128. W. Lindner and C. Pettersson, in *Liquid Chromatography in Pharmaceutical Development,* W. Wainer, Ed., Aster, Springfield, Mass., 1985, p. 63.
129. R. Dappen, H. Arm, and V. R. Meyer, *J. Chromatogr., 373:* 1 (1986).
130. A. C. Mehta, *J. Chromatogr., 426:* 1 (1988).
131. K. Iwaki, S. Yoshida, N. Nimura, T. Kinoshita, K. Takeda, and H. Ogura, *J. Chromatogr., 404:* 117 (1987).
132. K. Iwaki, S. Yoshida, N. Nimura, T. Kinoshita, K. Takeda, and H. Ogura, *Chromatographia, 23:* 272 (1987).
133. K. Iwaki, N. Nimura, T. Kinoshita, K. Takeda, and H. Ogura, *Anal. Chem., 58:* 2372 (1986).
134. K. Iwaki, S. Yoshida, N. Nimura, T. Kinoshita, K. Takeda, and H. Ogura, *Chromatographia, 23:* 899 (1987).
135. K. Iwaki, M. Yamazaki, N. Nimura, and T. Kinoshita, *J. Chromatogr., 625:* 353 (1992).
136. K. Iwaki, M. Yamazaki, N. Nimura, T. Kinoshita, R. Matsuda, and Y. Hayashi, *Chromatographia, 37:* 156 (1993).
137. A. G. Wright, A. F. Fell, and J. C. Berridge, *J. Chromatogr., 458:* 335 (1988).
138. A. G. Wright, A. F. Fell, and J. C. Berridge, *J. Chromatogr., 464:* 27 (1989).
139. J. A. Crow and J. P. Foley, *Anal. Chem., 62:* 378 (1990).
140. S. N. Deming and S. L. Morgan, *Anal. Chem., 45:* 278A (1973).

141. D. E. Long, *Anal. Chim. Acta, 46:* 193 (1969).
142. J. A. Nelder and R. Mead, *Comput. J., 7:* 308 (1965).
143. M. W. Routh, P. A. Swartz, and M. B. Denton, *Anal. Chem., 49:* 1422 (1977).
144. B. G. M. Vandeginste, *Trends Anal. Chem., 1:* 210 (1982).
145. Y. Hayashi and R. Matsuda, *J. Chromatogr., 585:* 187 (1991).
146. S. Terabe, K. Otsuka, K. Ichikawa, A. Tsuchiya, and T. Ando, *Anal. Chem., 56:* 111 (1984).
147. S. Terabe, K. Otsuka, and T. Ando, *Anal. Chem., 57:* 834 (1985).
148. S. Terabe, K. Otsuka, and T. Ando, *Anal. Chem., 61:* 251 (1989).
149. W. G. Kuhr, *Anal. Chem., 62:* R403 (1990).
150. S. Terabe, *Kikan Kagaku Sosetu, 9:* 188 (1990).
151. S. Terabe, *Micellar Electrokinetic Chromatography,* Beckman, Calif., 1992.
152. S. Terabe, *J. Pharm. Biomed. Anal., 10:* 705 (1992).
153. J. Vinderogel and P. Sandra, *Introduction to Micellar Electrokinetic Chromatography,* Hüthig, Heidelberg, 1992.
154. M. G. Khaledi, S. C. Smith, and J. K. Strasters, *Anal. Chem., 63:* 1820 (1991).
155. J. Vindevogel and P. Sandra, *Anal. Chem., 63:* 1530 (1991).
156. J. P. Foley, *Anal. Chem., 62:* 1302 (1990).
157. S. C. Smith and M. G. Khaledi, *Anal. Chem., 65:* 193 (1993).
158. Y. Hayashi, R. Matsuda, and S. Terabe, *Chromatographia, 37:* 149 (1993).
159. Y. Hayashi and R. Matsuda, *Fresenius J. Anal. Chem., 338:* 597 (1990).
160. T. Nauyen, G. K. Shiu, W. N. Worsley, and J. P. Skelly, *J. Pharm. Sci., 79:* 163 (1990).
161. J. R. Strimaitis and G. L. Hawk, Ed., *Advances in Laboratory Automation Robotics,* Vol. 4, Zymark Corp., Hopkinton, Mass., 1988.
162. R. L. Sharp, R. G. Whitfield, and L. E. Fox, *Anal. Chem., 60:* 1056A (1988).
163. R. Matsuda, Y. Hayashi, M. Ishibashi, and Y. Takeda, *J. Chromatogr., 438:* 319 (1988).
164. Y. Hayashi, R. Matsuda, S. Yoshioka, and Y. Takeda, *Anal. Chim. Acta, 209:* 45 (1988).
165. R. Matsuda, Y. Hayashi, M. Ishibashi, and Y. Takeda, *J. Assoc. Off. Anal. Chem.,* in press.
166. P. Haefelfinger, *J. Chromatogr., 218:* 73 (1981).
167. J. Weiling, P. M. J. Coenegracht, C. K. Mensink, J. H. G. Jonkman, and D. A. Doornbos, *J. Chromatogr., 594:* 45 (1992).
168. Y. Hayashi and R. Matsuda, *Chromatographia, 30:* 367 (1990).
169. Y. Hayashi, T. Shibazaki, R. Matsuda, and M. Uchiyama, *J. Chromatogr., 407:* 59 (1987).

170. P. R. Brown, *Adv. Chromatogr., 12:* 1 (1975).
171. E. Grushka and D. Isreali, *Anal. Chem., 62:* 717 (1990).
172. A. G. Wright, J. C. Berridge, and A. F. Fell, *Analyst, 114:* 53 (1989).
173. A. Said, G. S. Aly, and A. M. Jarallah, *J. High Resolut. Chromatogr. Chromatogr. Commun., 11:* 681 (1988).
174. R. Matsuda, Y. Hayashi, T. Suzuki, and Y. Saito, *Chromatographia, 32:* 233 (1991).
175. D. Ishii and T. Takeuchi, *Trends Anal. Chem., 9:* 152 (1990).
176. M. Goto, T. Takeuchi, and D. Ishii, *Adv. Chromatogr., 30:* 167 (1989).
177. R. P. W. Scott, *J. Chromatogr., 517:* 297 (1990).
178. R. T. Kennedy and J. W. Jorgenson, *Anal. Chem., 61:* 1128 (1989).
179. K. Jinno, S. Ogura, and K. Takayama, *J. High Resolut. Chromatogr., 13:* 323 (1990).
180. N. Vonk, W. P. Verstraeten, and J. W. Marinissen, *J. Chromatogr. Sci., 30:* 296 (1992).
181. Y. Hayashi and R. Matsuda, *Fresenius J. Anal. Chem., 347:* 225 (1993).
182. K. Kafadar and K. R. Eberhardt, *Adv. Chromatogr., 24:* 1 (1984).
183. Y. Hayashi and R. Matsuda, *Anal. Chim. Acta, 277:* 325 (1993).
184. M. R. Detaevernier, Y. Michotte, L. Buydens, M. P. Derde, M. De Smet, L. Kaufman, G. Musch, J. Smeyers-Verbeke, A. Thielemans, L. Dryon, and D. L. Massart, *J. Pharm. Biomed. Anal., 4:* 297 (1986).
185. M. De Smet, A. Peeters, L. Buydens, and D. L. Massart, *J. Chromatogr., 457:* 25 (1988).
186. H. Yuzhu, A. Peeters, G. Musch, and D. L. Massart, *Anal. Chim. Acta, 223:* 1 (1989).
187. L. Peichang and H. Hongxin, *J. Chromatogr., 452:* 175 (1988).
188. J. A. van Leeuwen, B. G. M. Vandeginste, G. J. Postma, and G. Kateman, *Chemometr. Intel. Lab. Syst., 6:* 239 (1989).
189. P. Conti, T. Hamoir, M. De Smet, H. Piryns, N. Vanden Driessche, F. Maris, H. Hindriks, P. J. Schoenmakers, and D. L. Massart, *Chemometr. Intel. Lab. Syst., 11:* 27 (1991).
190. J. A. van Leeuwen, B. G. M. Vandeginste, G. Kateman, M. Mulholland, and A. Cleland, *Anal. Chim. Acta, 228:* 145 (1990).
191. J. A. van Leeuwen, L. M. C. Buydens, B. G. M. Vandeginste, G. Kateman, P. J. Schoenmakers, M. Mulholland, et al., *Chemometr. Intel. Lab. Syst., 10:* 337 (1991).
192. J. A. van Leeuwen, L. M. C. Buydens, B. G. M. Vandeginste, G. Kateman, P. J. Schoenmakers, M. Mulholland, et al., *Chemometr. Intel. Lab. Syst., 11:* 37 (1991).
193. J. A. van Leeuwen, L. M. C. Buydens, B. G. M. Vandeginste, G. Kateman, A. Cleland, M. Mulholland, C. Jansen, and F. A. Maris, P. H.

Hoogkamer, J. H. M. van den Berg, et al., *Chemometr. Intel. Lab. Syst.,* *11:* 161 (1991).
194. A. Bertha and G. Vigh, *J. Chromatogr., 485:* 383 (1989).
195. R. M. Smith and C. M. Burr, *J. Chromatogr., 485:* 325 (1989).
196. J. Zupan, *Anal. Chim. Acta, 235:* 53 (1990).
197. P. Vankeerberghen and D. L. Massart, *Trends Anal. Chem., 10:* 110 (1991).
198. *United States Pharmacopeia XXII,* ⟨1225⟩ Validation of compendial methods (1990).

Index

Albumins, concentration in human plasma and serum of, 8
Alcohols, infrared data on, 88
Alkanes, MI/FT-IR spectra of, 89
Alkenes, MI/FT-IR spectra of, 77
Amides, infrared data on, 83–84
Anion-exchange chromatography of organic ions, 329–340
  effects of temperature, 335–336
  injection solvent effects, 336–340
  solvophobicity and retention, 329–335
Atomic emission detection, pesticides and herbicides and, 273–275

Blood plasma (*see* Plasma)
Branched oligosaccharides, HPCE separation of, 221–224

Capillary electrophoresis (CE), 2–7
  errors from measurement and signal processing steps in, 348, 349 (*see also* HPCE of serum and plasma proteins)
Capillary electrophoresis of carbohydrates, 177–250
  electrophoretic system, 179–207
    capillary columns, 190–196
    detection systems, 197–207
    electrolyte systems, 179–190
  separation methodologies and applications, 207–245
    glycolipids, 242–245
    glycoproteins, 224–237
    glycosaminoglycans, 238–242
    monosaccharides, 207–214
    polysaccharides, 214–224

Capillary gel electrophoresis (CGE), 6–7
  separation of serum and plasma proteins by, 31–34, 42–44
Capillary isoelectric focusing (CIEF), 5
  separation of serum and plasma proteins by, 28–31, 40–41
Capillary isotachophoresis (CITP), 4–5
  separation of serum and plasma proteins by, 26–28, 39–40
Capillary zone electrophoresis (CZE), 3–4
  separation of serum and plasma proteins by, 26, 35–39
Carbohydrates:
  carbohydrate–borate complexes, 180–185
  carbohydrate–metal cation complexes, 188–190 (*see also* Capillary electrophoresis of carbohydrates)
Carbonyl compounds, MI/FT-IR spectra of, 77–82
Cation-exchange chromatography of organic ions, 325–329
Chromatography/detector interfaces, 59
Chromatography/IR interfaces, 59–60
Commercially available coated capillaries, 196
Computer simulation:
  testing of one-dimensional overlap theories by, 131–140
    extended Poisson model, 139–140
    Fourier analysis, 134–139
    Poisson model, 131–134

[Computer simulation:]
  testing of two-dimensional and n-dimensional overlap theories by, 164–166

Detection systems, 197–207
  direct detection, 203–207
  indirect detection, 200–203
  precolumn derivatization, 197–200 (*see also* names of detection systems)
Disaccharides, HPCE separation of, 215–216

Electron capture detection (ECD), pesticides and herbicides and, 269–270
Electrophoretic system, 179–207
  capillary columns, 190–196
    commercially available coated capillaries, 196
  detection systems, 197–207
    direction detection, 203–207
    indirect detection, 200–203
    precolumn derivatization, 197–200
  electrolyte systems, 179–190
    carbohydrate–borate complexes, 180–185
    carbohydrate–metal cation complexes, 188–190
    highly alkaline pH electrolytes, 185–187
Environmental applications of SFC, 251–308
  pesticides and herbicides, 259–281
    atomic emission detection, 273–275

[Environmental applications of SFC]
  electron capture detection, 269–270
  flame ionization detection, 259–265
  Fourier transform infrared detection, 275–277
  ion mobility detection, 280
  mass spectrometric detection, 277–280
  photometric detection, 273
  SFE/SFC, 280–281
  thermionic detection, 270–273
  UV detection, 265–269
  phenols, 281–288
    flame ionization detection, 286–287
    miscellaneous detection, 287–288
    UV detection, 282–285
  polychlorinated biphenyls, 253–259
    flame ionization detection, 255
    mass spectrometric detection, 258–259
    UV detection, 255–258
  polynuclear aromatic hydrocarbons, 288–301
    mass spectrometric detection, 293–295
    miscellaneous detection, 295–300
    multidimensional techniques, 300–301
    UV detection, 288–293
Extended Poisson model, testing of one-dimensional overlap theories via computer simulation, 139–140

Flame ionization detection:
  pesticides and herbicides and, 259–265
  phenols and, 286–287
  polychlorinated biphenyls and, 255
Fluorescence detections in CE, 17–19
Fourier transform infrared (FT-IR) spectrometry, 58–59
  pesticides and herbicides and, 275–277
Function of mutual information (FUMI), 351
  applications of FUMI and MEI, 378–396
    evaluation of columns for optical resolution, 391–393
    merit of short columns, 390–391
    model optimization, 378–383
    optimal mobile-phase composition in pesticide analysis, 383–387
    optimization of MEEC analysis, 394–396
    optimization of wavelength and amount of internal standard, 387–389
    simplex optimization for HPLC, 392–394
  simplified TOCO with, 406–407
  TOCO with FUMI and MEI for antipyretic mixture, 396–399
FUMI theory, 351

Gas chromatography/matrix isolation/infrared (GC/MI/IR) spectrometric analysis of natural products, 57–108
  applications, 94–103

[Gas chromatography/matrix isolation/infrared (GC/MI/IR) spectrometric analysis of natural products]
  chromatography/detector interfaces, 59
  chromatography/IR interfaces, 59–60
  hardware, 60–71
    GC/light pipe/Fourier transform infrared spectrometry, 64–66
    GC/MI/Fourier transform infrared spectrometry, 60–64
    GC/MI/FT-IR methodologies, 68–71
    multidimensional gas chromatography, 66–68
  special interpretation, 71–93
    characteristics of matrix isolation, 72–73
    IR spectra of matrix-isolated organic compounds and applications to natural products, 73–75
    IR spectral libraries, 90–91
    matrix isolation spectrometry, 71–72
    matrix shifts, 75–77
    matrix versus light pipe, 91–93
    MI/FT-IR spectra of selected compound types, 77–89
Globulins, concentration in human plasma and serum of, 8
Glycolipids, HPCE separation of, 242–245
Glycoproteins, HPCE separation of, 224–237
  glycans, 232–237
  glycoforms, 225–230
  glycopeptide mapping, 230–232

Glycosaminoglycans (GAGs), HPCE separation of, 238–242

Herbicides (see Pesticides and herbicides, SFC analysis of)
Highly alkaline pH electrolytes, 185–187
High-performance capillary electrophoresis (HPCE), 177–178, 179
HPCE of serum and plasma proteins, 1–56
  basic considerations in protein separation, 10–34
    capillaries and coatings, 22–25
    protein detection, 15–21
    sample injection, 11–15
    sample preparation, 10–11
    separation of serum and plasma proteins, 25–34
  capillary electrophoresis, 2–7
  physical and chemical properties of selected human serum and plasma proteins, 47–48
  register of human proteins separated by CE, 46
  serum and plasma proteins, 7–10
  use of CE in serum and plasma protein separation, 35–44
    capillary gel electrophoresis, 42–44
    capillary isoelectric focusing, 40–41
    capillary isotachophoresis, 39–40
    capillary zone electrophoresis, 35–39
    combined separation techniques, 44

High-performance liquid
  chromatography (HPLC),
  errors from measurement
  and signal processing steps
  in, 348, 349
HPLC of homologous organic
  anions and cations, 309–346
  effect of temperature, 314–315
  homologous series, 310–312
  ion-exchange chromatography of
    organic ions, 324–340
    anion exchange, 329–340
    cation exchange, 325–329
  ion-exclusion chromatography of
    organic ions, 340–344
  methylene selectivity, 312
  micellar chromatography, 315–316
  mobile phase and selectivity, 312–313
  reversed-phase chromatography
    of organic ions, 316–324
    bonded silica stationary phases, 318–320
    ion exchange on reversed-phase columns, 320–324
    polymeric stationary phases, 316–317
  stationary phase and selectivity, 313
Hydroxyls, MI/FT-IR spectra of, 82–89

Indirect detection, 200–203
Information theory, 350–351
  quantitative analysis and, 371–375
    precision and information, 371–375
    resolution and information, 374–375

[Information theory]
  throughput of analysis, 373–374
Ion-exchange chromatography of
  organic ions, 324–340
  anion exchange, 329–340
  cation exchange, 325–329
Ion-exclusion chromatography of
  organic ions, 340–344
Ion mobility detection, pesticides
  and herbicides and, 280

Linear oligosaccharides, HPCE
  separation of, 217–221

Mass spectrometric (MS) detection, 58
  pesticides and herbicides and, 277–280
  polychlorinated biphenyls and, 258–259
  polynuclear aromatic
    hydrocarbons and, 293–295
Matrix-isolated organic compounds,
  IR spectra of, 73–75
Matrix isolation/Fourier transform
  infrared (MI/FT-IR) spectra, 60–64
  schematic representation of
    relationship between mass
    spectral information and MI/
    FT-IR information for
    compound identification, 103, 105
  of selected compound types, 77–89
    alkanes, 89
    alkenes, 77
    carbonyl compounds, 77–82
    hydroxyls, 82–89

430 / Index

Matrix isolation (MI) spectrometry, 71–72
  characteristics of, 72–73
Measurement-elicited information (MEI), 351 (see also Function of mutual information (FUMI), applications of FUMI and MEI)
Micellar chromatography of homologous organic anions and cations, 315–316
Micellar electrokinetic capillary chromatography (MECC), 7, 178
  optimization of analysis of, 394–396
Monosaccharides, HPCE separation of, 207–214
Multidimensional gas chromatography (MDGC), 66–68
  applications of, 97–101
Multidimensional separations, modeling of overlap in, 158–168
  applications of n-dimensional overlap theory in two-dimensional experimental data, 166
  number of dimensions necessary to resolve m components, 166–168
  testing of two-dimensional and n-dimensional theories by computer simulation, 164–166
  theories, 158–163
    application of S.A. Roach theory, 161–162
    early work, 159

[Multidimensional separations, modeling of overlap in]
  Poisson model in two dimensions, 159–160
  probability of total resolution in two-dimensional and three-dimensional separations, 160–161
  theory for overlap in n-dimensional separation, 163
Multidimensional SFC for polynuclear aromatic hydrocarbons, 300–301

Oligosaccharides, HPCE separation of:
  branched, 221–224
  linear, 217–221
One-dimensional (1-D) separations, modeling of overlap in, 111–158
  amplitude distribution of SCPs in one-dimensional separations, 156
  application and testing of one-dimensional overlap theories by experiment, 140–151
  basis for assumption of randomness, 130–131
  determination limits, 156–158
  resolution factors for one-dimensional models of overlap, 151–156
  testing of one-dimensional theories by computer simulation, 131–140
  theories, 111–130
    alternative means to estimate m by Poisson model, 118

[One-dimensional (1-D) separations, modeling of overlap in]
  analogy between overlap and depolymerization, 115–118
  combinatorial analysis, 111–113
  early work, 111
  extended Poisson model: density variations of SCPs, 129–130
  Fourier analysis, 123–128
  Poisson model, 115–116
  probability of distribution of observed peaks and SCPs, 119–121
  saturogram, 128
  statistical basis for eluent optimization, 122–123
  variance associated with Poisson model, 118–119
  variation of overlap probability with peak area, 113–115
Optimization of separation science, 350
  applications of FUMI and MEI, 378–396
    evaluation of columns for optical resolution, 391–392
    merit of short columns, 390–391
    model optimization, 378–383
    optimal mobile-phase composition in pesticide analysis, 383–387
    optimization of MEEC analysis, 394–396
    optimization of wavelength and amount of internal standard, 387–389
    simplex optimization for HPLC, 392–394

[Optimization of separation science]
  factors affecting precision and throughput, 397–415
    area ratio of adjacent peaks, 410–414
    interference of unnecessary peaks, 414
    peak-resolving power of data processing, 407–410
    sample size, 407
    system miniaturization, 414–415
    unresolvable peaks, 414
  fundamentals of optimization, 375–378
    nonlinear programming problems, 376–378
    practical optimization strategies, 375–376
  total chromatographic optimization (TOCO), 396–407
    simplified TOCO with FUMI, 406–407
    throughput of dissolution test using robotics, 399–402
    TOCO with FUMI and MEI for antipyretic mixture, 396–399
    validity of internal standard, 402–406
Organic anions and cations, HPLC of, 309–346
  effect of temperature, 314–315
  homologous series, 310–312
  ion-exchange chromatography of organic ions, 324–340
    anion exchange, 329–340
    cation exchange, 325–329
  ion-exclusion chromatography of organic ions, 340–344

432 / Index

[Organic anions and cations, HPLC of]
  methylene selectivity, 312
  micellar chromatography, 315–316
  mobile phase and selectivity, 312–313
  reversed-phase chromatography of organic ions, 316–324
    bonded silica stationary phases, 318–320
    ion exchange on reversed-phase columns, 320–324
    polymeric stationary phase, 316–317
  stationary phase and selectivity, 313

Peak overlap, statistical theories of, 109–175
  statistical modeling of overlap in multidimensional separations, 158–168
    application of n-dimensional overlap theory to two-dimensional data, 166
    number of dimensions necessary to resolve m components, 166–168
    testing of two-dimensional and three-dimensional overlap theories by computer simulation, 164–166
    theories, 158–163
  statistical modeling of overlap in one-dimensional separations, 111–158
    amplitude distribution of SCPs in one-dimensional separations, 156

[Peak overlap, statistical theories of]
    application and testing of one-dimensional overlap theories by experiment, 140–151
    basis for assumption of randomness, 130–131
    determination limits, 156–158
    resolution factors for one-dimensional models of overlap, 151–156
    testing of one-dimensional theories by computer simulation, 131–140
    theories, 111–130
Pesticides and herbicides, SFC analysis of, 259–281
  atomic emission detection, 273–275
  electron capture detection, 269–270
  flame ionization detection, 259–263
  Fourier transform infrared detection, 275–277
  ion mobility detection, 280
  mass spectrometric detection, 277–280
  photometric detection, 273
  SFE/SFC, 280–281
  thermionic detection, 270–273
  UV detection, 265–269
Phenols:
  infrared data on, 88
  SFC analysis of, 281–288
    flame ionization detection, 286–287
    miscellaneous detection, 287–288
    UV detection, 282–285
Photometric detection, pesticides and herbicides and, 273

Plasma, 7–10 (*see also* Serum and plasma proteins)
Poisson model of overlap in one-dimensional separations, 115–116, 118
　application and testing of, 140–148, 150–151
　density variations of SCPs, 129–130
　resolution factors for, 151–156
　testing by computer simulations, 131–134, 139–140
　variance associated with, 118–119
Poisson model of overlap in two dimensions, 159–160
Polychlorinated biphenyls (PCBs), SFC analysis of, 253–259
　flame ionization detection, 255
　mass spectrometric detection, 258–259
　UV detection, 255–258
Polynuclear aromatic hydrocarbons (PAHs), SFC analysis of, 288–301
　mass spectrometric detection, 293–295
　miscellaneous detection, 295–300
　multidimensional techniques, 300–301
　UV detection, 288–293
Polysaccharides, HPCE separation of, 214–224
　branched oligosaccharides, 221–224
　disaccharides, 215–216
　linear oligosaccharides, 217–221
Precolumn derivatization, 197–200
Protein separation, 10–34
　capillaries and coatings, 22–25
　protein detection, 15–21
　sample injection, 11–15

[Protein separation]
　sample preparation, 10–11
　separation of serum and plasma proteins, 25–34

Quantitative analysis:
　information theory and, 371–375
　　precision and information, 371–372
　　resolution and information, 374–375
　　throughput of analysis, 373–374
　uncertainty structure of, 361–371
　　error prediction in HPLC and CE, 361–366
　　literature on precision and reproducibility, 368–371
　　uncertainty principle in separation science, 367–368
　　uncertainty structure in analytical apparatus, 366–367

Raman spectroscopy as detection technique in CE, 20–21
Reversed-phase chromatography of organic ions, 316–324
　bonded silica stationary phases, 318–320
　ion exchange on reversed-phase columns, 320–324
　polymeric stationary phases, 316–317

Serum and plasma proteins, 7–10
　separation by CE, 25–34
　　capillary gel electrophoresis, 31–34, 42–44

[Serum and plasma proteins]
  capillary isoelectric focusing, 28–31, 40–41
  capillary isotachophoresis, 26–28, 39–40
  capillary zone electrophoresis, 26, 35–39
  combined separation techniques, 44
Simplex optimization for HPLC, 392–394
Spectrometric analysis by GC/MI/IR (*see* Gas chromatography/matrix isolation/infrared (GC/MI/IR) spectrometric analysis of natural products)
Statistical theories of peak overlap (*see* Peak overlap, statistical theories of)
Stochastic properties of signals, 351–361
  accuracy and precision, 357–361
  precision of independent peaks, 351–355
  precision of interfered peaks, 355–357
Supercritical fluid chromatography (SFC), environmental applications of, 251–308
  pesticides and herbicides, 259–281
    atomic emission detection, 273–275
    electron capture detection, 269–270
    flame ionization detection, 259–265
    Fourier transform infrared detection, 275–277
    ion mobility detection, 280

[Supercritical fluid chromatography (SFC), environmental applications of]
    mass spectrometric detection, 277–280
    photometric detection, 273
    SFE/SFC, 280–281
    thermionic detection, 270–273
    UV detection, 265–269
  phenols, 281–288
    flame ionization detection, 286–287
    miscellaneous detection, 287–288
    UV detection, 282–285
  polychlorinated biphenyls, 253–259
    flame ionization detection, 255
    mass spectrometric detection, 258–259
    UV detection, 255–258
  polynuclear aromatic hydrocarbons, 288–301
    mass spectrometric detection, 293–295
    miscellaneous detection, 295–300
    multidimensional techniques, 300–301
    UV detection, 288–293

Thermionic detection (TID), pesticides and herbicides and, 270–273
Total chromatographic optimization (TOCO), 351, 396–407
  simplified TOCO with FUMI, 406–407
  throughput of dissolution test using robotics, 399–402

[Total chromatographic optimization (TOCO)]
  TOCO with FUMI and MEI for antipyretic mixture, 396–399
  validity of internal standard, 402–406

Uncertainty structure of quantitative analysis, 361–371
  error prediction in HPLC and CE, 361–366
  literature on precision and reproducibility, 368–371

[Uncertainty structure of quantitative analysis]
  uncertainty principle in separation science, 367–368
  uncertainty structure in analytical apparatus, 366–367
UV detection:
  pesticides and herbicides and, 265–269
  phenols and, 282–285
  polychlorinated biphenyls and, 255–258
  polynuclear aromatic hydrocarbons and, 288–293